Applied Geostatistics for Reservoir Characterization

Mohan Kelkar
Professor
U. of Tulsa

Godofredo Perez
Staff Reservoir Engineer
BP Indonesia
Jakarta

Society of Petroleum Engineers

Book Editor

Printed in the United States of America.

The paper used in this book meets the minimum requirements of ANSI/NSIO Z39.48-1992 (R1997). ∞

ISBN 1-55563-095-2

08 09 10 11 12 13 14 / 9 8 7 6 5 4 3 2

Society of Petroleum Engineers
222 Palisades Creek Drive
Richardson, TX 75080-2040 USA

http://store.spe.org/
service@spe.org
1.972.952.9393

Half Price Books
1835 Forms Drive
Carrollton, TX 75006
OFS OrderID 38620510

Thank you for your order, MyFBAPrep Sarasota (AVI: C2906432)!

Thank you for shopping with Half Price Books! Please contact Support@hpb.com. if you have any questions, comments or concerns about your order (731479328)

SKU	ISBN/UPC	Title & Author/Artist	Shelf ID	Qty	OrderSKU
S461684100	9781555630959	Applied Geostatistics for Reservoir Charac... Kelkar, Mohan	DB 26.13.3	1	

SHIPPED STANDARD TO:
MyFBAPrep Sarasota (AVI: C2906432)
1657 West University Parkway Unit A
Sarasota FL 34243
abebooks.red@ergoder.com

ORDER# **731479328**
Abebooks

Preface

The use of geostatistics as a tool for reservoir characterization started gaining popularity in the petroleum industry in the mid-1980s. It provided a new approach to quantify and describe the spatial dependence of reservoir attributes. A course in geostatistics was first taught at the U. of Tulsa in 1989, and the present book is the evolution of notes that were used to teach that first course.

As a new technology adopted from other industries, geostatistics came with unfamiliar terminology and examples not easily applicable to the petroleum industry. This has been the main reason that made it difficult for petroleum geoscientists to understand this newly introduced technology and perhaps also made them hesitant to apply it for reservoir characterization. Therefore, the main goal of this book is to introduce the key concepts of geostatistics with terminology more familiar to petroleum professionals and to include actual field examples from petroleum reservoirs.

Although this book can be used as a text for either undergraduate or graduate level courses, it is primarily designed as a self-learning manual. The intuitive understanding is emphasized over mathematical details, and several numerical and field examples are provided to explain the actual calculations. Reading of the book does not require formal background in any advanced subject. We realize that integration of various types of data is becoming a key to a representative reservoir description. Therefore, the book is written such that any geoscientist can read it and follow it. For the truly initiated, additional mathematical details are provided in the appendices.

We hope that this book can provide the necessary background in geostatistics from a petroleum industry perspective. We know that many of the methods discussed in the book are routinely used in the industry. We expect that the usage will grow in the future with new and improved techniques constantly emerging.

Acknowledgments

We would like to thank several people who contributed to the success of this book. First and foremost are those who took the course from us and offered valuable suggestions and corrections. These participants include students at the U. of Tulsa, as well as participants in the OGCI courses that Mohan Kelkar teaches on a regular basis. We would like to thank Dr. Anil Chopra for agreeing to be the editor of the book and encouraging Godo Perez to be the coauthor. We appreciate Dr. Kirk Hird's efforts in making available the field data used in the examples for this book. Mohan Kelkar would like to thank the U. of Tulsa for providing the time and resources to make this book a reality.

We thank Shelley Nash for editing and improving the manuscript. Many members of the SPE Publications staff worked diligently to make this project a success. We would like to thank them all. There are those whose names may not have been mentioned because of oversight on our part. However, please accept our sincere gratitude for your contributions.

Mohan Kelkar is a professor of petroleum engineering at the U. of Tulsa. After receiving a BS degree in chemical engineering from the U. of Bombay and an MS degree in petroleum engineering and PhD degree in chemical engineering from the U. of Pittsburgh, he joined the U. of Tulsa as an assistant professor in 1983. His primary research interests are developing tools for integrated reservoir description and applying them to field studies. He has received funding from government agencies and international oil companies to conduct his research. He currently teaches a geostatistics course at the U. of Tulsa and for many oil companies around the world. Kelkar was a faculty adviser for the SPE Student Chapter at the U. of Tulsa for 5 years. He has served on the SPE Education and Accreditation Committee and has been a Review Chairman for *SPE Reservoir Evaluation & Engineering.*

Godofredo Perez works for BP Indonesia in Jakarta as a staff reservoir engineer. He is involved in field development planning and reservoir modeling. Previously, he worked for Arco Exploration and Production Technology in Plano, Texas, as a principal research engineer. At Arco, he worked in several integrated reservoir-modeling projects and conducted research on geostatistics, integration of seismic data in reservoir models, scaleup, and effect of heterogeneities on gas injection. He also taught internal geostatistics courses. Perez holds BS, MS, and PhD degrees in petroleum engineering from the U. of Tulsa. In 2001, he served as the Program Vice-Chairman for the Indonesian SPE section.

Dedication

MK
To my parents, who taught me the value of education; to my wife and children, who taught me the value of love.

GP
To my wife, Michele, for her extraordinary support and encouragement through the years while writing this book, and to Sebastian and Andres, my sons, for sharing their time.

Contents

Chapter 1
Introduction

1.1 Introduction

Reservoir characterization can be broadly defined as the process of describing various reservoir characteristics by use of all available data. The description can be qualitative or quantitative. Reservoir characteristics include pore- and grain-size distributions, reservoir permeability and porosity, facies distribution, depositional environment, and basin description. Considering the variety of characteristics that need to be described, the available data have to come from a broad spectrum of sources. These types of data may include core and log data; well-test data; tracer and production data; and, in some instances, 2D- and 3D-seismic data, vertical seismic profiles, wellbore tomography, and outcrop analogs.

Ideally, all the different sources of data should be included in a final reservoir description. If more information is used, the reservoir description is better. In practice, however, several problems are associated with the simultaneous use of all the data. Briefly, these problems include the following.

• All the data are not available at the same time. As exploration and delineation of the reservoir begins, additional data become available. Once production starts, more production-related information is collected. At the time of exploration, however, very few types of data are available. Typically, geophysical data (seismic surveys), regional depositional models, and data from analogous reservoirs are available at the beginning. Obviously, with the limited amount of information, the reservoir description has more uncertainty.

• The quality of data collected from various sources may differ. Accordingly, the weight and importance assigned to each set of data has to be different. For example, seismic data are collected with nonintrusive techniques. As a result, several steps of interpretations and qualitative judgments have to be included before those data can predict properties that are of interest for reservoir exploration or exploitation. Owing to these additional steps, more uncertainty is associated with the seismic interpretation. On the other hand, permeability and porosity measurements with sample cores are direct measurements of reservoir properties. Because the analysis requires a minimal number of assumptions, the data are highly accurate and less uncertain. All the other examples of collected data also have varying degrees of accuracies and uncertainties associated with them. When these different sources are used, it is important to account for the uncertainties in the sampled data in addition to the actual measurements, which may be difficult.

• Information collected from different sources may not be measured at the same scale. In addition, the resolution of measured values can differ significantly. As data from different sources are gathered, the scales over which the data are collected have to be accounted for properly. For example, permeability values are measured with cores as well as well tests. Although these two techniques measure the same variable, the scales over which these two measurements are made differ significantly. When information from both sources is used, the differences in scales must be included in the analysis.

In addition to differences in scale, the resolution of different measurements also can differ significantly, resulting in uncertainties at different scales. For example, seismic data that may be related to porosity data may have a resolution of ± 50 ft, whereas the core data or some log data may have a resolution of a few inches. In analyzing measurements of the same variable made with two or more different techniques, it is important to account for the resolution of different techniques.

• Data interpretation may be qualitative. An example would be tracer data collected during the production phase. Although tracer data are quantitative in nature, their interpretations can only be used in a qualitative manner in most instances. This may include knowledge of the direction of reservoir continuity and fracture directions, indication of thief zones, and the effectiveness of expensive injection fluids during an enhanced-oil-recovery (EOR) process.

Another good example is a geological description of a basin. Similar to tracer data, a geological description, although qualitative, can provide valuable insight into the framework of the reservoir. Typically, the descriptions are constructed by analyzing core and log data as well as from a geologist's prior experience with similarly situated reservoirs. The description of rock types and depositional environment can provide a valuable constraint in describing a reservoir.

Unfortunately, for many reservoir applications, a quantitative description of reservoir characteristics is necessary for

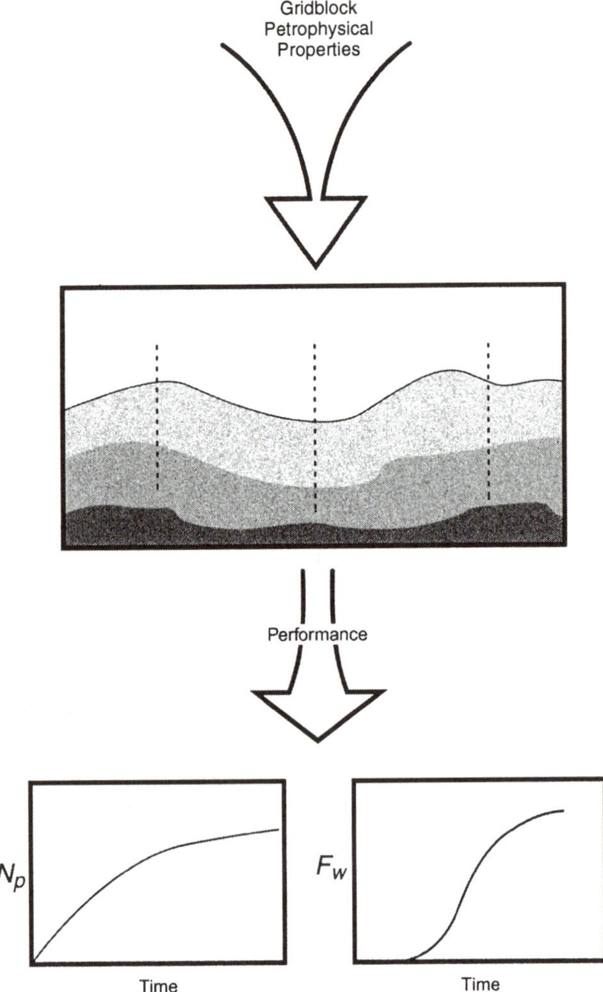

Fig. 1.1—Typical reservoir-simulation process.

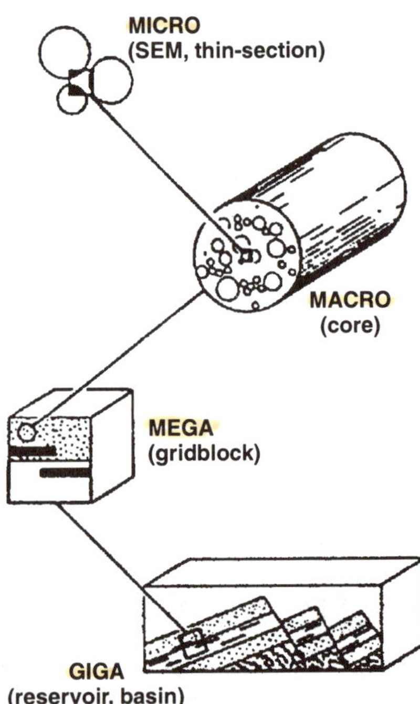

Fig. 1.2—Heterogeneity scales; SEM = scanning electron microscopy (courtesy of Academic Press).

proper reservoir analysis. Attempts have been made to quantify some qualitative data, but adequate procedures do not yet exist to use qualitative data that describe reservoir properties.

• Production data are probably the most valuable pieces of data from a petroleum engineer's point of view. Once production begins, production data are collected on a regular basis and are easily available. From an engineer's perspective, no matter what reservoir description can be constructed, it is worthless if it cannot reproduce the production data. Unfortunately, the use of production data as an input in reservoir description poses some important problems because of the complex mathematical relationship between the reservoir description and the production data. As **Fig. 1.1** shows, unlike simulation of other data, simulating production data are the result of the particular reservoir description entered as well as the reservoir production mechanism. With known reservoir description and reservoir mechanisms, future performance can be predicted. The model used for reservoir simulation is invariably complex and nonlinear with respect to input parameters. This nonlinearity makes it very difficult to predict a unique reservoir description from historical performance. This is a classic example of an "inverse problem," in which multiple configurations of input parameters can result in the same output. Currently, no single procedure adequately addresses the integration of production data in reservoir description.

As a result of these limitations, the reservoir-description methods discussed in this book may be able to use only part of the available data. While these methods are a considerable improvement over traditional methods, more work must be done to develop a more comprehensive scheme of reservoir description.

In this chapter, we first introduce the definition of heterogeneity and its associated scales. We also discuss the importance of scales in the reservoir description as well as in performance prediction. The next section briefly describes modeling a reservoir with conventional methods, enumerates the limitations of these methods, and proposes possible improvements in these methods. The last section introduces the importance of geostatistics in reservoir description. We define the advantages and the disadvantages of the geostatistical techniques and briefly state the procedures involved in applying geostatistical methods for reservoir description.

1.2 Reservoir Heterogeneity

Reservoir heterogeneity is defined as a variation in reservoir properties as a function of space. Ideally, if reservoirs were homogeneous, measuring a reservoir property at any location would allow a full description of the reservoir. The properties may include permeability, porosity, thickness, saturation, faults and fractures, rock facies, and rock characteristics. For a proper reservoir description, the variation in these reservoir properties as a function of spatial location must be predicted.

1.2.1 Definition of Scales. Reservoir heterogeneity can be measured on different scales. Heterogeneities can be measured from small-scale measurement of grain-size distribution at a microscopic level to large-scale geological description at a basin level. Although reservoir properties vary over many different continuous scales, for convenience, heterogeneity scales can be divided into the four levels shown in **Fig. 1.2,** which are defined, rather arbitrarily, by the following.

Microscopic Heterogeneities. Microscopic heterogeneities are the heterogeneities measured at a micro level. They can also be called pore-level heterogeneities. These heterogeneities include pore- and grain-size distributions, throat openings, and rock lithology and mineralogy.

TABLE 1.1—SCALES OF RESERVOIR HETEROGENEITIES (from Ref. 1)

Type	Level	Measurement Scale	Measurements	Geologic Observations/ Measurements	Flow Performance Effect	Flow Process Effect
Micro	Pore	≈μm	Pore geometry Grain size Mineralogy	Texture Mineralogy Fractures	Displacement efficiency (trapped oil)	Complex EOR process
Macro	Core	≈inch	k k_r ϕ P_c Wettability Saturation	Lamination Crossbedding baffles within genetic units	Sweep efficiency (bypassed oil)	Secondary recovery (waterflooding)
Mega	Gridblock	≈foot	Logs RFT's Single-well tracer Seismic	Boundaries of genetic units Permeability zonation within units	Sweep efficiency (bypassed oil)	Secondary recovery (waterflooding)
Giga	Interwell	≈mile	Well test Surface seismic interwell tracer tests	Sealing/nonsealing faults	Extraction efficiency (untrapped oil)	Primary recovery

Macroscopic Heterogeneities. Macroscopic heterogeneities are those measured at a core level. Hence, they are also called core-level heterogeneities. They include core measurements (such as permeability, porosity, saturation and wettability of the rock), petrophysical properties (such as relative permeability and capillary pressure), and some log data.

Megascopic Heterogeneities. Megascopic heterogeneities are the heterogeneities that have the same order of magnitude as a reservoir simulator gridblock, which is typically several feet in breadth and width. The properties measured on this scale include some log data; pressure-transient data, such as repeat formation tests (RFTs); and residual-oil-saturation measurements with single well-tracer tests. Some seismic data can also be considered to be on this scale.

Gigascopic Heterogeneities. Gigascopic heterogeneities are those measured at an interwell reservoir scale. The properties measured on this scale include permeability measurements from well-test data and interwell tracer tests. In addition, surface seismic data and major fault locations also can be considered part of gigascopic heterogeneities.

1.2.2 Effect of Various Scales. Table 1.1 gives brief descriptions of various levels of heterogeneities and their effect on reservoir flow performance. The following explains the table columns.

Measurement Scale. The measurement scales indicated in Table 1.1 are arbitrary. Reservoir properties change continuously at every scale. Therefore, the division of scales is made largely on the basis of the type of measurements made at various levels. Obviously, some overlap exists between different measurements, and some measurements can fall between two possible scales. For example, some types of logs (such as formation microscanner) fall under macro level, and some short-duration well-tests can fall under megascopic level.

Measurements. As Table 1.1 shows, different types of measurements are conducted at different levels. Typically, every property is not measured at every level. In describing the reservoir properties at a particular location, it is important to know not only the location in the reservoir but also the scale over which these properties are defined.

Any measurement over a microscopic scale shows significant variations in that property at that scale (for example, porosity at a microscopic level can be either one in the pore or zero in the grain). When the measurement is made at larger scales, however, the variation in the properties observed at a smaller level is not observed because of averaging of smaller-scale properties. Eventually, at the other extreme, when properties are measured on a reservoir scale, it may be possible to measure one single porosity value that represents the reservoir. In short, the variation of a given measured property decreases as the scale of measurement increases. This scale dependence is called a support effect.

For example, consider permeability. Typically, core permeability is measured over a 2-in.-diameter by 2-in.-long core. The volume of the core over which the permeability is measured is 0.004 ft^3. On the other hand, well-test permeability is measured over a volume of 10^7 to 10^9 ft^3. There are at least 10 orders of magnitude difference between the volume over which permeability is measured on a core scale vs. the volume over which it is measured with well-test data. This difference in measurement volumes has a significant impact on the measured value.

Fig. 1.3 compares well-test and core permeability data for an oil reservoir. As the figure indicates, where permeability values are plotted as functions of percentage of probability, the well-test data show less variability than the core data; this is consistent with the support effect. As the well-test data are collected over a much larger volume, the permeability values obtained are some representative averages of all the small-scale (core-scale) permeability values located within the volume represented by the well tests. As the averaging volume increases, the variability between measured values

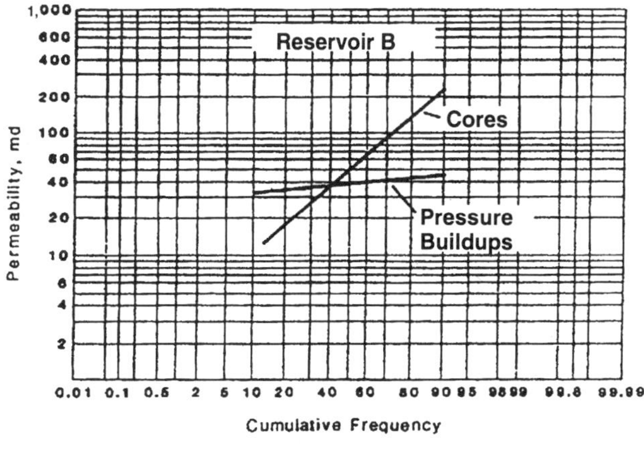

Fig. 1.3—Comparison between well test and core data (courtesy of Academic Press).

TABLE 1.2—ROCK VOLUMES INVESTIGATED FOR DIFFERENT LOGS (from Ref. 3)

Observation	Vertical Resolution (in.)	Radial Depth of Investigation (in.)	Radial Depth of Investigation (%)	Angular Dispersion (degrees)	Rock Volume Investigated (in.^{3})
Density	15	5		30	245
Neutron (mandrel)	24	16		360	28,952
Microelectrical log	2	1.5	50	30	7
Gamma ray	20	15		360	21,676
Unaided human eye focusing on a one-third slab of 4-in. whole core from a distance of 24 in.	0.2	NA		NA	NA
Dipmeter (one button)	0.5	6	90	14	5
Borehole televiewer	0.3	0.1		360	0.7/Scan
Correlation core gamma ray of a 4-in core by use of 3-in.-diameter detector	6	4		NA	75
6FF40	96	50	90	360	824,619
1 ft of 4-in. core	12	NA		360	150
1-× 2-in. plug	1	NA		360	1.6
MWD resistivity	6	32	90	360	24,127
MWD bit resistivity (oil-based mud)	1	32	90	360	4,021

decreases. Therefore, permeability values, based on well tests, show less variability than those from the core data.[2] From the data in Fig. 1.3, the standard deviation observed for the well-test data is 4.26 md, whereas the standard deviation for the core data is 186.4 md. A difference of almost two orders of magnitude exists between the variability observed for the core vs. the well-test data.

In describing reservoir heterogeneities, the heterogeneity has to be defined on a proper scale to be truly representative of that scale. Unfortunately, measurements of every reservoir property may not be available at every scale. For example, if we need to define the permeability value for a gridblock, the value should be representative of that scale. Permeability can be measured on a core level, which is a scale smaller than a typical gridblock, or on a reservoir level (well-test data), which is a scale much larger than a typical gridblock. Intuitively, the variability of the permeability values at a simulator gridblock level should be in between the variabilities of permeabilities at the core level and at the well-test data level. That is, for the data in Fig. 1.3, the standard deviation for estimated permeability values for the gridblocks should be between 4 and 200 md. Proper reconciliation of the two permeability measurements by accounting for the support effect is needed before simulator gridblock values can be represented correctly.

The importance of the support effect is not restricted only to permeability values. For porosity, data can be measured with core and log data. Core data typically show a lot more variability than log data. Even for the log data, as **Table 1.2** indicates, different log measurements are based on different support volumes. The resolution volume can be as small as 5 in.^{3} and as large as 800,000 in.^{3}. This difference has to be reconciled before the log measurements are used. A proper reservoir description requires that the support effect be properly accounted for in describing the reservoir properties.

Geologic Observations/Measurements. As Weber and van Geuns[4] explained, geologic heterogeneities (explaining the rock type and rock characteristics) vary depending on the scale of measurements. Similar to other quantitative information, different types of heterogeneities become evident as the scale over which a geological observation is made changes.

On a large (gigascopic) scale, a geologist may be able to identify large-scale heterogeneities that complicate the reservoir's internal architecture. These include main depositional units and the boundaries between them, which may be separated by sealing or nonsealing faults. On the next (megascopic) scale, the boundaries between the genetic units and possible permeability barriers between them can be identified.

At the macroscopic level, for clastic reservoirs, for example, baffles (small mud clasts or shale barriers) within individual units (which may affect effective vertical permeability[5]), laminations, and crossbedding within units can be identified. These observations can be made by studying whole cores and are based on prior experience with outcrop analogs. Laminations and crossbedding also can have a significant impact on the effective permeabilities of a gridblock.

At the smallest scale (microscopic), rock mineralogy as well as pore and grain structure may be studied. These observations are important for correctly identifying the depositional history as well as for understanding possible relationships between permeability and porosity, especially for carbonate reservoirs.[6] In addition, small fractures can be identified through microscopic examination. These fractures are, however, difficult to incorporate in conventional simulation models.

It is important to understand that the reservoir properties (such as permeability and porosity), even if measured on a microscopic level, are difficult to incorporate in a conventional reservoir simulator because of computer-memory limitations. Microscopic heterogeneities can be incorporated only indirectly as part of the reservoir description by understanding their impact on a larger scale. For example, minerology can be incorporated through its implication with respect to the large-scale depositional environment, or pore- and grain-size distributions can be incorporated through their impact on residual oil saturation.

Effect on Reservoir Performance. The scales of heterogeneities are also important because different heterogeneities affect reservoir performance differently. Microscale heterogeneities create preferential-flow-path channels. These are pore-level heterogeneities. Because of pore-level heterogeneities, displacing fluids may take preferential paths and bypass some hydrocarbons. The hydrocarbons left behind result

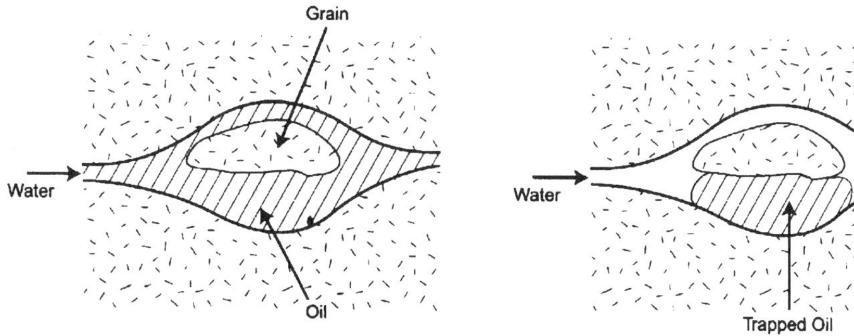

Fig. 1.4—Origin of trapped oil.

in a reduced displacement efficiency: the higher the amount of residual hydrocarbons, the lower the displacement efficiency. The displacement efficiency directly impacts the amount of oil recovered from the reservoir. The oil left behind is called trapped or residual oil.

For a typical water-wet reservoir in a dual-pore system, water enters the smaller pore, displaces the oil from it, and leaves oil behind in the larger pore (**Fig. 1.4**). Once water forms a continuous path, the discontinuous oil left behind can be displaced only by modifying the capillary forces between oil and the displacing fluid.

Macroscale and megascale heterogeneities result in preferential paths for displacing fluids on a larger scale. **Fig. 1.5** shows some simplified examples of these heterogeneities. Fig. 1.5a shows the impact of vertical heterogeneities on vertical sweep efficiency. Owing to stratification in the vertical direction, oil is preferentially displaced by the displacing fluid. The displacing fluid moves preferentially through high-permeability regions, leaving behind significant amounts of oil in the lower-permeability strata. Fig. 1.5b shows the effect of areal heterogeneities on areal sweep efficiency. Because of preferential flow paths, the displacing fluid reaches the producing well without reaching all parts of the reservoir. Once preferential paths are established, the displacing fluid short-circuits through these paths, leaving behind bypassed oil.

Gigascopic heterogeneities cause some oil sources to remain uncontacted. **Fig. 1.6** indicates that hydrocarbons may be stored in isolated sand lenses. These sand lenses are not connected with each other. If a drilled well intersects any of these lenses, the hydrocarbons can be produced from these lenses. However, if the lens is not in contact with any of the drilled wells (like the one in the middle of Fig. 1.6), the oil in that lens cannot be extracted unless an infill well is drilled. The oil left behind in the isolated parts of the reservoir is called untapped oil. Extraction efficiency of these reservoirs can be improved by drilling infill wells.

Effect of Heterogeneities on Flow Process. In a typical oil reservoir, different flow processes may be used to extract oil. During the initial stages of production, primary oil recovery uses the natural energy of the reservoir. As the natural energy is depleted, the reservoir may be subjected to a waterflooding process. If economically feasible, waterflooding may be followed by tertiary-oil-recovery processes, such as CO_2 flooding, surfactant flooding, or polymer flooding.

A clear relationship exists between the type of flow process taking place in the reservoir and the types of heterogeneities that play a significant role in determining the flow performance. On one extreme, in single-phase depletion of a gas reservoir, only gigascopic (and to some extent megascopic) heterogeneities impact overall reservoir performance. As long as the correct genetic units and the flow barriers between them are identified, the reservoir performance can be reproduced. Primary recovery is "forgiving" to mistakes with respect to small-scale heterogeneities or macroscale heterogeneities. Any reasonable representation of these heterogeneities is sufficient to reproduce the primary-recovery performance.

Mistakes with respect to large-scale heterogeneities, however, can result in an incorrect recovery prediction even under primary recovery. **Fig. 1.7,** which plots estimated reserves as functions of time for several North Sea reservoirs, shows an example. Many of these reservoirs are still under primary depletion or under natural water-influx drive. In tectonically quiet, relatively unfaulted areas, the estimates tend to go up because the original forecasts were conservative. In highly faulted fields, estimates go down as additional faults and flow barriers are discovered during the depletion phase.

As the flow process changes from primary depletion to secondary flooding, additional levels of heterogeneities can affect reservoir performance. Mega- and macroscopic levels especially have significant impact on prediction of reservoir performance under secondary-recovery performance. These heterogeneities include internal boundaries within units, areal reservoir connectivity, and the presence of contrasting zones.[7] Complex lithological structures can also affect secondary-recovery performance. Proper identification of mega- and macroscopic heterogeneities is critical before flow performance under secondary recovery can be honored.

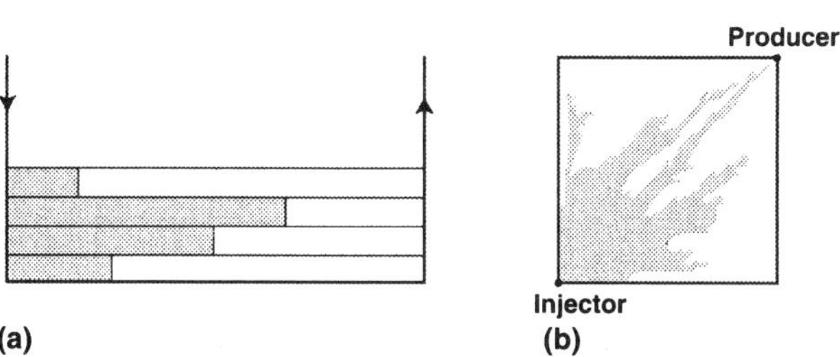

Fig. 1.5—Effect of heterogeneities on sweep efficiency.

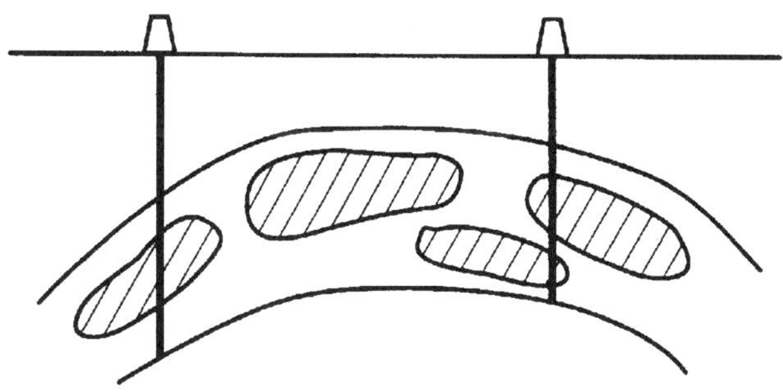

Fig. 1.6—Effect of gigascopic heterogeneities on performance.

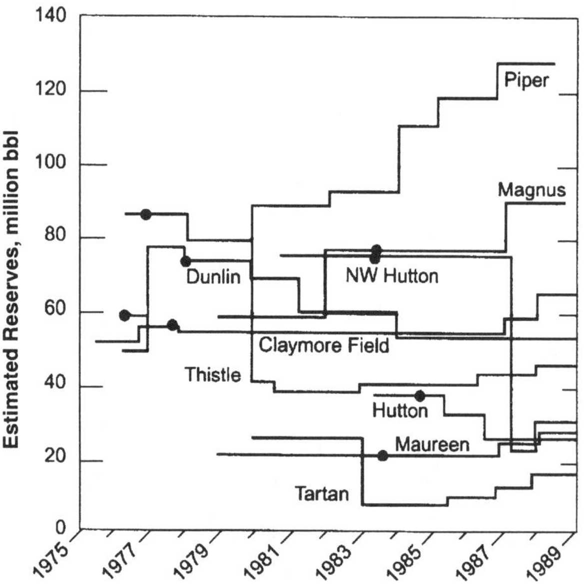

Fig. 1.7—Reserve estimates for North Sea fields.

The requirement that heterogeneities be represented becomes even more critical as the flow process becomes more involved. In an EOR process, owing to the complexity of the flow process, even the microlevels of heterogeneity can affect flow performance. These small levels of heterogeneity can affect the flow behavior of displacing fluids as well as affecting the way in which the interface between the displacing fluid and oil is established and moved.

On the basis of this discussion, we can make a few observations. First, as the flow process becomes more complex, the level of detail required with respect to heterogeneity increases. The more complex the flow process, the more small-scale detail is required. However, the effect of large-scale heterogeneities is not minimized, even for complex flow processes. In terms of data hierarchy, large-scale heterogeneities still have the greatest impact on overall flow performance, even for complex flow process.[7] As the level of heterogeneities decreases, the relative impact on flow performance also decreases. Therefore, more importance should always be placed on representing the largest-scale heterogeneities correctly, followed by relatively smaller levels.[8] The point to remember is that correct representation of smaller-level heterogeneities becomes increasingly important as the flow process becomes more complex.

A second observation relates to the practical difficulty in simulating reservoir performance. As the flow process becomes more complex, the physics of the flow process also becomes complex and must be represented correctly in a typical reservoir simulator. As a result, the computational requirements, even for the same number of gridblocks, increases for a more complex flow process. Obviously, to be able to simulate flow performance for a more complex flow process, it is preferable to use a smaller number of gridblocks to maintain control of the computational time. Unfortunately, it seems that, to represent the heterogeneities properly as the flow process becomes more complex, more gridblocks might have to be used, which may be difficult in practice. As an alternative, the representation of the flow process or the scale of heterogeneities has to be simplified; which one is more critical is a debatable issue that has yet to be resolved.

The third observation is a direct continuation of the second one. Knowing that different levels of heterogeneity play different roles, depending on the type of flow process, makes it obvious that simply matching flow performance under primary production may not guarantee that the reservoir description can predict secondary- or tertiary-oil-recovery performances correctly. For example, primary-recovery performance is forgiving to misrepresentation of vertical barriers between layers of heterogeneities. However, these barriers can play an important role in a waterflooding process. This may explain why some reservoir-performance-prediction studies underpredict breakthroughs or overpredict sweep efficiency, even after matching primary performance. This indicates that as many small-scale heterogeneities as possible should be represented, even though they may not affect the current flow process, because it may be valuable when the same reservoir description is used to predict future performance under a more complex process.

To summarize, different levels of heterogeneities have different impacts on oil recovery. Proper identification and knowledge of heterogeneities on various scales, therefore, is necessary for optimum reservoir production performance.

1.3 Reservoir Modeling in the Presence of Heterogeneities

For the purposes of optimizing production from a reservoir, reservoir behavior must be modeled properly to predict its performance. In modern times, the technique often used for modeling the reservoir behavior is the use of a reservoir simulator. Simulation has several advantages over analytical tools, including the following.

- The ability to test new procedures and new concepts without actually testing them in the field.[9]
- The ability to represent complex physical processes in a rigorous manner.
- The ability to incorporate reservoir heterogeneities at various scales.
- The ability to forecast future performance under existing, as well as modified, conditions.
- The ability to interpret performance under varied operating conditions.[9]

Compared with analytical methods, reservoir flow simulations require significantly more input information for proper application. In addition, simulation exercises are constrained by the computational abilities of individual computers. These limitations become more evident in the following discussion of the conventional application of reservoir simulation.

This section first discusses the conventional approach followed in reservoir simulation to predict reservoir performance, then covers some of the associated drawbacks and suggests possible improvements.

1.3.1 Conventional Approach. The most common application for reservoir simulation in a conventional approach is future-performance prediction. Because significant uncertainty exists in reservoir parameters used as input in reservoir simulation, the simulator model has to be validated before it can be used to predict future performance. As Mattax and Dalton[10] stated, "the most useful way to determine whether a model adequately describes a reservoir is to simulate past performance and to compare the simulation with the reservoir's actual performance history." The step of comparing prior actual performance with simulated performance is a process called "history matching." Once prior data have been matched, the modeler has more confidence in the simulation of future performance.

The commonly used model for history matching is a layer-cake model. This approach involves defining the reservoir in terms of several layers that have different physical properties and numerically simulating reservoir behavior in either two or three dimensions. This simple approach has several advantages.

- It is easy to understand the physical architecture of the reservoir in terms of a layer cake.
- It conforms with geology in many instances. For example, Weber and van Geuns[4] note that the layer-cake model can be applied to certain geological environments, including single or amplified barrier bars, shallow marine sheet sands, transgressive sands, offshore bars, certain turbidites, braided rivers, and aeolin deposits.
- It is easy to simulate because the numerical simulators are easily adaptable to orthogonal, preferably rectangular, gridblocks. (From the simulation point of view, orthogonal grids are preferable to nonorthogonal grids[11]).
- It can mimic historical performance within a few iterations.

In most field-scale simulators, use of a layer-cake model involves the following steps.

1. Establish the number of layers by use of the cross correlation between wells and/or on the basis of similar values of permeability-thickness product, porosity-thickness product, or permeability/porosity ratio. The initial correlation between wells is done by a geologist from log and core observations as well as from depositional history.

2. Define petrophysical properties for individual layers with core and/or well-test data. These properties include rock properties, such as permeability and porosity, and fluid saturations. They also include relative permeabilities and capillary pressure values if the simulation involves multiphase flow.

Rock properties at well locations are estimated with either core or log values; in the case of permeabilities, they are based on a relationship between permeability and log-derived porosity values. Permeabilities of individual layers may be adjusted to reflect the observed well-test-derived permeability at a particular well. Once the properties are defined at individual well locations for a particular layer, these values are interpolated to assign the values at the gridblocks at interwell locations.

Coreflooding data are used to assign relative permeabilities and capillary pressure values. In the laboratory, relative permeabilities and capillary pressure data are obtained for several core samples. Some type of averaging scheme is used to assign a single set of relative permeabilities as well as capillary pressure data for the reservoir.

3. Numerically simulate the reservoir performance. In numerically simulating the reservoir, every attempt is made to simulate the physics of the process. In some instances, however, the physics of the process may have to be simplified because of computational limitations.

4. If the simulated performance does not match the historical performance, adjust the reservoir parameters until a match with the production data are obtained. The parameters that are normally adjusted include relative permeability data, number of layers, and individual layer properties. For example, to match early breakthroughs, permeabilities of some layers may be increased, while it may be decreased for the other layers so that the averaged permeability still compares favorably with the well-test-derived permeability value.

Typically, a reasonable match between historical and simulated performance can be obtained within three to four iterations. After a match is obtained, future performance can be predicted with the same reservoir description. Prediction of future performance can be under existing conditions, but it is more likely to be under modified conditions that may include a sensitivity analysis of various operating parameters.

1.3.2 Limitations of the Conventional Approach. The conventional approach has four important drawbacks.

Series Modeling. A series (or sequential) modeling approach has five stages.[12]

1. Analysis of the raw data. In this stage, geologists and geophysicists analyze the raw data (e.g., cores, logs, seismic surveys, and depositional histories). The information may also include certain qualitative observations based on prior experience. On the basis of available information, the geologist may describe the reservoir in terms of several correlated or uncorrelated layers.

2. Construction of the model. In this step, once the information reaches the engineer, depending on the computing limitations and any other constraints, the engineer may lump together several layers to describe the reservoir in terms of a limited number of layers. Lumping may involve adding layers that may have similar petrophysical properties, such as permeability and porosity, but that may not be geologically similar.

3. History matching of prior performance. During this trial-and-error history-matching procedure, changes may be made in the reservoir description to match performance. These changes include changing the number of layers, adjusting individual-layer properties, or adjusting petrophysical properties.

4. Calibration of the model to match the results better. This step goes together with Step 3, in which the model is calibrated with prior production data and modified to account for any changes in the flow process as well as in well configurations.

5. Prediction of future performance. In this last step, future performance is predicted under desired conditions with the calibrated model.

This series modeling approach has the following pitfalls.

- Because of its nature, the description process does not become interdisciplinary. Once the initial description is created, the geologist and geophysicist may rarely get involved in the modeling process. The changes in reservoir description made by an engineer may be consistent with the engineering information but may violate some of the "soft information" used in constructing the correlation. Also, use of the series approach contains the implicit assumption that the model created by the geologists is static; that is, it does not change or get modified on the basis of insights gained during the flow-simulation process.
- Sequential planning assumes that each of the five steps is completed in a timely manner. If any of the steps is delayed, all subsequent steps are also delayed. In short, any delay in the process can cause a corresponding delay in the delivery of the final product. Most simulation studies take a longer time than originally anticipated.
- There is no flexibility for incorporating additional data or knowledge that becomes available while the study is in progress. If such knowledge is incorporated, it delays the project even further. Delays, in some instances, are translated into missed opportunities.

Lack of Scale Dependence on Reservoir Properties. The inclusion of scale dependence is important in reservoir description from two perspectives: input and output. From the input perspective, it is important to account for the scale over

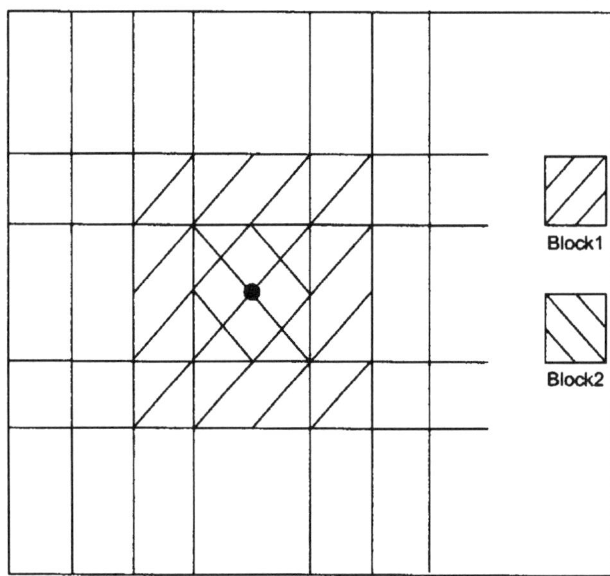

Fig. 1.8—Gridblock configurations.

which input data are gathered. As discussed in the previous section, all the reservoir properties are not necessarily measured at a desired scale. If the properties are measured on a different scale, we must reconcile the information collected on various scales and ensure that the differences in the scales of observed data are accounted for properly.

A commonly used method that violates the principle of scaling is matching well-test data with core-derived permeability data at individual well locations. In many conventional simulation studies, the log-derived porosities are transformed to permeability values. These permeability values (vertically defined) are then averaged (either arithmetically or geometrically) for comparison with the well-test data. If they do not compare well, the log-derived values may be calibrated to match the well-test data. What is ignored in this analysis is that the well-test-derived permeability represents not only the values at the wellbore but also those at the region surrounding the wellbore.

The second perspective, regarding the scale of reservoir properties, arises from the requirement that reservoir properties be estimated at locations where no data are available. Specifically, if a reservoir simulator is being used, gridblock properties must be defined at the interwell scale. The conventional method defines the gridblock-property value at the center of the gridblock (**Fig. 1.8**). This figure shows two potential gridblock configurations. One is indicated by solid lines, and the other is indicated by dashed lines. By coincidence, the center of the gridblocks (Blocks 1 and 2) for both configurations is at the same location. Using a conventional method, we interpolate the values at well locations and assign a value at the center of the gridblock. That value represents the block value.

In reality, the two blocks should have different physical properties because their sizes differ. Each block is made up of different small-scale heterogeneities ("small cores" present within each block). The effective physical property defined for a block should account for the smaller-scale heterogeneities that are present within the gridblock. Obviously, Blocks 1 and 2 do not contain the same small-scale heterogeneities. The estimated physical properties, therefore, should be different rather than identical. Also, overall variance observed between the values for the smaller gridblocks should be greater than those for the larger gridblocks. By simply interpolating the values at the center of the gridblock, the conventional method fails to account for these effects.

Uncertainty Estimations. History matching of prior production performance is used primarily to predict future performance. The implicit assumption in predicting future performance is that the reservoir description used in matching prior performance uniquely defines the reservoir description; therefore, it can be used to predict future performance.

In reality, the reservoir description used for matching historical performance is a result of several subjective judgments, which include decisions regarding type of data that need to be honored, scaleup of certain petrophysical properties (core relative permeabilities vs. pseudofunctions), individual layer properties, and appropriate number of layers. It is easy to show that two people working with the same data set will come up with different reservoir descriptions that still match the historical performance. Considering the uncertainties involved in the estimation of reservoir properties, it is reasonable to assume that the reservoir description that matches the historical performance is nonunique.

It is unreasonable to assume that simply duplicating past performance validates the model. In fact, the match may reflect compensating errors among various input parameters. Further, recent studies have shown that it is relatively easy to match fieldwide performance but very difficult to match individual well performances without making significant subjective assumptions. If we know that the reservoir description is nonunique, it is imperative to develop a future-performance profile that indicates the nonuniqueness of the history match and therefore presents the uncertainties in future performance. The conventional method, unfortunately, does not predict uncertainties in future performance. Instead, after "validating" the model with past performance, it predicts a single future performance.

Inclusion of Geological Features. Conventional layer-cake models may be appropriate for many types of reservoirs that exhibit similar geological architecture. However, in many instances the geological features exhibit complex patterns that may not conveniently fit in a layer-cake model. Weber and van Geuns[4] describe three reservoir categories for clastic reservoirs.

Layer-Cake Reservoirs. These reservoirs have structures continuous enough that they can be represented with a conventional layer-cake model.

Jigsaw-Puzzle Reservoirs. These reservoirs are composed of series of sand bodies that fit together without major gaps. Some nonpermeable baffles may exist between superimposed sand bodies. Examples of these reservoirs include point bars and turbiditic-fan sand bodies.

Labyrinth Reservoirs. These reservoirs consist of complex arrangements of sand bodies and lenses. They appear discontinuous, and detailed correlations are possible only when the well spacing is small. Examples of these reservoirs include tidal-channel fill embedded in a barrier-bar sand and lacustrine sheet sands combined with fluvial channel sands.

Fig. 1.9 shows schematic representations of these three types of reservoirs. The increase in lack of continuity can be observed as we go from layer-cake reservoirs to labyrinth-type reservoirs. In constructing an appropriate model, it is important that both the "form" and "function" of the reservoir is honored as much as possible.* The form represents the geological environment, rock types, and the reservoir architec-

*Personal communication with D. Kerr, U. of Tulsa, Tulsa, Oklahoma (1992).

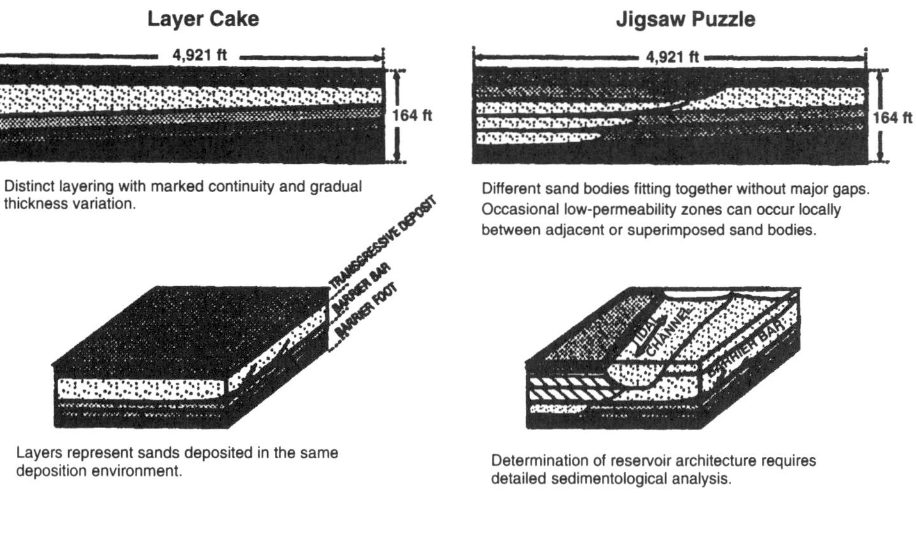

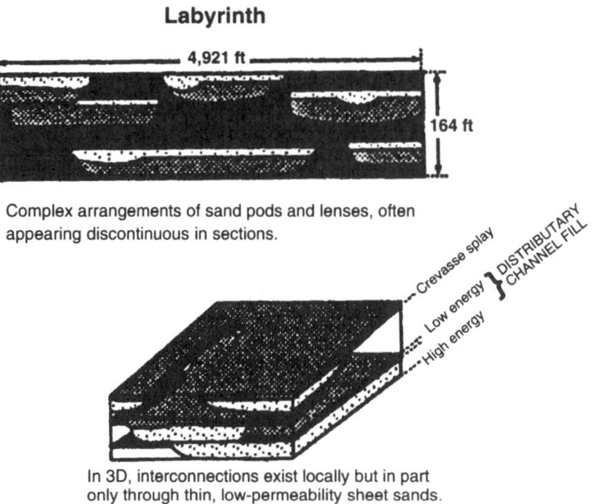

Fig. 1.9—Reservoir models.

ture or building blocks. The function represents the petrophysical properties that result in the flow performance. In conventional history matching, the focus is on matching the function or the historical performance. However, the reservoir architecture (or the form) may be violated in that process.

1.3.3 Possible Improvements. The conventional simulation method can be improved. The possible improvements suggested in this section address some of the drawbacks discussed in the previous section.

Parallel Modeling. Even in the presence of extensive data, a lot of uncertainties remain with respect to reservoir description. These uncertainties can be minimized if all the available data are properly used and are accounted for. In a series-modeling approach, some constraints imposed by the geologists or geophysicists may be ignored during the simulation process, possibly creating more uncertainties in the reservoir description. The process may also be delayed because of delays in any of the steps.

In contrast, parallel modeling offers an alternative for integrating the data in a consistent manner. In parallel modeling, a multidisciplinary team, which may include geologists, geophysicists, reservoir engineers, and production engineers, studies a single reservoir. Instead of taking steps in series, the steps go in parallel. An attempt should be made to construct an integrated description with all available data, as shown in **Fig. 1.10.** As a possible process, all five steps (analysis, construction, history matching, calibration, and future-performance prediction) can occur almost simultaneously. As a start, a model that honors large-scale heterogeneities can be constructed from geological/geophysical information and can be flow simulated to gauge predicted flow performance by use of average petrophysical properties. The predicted flow performance can be fed back to a geoscientist to investigate any possible modifications at a large scale in the original model. The refined model can be flow simulated again for calibration purposes. As the model gets progressively closer to the observed performance, a higher level of detail with respect to heterogeneities can be added to give a better picture of the reservoir. Because the description is created by a combined effort, much of the information (whether soft or hard) is honored in constructing the final description. In addition, because the process involves significant feedback, new information that becomes available during the process can be incorporated easily.[12]

Fig. 1.11 shows an example of a successful integrated, parallel approach in constructing a reservoir description. Fig. 1.11a shows the history-match results between the simulated and the observed data for a large Middle Eastern reservoir.[13]

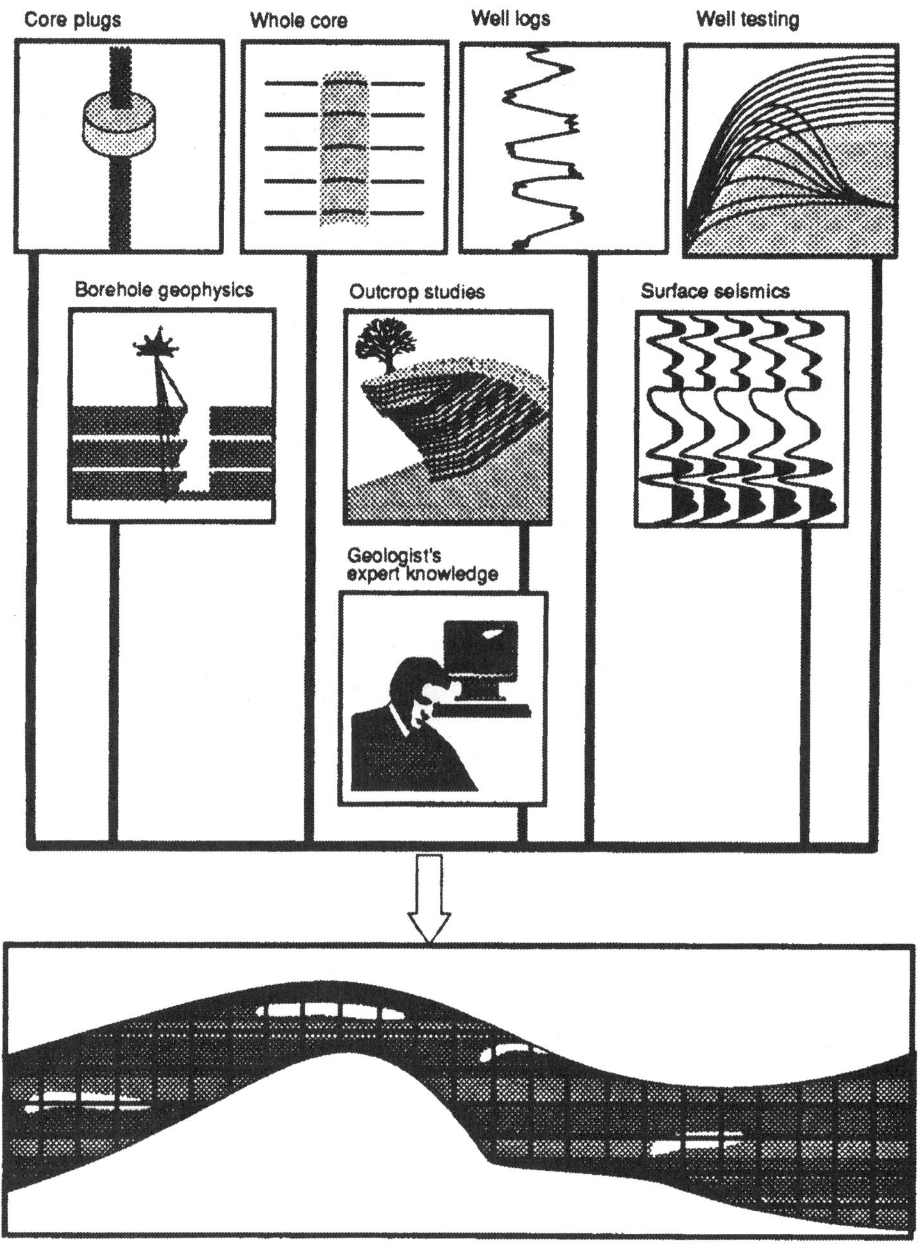

Fig. 1.10—Reservoir description.

This match was based on a series-modeling approach. First, a geologic model was developed by a team of geologists from two partner companies. Once approved, the geologic model could not be changed. After making changes in petrophysical properties and aquifer strength, the engineers obtained the best match (Fig. 1.11a). The total effort took three years. The alternative approach of parallel modeling was used starting in 1992. A team was formed consisting of a geophysicist, two geologists, a production engineer, two reservoir engineers, and a computer scientist. During the history-matching phase, the geologists and geophysicist updated the geologic model in areas that lacked geologic data. Also, additional information, which had to be left out in the beginning, was added as the history matching progressed. The total effort took nine months. Fig. 1.11b shows the results. As is evident, a much better match between simulated and observed performance was obtained with less effort.[13]

Inclusion of Scale Effect. Inclusion of the appropriate scale in describing the reservoir properties is important for a proper reservoir description. Especially in numerical simulators, gridblock properties should depend not only on the relative gridblock location but also on gridblock size.

Many ways have been proposed to define the physical properties on a particular scale appropriately. Although core values may be known, it is impossible to define small enough gridblocks that have the same scale as the cores. Therefore, in describing gridblock properties, some scaleup process may be needed. The properties involved are porosity, permeability, relative permeability, and capillary pressure.

Scaleup of porosity is relatively easy. Arithmetic average of small-scale porosities results in the effective porosity of the gridblock. For permeability, scaling up is much more difficult. Different methods have been proposed to scale up permeability values ranging from geometric averaging to effective permeability tensors. For scaleup of relative permeabilities, a common approach is to use pseudofunctions. In this approach, the gridblock is simulated in the presence of small-scale heterogeneities, and the effect of the heterogeneities is captured in terms of effective permeabilities. These effective properties are used in subsequent simulations on a large

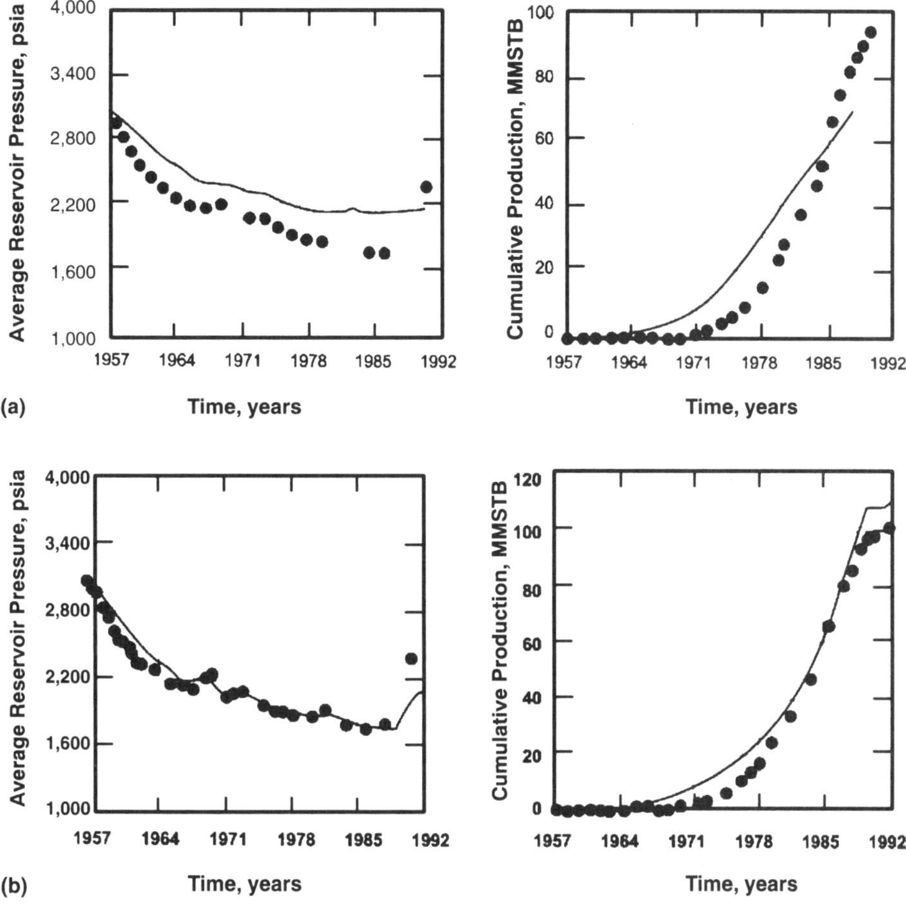

Fig. 1.11—(a) Series vs. (b) parallel modeling.

scale.[14] Although use of pseudofunctions is promising, the approach still has some limitations. Chief among them is the inability of pseudofunctions to describe gridblock properties independent of the imposed boundary conditions.[15] It should be remembered that any upscaling process, no matter how accurate, results in homogenization of small-scale properties. As properties are scaled up, some flow-process details are lost. The question of the extent of scaling up and its effect on the flow process has not been fully resolved. Chap. 8 discusses the entire upscaling process in detail.

Inclusion of Uncertainties. Correct modeling of prior historical performance does not ensure correct modeling of future performance. As explained in the previous section, the effect of scales of heterogeneities is flow-process dependent. As a result, matching primary performance may not lead to correct prediction of the waterflooding performance. What is adequate in terms of heterogeneity representation for primary processes may not be adequate for secondary processes. In addition, although overall field performances can be reasonably matched with 10 to 40% accuracy,[12] matching individual well performance is difficult because of both the limited amount of information on individual well variables and a lack of accuracy. Considerable uncertainty surrounds individual well performance matching. Also, all the reservoir properties are not available at every desired scale. Some subjective assumptions have to be made with respect to how properties are defined at individual gridblock levels.

In light of these uncertainties in the prediction of future performance, predicting a unique performance on the basis of prior history matching is misleading and, in some cases, may be inappropriate. Instead, accepting the uncertainty in the reservoir description allows prediction and quantification of the uncertainties in future performance. **Fig. 1.12** shows that, rather than predicting one unique future performance, a better procedure is to predict an array of future performances, all reasonably matching the historical perfor-

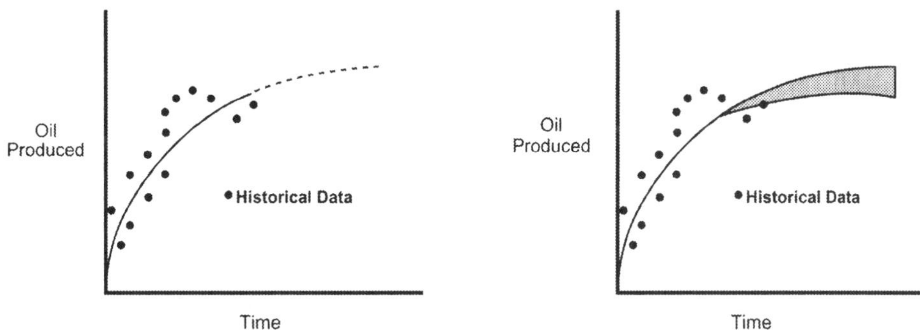

Fig. 1.12—Prediction of future performance.

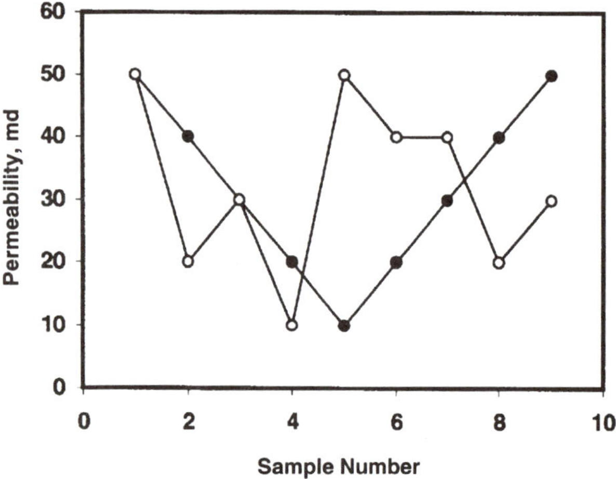

Fig. 1.13—Permeability values as function of distance.

mance as well as the available input data. By investigating the range of possible future performances, the uncertainty with respect to future production and the resulting economic evaluation can be represented better.

Representation of Geological Features. Many conventional methods create reservoir descriptions that may not strictly honor the reservoir's geological architecture. In emphasizing function (performance) over form (architecture), some key architectural features may be violated in the final description. One way to overcome this problem is through parallel modeling. Ensuring that the created reservoir description honors the geologist's view ensures that some of the qualitative information that goes into the geological model is retained. Another approach is to use a two-step modeling process. In this process, the reservoir description is created first by use of large-scale geological features. Once the architecture is in place, the petrophysical properties that are consistent with the observed data are estimated within each feature. Chap. 7 discusses this procedure in greater detail.

1.4 Use of Geostatistics in Reservoir Description

In the presence of limited information, any reservoir description involves the use of statistics. Even in conventional reservoir simulations, different interpolation schemes are used to estimate reservoir properties at interwell locations. These schemes assume some type of implicit relationship between the sample data.

In this book, we concentrate on the application of geostatistical techniques to reservoir description. We define geostatistics in a broad sense as a statistical technique that accounts for spatial relationships of variables in estimating values of the variables at unsampled (unobserved) locations. These variables include conventional petrophysical properties, such as permeability and porosity; seismic information, such as impedance values; or geological variables, such as sands and shales. The relationships may include information concerning how neighboring values of the same variable are related to each other or information concerning what the chances are that two different variables are close to each other (e.g., channel fill and crevasse splay).

Geostatistics takes advantage of the fact that, in many natural phenomena, variable values measured close to each other are similar. As the distance between the measured values increases, the similarity between the two measurements decreases. On the basis of this similarity, geostatistics captures the spatial relationship through certain correlation functions.

Consider an example of two permeability measurements in the wellbore.[16] Both measurements are conducted at uniform intervals of 1 ft. Close observation of these two measurements reveals that both measurements have the same number of sample points and the same values. From a conventional statistical measurement (such as mean and variance), there is no difference between these two samples. However, plotting these two samples as a function of distance (**Fig. 1.13**) shows that a difference does exist between the two measurements. Measurement 1 shows more continuity in the changes than Measurement 2, where the changes are rather abrupt. The type of continuity observed in Measurement 1 cannot be captured in conventional statistics. However, geostatistics can capture and quantify this continuity and use the information to predict values at unsampled locations.

1.4.1 Applications.

Geostatistical procedures are versatile enough to be used for several purposes related to reservoir description.

Interpolation and Extrapolation. Geostatistics can be used to interpolate and extrapolate the values of reservoir variables at unsampled locations. Contour maps can be created from the estimated values

Spatial-Distribution Analysis. Geostatistics can provide the quantitative relationship describing the spatial variability of a reservoir property. This includes the distance over which a given variable is related as well as how that variable is spatially related to other variables.

Risk Analysis/Uncertainty Estimates. In any estimation process, in addition to the estimated value, it is desirable to know the relative uncertainties in the estimation. Geostatistics can provide different ways of defining uncertainties in estimated values at unsampled locations. These uncertainty estimates can be useful in risk analysis. A good example of these estimates is the determination of hydrocarbons in place in a reservoir. Knowing the uncertainties involved in such a determination, a proper economic evaluation that incorporates the associated risks of uncertainties can be conducted.

Use of Intercorrelated Attributes. In a reservoir description process, utilization of all available sources of data is very important. Different sources of data are collected over different regions, and some data are collected much more extensively than other data. If we are interested in estimating values of a variable for which only a few samples are available, it may be useful to use the sample data from another extensively sampled variable if a spatial relationship exists between the two variables. A good example of such an application would be estimation of porosity values during the exploration stage. Core and log data may be available from only a few drilled wells, while seismic data may be extensively collected. If a spatial relationship can be established between the well data and the seismic data, the extensively sampled seismic information can be used to estimate the porosity values at unsampled locations. Another example is the use of log porosity data from various wells to estimate permeability values. Using limited core permeability data, we can generate permeability values at unsampled locations if a spatial relationship between porosity and permeability can be established. Geostatistics allows the development of a cross relationship between any two variables; therefore, one variable can be effectively used to estimate the other.

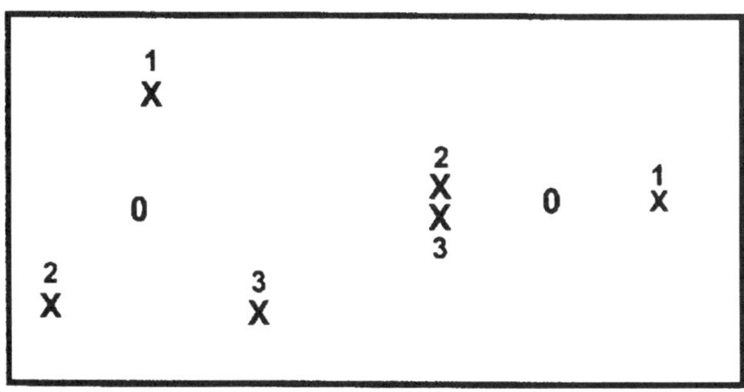

Fig. 1.14—Data configuration of sampled data.

1.4.2 Advantages. Geostatistical methods have several advantages over conventional methods.

• Most estimation methods used for interpolation and extrapolation purposes are highly empirical in nature. Although based on sound mathematical principles, these methods may not provide a physical reasoning for using sample values in certain ways to estimate values at unsampled locations. For example, an inverse-square-distance method assumes that the influence of a sample point on an estimated value at an unsampled location is inversely proportional to the distance squared between those two points. No physical basis exists for this particular assumption. In contrast, methods that use geostatistics are based on spatial relationships that are established by use of the sampled data. Depending on the available sample data and their configuration, the relationships change. In essence, spatial relationships are customized for a particular data set, unlike when a generalized relationship is used for all the data sets. This way, the influence of the surrounding sample data on the unsampled location is calibrated properly.

• Most estimation methods do not provide the estimation errors along with the estimated value. However, it is important that the uncertainties associated with the estimation be established properly. Depending on the availability of the sampled data, the uncertainties with respect to the estimate can vary significantly. In addition to providing the estimated values, geostatistical procedures also provide associated uncertainties in the estimation.

• Most estimation methods estimate only the point values at unsampled locations. In contrast, geostatistical methods can account for the support or volume of the location at which the variable needs to be estimated. This may be especially useful in estimating simulator gridblock values. Although certain limitations exist with respect to area- or volume-based estimates of a variable, in principle, geostatistical methods can account for the volume or area of a location in estimating a variable value at that location.

• Geostatistical techniques honor sampled data. This means that values calculated by these methods are identical to sampled values at sampled locations.

• Many estimation techniques do not account for data redundancy. Although the same amount of information is available, depending on the data configuration, some data may be more valuable in the estimation procedure than others. **Fig. 1.14** shows two possible configurations of three sampled points to illustrate this. Configuration A shows that the three sample points (Xs) are uniformly distributed surrounding the unsampled point (O). Estimation of a value at Location O is influenced equally by all three sampled points. In contrast, for Configuration B, Sample Point 1 is on one side of the unsampled location and Sample Points 2 and 3 are so close to each other such that, individually, they would not provide the same amount of information as Point 1. This intuitive conclusion is borne out by geostatistical methods, where, in Configuration A, equal weights are assigned to all three sampled points in estimating the value at the unsampled location and, in Configuration B, Sample Point 1 is assigned a higher weight than Sample Points 2 and 3. In addition, the uncertainty estimate under Configuration A is less than that under Configuration B. This is consistent with the fact that the three sample points are uniformly distributed for Configuration A, thereby providing more information than the three sample points in Configuration B.

• Most estimation techniques do not allow the use of qualitative information in estimation of a variable at unsampled locations. If proper procedures are used to quantify the qualitative information, some geostatistical techniques do allow the use of qualitative information to estimate values at unsampled locations.

• As described in the previous section, some geostatistical methods allow the use of extensively sampled variables to estimate the values of other sparsely sampled variables.

1.4.3 Disadvantages. Although the geostatistical procedure offers several advantages over conventional methods, one must be aware of certain disadvantages.

Additional Data. Geostatistical techniques allow manipulation of only available data. The data are analyzed in a physically reasonable manner, and the variable values are estimated at unknown locations. Just because estimates have been obtained at every unsampled location should not give a false sense of security because all the estimated values are associated with uncertainties. These uncertainties can be reduced by collecting additional data. In addition, the spatial relationship model can also be improved if additional data are collected. Massaging of available data never replaces the value of additional data. Therefore, any additional data should always be included in the analysis, although this may involve modifying spatial relationships and estimating values again.

In collecting additional data, more time should be spent on data that have a greater effect on the quality of the solution.[8] Data hierarchy in terms of data collection is extremely important in terms of cost/benefit analysis. It seems logical that additional data that represent large-scale heterogeneities will have a bigger impact on reducing uncertainties in overall performance than data collected on small-scale heterogeneities.

Subjective Decision Making. Similar to any other statistical procedure, geostatistical techniques involve subjective decision making. During the analysis, each step requires a certain amount of subjective decision making. Depending on the decision made, the results are consistent with that decision. Considering the importance of each decision, these decisions should be made in a consistent manner and on an informed basis; i.e., they should be consistent with the available information and should be made by an expert in that area.

Hard Work. Geostatistical methods allow a lot more flexibility in adopting various sources of data in describing the reservoir. Using these various sources of information in a consistent manner requires the user to work harder than with a conventional method. Inconsistent data sets can no longer be reconciled by favoring one data set over the other; instead, the data sets should be reconciled by accounting for the different supports over which the samples are collected. In addition, precision in a data set should also be taken into account. Qualitative information submitted by the geologist in creating the reservoir description can no longer be ignored; instead, the description has to be consistent with the qualitative constraints. Other, similar constraints make geostatistical methods more difficult and time-consuming to use than conventional methods.

1.4.4 Procedure. Application of geostatistics is a three-step procedure: (1) assumption of stationarity, (2) spatial modeling of sample data, and (3) estimation of a variable value at unsampled locations.

Assumption of Stationarity. The assumption of stationarity is probably the most important assumption in geostatistical analysis. It requires that the model developed on the basis of the sampled data is applicable over the region of interest; this region defines the region of stationarity.

Any statistical analysis requires an assumption of stationarity; geostatistics is no exception. Any statistical process requires that the inference made on the basis of the sample data is applicable to the population of interest. In most instances, the assumption of stationarity can be made intuitively. For example, calculating the average age of all children going to a single elementary school and stating that this average represents the average age of all children going to elementary school in that particular city may give a reasonable assumption of stationarity. The assumption is that the data from one elementary school sample can be extended to the population of all the elementary schools in the city. As the population is extended, however, the assumption of stationarity may get weaker. If we state that this average represents the average of all children going to the elementary school within the entire state, we may not be inferring correctly because of, for example, differences in rural and city populations or differences in inner-city vs. suburban populations. The point is, the narrower the definition of the region of stationarity, the closer it is to the assumption of stationarity. As the region of stationarity is broadened with the same number of samples, the assumption of stationarity may be violated.

In this example, defining the region of interest is relatively easy. In the case of reservoir description, it may be difficult. In many instances, the region over which the model based on the wellbore data is applicable has to be defined from limited wellbore data. We can never be certain about this region; the decision is necessarily subjective. Because of the scarcity of samples, the region of stationarity is extensive during the early stages of E&P. As more information is collected, the region of stationarity can be defined more narrowly.

The uncertainty in the definition of the region of stationarity does not minimize its importance. The decision must be made carefully and has to be based on available data. The region of stationarity cannot be proved or disproved in the absence of additional data. The only thing we can state is that our analysis is consistent with the assumption of stationarity.

Modeling of Spatial Relationship. The second step in a geostatistical process is modeling of a spatial relationship on the basis of the sampled data. In essence, we evaluate the spatially collected sampled data for a given variable and estimate the extent of a spatial relationship between the sample data. As the distance between the sampled data increases, the similarity between the data decreases. This similarity or lack thereof is mathematically defined with a statistical function called the variogram, and the estimated values of this similarity is modeled with analytical functions. The variogram quantifies how closely neighboring sample values are related to each other, and the technique can be extended to establish a spatial relationship between two variables. Other functions can be used to quantify spatial relationships; however, the variogram is the one most commonly used in geostatistics for this purpose.

Estimation. The third and last step in a geostatistical process is estimation of values at unsampled locations. In conventional geostatistics, the estimation procedure is called kriging. Different kriging procedures are used, depending on the type of estimation procedure. For example, if point estimates are required, punctual or point kriging is used; if estimates of block values are required, block kriging is used. Several other methods have been proposed to estimate values at unsampled locations, depending on the specific type of application.

In addition to conventional kriging techniques is the conditional-simulation method, which holds greater promise in the area of reservoir description. This method allows creation of alternative reservoir images, each having an equal probability of existence. By creating several possible images of a reservoir description, conditional simulation allows the uncertainties in the estimation to be accounted for appropriately.

1.5 Outline

This section is a brief outline of the chapters included in this book.

Chap. 2, "Principles of Statistics," introduces the basic principles of conventional statistics that are necessary for understanding geostatistical procedures and analysis. Starting with frequency distribution, it covers summary statistics, probability definitions, cumulative distribution and probability density functions, expected values, inference techniques, and covariance and variogram definitions.

Chap. 3, "Spatial Relationships and Modeling," discusses the procedures used to estimate and to quantify spatial relationships. Although the emphasis is on estimation of the variogram, the chapter also discusses other methods (e.g., the covariance and the two-point histogram) and other unconventional methods (e.g., the madogram and the radogram). The chapter also covers modifications used for estimating conventional variograms to obtain more robust estimates. It further discusses various nonlinear transformations commonly used to transform sample data. These transformations include log, Gaussian, and indicator transformations. The second part of Chap. 3 deals with modeling spatial relationships. Different models, including fractal models, that allow modeling of these relationships are discussed along with ways of modeling the estimated values. In addition, modeling of relationships that exhibit anisotropy and relationships between two attributes (cross variograms) are presented.

Chap. 4, "Conventional Estimation Techniques," addresses conventional geostatistical techniques used for estimation purposes. The type of procedure used depends on the type of estimation required, and different procedures have been developed for this purpose. The chapter covers simple kriging, point or ordinary kriging, block kriging, cokriging, universal kriging, and nonlinear kriging. The utility and limitations of these various techniques are presented along with applications.

Chap. 5, "Simulation Methods," introduces the reader to simulation methods, which allow a user to create alternative images of the reservoir, each equally likely. The chapter discusses the advantages of simulation methods and presents a preview of various simulation techniques.

Chap. 6, "Grid-Based Conditional-Simulation Methods," discusses various techniques that have been recently introduced in the literature to generate reservoir properties at the grid scale. This is the scale that is defined by grids (preferably orthogonal) and that can be used as direct input to reservoir flow simulators. The first section of the chapter discusses sequential-simulation techniques; these include sequential Gaussian simulation, sequential indicator simulation, and probability simulation. The second section discusses cosimulation of multiple attributes, accounting for interrelationships among various attributes. The last section discusses use of simulated annealing to describe reservoir properties.

Chap. 7, "Object-Based Conditional Simulation Methods," presents conditional simulation techniques based on honoring the shapes of various objects rather than grid scales. Typically, these shapes include geological sand bodies (such as channel fill) and other large- and small-scale features (such as faults and fractures). The methods discussed include Boolean methods and Markov point processes.

Chap. 8, "Scaling of Reservoir Properties," introduces methods used to scale up reservoir properties generated on a very fine scale. Conditional-simulation methods discussed in Chaps. 6 and 7 are relatively fast and can generate millions of gridblock values in a relatively short period. To use them as input in flow simulation, these properties need to be scaled up. This chapter covers various methods used for upscaling permeability, relative permeability, and capillary pressure.

Chap. 9, "Looking Ahead," summarizes accomplishments in the area of geostatistical applications to reservoir description, discusses the advantages and limitations of these techniques, and concludes with potential areas of future research interest and applications.

The chapters all contain several examples to illustrate the concepts. Two types of examples are provided: numerical examples show the calculation procedure for a particular method with limited, synthetic data sets, and field examples show application of the same procedure to a field data set.

To maintain consistency and avoid repetition of explaining the background of the field data, a single field data set illustrates all the field examples. Appendix A provides the details about the field data.[17] Briefly, the data are collected from the Burbank field, located 90 miles northwest of Tulsa. A total of 168 wells, either cored or logged, were used to evaluate the 12,320-acre study area. On the basis of 30 cross sections across the study area, the reservoir was divided vertically into 10 flow units. In most field examples, data are used from individual wells or individual flow units. Although alternative descriptions of flow units may be possible, we assume that our field-unit description is correct for the examples in this book.

The rationale for choosing these field data is that, in practice, this is the approach one might follow. Starting with the individual well data, the data can be divided vertically according to geological units, and each unit (which need not be areally continuous) can be described to capture the areal variability. The division of individual geological units is subjective. Because it is not the purpose of this book to describe methods for drawing isochronal lines to divide the reservoir in subregions, we assume that the division into 10 units is a correct description for our field data. We realize that alternative ways exist by which the same field can be geologically described; however, we believe that the geostatistical techniques discussed in the subsequent chapters are valid even if the geological interpretation changes. The descriptions we create throughout the subsequent chapters are consistent with the individual-unit characterization we have assumed to be correct. If that description is modified, the same techniques can be used to create consistent descriptions with a modified unit characterization.

Many of the programs we used to analyze the field data are from Ref. 18. This reference contains a wealth of geostatistical programs as well as many utility programs. Most important, these programs are in the public domain and can be modified to suit any specific purpose. Ref. 18 also provides brief descriptions of many of the geostatistical tools.

Before we explain the various geostatistical tools, we want to emphasize the importance of understanding the usefulness and the limitations of these tools before applying them. Geostatistical applications, without a doubt, are one of the most promising areas in the reservoir characterization process. We hope that this book helps the reader evaluate and weigh different applications carefully before choosing the most appropriate technique.

Nomenclature

F_w = fractional flow of water
k = permeability, L^2, md
k_r = relative permeability
N_p = cumulative oil produced, L^3, bbl
P_c = capillary pressure, m/Lt^2, psi
ϕ = porosity

References

1. "Trends in Reservoir Management," *Oilfield Review* (January 1992) 8.
2. Ahmed, U., Crary, S.F., and Coates, G.R.: "Permeability Estimation: The Various Sources and Their Interrelationships," *JPT* (May 1991) 578; *Trans.*, AIME, **291.**
3. Enderlin, M.B., Hansen, D.K.T., and Hoyt, B.R.: "Rock Volumes: Considerations for Relating Well Log and Core Data," *Reservoir Characterization II*, L.W. Lake, H.B. Carroll Jr., and T.C. Wesson (eds.), Academic Press, San Diego, California (1991) 277–88.
4. Weber, K.J. and van Geuns, L.C.: "Framework for Constructing Clastic Reservoir Simulation Models," *JPT* (October 1990) 1248; *Trans.*, AIME, **289.**
5. Journel, A.G.: "Use of Geostatistics for Reservoir Characterization," course notes, Stanford U., Stanford, California (December 1987).
6. Lucia, F.J.: "Petrophysical Parameters Estimated From Visual Descriptions of Carbonate Rocks: A Field Classification of Carbonate Pore Space," *JPT* (March 1983) 629.
7. Saleri, N.G., Toronyi, R.M., and Snyder, D.E.: "Data and Data Hierarchy," *JPT* (December 1992) 1286; *Trans.*, AIME, **293.**
8. Raza, S.H.: "Data Acquisition and Analysis for Efficient Reservoir Management," *JPT* (April 1992) 466.
9. Lake, L.W.: "Predicting vs. Interpreting: A Survey of Oil Recovery Modeling," paper presented at the 1993 SIAM Conference on Math and Computational Issues in Geosciences, Houston, 19–21 April.
10. Mattax, C.C. and Dalton, R.L.: "Reservoir Simulation," *JPT* (June 1990) 692–95.
11. Aziz, K.: "Reservoir Simulation Grids: "Opportunities and Problems," *JPT* (July 1993) 658.
12. Saleri, N.G.: "Reservoir Performance Forecasting: Acceleration by Parallel Planning," *JPT* (July 1993) 652.
13. Hazlett, W.G. *et al.*: "The Integrated Approach to Performing Reservoir Simulation Studies," paper SPE 26224 available from SPE, Richardson, Texas (1993).
14. Lasseter, T.J., Waggoner, J.R., and Lake, L.W.: "Reservoir Heterogeneities and Their Influence on Ultimate Recovery," *Reservoir Characterization*, L.W. Lake and H.B. Carroll Jr. (eds.), Academic Press, Orlando, Florida (1986) 545–60.

15. Hewett, T.A. and Behrens, R.A.: "Conditional Simulation of Reservoir Heterogeneity With Fractals," *SPEFE* (September 1990) 217; *Trans.*, AIME, **289.**
16. Kelkar, M.: "Introduction to Geostatistics," tutorial paper presented at the 1991 Intl. Reservoir Characterization Technical Conference, Tulsa, 11–13 November.
17. Hird, K.B.: "A Conditional Simulation Method for Reservoir Description Using Geological and Well Performance Constraints," PhD dissertation, U. of Tulsa, Tulsa (1993) Chap. 2.
18. Deutsch, C.V. and Journel, A.G.: *GSLIB: Geostatistical Software Library and User's Guide,* Oxford Press, New York City (1992).
19. Haldorsen, H.H.: "Simulator Parameter Assignment and the Problem of Scale in Reservoir Engineering," *Reservoir Characterization*, L.W. Lake and H.B. Carroll Jr. (eds.), Academic Press, Orlando, Florida (1986) 293–340.
20. Weber, K.J.: "How Heterogeneity Affects Oil Recovery," *Reservoir Characterization*, L.W. Lake and H.B. Carroll Jr. (eds.), Academic Press, Orlando, Florida (1986) 487–544.
21. Worthington, P.F.: "Reservoir Characterization at the Mesoscopic Scale," *Reservoir Characterization II*, L.W. Lake, H.B. Carroll Jr., and T.C. Wesson (eds.), Academic Press, San Diego, (1991) 123–65.
22. Corrigan, T.: "Factors Controlling Successful Reserve Prediction: A Cautionary Tale from the UK North Sea," paper presented at 1988 Norwegian Petroleum Soc. Conference on Reservoir Management in Field Development and Production, Stavanger, 14–15 November.
23. "Reservoir Characterization Using Expert Knowledge, Data and Statistics," *Oilfield Review* (January 1992) 25.
24. Haldorsen, H.H. and Damsleth, E.: "Challenges in Reservoir Characterization," *AAPG Bulletin* (April 1993) **77,** No. 4, 541.
25. Journel, A. G.: "Geostatistics for Reservoir Characterization," paper SPE 20750 presented at 1990 SPE Annual Technical Conference and Exhibition, New Orleans, 23–26 September.

SI Metric Conversion Factors

acre	$\times$ 4.046 856	E − 01	= ha
bbl	$\times$ 1.589 873	E − 01	= m^3
ft	$\times$ 3.048*	E − 01	= m
md	$\times$ 9.869 233	E − 04	= μm^2
mile	$\times$ 1.609 344*	E + 00	= km

*Conversion factor is exact.

Chapter 2
Principles of Statistics

2.1 Introduction

The principles of statistics must be understood so the application of geostatistics can be understood. Principles of statistics form the backbone of geostatistics. In addition, many statistical principles can be directly applied to provide a better understanding of the sample data. This chapter focuses on the principles necessary to understand and analyze spatially distributed data. Only those principles are covered. Refs. 1 through 5 provide more general information on statistical principles.

The chapter is divided into two broad sections. The first section deals with a branch of statistics called descriptive statistics. This branch deals with the organization, presentation, and summarization of data. These techniques are particularly useful in developing a better understanding of the type of information currently available and the peculiar characteristics it possesses. Understanding the characteristics of the information allows us to use it more productively.

The second section deals with a branch of statistics called inferential statistics. This branch deals with the procedures for deriving conclusions about a population on the basis of sample data. Here, sample data are the collected information and population is all the information, we would like to have within the region of interest. Most geostatistical applications use principles of inferential statistics in estimating values at unsampled locations. This chapter, however, concentrates only on the inferential statistical principles used in conventional statistics.

2.2 Descriptive Statistics

Descriptive statistics, which has been in use for more than 200 years, is the oldest branch of statistics. Its origins lie in surveys and census activity, where investigators tried to capture details about the activity in as few numbers as possible. With modern computer technology, the field of descriptive statistics has become even more powerful; general trends, anomalies, and variations in data can be illustrated with contrasting colors or gray-scale maps. In general, descriptive statistics can be applied to either sample data or population. In practice, however, the population is rarely known. Therefore, we restrict the applications to sample data only.

This section covers several techniques for analyzing and characterizing sample data, starting with a frequency-distribution technique, probably the oldest statistical technique for analyzing samples. Then, we discuss summary statistical techniques, which try to capture basic features of sample with a limited number of parameters. A separate section covers techniques useful for understanding spatial data sets. Finally, we briefly address techniques used to understand bivariate data sets and try to relate two variables to each other.

2.2.1 Frequency Distribution. The frequency-distribution method is one of the simplest ways to analyze sample data. It summarizes the data in a more compact form than original sample observations. To construct a frequency distribution, the range of the data is divided into intervals called class intervals. It is a common practice to use class sizes of equal width, but this is not necessary. The number of measurements falling within a particular class, i, is called a class frequency, f_i.

The number of classes should be chosen so that the sample characteristics, or "signature," are clearly visible from the display of class frequencies. Too few or too many class intervals do not provide the necessary information. If too few classes are used, some details may be lost; if too many classes are used, too few values may fall within each class, giving very little information about sample data tendencies. Typically, the number of class intervals depends on the number of sample data points. One way to estimate is to use the square root of the total number of sample points to approximate the number of class intervals. In most cases, 5 to 20 intervals are sufficient.

On the basis of frequency values within each class, we can calculate the relative-frequency values for each class. If n = the total number of samples, we can calculate the relative frequency, f_{Ri}, for Class i as

$$f_{Ri} = \frac{f_i}{n}. \quad \text{(2.1)}$$

TABLE 2.1—CLASS FREQUENCY VALUES FOR NUMERICAL EXAMPLE 2.1

Class	i	f_i	f_{Ri}	F_i
$0.10<\phi\leqq 0.12$	1	5	0.1	0.1
$0.12<\phi\leqq 0.14$	2	9	0.18	0.28
$0.14<\phi\leqq 0.16$	3	20	0.4	0.68
$0.16<\phi\leqq 0.18$	4	10	0.2	0.88
$0.18<\phi\leqq 0.2$	5	6	0.12	1.0

Normalizing the class frequency by the total number of samples ensures that all the relative class frequencies add to one. In other words,

$$\sum_{i=1}^{N} f_{Ri} = 1, \quad \text{.............................. (2.2)}$$

where N = total number of classes.

In addition to relative class frequency, cumulative relative class frequency can be calculated by

$$F_j = \sum_{i=1}^{j} f_{Ri} \quad \text{.............................. (2.3)}$$

for $j = 1,\ldots,$ where F_j = cumulative frequency for class j that results from addition of all the relative frequencies up to and including class j.

Numerical Example 2.1. The following porosity samples are measured in a wellbore: 0.141, 0.124, 0.152, 0.156, 0.113, 0.167, 0.194, 0.142, 0.133, 0.149, 0.106, 0.137, 0.147, 0.159, 0.174, 0.129, 0.153, 0.173, 0.189, 0.16, 0.193, 0.156, 0.149, 0.135, 0.145, 0.171, 0.101, 0.151, 0.176, 0.191, 0.121, 0.148, 0.153, 0.171, 0.183, 0.108, 0.123, 0.169, 0.185, 0.153, 0.117, 0.127, 0.145, 0.141, 0.165, 0.14, 0.143, 0.178, 0.179, 0.157. Analyze these porosity, ϕ, values using a frequency-distribution analysis.

Solution. For these 50 values, we divide the data into five classes. **Table 2.1** lists the class-frequency values: column 1 lists the porosity, column 2 lists the number of values (class frequencies) within each class, and column 3 lists the relative-frequency values for each class calculated with Eq. 2.1. For example, for class values between 0.12 and 0.14,

$$f_{Ri} = \frac{f_i}{n} = \frac{9}{50} = 0.18.$$

Column 4 gives the cumulative relative class frequency values calculated with Eq. 2.3. For example, for class values between 0.14 and 0.16,

$$F_3 = \sum_{i=1}^{3} f_{Ri} = 0.1 + 0.18 + 0.4 = 0.68.$$

We can plot relative frequencies as well as cumulative relative class frequencies as functions of the variable values (see **Fig. 2.1**). The plot of relative frequency vs. the variable is called a relative-frequency histogram.

These plots are useful in characterizing samples. For example, from this plot, we can conclude that 40% of the porosity values fall between 0.14 and 0.16 (or that 10% of the porosity values fall between 0.1 and 0.12). The cumulative relative histogram plot shows that 68% of the porosity values are less than 0.16. For equal-sized class distributions, the area of the rectangle for a particular class is proportional to the chance that a porosity value will fall within that class (**Fig. 2.2**). In Fig. 2.2, the chance of porosity falling between 0.14 and 0.16 is proportional to the crosshatched rectangle. Fig. 2.2 also shows that the chance that porosity is less than 0.16 is 68%.

In defining frequency distributions, we assume a discrete distribution for a variable. That is, once the data are divided into several classes, we do not distinguish among the values within a class; all values are treated the same. The advantage of using such an approach is that the sample can be characterized with fewer parameters. Instead of knowing all the porosity values, we may need to know only the individual classes and the distribution of values within all the classes. Such information may be adequate for understanding the sample characteristics.

In reality, porosity is not a discrete variable; it is a continuous one. Just because a porosity value of 0.102 is not observed does not mean that a value of 0.102 does not exist. It may be obtained in the next sample. Porosity can take any value within the extreme values.

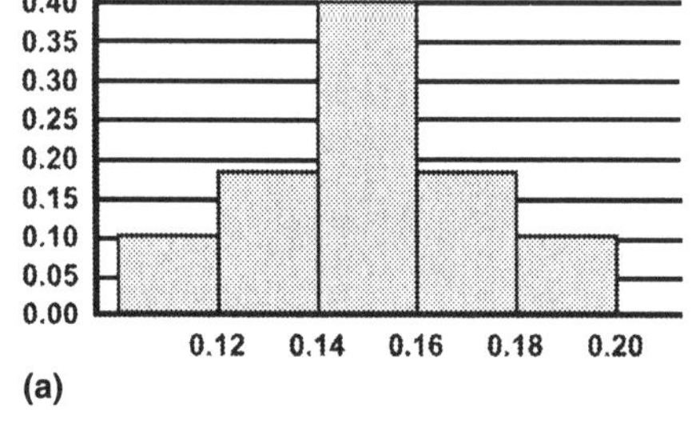

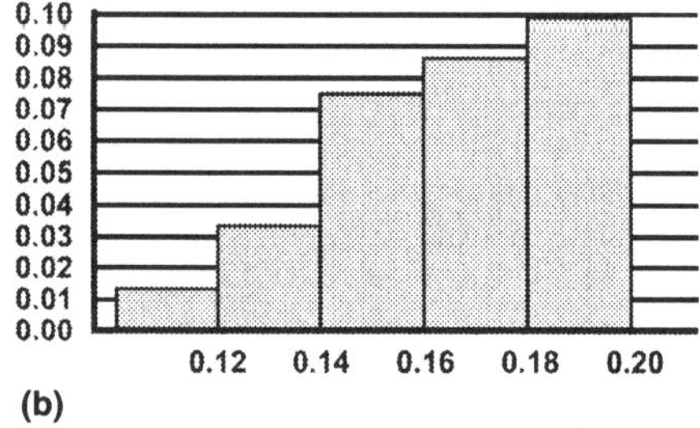

Fig. 2.1—(a) Relative frequency and (b) cumulative relative frequency.

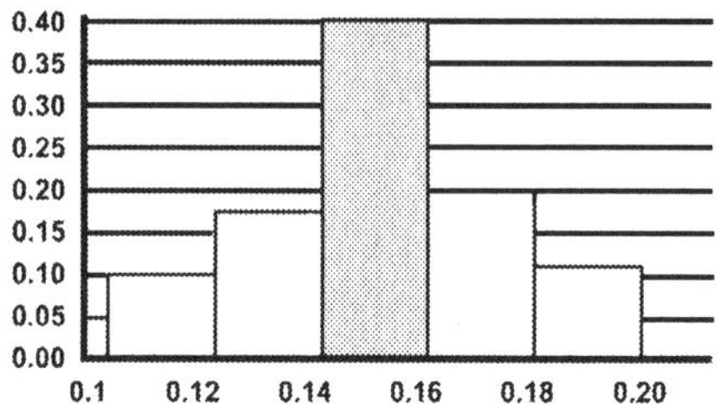

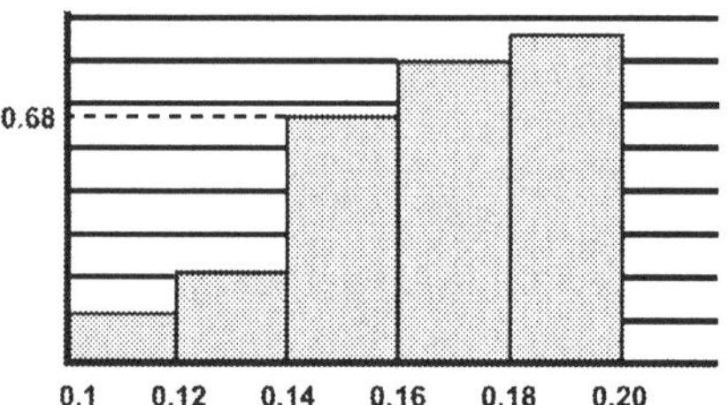

Fig. 2.2—Characterizing the frequency function.

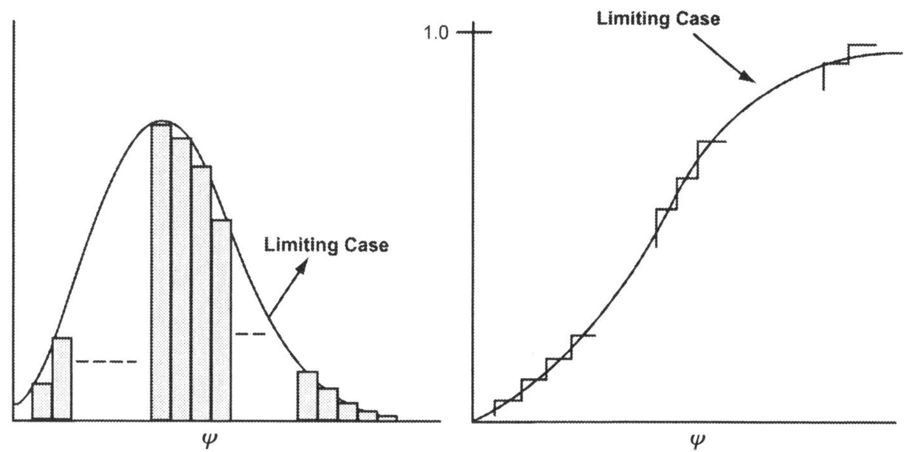

Fig. 2.3—Frequency distribution for a large number of classes.

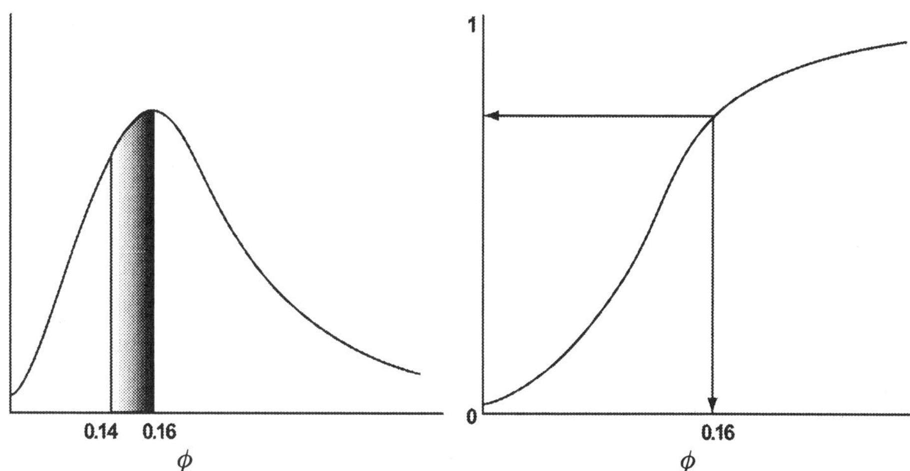

Fig. 2.4—Characterizing continuous distribution.

Collecting 5,000 samples instead of 50 may allow division of the data set into additional classes. Under these circumstances, Fig. 2.1 may look like **Fig. 2.3.** Within the limits, the discrete distribution may look like a continuous distribution as shown in Fig. 2.3 for a very large sample set approaching the population. Applying the principles used in Fig. 2.2, we can say that the chance that porosity will fall between 0.14 and 0.16 is proportional to the area under the relative-frequency histogram curve between these limits. The cumulative relative-frequency diagram also shows the chance that the porosity will be less than 0.16 (**Fig. 2.4**).

Although the analysis presented in Fig. 2.4 is not strictly correct mathematically, it provides an intuitive understanding of how the frequency distribution can eventually lead to continuous distribution. This section provides a basis for understanding the conceptual framework of the probability-density function (analog of f_{Ri}) and cumulative-distribution function (analog of F_i), which is discussed in Sec. 2.3. A very clear relationship exists between the frequency distributions and the statistical functions for a continuous distribution.

Field Example 2.1. As for all field examples, Appendix A provides the field data. This example uses histogram plots for porosity and permeability data for Well 34-29 in the Burbank field (**Fig. 2.5**). The well is in the north-central portion of the reservoir. On the basis of initial potential and permeability and porosity data, the well is in a better part of the reservoir.

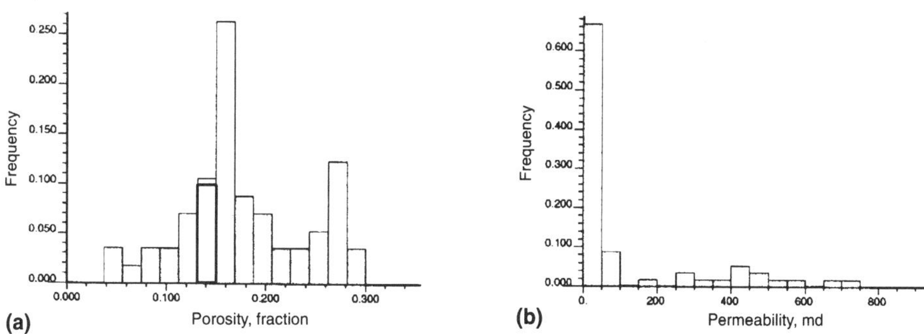

Fig. 2.5—Histogram for (a) porosity and (b) permeability data for Well 34-29.

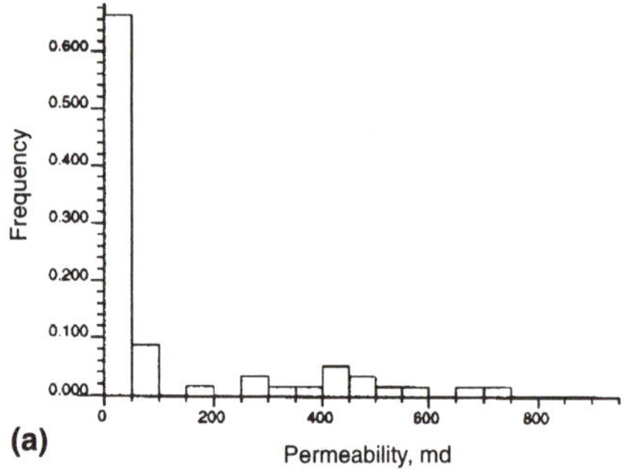

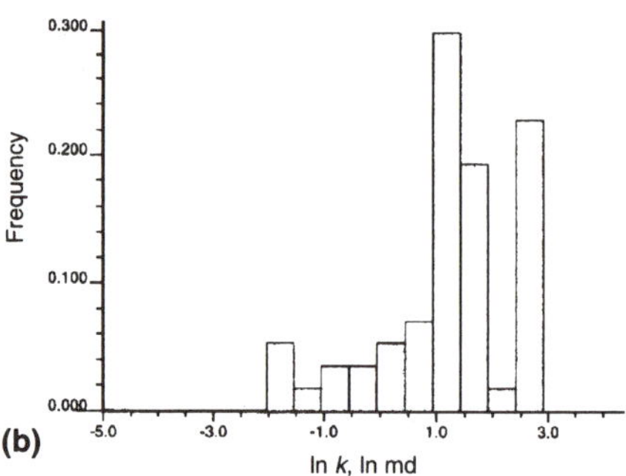

Fig. 2.6—Effect of log transform on permeability histogram: (a) permeability data and (b) log *k* data for Well 34-29.

Fig. 2.5a, the porosity histogram, shows a reasonably symmetric distribution with an exception at the high end of the values. Other than in the middle, another peak of frequency appears at a porosity value of 0.27. This may indicate a mixing of porosity distributions from two different geological units. Appendix A indicates that the Burbank sand is divided into several geological units. Although some units are hard to distinguish from each other, many units are separated by continuous shales that are extensive in an areal direction. Specifically, the lower unit (Unit 10) is a relatively thick sand with low porosity values. This unit can significantly influence the overall distribution of porosity values.

The permeability distribution at Well 34-29 (Fig. 2.5b) is skewed positively. Although the permeability values range from as low as 0.01 md to 750 md, the majority of the values are at the lower end of the region. This type of histogram is rarely useful for characterizing a sample because the values are clustered at one end.

One way to overcome this problem is to transform the sample data in some way so that some sample characteristics are evident from the histogram plot. The most commonly used approach for permeability values is the log transform. **Fig. 2.6** shows the original permeability histogram and the log of permeability, log k, histogram. The log k distribution is much more symmetric than the permeability distribution. In addition, the log k and ϕ histograms are remarkably similar. Both show similar trends with two peaks in the histogram plot, one of which is at the higher end of the values. Although this needs to be validated, such characteristic similarity may indicate a relationship between log k and ϕ.

Fig. 2.7 shows histogram plots for two additional wells in the field: Well 31-W23 in the northwest portion of the reservoir and Well 36-16 in the northeast portion of the reservoir. Both wells are thought to be in poor regions of the reservoir. On both the east and west sides of the study region, the amount of shale increases, and the permeability values become smaller. Core porosities reflect this transition in the histogram. Both wells show a significant portion of shale, indicated by a high fraction of very low porosity values (almost 40%). The overall range of porosity values is also smaller compared with that in Well 34-29.

These types of histogram maps for wells located in different regions of the reservoir serve several purposes. First, changes in the sample distribution of a physical property can be visually observed. This may confirm some geological boundaries and sand units that are established on the basis of log interpretation. Also, when a particular range of porosity values is missing at certain wells (as in the case of Well 36-16, where no values greater than 18% or between 2 and 10% are observed), it may indicate that a particular geological unit may not extend over a region. Additional analysis can confirm this. Another benefit of generating such histogram maps is to define the region of stationarity. As Chap. 1 discussed, this region represents an area where the proposed model can be applied on the basis of the sample data. If the histogram characteristics vary over a wide range within an area, the region of stationarity may have to be re-examined and redefined.

2.2.2 Summary Statistics for Univariate Distribution.

In addition to the frequency distribution, which characterizes the sample visually and numerically, characterizing the sample through summary information is also beneficial. This section addresses the sample of one variable: a univariate distribution. Several types of summary statistics for univariate distributions are available.

Mean. The sample mean represents the arithmetic mean of the sample data points. Mathematically,

$$\bar{x} = \frac{1}{n}\sum_{i=1}^{n} x_i, \quad (2.4)$$

where n = total number of samples and x_i = the value of Sample i. The arithmetic mean represents the central tendency of the sample.

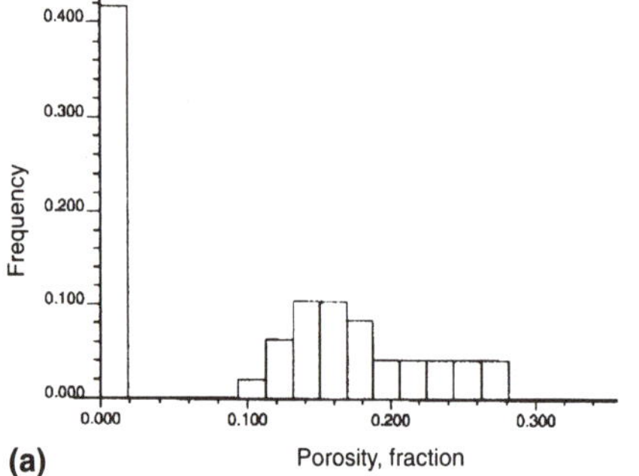

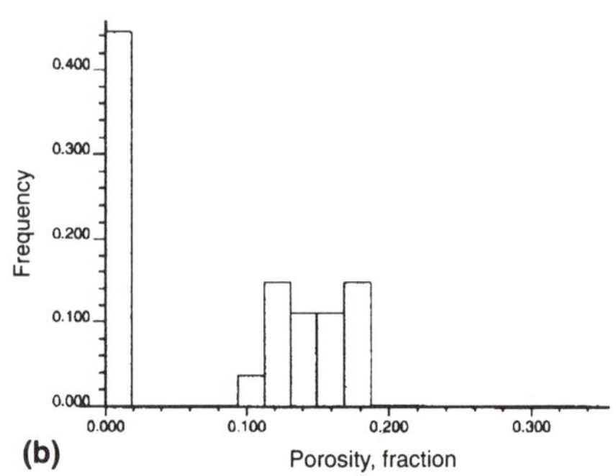

Fig. 2.7—Porosity histograms for Wells (a) 31-23 and (b) 36-16.

Median. Another measure of central tendency is the median, which is the sample point that divides the sample into equal halves. If all the samples are arranged in an ascending order so that $x_1 < x_2 \ldots < x_n$, the sample median, $\tilde{x}$, is calculated by

$$\tilde{x} = x_{(n+1)/2} \qquad (2.5a)$$

when n is odd and by

$$\tilde{x} = \left(\frac{x_{n/2} + x_{n/2+1}}{2}\right) \qquad (2.5b)$$

when n is even. An advantage of using median rather than mean is that median is not influenced by extreme values. In contrast, because the mean is an arithmetic average, it can be affected by one or two extreme values.

Mode. The mode, which is another measure of a central tendency, is an observation that occurs most frequently in the sample. The value of the mode obviously depends on the precision of the data, especially for naturally occurring variables. If the data are very precise, each value is unique and none is repeated. One option is to use the frequency-distribution classes as a range of precision and define the mode as the class that contains the highest number of values.

Mean, median, and mode coincide with each other if the distribution is symmetric. If the distribution is skewed (e.g., the permeability histogram in Fig. 2.5b), these three tendencies exhibit different values. If the distribution is skewed positively (to the right), mode < median < mean. If the distribution is skewed negatively (to the left), mode > median > mean.

Extremes. In addition to the mean, median, and mode, we can also define the minimum and maximum of the sample values, where the minimum represents the smallest value and the the maximum represents the highest value.

Percentile. Percentile values represent sample values that are greater than a certain percentage of the sample values. The median is an example of the 50th percentile value because 50% of the values are smaller than the median. If the values are arranged in ascending order, x^p represents a value where p percent values are smaller than x^p. For example, x^{10} represents a sample value that is greater than 10% of the total sample points.

Certain types of percentiles are commonly used in describing sample data. For example, the first quartile represents x^{25}, where 25% of the sample values are less than x^{25}. x^{75} represents a value where 25% of the sample values are greater than x^{75}. In a similar manner, deciles describe the data in terms of 10ths. The first decile represents the value greater than 10% of the sample data. The fifth decile represents the median. Overall, percentile distribution is simply another way of looking at data distribution.

Variance. The sample variance represents the spread of the data. It is a quantitative measure of how widely the data are distributed. Mathematically, variance is calculated as

$$s^2 = \frac{\sum_{i=1}^{n} (x_i - \bar{x})^2}{n - 1}, \qquad (2.6)$$

where s^2 = sample variance, $\bar{x}$ = sample mean, and n = total number of samples. Variance can also be calculated as

$$s^2 = \frac{\sum_{i=1}^{n} x_i^2 - n\bar{x}^2}{n - 1}. \qquad (2.7)$$

Although Eqs. 2.6 and 2.7 should give the same result, Eq. 2.7 is preferred because of its numerical precision. The square root of variance, s, is called the standard deviation. It has the same units as the variable being sampled.

Coefficient of Variation. The coefficient of variation, C_V, is defined as

$$C_V = \frac{s}{\bar{x}}, \qquad (2.8)$$

where s = standard deviation and $\bar{x}$ = sample mean. Because s and $\bar{x}$ have the same units, C_V is a dimensionless quantity; therefore, it provides a measure of the relative spread of a sample. When samples from two different variables are compared, the C_V value provides an indication of which variable has the relatively greater spread. For very tight samples, the value of C_V is small. For samples that exhibit several-orders-of-magnitude variations, C_V typically is > 1, sometimes in the range of 2 to 5. High C_V values are a warning sign that the sample data contain extreme values that may affect estimation of values at unsampled locations.

Range. Range is another quantitative measure of the spread. A simple definition of range, R, is

$$R = x_{max} - x_{min}, \qquad (2.9)$$

where x_{max} = the maximum value and x_{min} = the minimum value. Other definitions of range have also been used. For example, interquartile range represents the difference between two successive quartile values. We can define the first quartile range as

$$R_1 = x^{25} - x_{min}, \qquad (2.10)$$

where x^{25} = the 25th percentile value and x_{min} = the minimum value. Similar definitions can be used for other quartile ranges.

Numerical Example 2.2. The following data for pay-zone thickness (in feet) are collected from all available wells in a reservoir: 6, 10, 20, 12, 20, 10, 15, 32, 27, 10, 18, 29, 8, 17, 23, 36, 19, 13, 33, 10, 26. Calculate mean, median, mode, quartile values, variance, C_V, and range.

Solution. The total number of samples is 21.

Mean. With Eq. 2.4,

$$\bar{x} = \frac{\sum_{i=1}^{n} x_i}{n}$$

$$= \frac{(6 + 10 + 20 + 12 + \ldots + 33 + 10 + 26)}{21}$$

$$= 18.76 \text{ ft.}$$

Median. By arranging the values in ascending order (6, 8, 10, 10, 10, 10, 12, 13, 15, 17, 18, 19, 20, 20, 23, 26, 27, 29, 32, 33, 36), we can calculate the median. For an odd number of samples, with Eq. 2.5a,

$$\tilde{x} = x_{(n+1)/2} = x_{11} = 18 \text{ ft.}$$

TABLE 2.2—SUMMARY STATISTICS FOR FIELD EXAMPLE 2.2

Statistics	Variable ϕ	k	$\ln k$
Mean	0.1762	123.3	2.82
Median	0.16	24.2	3.19
x^{25}	0.1437	7.4	1.83
x^{75}	0.2203	97.0	4.31
Variance	0.0038	39,761.2	7.97
σ	0.0611	197.6	2.82
C_V	0.347	1.6	1.0
Range	0.238	748.0	11.23

Mode. The mode is the sample that occurs the most frequently. In our data set, mode is equal to 10 ft because it occurs the most frequently.

Quartiles. The 25th quartile represents a value ≧25% of the sample values. For 21 samples, this represents the sixth value, or 10 ft. The 75th percentile represents a value ≧75% of the sample values. This represents the16th value, or 26 ft.

Variance. Variance is calculated with Eq. 2.7.

$$s^2 = \frac{\sum_{i=1}^{n} x_i^2 - n\bar{x}^2}{n-1}$$

$$= \frac{(6^2 + 10^2 + \ldots + 33^2 + 10^2 + 26^2) - (18.76)^2}{21 - 1}$$

$$= 80.19 \text{ ft}^2.$$

Standard Deviation.

$$s = \sqrt{s^2} = \sqrt{80.19} = 8.95.$$

Coefficient of Variation. With Eq. 2.8,

$$C_V = \frac{s}{\bar{x}} = \frac{8.95}{18.76} = 0.477.$$

This value represents a relatively small variation within the sample. Typically, a value of $C_V < 1$ indicates a relatively narrow distribution.

Range. With Eq. 2.9,

$$R = x_{\max} - x_{\min} = 36 - 6 = 30 \text{ ft}.$$

We can also calculate the interquartile range.

$$R_1 = x^{25} - x_{\min} = 10 - 6 = 4 \text{ ft}.$$

and $R_4 = x_{\max} - x^{75} = 36 - 26 = 10$ ft.

Field Example 2.2. This example shows application of the same summary-statistics analysis to a field example with some summary statistics obtained for porosity and permeability data for Well 34-29. The total number of sample points is 57. **Table 2.2** provides the summary statistics.

For porosity data, the mean and median are very close to each other, indicating a symmetric distribution. In contrast, the median for the permeability data is much smaller than the mean, indicating a positively skewed distribution.

The coefficient of variation, C_V, shows that the variation in porosity is small, while the value of $C_V > 1$ for the permeability data indicates order-of-magnitude variations within the sample data. Such variations should be treated with care when estimating values at unsampled locations.

TABLE 2.3—POROSITY DATA FOR FIELD EXAMPLE 2.2

Statistics	Well 31-W23	34-29	36-16
Mean	0.1043	0.1762	0.0816
Median	0.132	0.16	0.113
x^{25}	0.000	0.1437	0.000
x^{75}	0.175	0.2203	0.153
Variance	0.0092	0.0038	0.0059
Standard Deviation	0.0948	0.0611	0.0754
C_V	0.91	0.347	0.92
Range	0.268	0.238	0.184

As discussed in Sec. 2.2.1, a wide variation in sample values can be overcome by use of some type of nonlinear transform, such as a log transform. Table 2.2 also shows the summary statistics for natural log of permeability ($\ln k$) values.

Note that the mean and median values are a lot closer to each other, indicating symmetry in the distribution. Also, the interquartile ranges are much more symmetric for the log transform than for the raw permeability values. For the raw permeability data, for example, the last quartile range is 651 md, which represents more than 80% of the total range of the sample data. C_V is also smaller for $\ln k$ data, indicating a smaller variation than in the sampled data values.

Dealing with samples that are symmetric and that show a narrow spread is always easier than dealing with highly skewed samples with a large spread. Some type of nonlinear transform is commonly used to convert raw sample data into transformed data that is "better behaved." Although this seems simple, such transformations should be used with caution. Discussions in subsequent chapters indicate that these transformations may result in erroneous estimated values at unsampled locations within the region of interest.

Analyzing the field data further, we examine data from the other two wells investigated in Field Example 2.1: Well 31-W23 in the northwest part of the reservoir and Well 36-16 in the northeast part of the reservoir. These two wells represent a relatively poor part of the reservoir, where the proportion of shale is much higher. **Table 2.3** provides the summary statistics for both these wells and also includes data from Well 34-29 for comparison. The table shows that the mean is much smaller for both Wells 31-W23 and 36-16, indicating a relatively large fraction of shale. The high C_V values at both wells indicate a skewed porosity distribution, which was evident from the histograms (Figs. 2.5 through 2.7).

Another visual display that is sometimes useful for comparing the two samples is the *Q-Q* plot. This plot represents quantile comparisons of the two data sets. For example, the 10th quantile value of one set is plotted vs. the 10th quantile of the other set, the 20th quantile value of one set is plotted vs. the 20th quantile value of the other set, and so on. If two samples have essentially the same distribution, the *Q-Q* plot shows a perfect 45° straight line. **Fig. 2.8** shows the comparison between Wells 34-29 and 31-W23 and between Wells 34-29 and 36-16. Both plots clearly show a big deviation from the 45° line, indicating significant differences in the distributions of the two data sets. All points fall below the 45° line, indicating that Well 34-29 is always represented by a higher value than the other two wells for the entire percentile range.

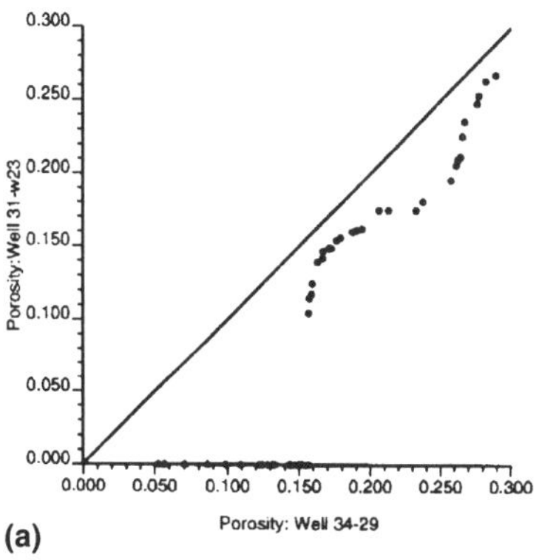

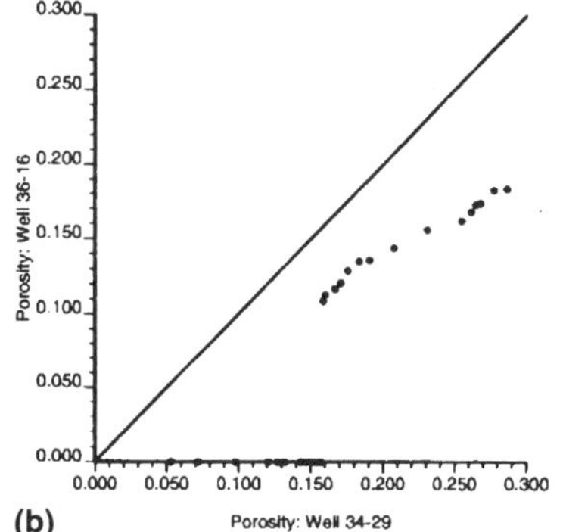

Fig. 2.8—Percentile plot (*Q-Q* plot) comparisons of porosities of two wells: (a) Well 34-29 vs. Well 31-w23 and (b) Well 34-29 vs. Well 36-16.

2.2.3 Spatial Data Sets. In this section, we apply some techniques discussed in the two previous sections to a spatially distributed data set. The discussion is limited to univariate (single variable) descriptions. Spatial data sets collected in petroleum reservoirs are unique in several ways.

1. The data sets do not represent random sampling. Random sampling represents a sampling where all samples have an equal likelihood of being selected. In practice, however, all the wells in a reservoir are not drilled at the same time. A few exploratory wells are drilled first. Then, on the basis of information gathered, additional wells are drilled. As more wells are drilled, the information collected from the previously drilled wells is used to drill the next set of wells. That is, all wells are not drilled on the basis of the same information available. Therefore, the likelihood that a particular well will be drilled differs, depending on the information available at the particular time.

2. The data sets result in biased sampling. The goal in drilling wells in petroleum reservoirs is to drill wells that have the maximum potential of economic success, not to collect a good sample set. When selecting locations of new wells, the primary purpose is always the most effective way to produce additional oil with minimum investment. Obviously, more wells are drilled in an area where greater potential for additional recovery exists. Subjective decisions based on potential as a criterion result in biased sampling. That is, more wells are selectively drilled in an area where either the pay zone is thicker, the porosity is higher, or the reservoir is more permeable. Fewer wells are drilled in regions with low permeability, low pore volume, or both. Treating these sampled wells without accounting for the sampling bias may paint an overly optimistic picture of the reservoir because more samples selectively come from regions with a higher potential.

3. The data sets may show local variability. A typical reservoir has several wells drilled, which may intersect several geological horizons. In addition, even areally, a reservoir may contain several different geological units. In a fluvial reservoir, an area may include a channel sand, a crevasse splay, point bars, and a flood plain. Because of the differences in the environment, there are local variations in the statistical properties (e.g., some areas show higher porosity than others). In considering such a sample, we need to understand how these local variabilities affect the overall estimation procedure.

One may argue that this problem can be avoided by simply defining a small enough region of interest so that these types of local variabilities are not evident. Unfortunately, as a region is made smaller and smaller, fewer sample points become available in a particular region. For statistically valid results, a data set must have a sufficient number of samples. A compromise is to consider local variability as part of the data and account for these variabilities through various procedures. Later chapters discuss some of these procedures.

Here, we discuss two procedures commonly used to analyze and understand spatial data sets. The first procedure can be used to remove bias in the sampling. The second procedure allows us to understand the local variability in a sample set.

Sample Declustering. Sample declustering is one of the simplest ways to remove the undue influence of biased sampling. Originally proposed by Journel,[7] the method requires that spatial data be divided into several small subareas (**Fig. 2.9**). These subareas typically are rectangles that cover the entire region. Once the data are divided into small subregions, the number of data points within in each subarea is calculated. For example, Subarea 1 in Fig. 2.9 has three sample points, and Subarea 2 has six sample points. Depending on the number of sample points within each subarea, an appropriate weight is assigned to each sample point within that subarea. One logical choice is to assign a weight to each point equal to the inverse of the total number of points within each subarea. For example, if Point x_i falls within Subarea l where n_l sample points are located, then

$$w_i = \frac{1}{n_l}, \qquad (2.11)$$

where w_i = weight assigned to Sample Point x_i. For example, in Fig. 2.9, each point in Subarea 1 is assigned a weight

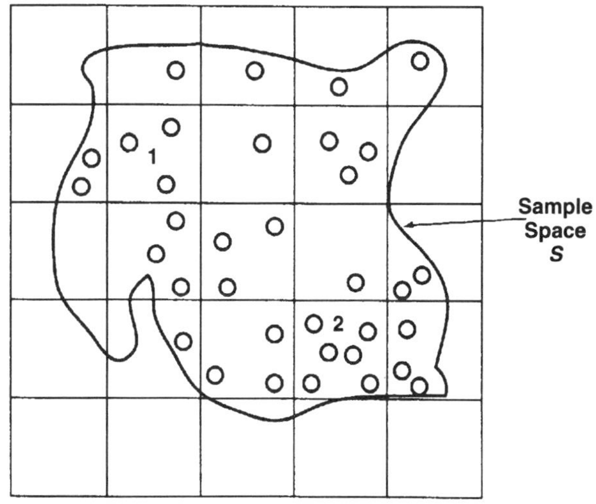

Fig. 2.9—Declustering of data.

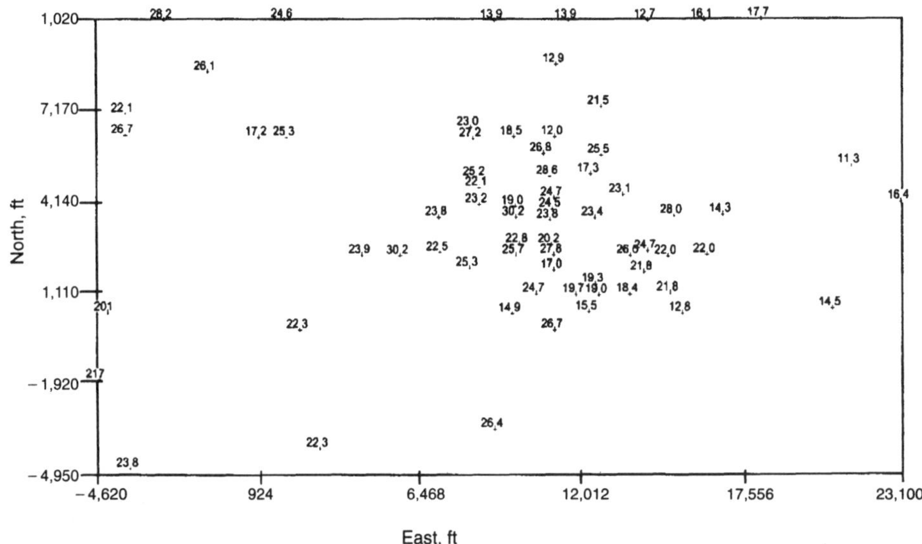

Fig. 2.10—Areal distribution of porosity values.

of one-third, and each point in Subarea 2 is assigned a weight of one-sixth. If only a single point falls within an area, a weight of one is assigned to that point.

The idea of assigning weight is simple. The greater the number of data points within an area, the more clustered the sample is within that area. Assigning a smaller weight to those points reduces any undue influence of those points.

Once weights of individual points are calculated, the arithmetic mean of the samples is calculated as

$$\bar{x} = \frac{\sum_{i=1}^{n} w_i x_i}{\sum_{i=1}^{n} w_i}, \qquad (2.12)$$

where $\bar{x}$ = declustered sample mean and w_i = weight assigned to individual Sample Point i. The variance can be calculated as

$$s^2 = \frac{\sum_{i=1}^{n} w_i (x_i - \bar{x})^2}{\sum_{i=1}^{n} w_i}, \qquad (2.13)$$

where s^2 = declustered variance of the sample. Once the mean and variance are calculated, the standard deviation and coefficient of variation can be estimated.

One unknown in these calculations is what size the subarea should be. As a guess, one option is to use

$$A_{\text{sub}} \approx \sqrt{\frac{A_t}{n}}, \qquad (2.14)$$

where A_{sub} = subarea size = size of the individual subareas, n = total number of sample points, and A_t = total area = area of the study region. The shape of the subarea can be the same as that of the total area.

To be precise, different subarea sizes should be tried, and the declustered mean plotted as a function of the subarea size. A size should be chosen so that the declustered mean reaches a minimum or maximum value. When analyzing most reservoir properties, the declustered mean reaches a minimum at a desired subarea size because sampling is always biased toward high values, such as high porosity, high pay-zone thickness, or high permeability. That is, the sample contains a more of high values than it would contain with random sampling. Eliminating the undue influence of these high-valued samples should make the mean lower than what the sample represents. The mean is estimated by examining the declustered mean for various subarea sizes. The correct size is the one that represents the minimum because that size represents the maximum effect that declustering can achieve. The Field Example illustrates this further.

Field Example 2.3. Data from Flow Unit 3 are used in this example. Appendix A indicates that the reservoir is divided into ten flow units; Flow Unit 3 is one of the units with relatively high porosity and permeability values. At each well, the log porosity values were arithmetically averaged over the pay-zone thickness for that unit to estimate the average porosity for the unit at that location. **Fig. 2.10** shows the areal distribution of porosity values over the study area. The values vary over a wide range, from as low as 11% to as high as 30%. As the map shows, the wells are clustered in areas that have high porosity values.

We can apply the declustering technique and, with Eq. 2.11, calculate the declustered mean for various cell sizes. **Fig. 2.11** shows a plot of declustered mean as a function of cell size. As expected, the declustered mean starts with a high value for a large cell size. It reaches a minimum value at a cell size of approximately 3,000 × 3,000 ft. The sample mean is 21.6 vs. the declustered mean of 20.7. Although the difference between the two may not seem significant, remember that the porosity variation is not that significant in this case. For variables that do exhibit an order-of-magnitude variation, the difference between sample and declustered means can be significant. More relevant, however, is the fact the sample bias can be removed with a very simple technique.

Fig. 2.12, which compares the histogram of the sample data with the histogram of the declustered data, also shows the effect of declustering. Although the maximum and minimum values have not changed, note that the quartile values for the declustered data are smaller than those for the sample data. Sample data, being biased, tend to exaggerate the percentage

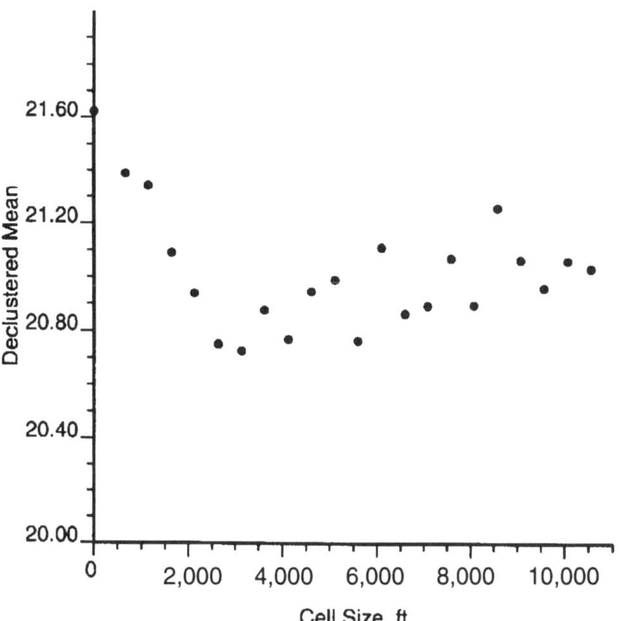

Fig. 2.11—Effect of declustering on sample mean.

of high values, which, in turn, increases quartile values. Declustering the sample minimizes this influence. In the histogram plot, the relative-frequency values for high porosities are smaller in the declustered data than in the clustered data. The reverse is true for low porosity values. For the low porosity values, the standard deviation for the declustered data is higher, indicating a more uniform spread over the data range than that indicated by clustering the data. The *Q-Q* plot in **Fig. 2.13**, which plots clustered vs. declustered data, further emphasizes this effect. All points fall below the 45° line, indicating that, for clustered (sample) data, the value for a given percentile is higher than for the declustered data.

In general, for data sets that indicate preferential clustering of the samples, it is safe to test them for declustering analysis. If any bias is present, it can be removed easily. If there is no bias or if the bias is very small, the declustered properties will be very similar to the sampled data set.

Moving-Window Statistics. Moving-window statistics is a technique by which local variations in statistical properties within a study region can be investigated. The technique is relatively simple but very powerful. A small window of a desired size is overlaid on the study region, and all sample points falling within that window are used to calculate local summary statistics. The two statistical properties most commonly used are the mean and the variance or standard deviation. The procedure is repeated by moving the window over the entire region. If the local means and variances are relatively uniform, the region is considered homoscendastic. If the local means and variances show significant variation, the study region is considered heteroscendastic.

The size and shape of the window is a subjective decision. A rectangular window is the normal choice because of computational efficiency. The size of the window is determined on the basis of the number of samples available within the study region. The window should be large enough so that each window contains sufficient samples to provide representative local statistics and small enough so that local variabilities are preserved and exhibited through moving-window statistics. One possible compromise is to select large windows to ensure sufficient samples but overlap the windows with a certain lag. This way, a sufficient number of local variations across the region is obtained without too much reduction in the size of the window. In this approach two adjacent windows partially overlap each other, resulting in some sample points being used more than once in estimation of local statistics.

In examining local properties, it is important to understand both local mean and local variance. Several possibilities may exist with respect to the variation of these properties. Local mean and variance can both vary; local mean can vary while local variance remains fairly constant; or local mean can remain constant while local variance varies. If both vary, a possibility exists that they may be related to each other. In many earth science data sets, local means are observed to be proportional to local standard deviations; this is called a proportional effect. Observation of such a relationship can be useful in estimating values at unsampled locations.

Field Example 2.4 illustrates application of moving-window statistics.

Field Example 2.4. The data set used in Field Example 2.3 is used in this example. The flow-unit porosity data are divided with windows of approximately 7,000 × 7,000 ft. The windows are defined so that half the window overlaps the preceding window. **Fig. 2.14** displays the window statistics. The mean and variance are shown at the centers of the windows. There is significant variation in the mean porosity, from a low of 15% to a high of up to 25%. The standard deviation varies from a low of 1% to a high of 6%. Understanding this type of areal variability in statistical properties is important for estimating values at unsampled locations. It is also helpful in deciding the type of technique to be used in the estimation procedure. These aspects are discussed further in later chapters.

Fig. 2.15 shows a plot of local mean vs. local standard deviation for a moving window that has at least seven sample points. The choice of seven as the minimum number of data is somewhat arbitrary. The figure shows that the relationship between the standard deviation and the local mean is rather

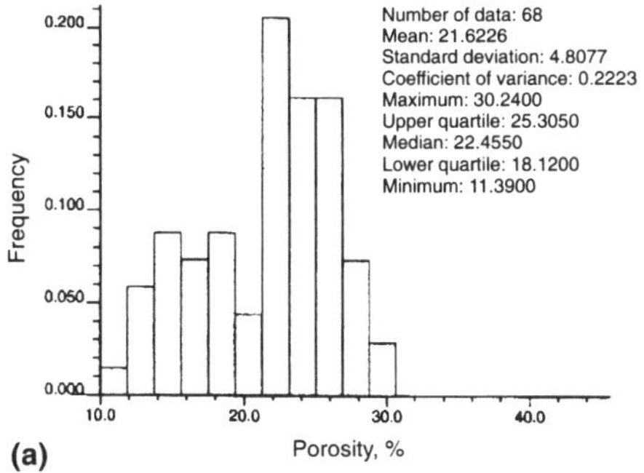

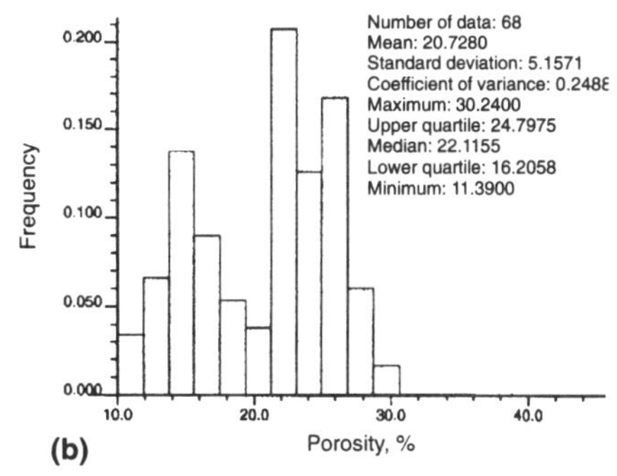

Fig. 2.12—Comparison of (a) clustered and (b) declustered porosity distributions for Flow Unit 3.

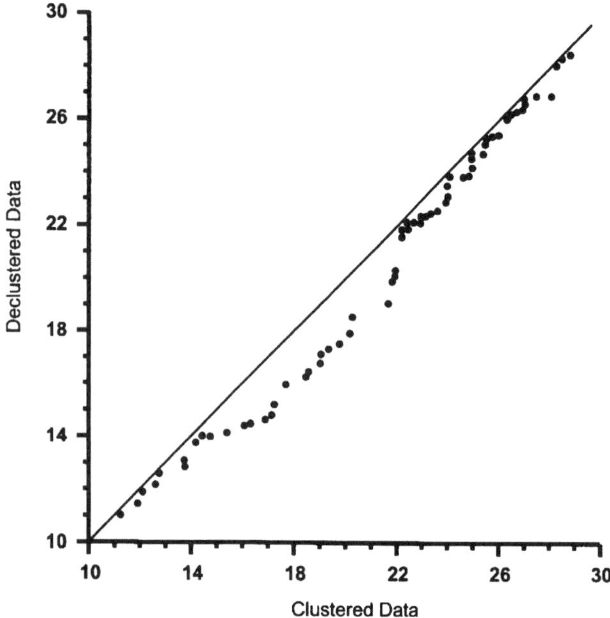

Fig. 2.13—Percentile plot comparisons between clustered and declustered data.

weak. If there is any relationship, it shows that the local mean is negatively related to the standard deviation; that is, as the local mean increases, the standard deviation decreases. Such a relationship generally is observed for a negatively skewed histogram. Note that the porosity histogram for Flow Unit 3 is slightly negatively skewed. For a positively skewed histogram, the local standard deviation normally increases as the local mean increases. Use of this type of relationship may aid the estimation process.

In analysis of spatial data sets, it is important to remember that spatial data sets collected from a reservoir may exhibit some unique features. Before these data sets are used as samples, it is important to conduct some exploratory analysis. Exploratory analysis may eliminate some of the bias in the data sets and also may indicate local variations in the observed properties.

2.2.4 Bivariate Statistics. This section analyzes statistics of two variables. We first discuss the conditional frequency distribution of one variable with respect to the other variable. We then examine various summary-statistics tools used to analyze the relationship between two variables and discuss application of these tools to describe and analyze spatial data sets. Again, both numerical and field examples illustrate the principles.

Conditional Frequency Distribution. Sec. 2.2.1 discussed the frequency-distribution analysis. Under that analysis, we divide data sets into several classes and consider the number of values falling within each class. We can extend this analysis to two variables. The concept of conditional distribution is used to present frequency distribution of two variables in a suitable format.

A conditional distribution is a distribution of one variable that is *conditional* on the distribution of the other variable. For example, if we are considering two variables, permeability and porosity, we can generate a conditional distribution of permeability for a given porosity value. Obviously, for a large number of sample values for any given porosity value, there may be multiple observations of permeability values. We can consider the frequency distribution of only these permeability values. This distribution is conditional to that value at which the porosity is fixed.

In practice, there may not be sufficient samples to generate a conditional distribution for one variable for a given value of the other variable. As an alternative, we can consider a range of values for one variable and the conditional distribution of the other variable for that range. As this range becomes smaller, the conditional distribution comes closer to a distribution corresponding to a point value. The following example illustrates this approach.

Field Example 2.5. The data from Well 34-29 are used for this example. This well has 57 porosity and permeability values, with porosity ranging between 0.052 and 0.29. The porosity values are divided arbitrarily into five classes; **Table 2.4** shows class ranges and mean of each class.

For convenience, assume that each class is represented by the mean value. The permeability values corresponding to each porosity class also are divided into five classes; **Fig. 2.16** shows the histograms of permeability values for each porosity class. These histograms are conditional distributions of

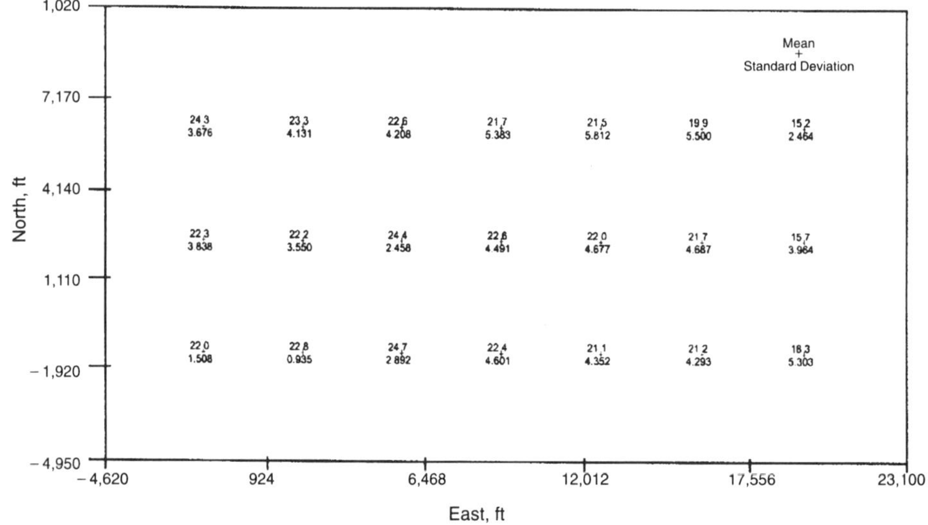

Fig. 2.14—Areal distribution of mean/standard deviation.

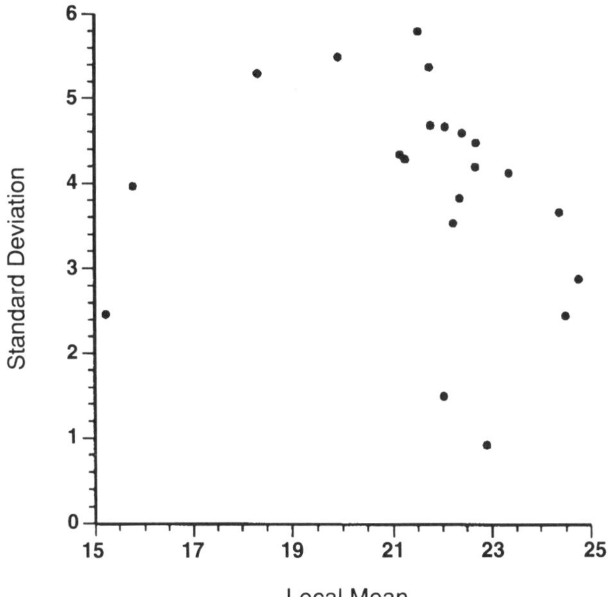

Fig. 2.15—Local mean vs. standard deviation for Flow Unit 3 porosity data.

permeability values for a given porosity class. Although the number of permeability values in each histogram is rather limited, a trend is clearly evident. Low porosity values are associated with low permeability values. As the average porosity increases, so do the permeability values. With the assumption that the mean porosity value for a class represents the class, the permeability distribution within each class can be associated with that particular value. For a large number of porosity values, class sizes can be made smaller. Eventually, in the limit, a conditional distribution of permeability values can be associated with each porosity value.

This type of conditional information is useful in associating the uncertainty of one variable with respect to the value of another variable. For example, in the geological description of a channel sand, the width of a channel and the thickness (depth) of a channel are related to each other; however, for a given width, several thickness values are possible. When constructing geological descriptions, once a channel width is chosen, we can use the conditional distribution of thickness to select a particular thickness from that distribution. This allows us to capture the nonunique relationship between the channel width and the thickness. Chap. 7 provides a detailed discussion of this particular method.

Summary Statistics for Bivariate Distribution. Similar to univariate statistics, several tools are used to summarize the statistics between two variables. The covariance is defined as

$$c(x,y) = \frac{1}{n}\sum_{i=1}^{n} x_i y_i - \frac{1}{n}\sum_{i=1}^{n} x_i \frac{1}{n}\sum_{i=1}^{n} y_i, \quad \text{........ (2.15)}$$

where x_i and y_i = samples of the variables x and y, respectively, and n = total number of sample pairs. Note that covariance reduces to variance if $x = y$.

Covariance is a measure of the relationship between two variables. If x and y are positively related (i.e., as x increases, y increases), the covariance has a positive value. If x and y are negatively related (i.e., as x increases, y decreases), covariance has a negative value. In the same manner, if x and y are not related, the covariance has a value close to zero.

TABLE 2.4—POROSITY CLASSES FOR FIELD EXAMPLE 2.5

Class	Mean
0.052 to 0.132	0.098
0.132 to 0.157	0.150
0.157 to 0.175	0.166
0.175 to 0.261	0.212
0.261 to 0.290	0.273

Similar to variance, covariance is defined in units that depend on the units of x and y, and it can be made dimensionless by defining a correlation coefficient.

$$r(x,y) = \frac{c(x,y)}{s_x s_y}, \quad \text{........................ (2.16)}$$

where $r(x, y)$ = correlation coefficient, $c(x, y)$ = covariance between x and y, s_x = standard deviation of the x variable, and s_y = standard deviation of the y variable. Sec. 2.2.2 discussed the calculation of standard deviation, which is the square root of variance.

By making the correlation coefficient dimensionless, we can define its limits. The value of the correlation coefficient always falls between the limits of +1 and −1. If x and y are positively related, the correlation coefficient falls between 0 and +1. The stronger the relationship, the closer the value will be to +1. If x and y are negatively related, the correlation coefficient falls between 0 and −1. The stronger the relationship, the closer the value will be to −1. If x and y are not related, the correlation coefficient is zero.

In some instances, the square of the correlation coefficient, $r^2(x, y)$, is used instead of the correlation coefficient to describe the relationship between the two variables. One advantage of using this value (sometimes called the r^2 statistic) is that it always falls between zero and one, whether x and y are positively or negatively related. This is the term most commonly used in describing the "goodness of fit" in a linear-regression analysis between two variables. The following examples illustrate the calculation procedure.

Numerical Example 2.3. Table 2.5 provides core permeability vs. core porosity data from a well. Calculate the covariance and the correlation coefficient between log k and ϕ data.

Solution. In practice, we assume that log k is related to ϕ values. Therefore, we first calculate the log k values; **Table 2.6** shows the calculated values.

Calculate the covariance with Eq. 2.15.

$$c(x,y) = \frac{1}{n}\sum_{i=1}^{n} x_i y_i - \frac{1}{n}\sum_{i=1}^{n} x_i \frac{1}{n}\sum_{i=1}^{n} y_i,$$

where $n = 9$ because we have nine pairs of data. Treating x as log k and y as ϕ, we can calculate

$$c(\log k, \phi) = \frac{1}{9}\big[3.063 \times 29.49 + 2.725 \times 26.79 + \ldots + 2.559 \times 25.54\big]$$
$$- \frac{1}{9}\big[3.063 + 2.725 + \ldots + 2.559\big] \times \frac{1}{9}\big[29.49 + 26.79 + \ldots + 25.54\big].$$
$$= 0.8875.$$

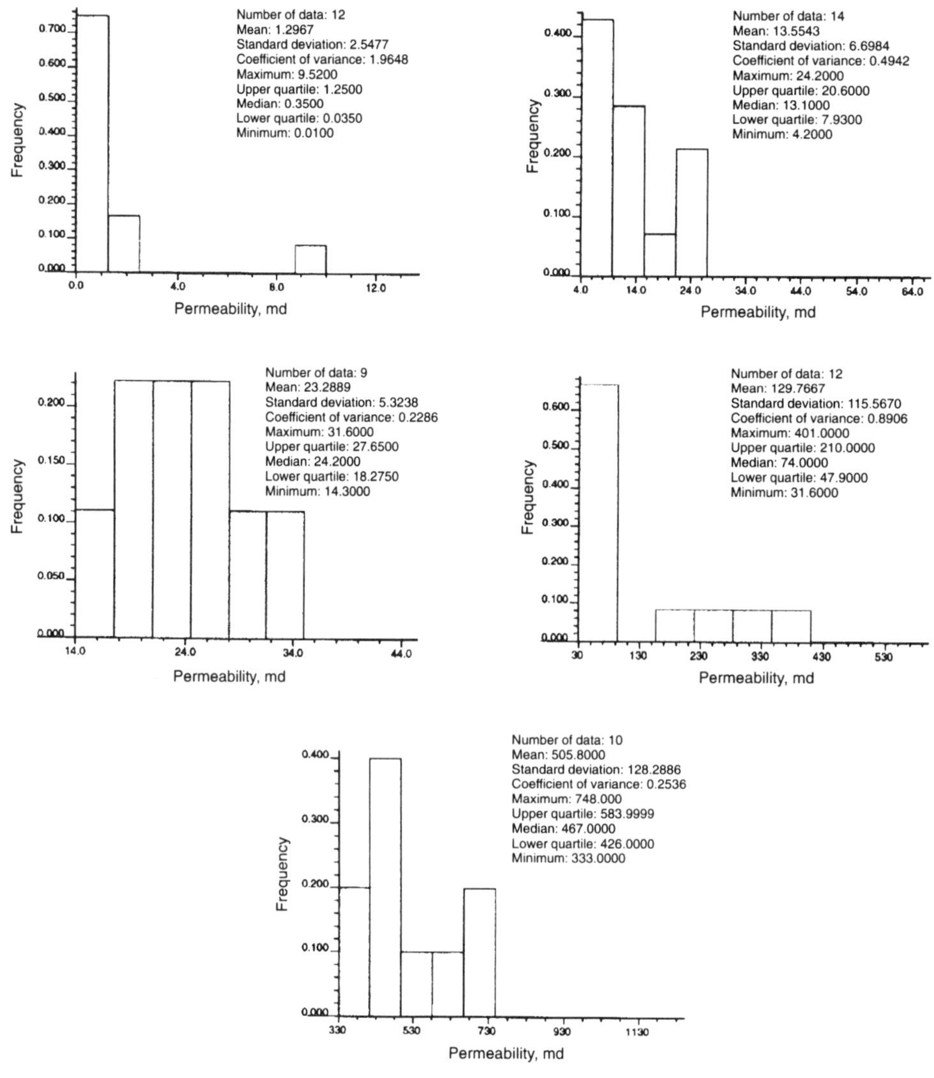

Fig. 2.16—Conditional distribution of permeability for Well 34-29: (a) $\phi = 0.052$ to 0.132, (b) $\phi = 0.132$ to 0.157, (c) $\phi = 0.157$ to 0.175, (d) $\phi = 0.175$ to 0.261, and (e) $\phi = 0.261$ to 0.290.

To calculate correlation coefficient, we must calculate standard deviation for both log k and ϕ. Use Eq. 2.7 with n instead of $(n-1)$ as the denominator.

$$s^2 = \frac{\sum_{i=1}^{n} x_i^2 - n\bar{x}^2}{n}.$$

Modification of the denominator is necessary to make the definition consistent with the covariance definition. For log k, the mean is 2.6322; therefore,

$$s^2 = \frac{[3.063^2 + 2.725^2 + \ldots + 2.559^2] - 9 \times 2.6322^2}{9}$$

$$= 0.1489,$$

TABLE 2.5—PERMEABILITY AND POROSITY DATA FOR NUMERICAL EXAMPLE 2.3

Sample	ϕ	k (md)
1	29.49	1,156.0
2	26.79	531.0
3	28.74	1,059.0
4	27.65	822.0
5	27.69	1,014.0
6	22.69	109.0
7	23.3	138.0
8	23.81	166.0
9	25.54	362.0

TABLE 2.6—log k VALUES FOR NUMERICAL EXAMPLE 2.3

ϕ	log k	k (md)
29.49	3.063	1,156.0
26.79	2.725	531.0
28.74	3.025	1,059.0
27.65	2.915	822.0
27.69	3.006	1,014.0
22.69	2.037	109.0
23.3	2.14	138.0
23.81	2.22	166.0
25.54	2.559	362.0

TABLE 2.7—ORIGINAL POROSITY AND PERMEABILITY VALUES AND ASSOCIATED RANKINGS FOR NUMERICAL EXAMPLE 2.4

ϕ	log k	R_ϕ	$R_{\log k}$
29.49	3.063	9	9
26.79	2.725	5	5
28.74	3.025	8	8
27.65	2.915	6	6
27.69	3.006	7	7
22.69	2.037	1	1
23.3	2.14	2	2
23.81	2.22	3	3
25.54	2.559	4	4

which gives $s_{\log k} = 0.3859$. The standard deviation for ϕ values is calculated similarly and results in $s_\phi = 2.329$. Using Eq. 2.16 gives

$$r(\log k, \phi) = \frac{c(\log k, \phi)}{s_{\log k} s_\phi} = \frac{0.8875}{0.3859 \times 2.329} = 0.988.$$

As is evident, a positive correlation-coefficient value indicates a positive relationship between k and ϕ. Furthermore, a value close to one indicates that a strong relationship exists between k and ϕ values.

In addition to correlation coefficient, rank correlation coefficient is another measure that indicates the relationship between two variables. To estimate the rank correlation coefficient, all data values are first sorted in ascending order. Then, each value is assigned a rank, depending on where it falls. The smallest value receives the lowest rank, and the largest value has a rank of n, where n = total number of samples. Both variables are ordered this way, and ranks are assigned to both. Then, the rank correlation coefficient is calculated as

$$r(R_x, R_y) = \frac{c(R_x, R_y)}{s_{R_x} s_{R_y}}, \qquad (2.17)$$

where $r(R_x, R_y)$ = rank correlation coefficient, $c(R_x, R_y)$ = covariance between the rank values of the two variables, and s_{R_x} and s_{R_y} = standard deviations for the rank values for the two variables. When each variable has the same number of data values, $s_{R_x} = s_{R_y}$.

The rank correlation coefficient is a useful statistical tool for comparing two variables. Unlike the correlation coefficient, which can be influenced by extreme values within the data set (extreme values can change the mean and variance), the rank correlation coefficient is not affected significantly. Therefore, it is a relatively robust measure and may allow detection of any measurement errors, especially if there is a noticeable difference between the values of the correlation coefficient and the rank correlation coefficient. A high value of $r(R_x, R_y)$ and a low value of $r(x, y)$ may indicate that some erratic data pair need to be examined to ensure that there is no measurement error.

Alternatively, a high value of $r(R_x, R_y)$ and a low value of $r(x, y)$ may indicate that the relationship between the two variables is monotonic (trends are the same) but not necessarily linear. As noted previously, $r(x, y)$ is a measure of the goodness of fit between two variables when x and y are linearly related. If the relationship is nonlinear, $r(R_x, R_y)$ can still indicate a high value but $r(x, y)$ may show a low value.

TABLE 2.8—MODIFIED POROSITY AND PERMEABILITY VALUES AND ASSOCIATED RANKINGS FOR NUMERICAL EXAMPLE 2.4

ϕ	log k	R_ϕ	$R_{\log k}$
29.49	3.063	9	9
26.79	2.725	5	5
28.74	3.025	8	8
27.65	2.915	6	6
27.69	3.006	7	7
22.69	2.037	1	2
23.3	1.14*	2	1
23.81	2.22	3	3
25.54	2.559	4	4

*log (13.8) = 1.14

Numerical Example 2.4. Using the same data as in Numerical Example 2.3, calculate the rank correlation coefficient between log k and ϕ values.

Solution. To calculate the rank correlation coefficient, first rearrange the porosity and permeability values in ascending order and rank them. **Table 2.7** shows original values and associated rank for each pair. The ranks are identical for both porosity and permeability values. Therefore, the relationship between the two variables is perfect.

$$c(R_{\log k}, R_\phi) = \tfrac{1}{9}[9 \times 9 + 5 \times 5 + \ldots + 4 \times 4]$$
$$- \tfrac{1}{9}[9 + 5 + \ldots + 4]$$
$$\times \tfrac{1}{9}[9 + 5 + \ldots + 4] = 7.5,$$

$$s_{R\log k} = 2.7386,$$

and $s_{R_\phi} = 2.7386$.

Therefore,

$$r(R_{\log k}, R_\phi) = \frac{7.5}{2.7386 \times 2.7386} = 1.0.$$

The value of the rank correlation coefficient is consistent with the perfect relationship.

Now, assume that the seventh sample permeability is 13.8 md instead of 138 md. Using the modified value, we can rewrite the ranks (**Table 2.8**). The new value changes the ranking slightly. Calculating the correlation coefficient gives $r(\log k, \phi) = 0.868$; whereas, $r(R_{\log k}, R_\phi) = 0.983$. While simply changing one value does not affect the correlation coefficient for rank index significantly, it does significantly affect the correlation coefficient for the actual values (from 0.99 to 0.87). The discrepancy between the two correlation coefficients should prompt a careful investigation of the sample data for any errors in reporting the values.

Linear Regression. A logical extension for applying the summary statistics principles is establishing a linear relationship between the two variables. A linear relationship is useful in predicting a value of one variable when the value of the other variable is known. The simplest type of this relationship is

$$y = mx + b, \qquad (2.18)$$

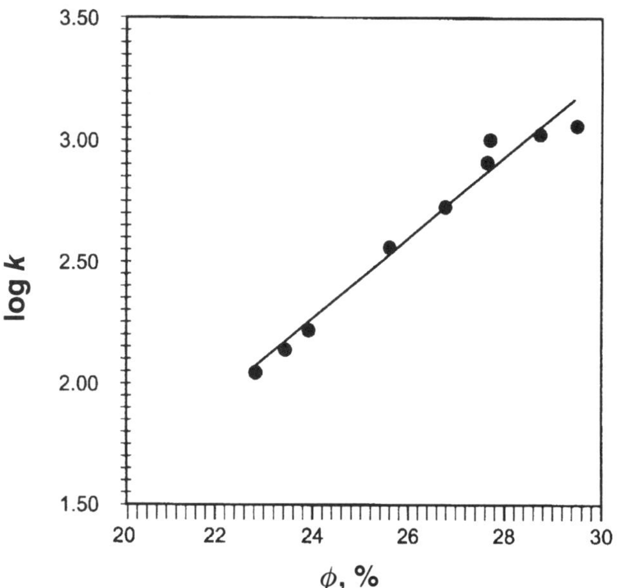

Fig. 2.17—Relationship between log *k* and *φ* for Numerical Example 2.5.

where y = the variable to be estimated; x = the known variable, m = the slope of the straight line, and b = an intercept on the y axis. To estimate the values of m and c, we first use the available sample pair of x and y, and obtain the "best" fit between the two variables. We can show that the best fit can be obtained by defining the values of m and c as

$$m = \frac{c(x,y)}{s_x^2} \quad \text{(2.19a)}$$

and $b = \bar{y} - m\bar{x}$, ... (2.19b)

where $c(x, y)$ = covariance between x and y, s_x^2 = variance of x, and $\bar{y}$ and $\bar{x}$ = arithmetic means of the y and x variables, respectively. As stated before, the goodness of fit is indicated by the correlation coefficient between the two variables.

Numerical Example 2.5. Using the same data set used for Numerical Example 2.3, obtain the best-fit line between log k and ϕ values.

Solution. In this case, assume that the y variable is log k and the x variable is ϕ. Therefore, the relationship is

$$\log k = m\phi + b.$$

From Numerical Example 2.3,

$$c(\log k, \phi) = 0.8875,$$

$$s_\phi = 2.329,$$

and $s_\phi^2 = 5.424$.

With Eq. 2.19a,

$$m = \frac{c(\log k, \phi)}{s_\phi^2} = \frac{0.8875}{5.424} = 0.1636.$$

To calculate the intercept, b, we need the arithmetic means of both k and ϕ.

With Eq. 2.4,

$$\bar{x} = \frac{1}{n}\sum_{i=1}^{n} x_i,$$

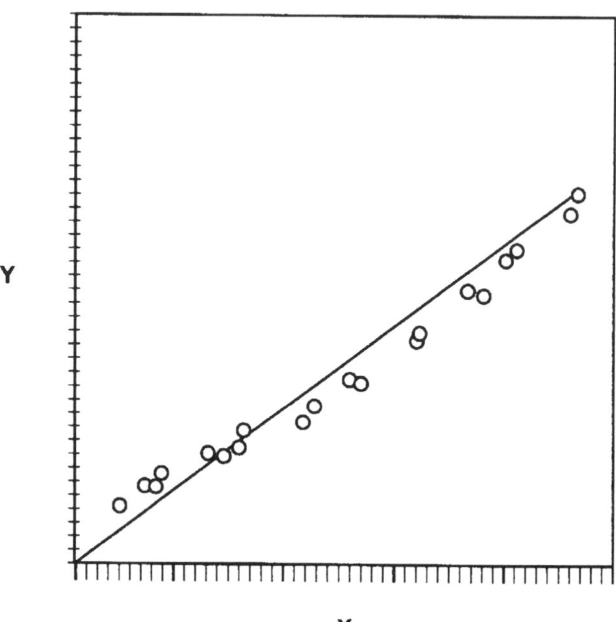

Fig. 2.18—Local bias in linear regression.

and we can calculate the arithmetic means of both variables: $\overline{\log k} = 2.6322$ and $\bar{\phi} = 26.19$. With Eq. 2.19b,

$$b = \overline{\log k} - m\bar{\phi} = 2.6322 - 0.1636 \times 26.19$$

$$= 2.6322 - 0.1636 \times 26.19 = -1.652;$$

therefore, the overall equation can be written as

$$\log k = 0.1636\phi - 1.652.$$

Fig. 2.17 shows the plot of log k vs. ϕ and the best-fit line. As expected, the relationship between log k and ϕ is excellent.

A linear relationship established between any two variables must be locally, as well as globally, unbiased. **Fig. 2.18** shows an example of a locally biased relationship. In this data set, the overall linear relationship between y and x is quite good and is globally unbiased; i.e., the data are spread evenly on both sides of the best-fit line. Careful examination shows, however, that the relationship is locally biased. The value of y is consistently underpredicted with the linear relationship at low values of x and consistently overpredicted at high values of x. This type of local bias may not allow correct prediction of y values given the value of x. If such a relationship is observed, the sample data must be divided further into different subsets and different correlations must be established for different regions.

Field Example 2.6. For this example, we use 57 porosity and permeability core values gathered from Well 34-26 to investigate the relationship between log k and ϕ for the well. **Fig. 2.19** shows a plot of log k vs. ϕ and the best-fit line for Well 34-26. The relationship for the best-fit line is given by $\log k = 18.567\phi - 2.048$, where ϕ is in decimal fractions and k is in md. The correlation coefficient is 0.934, and the rank correlation coefficient is 0.973. The small difference between the two correlation coefficients indicates that there is not a significant number of outlier values.

The high values of the correlation coefficient indicate that the fit is reasonable; however, local bias does exist in this relationship because several geological units are combined together in the vertical direction to develop a single relationship. We can remove this type of local bias if we can separate

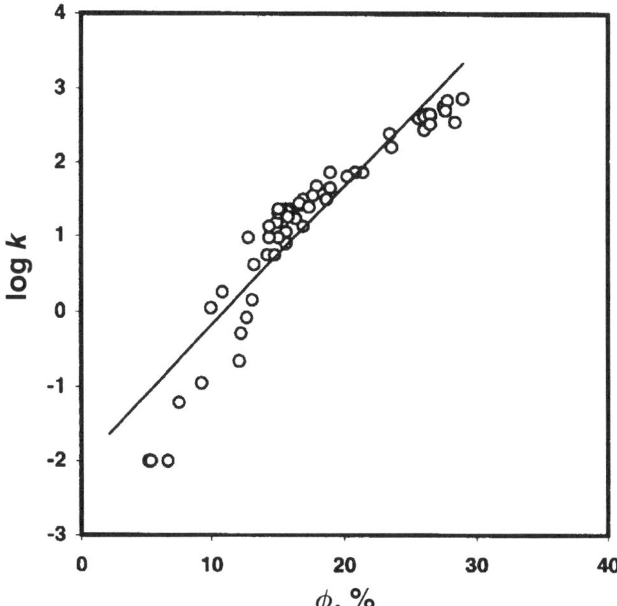

Fig. 2.19—Local bias in log k/ϕ relationship for Well 34-29.

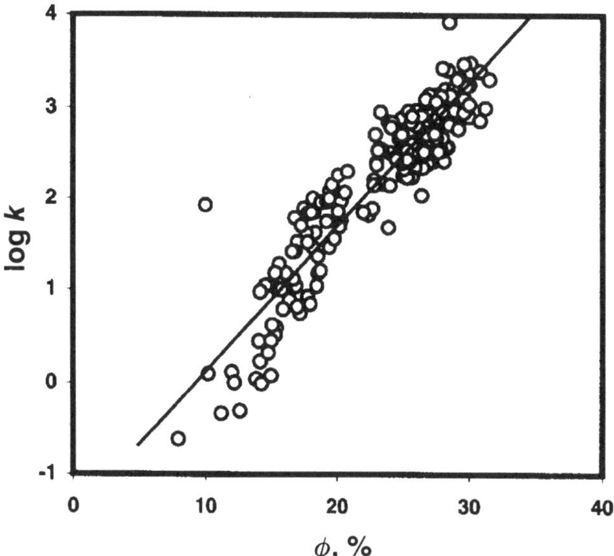

Fig. 2.20—Log k vs. ϕ for Flow Unit 3.

the data into several distinct, geological units and develop a relationship for each individual unit.

Field Example 2.7. This example investigates the log k vs. ϕ relationship for Flow Unit 3. Once the flow units at each well were identified, all cored data for that unit were collected together to develop such a relationship. **Fig. 2.20** shows a plot of log k vs. ϕ and the best-fit line for Flow Unit 3. The correlation coefficient is 0.927, and the rank correlation coefficient is 0.900. The equation for the best-fit line is given by $\log k = 0.158\phi - 1.457$, where porosity is in percent and permeability is in millidarcies.

There is very little local bias in this relationship, indicating good local and global fits. A close examination of the plot reveals that one sample data point can be considered as an outlier datum. This corresponds to a porosity value of 10% and log k of 1.924. Removing this data point and recalculating the best-fit line improves the correlation coefficient to 0.938. As expected, the rank correlation coefficient remains the same. The best-fit equation changes slightly. Note that the decision to remove a particular data point from the best-fit correlation is a subjective one. It should be done only after careful examination of the data set to ensure that such a data point is indeed erroneous and can be removed.

Bivariate Relationships for Spatial Data. In the previous sections, we examined the relationship between two variables and showed that covariance can be used as a statistical tool to quantify such a relationship. This section briefly introduces the application of bivariate summary statistics to spatial data sets. Chap. 3 discusses such relationships in detail.

An important distinction when establishing a bivariate relationship for spatial data is that the same variable is examined but at different spatial locations. It also is possible to develop a relationship between two different variables at different locations; Chap. 3 also discusses this.

Numerical Example 2.6. Table 2.9 shows porosity data collected from a vertical well at uniform intervals of 1 ft. Establish a relationship between porosity values at different locations as functions of distance between those values.

Solution. Although a lot more data typically are collected from a vertical well, we consider only seven data points for this example. The data are collected at uniform intervals. Recall that the covariance relationship (Eq. 2.15) states that

$$c(x,y) = \frac{1}{n}\sum_{i=1}^{n} x_i y_i - \frac{1}{n}\sum_{i=1}^{n} x_i \frac{1}{n}\sum_{i=1}^{n} y_i. \quad \text{........ (2.15)}$$

We use the same relationship, except that the x and y variables are the same variable at different locations. For example, if we denote variables $x(u)$ as a value of x at Location u and a variable $x(u+L)$ as a value of x at Location $u+L$, we can write Eq. 2.15 as

$$c[x(u), x(u+L)] = \frac{1}{n}\sum_{i=1}^{n} x(u_i)x(u_i+L) - \frac{1}{n}\sum_{i=1}^{n} x(u_i)\frac{1}{n}\sum_{i=1}^{n} x(u_i+L), \quad \text{.................. (2.20)}$$

where L = distance between the two variables, also called the lag distance. Note that the difference between the two variables in Eq. 2.20 is the lag distance. Eq. 2.20 still contains

TABLE 2.9—POROSITY DATA FOR NUMERICAL EXAMPLE 2.6

Depth (ft)	Porosity (%)
2,040	8.25
2,041	9.00
2,042	6.25
2,043	5.00
2,044	5.30
2,045	4.75
2,046	5.00
.	.
.	.
.	.

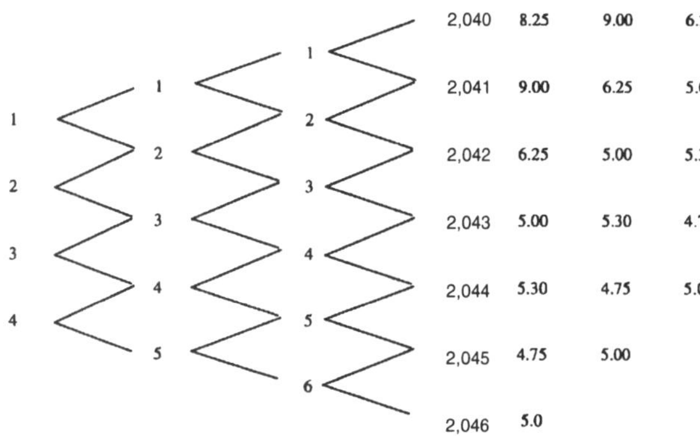

Fig. 2.21—Pairs for Numerical Example 2.6.

a term n, which represents the number of pairs. In this example, n = the number of pairs located a distance L apart.

To apply Eq. 2.20 to our the data set for a lag distance of 1 ft, we first need to find how many pairs are 1 ft apart. **Fig. 2.21** shows that, for a lag distance of 1 ft, we can gather six pairs; for a lag distance of 2 ft, we can gather five pairs; and so forth. With Eq. 2.20, we can calculate the covariance for a lag distance of 1 ft with $n = 6$.

$$\begin{aligned} c[\phi(u),\phi(u+1)] &= \tfrac{1}{6}\big[8.25 \times 9.0 + 9.0 \times 6.25 \\ &\quad + \ldots + 4.75 \times 5.0\big] \\ &\quad - \tfrac{1}{6}\big[8.25 + 9.0 + \ldots + 4.75\big] \\ &\quad \times \tfrac{1}{6}\big[9.0 + 6.25 + \ldots + 5.0\big] \\ &= 1.73. \end{aligned}$$

For covariance, we can simply write the left side of the equation as $c(1)$ because it reflects the covariance for a lag distance of 1 ft. Covariance for a lag distance of 2 ft is calculated in the same way. There are five pairs at that lag distance.

$$\begin{aligned} c(2) &= \tfrac{1}{5}\big[8.25 \times 6.25 + 9.0 \times 5.0 + \ldots + 5.3 \times 5.0\big] \\ &\quad - \tfrac{1}{5}\big[8.25 + 9.0 + \ldots + 5.3\big] \\ &\quad \times \tfrac{1}{5}\big[6.25 + 5.0 + \ldots + 5.0\big] \\ &= 0.43. \end{aligned}$$

There are four pairs for a lag distance of 3 ft, and the same equation is used to calculate $c(3) = 0.195$. The values of correlation coefficient at various lag distances can be calculated similarly. Recall that Eq. 2.16 states that

$$r(x,y) = \frac{c(x,y)}{s_x s_y}. \qquad \text{(2.16)}$$

For spatial data sets,

$$r[x(u),x(u+L)] = \frac{c[x(u),x(u+L)]}{s_{x(u)}s_{x(u+L)}}. \qquad \text{(2.21)}$$

As in the case of covariance, the correlation coefficient can be written simply as a function of the lag distance. That is, Eq. 2.21 can be written as

$$r(L) = \frac{c(L)}{s_{x(u)}s_{x(u+L)}}, \qquad \text{(2.22)}$$

where both the correlation coefficient and the covariance are functions of lag distance. For a lag distance of 1 ft, we can calculate $s_{x(u)}$ by calculating the variance of all data points used as a first data point in a given pair. The mean is

$$\begin{aligned} \overline{x(u)} &= \tfrac{1}{6}[8.25 + 9.0 + 6.25 + 5.0 + 5.3 + 4.75], \\ &= 6.425, \end{aligned}$$

and the variance is

$$\begin{aligned} s^2_{x(u)} &= \frac{\sum_{i=1}^{6} x^2_{(ui)} - 6\overline{x(u)}^2}{6} \\ &= \frac{\big[8.25^2 + 9.0^2 + \ldots + 4.75^2 - 6 \times 6.425^2\big]}{6} \\ &= 2.6823. \end{aligned}$$

Therefore, $s_{x(u)} = 1.638$. Similarly, for the second data point in each pair, $s_{x(u+1)} = 1.474$. Using Eq. 2.22 gives

$$\begin{aligned} r(1) &= \frac{c(1)}{s_{x(u)}s_{x(u+1)}} \\ &= \frac{1.73}{1.638 \times 1.474} = 0.7165. \end{aligned}$$

Similarly, for lag distances of 2 and 3 ft, respectively,

$$\begin{aligned} r(2) &= \frac{c(2)}{s_{x(u)}s_{x(u+2)}} \\ &= \frac{0.43}{1.595 \times 0.525} = 0.5135, \end{aligned}$$

and

$$\begin{aligned} r(3) &= \frac{c(3)}{s_{x(u)}s_{x(u+3)}} \\ &= \frac{0.195}{1.586 \times 0.195} = 0.631. \end{aligned}$$

A special case exists when the covariance and correlation-coefficient values are estimated at a lag distance of zero. At $L = 0$, the equation for covariance reduces to the corresponding equation for variance.

$$c(0) = \frac{1}{n}\sum_{i=1}^{n} x(u_i)x(u_i) - \frac{1}{n}\sum_{i=1}^{n} x(u_i)\frac{1}{n}\sum_{i=1}^{n} x(u_i) = s^2_{x(u)}.$$

In our example, $n = 7$ for $L = 0$, which gives $c(0) = s^2_{x(u)} = 2.548$.

We can easily show that $r(0) = 1$ because a perfect relationship exists between $x(u)$ and $x(u)$—they are identical. Also mathematically, from Eq. 2.22,

$$r(0) = \frac{c(0)}{s_{x(u)}s_{x(u)}} = \frac{s^2_{x(u)}}{s_{x(u)}s_{x(u)}} = 1.0.$$

Although the covariance and therefore the correlation-coefficient value can be calculated for lag distances of 4, 5, and 6 ft with our data, we stopped at a 3-ft lag distance primarily because the number of pairs decreases as lag distance increases. This makes estimates of covariance and correlation coefficient less reliable. It is similar to estimating a best-fit line with three data points. We can fit a line but may not trust it as much because of the limited data. Chap. 3 discusses the importance of having sufficient pairs for a given lag distance in more detail. Obviously, however, a sufficient number of pairs is necessary for a reliable estimate of spatial relationship.

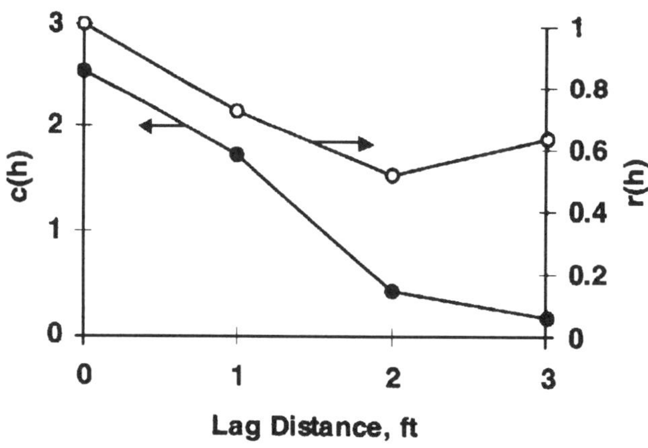

Fig. 2.22—Effect of lag distance on covariance and correlation coefficient.

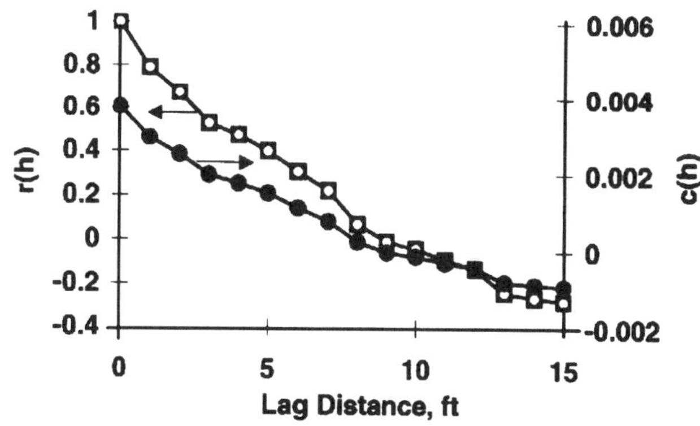

Fig. 2.23—Covariance and correlation coefficient for porosity data in Well 34-29.

If estimated covariance and correlation-coefficient values are plotted as functions of lag distance (**Fig. 2.22**), both the covariance and correlation coefficient decrease as functions of lag distance. This trend occurs because of the way in which most geoscience data are distributed. These types of data become significantly more similar as lag distance decreases. Recall that the correlation coefficient is a measure of how closely two variables are related. For a perfect relationship, the correlation coefficient is equal to one. As the relationship gets weaker, the correlation-coefficient value approaches zero. Starting at $L = 0$, the correlation coefficient equals one. As lag distance increases, the neighboring values become increasingly dissimilar, which is reflected in progressively smaller values of covariance as well as correlation coefficient. The correlation coefficient for a lag distance of 3 ft is slightly higher than one for a lag distance of 2 ft. This may be because of the limited number of pairs used for these calculations. The covariance, however, shows the expected behavior; as the lag distance increases, covariance decreases.

This type of spatial relationship can be quantified by another method called a variogram. Chap. 3 discusses the variogram, its computation, and applications in detail.

Field Example 2.8. This example uses data from Well 34-29 collected vertically at uniform intervals of 1 ft. Using the procedure established in Numerical Example 2.6, we can calculate the covariance and correlation coefficient for the vertical data; **Fig. 2.23** shows plots of both. As expected, the estimated values show a decreasing trend, eventually reaching a constant value at a lag distance of approximately 15 ft. Chap. 3 explains the significance of a constant value.

Fig. 2.24, which plots the first value of the pair vs. the second value of the pair for lag distances of 0, 1, 4, and 10 ft, reiterates what Fig. 2.23 tries to quantify. Recall that the correlation coefficient indicates how well two variables relate to each other. For a perfect relationship, data pairs should be identical, which is possible only at $L = 0$. Note that, as lag distance increases, the deviation from the 45° line increases, which is reflected in smaller correlation-coefficient values. At $L = 10$ ft, hardly any correlation exists, which is reflected by a correlation coefficient value close to zero.

This particular behavior of correlation coefficient or covariance is very useful in geostatistical analysis. In this section, we introduced the type of behavior one should expect in analyzing spatial data. Chap. 3 discusses the details regarding the best ways to capture the spatial relationships and the modeling of such relationships.

2.3 Inferential Statistics

Inferential statistics is a logical extension of descriptive statistics. Descriptive statistics most often deals with analyzing sample data sets. However, from the characteristics of the sample, conclusions (inferences) can be drawn about the population from which the sample was taken. Inferential statistics is the class of statistical techniques that deals with these types of problems. Only in use for approximately 80 years, inferential statistics is a relatively new branch of statistics. It is, however, a much more useful branch for petroleum-engineering-related problems than descriptive statistics because most problems that petroleum engineers deal with involve inferences and decision-making about unsampled locations in the reservoir.

We cover only topics related to inferential statistics that are necessary to understand the geostatistical principles. It is beyond the scope of this book to cover many of the details about inferential statistics; these are readily available from standard statistics books. Instead, we concentrate on covering some of the essential details that lay a foundation for understanding geostatistical principles and practice.

Sec. 2.3.1 covers the definition of a random experiment. Sec. 2.3.2 briefly reviews set-theory principles. Sec. 2.3.3 defines probability and the rules related to it. Sec. 2.3.4 discusses probability and cumulative-distribution functions. Sec. 2.3.5 presents the principles related to expected value and its applications. Sec. 2.3.6 covers some of the basic continuous-distribution functions that are widely used. Finally, Sec. 2.3.7 briefly describes some of the desirable characteristics of inference. We attempt to keep the discussion as simple as possible because of the range of the topics covered. Many concepts have to be introduced through mathematical equations, and we try to provide some intuitive feel for the equations used. Appendix B provides details about the equations, and the references can provide further aid in understanding the concepts.

2.3.1 Random Experiment. There is no rigorous mathematical definition for a random experiment. However, conceptually, it can be defined as an experiment whose outcome cannot be predicted with certainty in advance. Obviously, a random experiment has to result in more than one possible outcome. Another characteristic of a random experiment is that it can be repeated under controlled (unchanged) conditions with the outcomes appearing in a random manner.

An often cited example in statistics books of a random experiment is the tossing of a coin. Tossing a coin results in more

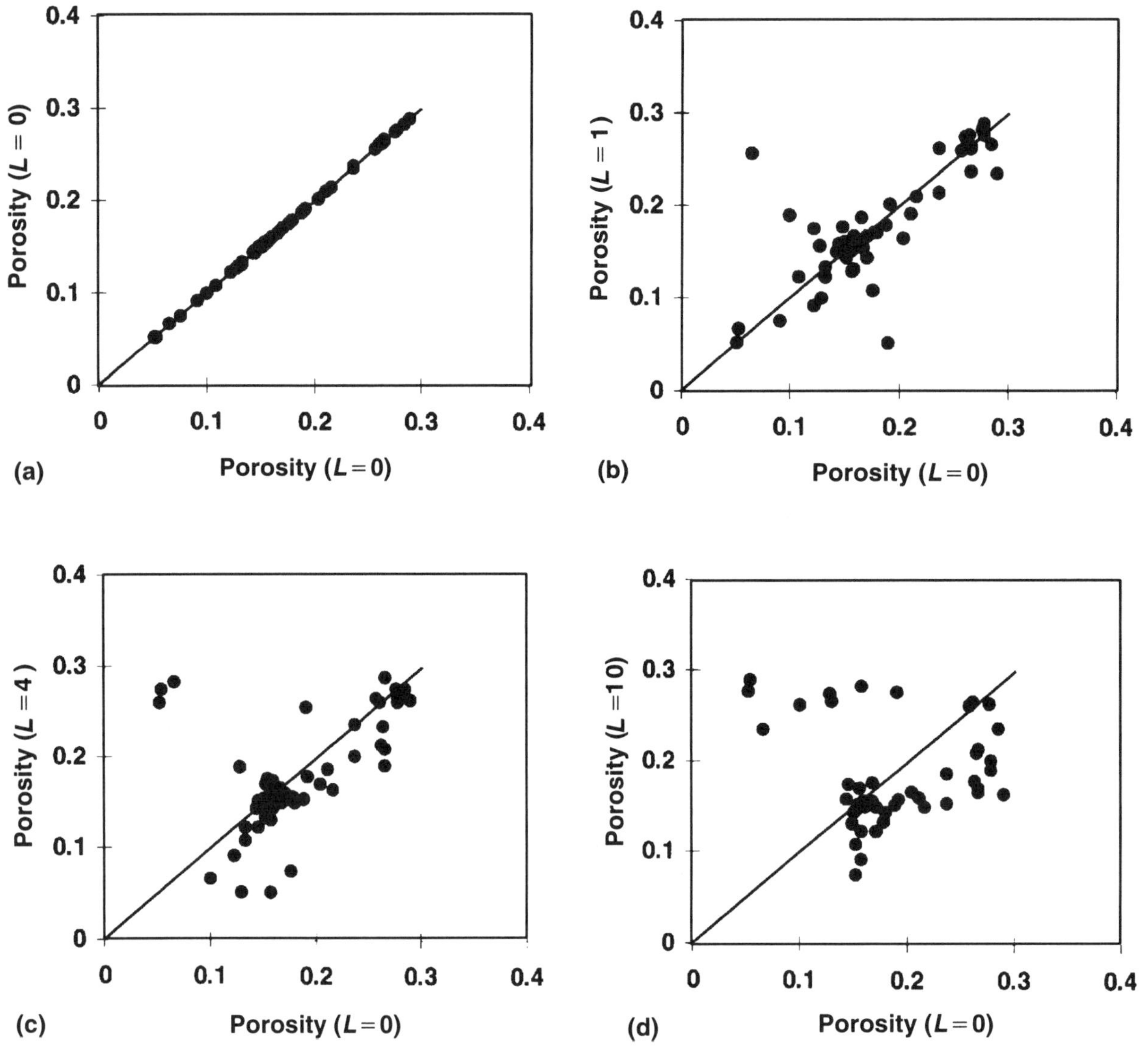

Fig. 2.24—Scatter plots for porosity data for Well 34-29: (a) $L=0$ ft, (b) $L=1$ ft, (c) $L=4$ ft, and (d) $L=10$ ft.

than one outcome: heads or tails. If the coin is true, we cannot predict the outcome with certainty, and we may be able to repeat the experiment under the same conditions so that the outcomes appear (heads or tails) in a random manner. Therefore, all characteristics of a random experiment are satisfied, and we can consider tossing a coin as a random experiment.

Another simple example of a random experiment is rolling a pair of dice. This results in multiple outcomes (36 possibilities to be exact), we cannot predict the outcome with certainty, and we can repeat the experiment under controlled conditions so that the outcomes appear in a random order. As in the case of tossing a coin, this satisfies all characteristics of a random experiment.

An example that is more difficult to justify as a random experiment, but closer to what we are interested in, is drilling a new well in a reservoir. It can be argued, with some justification, that drilling a well should not be considered a random experiment because, once a well location is chosen, drilling the well results in only one possible outcome, not multiple outcomes. The porosity and permeability values are distributed vertically in one particular way, the well is going to produce at a particular rate, and the well-test results (with a particular model) may provide a unique set of reservoir properties. If this is the case, why should drilling a well be considered to be a random experiment? If there is only one outcome, then drilling a well should be treated as a deterministic experiment where only one outcome is possible.

This argument has been used in the past to justify the use of a deterministic model to describe a reservoir. In a deterministic model, everything is known with certainty. The argument in favor of this is that each reservoir is unique and already defined. Although intuitively appealing, some problems are associated with this argument. First, understand that a deterministic model is the most desirable model to describe the reservoir. If we can describe every aspect of the reservoir with certainty, we do not need to know geostatistics or any other statistical principle to describe the reservoir. The question is whether we can do it.

In principle, with sufficient knowledge about the genesis of the reservoir, we can apply a deterministic model to describe the reservoir. This requires knowledge about how the reservoir was formed, under what circumstances the oil was trapped, what type of temperatures and pressures prevailed at the time the hydrocarbons were created, what type of geological processes took place subsequent to hydrocarbon presence, what chemical interactions took place after the hydrocarbons were trapped, and possibly some other factors. If we understand all these processes in sufficient detail and have the ability to put in a gigantic, complete, mathematical model to solve the model, we arrive at the deterministic reservoir description. The model will be able to predict all the necessary reservoir characteristics, including its size, shape, volume, physical properties, and the chemical properties of the hydrocarbons. This description has no uncertainty. We will be able to drill the wells in the most optimal way and produce the reservoir in the most efficient manner.

If this sounds impossible, it is because it is. What is fundamentally lacking in describing the reservoir in a deterministic manner is our knowledge. It is true that the reservoir is deterministic and that all the properties are uniquely defined. However, our understanding of the processes that resulted in these properties is so inadequate that they appear to be random. This does not imply that the processes are random, but simply that we have inadequate knowledge about them.

Therefore, should drilling a new well be considered a random experiment? Consider the characteristics of drilling a well with the characteristics of a random experiment. The first characteristic of a random experiment is that the outcome cannot be predicted with certainty. Because of our lack of knowledge, we do not know with certainty what the outcome will be of drilling a new well. The second characteristic is that a random experiment can result in several possible outcomes. Our lack of knowledge before drilling results in alternative outcomes; the most obvious is whether the well will be a producer or a dry hole. The third characteristic of a random experiment is that it can be repeated under controlled conditions so that the outcomes appear in a random manner. Although this may be a stretch of imagination, drilling a new well is a new experiment, and the outcome every time will appear in a random manner. Drilling each new well is based on prior experience; therefore, we have a little more information about the reservoir than we had before. As a result, we may be able to predict the possible outcomes with a little more confidence. However, because the outcome still cannot be predicted with complete certainty, the experiment is still considered random.

The concept of random experiment is extremely important. However, drilling a well as a random experiment is different from tossing a coin as a random experiment. Tossing a coin results in either of two outcomes; therefore, it is a random experiment. Drilling a well results in only one outcome; however, our lack of knowledge does not allow us to predict that outcome with certainty. Therefore, we treat it as a random experiment with multiple possibilities of outcomes.

This concept can be extended to the use of geostatistical principles to estimate values at unsampled locations. Although the properties at each unsampled location are uniquely defined, lack of knowledge does not allow us to predict them with certainty. Therefore, our prediction is a random experiment with several possible outcomes. Later chapters that cover the application of geostatistics discuss this in more detail.

To summarize, a random experiment can result in multiple outcomes, none of which can be predicted with certainty. When lack of knowledge prevents us from predicting a single outcome with certainty, we also treat that experiment as a random experiment.

2.3.2 Sample Space and Events. This section briefly reviews the principles of set theory. For convenience, rolling a die illustrates some of these principles. We already saw that rolling a die is a random experiment.

A sample space, S, is a set of all possible outcomes. For a rolling-a-die experiment, we can denote the sample space as

$$S = (1, 2, 3, 4, 5, 6) \qquad (2.23)$$

because the experiment has six possible outcomes.

An event is defined as a set consisting of some of the possible outcomes. If Event A consists of all the even-numbered outcomes of the rolling-a-die experiment, then Event A is

$$A = (2, 4, 6). \qquad (2.24)$$

If Event B consists of all the outcomes less than five for the rolling-a-die experiment, then Event B is

$$B = (1, 2, 3, 4). \qquad (2.25)$$

Using only two events, A and B, of a sample space, we can define the union of these two events, $A \cup B$, as consisting of all the outcomes present in either A or B. Therefore,

$$A \cup B = (1, 2, 3, 4, 6). \qquad (2.26)$$

Similarly, using Events A and B, we can define the intersection of these two events, $A \cap B$, as consisting of all outcomes that are present in both A and B. Therefore,

$$A \cap B = (2, 4). \qquad (2.27)$$

If the intersection of two events results in a null set (containing no outcome), these two events are mutually exclusive. For example, if Event C contains all the odd-numbered outcomes from a rolling-a-die experiment, then

$$C = (1, 3, 5). \qquad (2.28)$$

With the definition of intersection of events,

$$A \cap C = \Phi, \qquad (2.29)$$

where Φ = a null set and A is as defined by Eq. 2.24. Because $A \cap C$ contains no outcomes, A and C are considered mutually exclusive.

The best way to illustrate some of these definitions is with Venn diagrams. **Fig. 2.25** shows examples of Venn diagrams

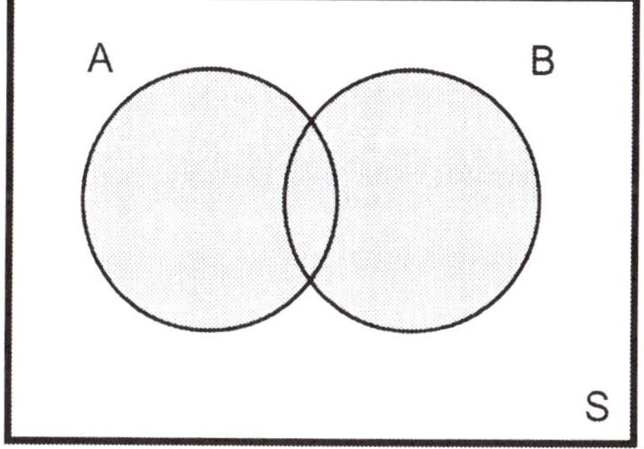

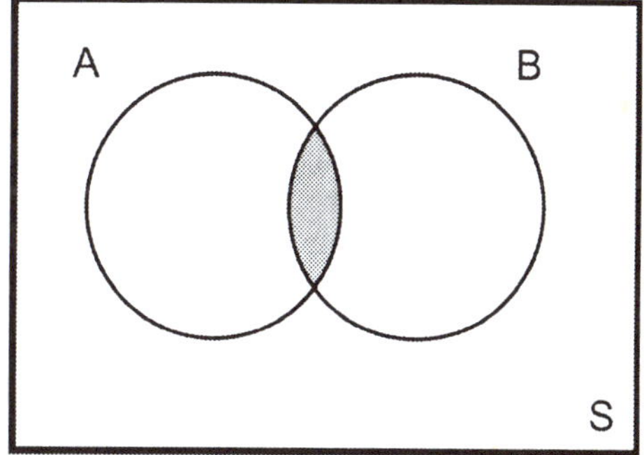

Fig. 2.25—Venn diagram.

for union and intersection of events within a sample space. The shaded region indicates the resulting event.

2.3.3 Probability. Probability normally is associated to a particular event of a random experiment. A geologist's statement that "there is 30% probability of finding oil at a location where a new well is to be drilled" can have two meanings. Both meanings are correct and are a result of the way we define the random experiment.

1. The first interpretation is that the geologist believes that, in reservoirs with a similar depositional environment, 30% of the wells will produce oil. That is, if several wells are drilled in very similar depositional environments, 30% will produce oil and, by inference, 70% will be dry holes.

2. The second interpretation is that the 30% probability is a measure of the geologist's subjective belief that the well will produce oil.

These two interpretations can be directly related to the description of random experiments. The first interpretation is closely related to the random experiment of rolling a die. That is, if we repeat the experiment a large number of times under controlled conditions, a pattern emerges about the outcomes. For example, for a true die, if we roll the die a large number of times, we observe that each of the six outcomes is equally likely. Under this interpretation, probability can be defined as

$$p(A) = \lim_{n_c \to \infty} \frac{n_A}{n_c}, \quad \ldots\ldots (2.30)$$

where $p(A)$ = probability of Event A, n_A = number of times outcome has occurred, and n_c = number of times the random experiment is conducted under controlled conditions. n_c has to be large to ensure that a correct pattern is captured. For the rolling-a-die experiment, the probability of each of the six outcomes is 1/6 or 0.1667. Going back to the geologist's statement, if we drill a large number of wells in a similar, depositional environment, we observe that $p(P)$, where P = producer is 30%, or 30% of the wells should produce oil.

Interpretation 2 is closely related to the random experiment of drilling a well. The geologist is simply using a subjective belief about the chance of success. Uncertainty exists is because of a lack of complete knowledge about the reservoir. However, the geologist is using his/her partial knowledge to assign a value to the probability of success.

Both interpretations are correct, and the mathematics of probability does not change with the interpretation applied. Deterministic events can be treated as random events if we lack sufficient knowledge about those events; however, with partial knowledge about the events, probabilities can be assigned to the likely outcomes (events) of that experiment.

Laws of Probability. On the basis of our definition of probability, we can write three basic laws related to probability. For Event A of a random experiment with Sample Space S,

$$0 \leqq p(A) \leqq 1. \quad \ldots\ldots (2.31)$$

That is, the probability value can never be less than zero or greater than one. Also,

$$p(S) = 1. \quad \ldots\ldots (2.32)$$

That is, the probability that the outcome will be part of the sample space is equal to one. Secondly,

$$\sum_{i=1}^{ne} p(A_i) = p\left(\bigcup_{i=1}^{ne} A_i\right), \quad \ldots\ldots (2.33)$$

where A_i = sequence of mutually exclusive events and n_e = number of mutually exclusive events. For $n_e = 3$,

$$p(A_1) + p(A_2) = p(A_1 \cup A_2), \quad \ldots\ldots (2.34)$$

and for $n_e = 3$,

$$p(A_1) + p(A_2) + p(A_3) = p(A_1 \cup A_2 \cup A_3). \quad \ldots\ldots (2.35)$$

That is, the probability of the union of mutually exclusive events is equal to the addition of the probabilities of the individual events.

For two events that are not mutually exclusive,

$$p(A \cup B) = p(A) + p(B) - p(A \cap B). \quad \ldots\ldots (2.36)$$

Numerical Example 2.7. The following three events are defined for a rolling-a-die experiment.

1. Event A. All even-numbered outcomes.
2. Event B. All outcomes greater than 3.
3. Event C. All odd-numbered outcomes less than 4

Calculate $p(A)$, $p(B)$, $p(C)$, $p(A \cup B)$, $p(A \cup C)$ and $p(A \cap B)$.

Solution. From the description of events, the events are $A = (2,4,6)$, $B = (4,5,6)$, and $C = (1,3)$. Knowing that the probability of individual outcomes for rolling a die is 1/6, we can calculate the probability of individual events on the basis of Eq. 2.35 as

$$p(A) = \frac{1}{6} + \frac{1}{6} + \frac{1}{6} = \frac{1}{2}.$$

If we consider Event A to consist of three mutually exclusive events with Outcomes 2, 4, and 6, then the probability of an event is the addition of the probabilities of these three events. Similarly,

$$p(B) = \frac{1}{6} + \frac{1}{6} + \frac{1}{6} = \frac{1}{2}$$

and $p(C) = \frac{1}{6} + \frac{1}{6} = \frac{1}{3}$.

Because $p(A \cap C) = \Phi$,

$$p(A \cup C) = p(A) + p(C) \quad \ldots\ldots (2.37)$$

$$= \frac{1}{2} + \frac{1}{3} = \frac{5}{6},$$

which can be confirmed because

$$A \cup C = (1,2,3,4,6).$$

Therefore,

$$p(A \cup C) = \frac{1}{6} + \frac{1}{6} + \frac{1}{6} + \frac{1}{6} + \frac{1}{6} = \frac{5}{6}$$

and $A \cap B = (4,6)$,

which results in

$$p(A \cap B) = \frac{1}{3}.$$

Using Eq. 2.36 gives

$$p(A \cup B) = p(A) + p(B) - p(A \cap B)$$

$$= \frac{1}{2} + \frac{1}{2} - \frac{1}{3} = \frac{2}{3},$$

which is confirmed because

$$A \cup B = (2,4,5,6).$$

Therefore,

$$p(A \cup B) = \frac{1}{6} + \frac{1}{6} + \frac{1}{6} + \frac{1}{6} = \frac{2}{3}.$$

Conditional Probability. As the name indicates, conditional probability is the probability of an event that is conditional on some information. This allows calculation of the probability of a given event when partial information regarding the result of the random experiment is available.

For example, rolling a pair of dice has 36 possible outcomes: (1,1), (1,2) , . . . , (1,6) , . . . , (3,3) , . . . , (6,6). The first number in the parentheses is the outcome of the first die, and the second number in the parentheses is the outcome of the second die. Because each die can take six possible values, the total number of outcomes is 36. If asked to calculate the probability that the addition of the outcomes is two, we immediately see that the only outcome that results in such addition is (1,1). Therefore, the probability that the addition will be two is 1 out of 36, or 1/36. If the first die lands on Side 1, what is the probability now that the addition of the two outcomes is two? We can calculate this probability easily because we know that the outcome of one of the dice is one. The second die can result in six possible outcomes. We want to calculate the probability that the outcome will be one because that is the only outcome that will result in an outcome of two when rolling both dice. The probability that the outcome of the second die will be one is one out of six, or 1/6. Therefore, the conditional probability of the addition of the outcomes of a pair of dice being two if the first die landed on Side 1 is 1/6. This value is considerably smaller than 1/36, which is what we calculated without any additional information.

The most common notation used to describe conditional probability is $p(A|B)$. This indicates the conditional probability of Event A occurring given that Event B has occurred. A general equation for calculating conditional probability is

$$p(A|B) = \frac{p(A \cap B)}{p(B)}. \quad \text{.......................} (2.38)$$

Intuitively, Eq. 2.38 can be explained as follows. Once Event B has occurred, we reduced the sample space corresponding to Event B only because no outcome outside Event B is possible. For Event A to occur, the outcome has to be common to both Events A and B because Event B has occurred. That is, it has to come from $A \cap B$. Therefore, the probability of Event A occurring given that Event B has occurred is equal to the ratio of the probability of Events A and B occurring divided by the probability of Event B.

Numerical Example 2.8. The probability of finding oil in an exploration well is estimated to be 0.2. One of the uncertainties in finding oil in this well is the presence of source rock. The probability of the presence of source rock is 0.7. After these preliminary calculations were made, a well drilled in a nearby area confirmed the presence of source rock in the region. What is the probability of finding oil in the first exploration well given that the presence of source rock is confirmed?

Solution. Let A be the event that the oil is found in the exploration well, and B be the event that the source rock is present. Using Eq. 2.38 gives

$$p(A|B) = \frac{p(A \cap B)}{p(B)}.$$

We know that $p(B) = 0.7$ and $p(A \cap B) = 0.2$ (because A is a subset of B); we have to have source rock to find oil in the well. Substituting gives

$$p(A|B) = \frac{0.2}{0.7} = 0.286.$$

The probability of finding oil improves to 0.286 because source rock is present.

We can rewrite Eq. 2.38 as

$$p(A \cap B) = p(A|B)p(B). \quad \text{..................} (2.39)$$

Eq. 2.39 allows us to derive a few more conclusions. First, for mutually exclusive events, $(A \cap B)$is a null set; therefore, $p(A \cap B) = 0$. Because the left side of Eq. 2.39 is zero, $p(A|B) = 0$, which is consistent with the idea that if B has occurred, A cannot occur. Therefore, the probability of A occurring given that B has occurred is zero. The same definition can be used to define independent events, which are events that are independent of each other. The occurrence of an independent event is not affected by whether any of the events from which it is independent has occurred. For example, in rolling a pair of dice, the outcome of one die does not affect the outcome of the other die. The outcomes of the two dice are completely independent of each other. In other words,

$$p(A|B) = p(A) \text{ and } p(B) = p(B|A), \quad \text{.........} (2.40)$$

where the probability that Event A will occur is not affected by the fact that B has occurred. The same thing can be said about Event B. The probability of Event B occurring is not affected by the fact that Event A has occurred.

Substituting Eq. 2.40 in Eq. 2.39 gives

$$p(A \cap B) = p(A)p(B) \quad \text{...................} (2.41)$$

for independent events. For more than two events, Eq. 2.41 can be extended as

$$p(A_1 \cap A_2, \ldots, \cap A_n) = p(A_1)p(A_2), \ldots, p(A_n), \quad \text{................} (2.42)$$

where $A_1, A_2, \ldots, A_n$ = independent events.

Numerical Example 2.9. The probability of success is estimated to be 0.2 for an exploration well in Basin 1. For another exploration well in Basin 2, the probability of success is estimated to be 0.3. If both wells are drilled, what is the probability that both will be successful?

Solution. Because these wells are drilled in different basins, they can be considered as independent events; the outcome of one well does not affect the outcome of the other. If Event A is a successful well in Basin1 and Event B is a successful well in Basin 2,

$$p(A) = 0.2 \text{ and } p(B) = 0.3.$$

With Eq. 2.41,

$$p(A \cap B) = p(A)p(B) = 0.2 \times 0.3 = 0.06.$$

The probability that both events will occur is 0.06 or 6%.

Another useful extension of Eq. 2.38 can be written. Recall that Eq. 2.38 states that

$$p(A|B) = \frac{p(A \cap B)}{p(B)}. \quad \text{......................} (2.38)$$

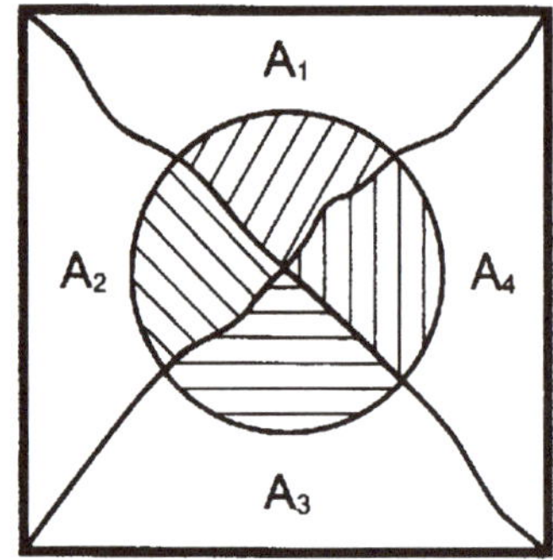

Fig. 2.26—Explanation of Bayes' rule.

With a sample space consisting of A_i mutually exclusive events so that

$$\sum_{i=1}^{n_e} p(A_i) = 1,$$

we can easily write

$$p(B) = p(A_1 \cap B) + p(A_2 \cap B) + \ldots + p(A_{n_e} \cap B). \quad \text{(2.43)}$$

In **Fig. 2.26**, which illustrates this, the sample space is divided into four mutually exclusive events, A_1 through A_4, and Event B is located in the sample space. As the figure shows, Event B can be written as

$$B = A_1 \cap B + A_2 \cap B + A_3 \cap B + A_4 \cap B. \quad \text{(2.44)}$$

This can be generalized for n_e mutually exclusive events.

Also, the numerator of Eq. 2.38 can be written as

$$p(A \cap B) = p(B|A)p(A) \quad \text{(2.45)}$$

because we know that

$$p(B|A) = \frac{p(A \cap B)}{p(A)}. \quad \text{(2.38)}$$

Using Eqs. 2.45 and 2.43 along with Eq. 2.38 gives

$$p(A_j|B) = \frac{p(B|A_j)p(A_j)}{\sum_{i=1}^{n_e} p(A_i \cap B)} \quad \text{(2.46a)}$$

or $$p(A_j|B) = \frac{p(B|A_j)p(A_j)}{\sum_{i=1}^{n_e} p(B|A_i)p(A_i)}. \quad \text{(2.46b)}$$

Eq. 2.46b is Bayes' theorem, which represents a generalized equation for conditional probability. An entire branch of statistics, Bayesian statistics, is based on the concept of determining conditional possibilities. The discussion of the conditional-simulation technique covers the usefulness of this principle in more detail. Numerical Example 2.10 illustrates Bayes' theorem.

Numerical Example 2.10. A 3D geophysical survey concludes that a new area has three potential regions where oil can be found. It is equally likely that oil could be found in any of the three regions. On the basis of the surrounding regions and the presence of source rock, it is possible for only one of the regions to contain oil. Geophysicists are certain that one of these regions should have commercial reserves. If only one exploration well is drilled in a region, there is a 40% chance that oil will not be discovered in that region although oil may indeed be present (overlook probability). If one well is drilled in Region 1 and is unsuccessful, what is the probability that oil is present in Region 1? What is the probability that the oil is present in the other two regions?

Solution. Let E_i, where $i = 1, 2, 3$, be an event that oil is in Region i. Let F be an event that the search of Region 1 is unsuccessful. With Bayes' theorem,

$$p(E_1|F) = \frac{p(E_1 \cap F)}{\sum_{i=1}^{3} p(F|E_i)p(E_i)} \quad \text{(2.47a)}$$

and $p(E_1 \cap F) = p(F|E_1)p(E_1)$. $\quad$ (2.47b)

The denominator of Eq. 2.47b can be written as

$$\sum_{i=1}^{3} p(F|E_1)p(E_i) = p(F|E_1)p(E_1) + p(F|E_2)p(E_2) + p(F|E_3)p(E_3). \quad \text{(2.48)}$$

Each region has an equal likelihood of success; therefore,

$$p(E_1) = p(E_2) = p(E_3) = 0.333.$$

Also, $p(F|E_1) = 0.4$, $p(F|E_2) = 1$, and $p(F|E_3) = 1$ because the exploration well in Region 1 results in failure if oil is in either Region 2 or 3. Substituting all these values gives

$$p(E_1|F) = \frac{0.4 \times 0.33}{0.4 \times 0.33 + 1 \times 0.33 + 1 \times 0.33} = 0.167.$$

The probability of finding oil in Region 1 is reduced to 0.167.

Similarly,

$$p(E_2|F) = p(E_3|F) = \frac{1 \times 0.33}{0.4 \times 0.33 + 1 \times 0.33 + 1 \times 0.33} = 0.417.$$

The probability of finding oil in Regions 2 and 3 improved to 0.417 because the search in Region 1 was unsuccessful.

2.3.4 Random Variables. A random variable is a variable whose values are generated by a random experiment on the basis of some probabilistic function. For example, the rolling-a-die experiment produces any one of the six possible outcomes randomly. If the random variable is letter X for this random experiment, for a true die,

$$p(X = 1) = \frac{1}{6} = p(X = 2) = p(X = 3) = p(X = 4) = p(X = 5) = p(X = 6). \quad \text{(2.49)}$$

That is, the probability that a random variable can take any one of the six values is 1/6.

It is important to maintain the distinction between a random variable and an actual outcome of a random variable. To make this distinction, we use an uppercase letter to denote a random variable (e.g., X) and a lowercase letter to denote the outcome (or realization) of a random variable (e.g., x). Conceptually, the difference between the random variable and its realizations can be explained with the same example of the rolling-a-die experiment. The random variable can take any of the six

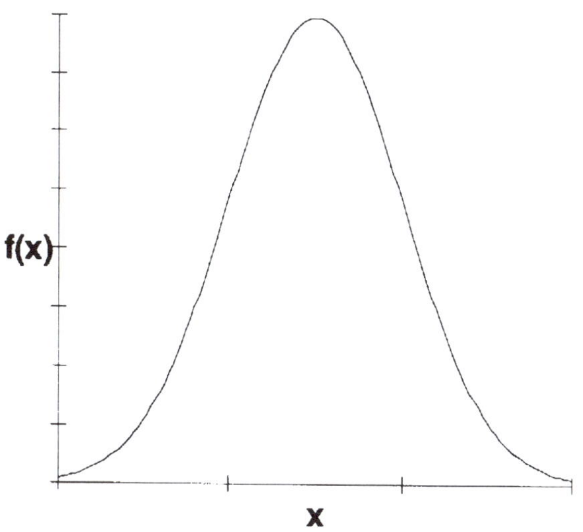

Fig. 2.27—Probability density function.

outcomes 1, 2, 5, 6, 3, 4, 3, 3, 4, 6, 1, 4, 2, ..., these are the realizations of the random variable, which we can denote as $x_1 = 1$, $x_2 = 2$, $x_3 = 5$, and $x_4 = 6$, where the subscripts denote the number of a particular realization.

Random variables are defined as two types: discrete and continuous. Discrete random variables can take a finite number of values. An example is the rolling-a-die experiment, where a random variable can take only six possible values. Continuous random variables can take a very large number of values (for example, a collection of porosity data from a reservoir), and a large number of outcomes are possible. The following two sections describe some functions of both discrete and continuous random variables.

Probability Function. The probability function describes the probability that a random variable will take a certain value. The probability function is closely related to the relative-frequency-distribution function, which describes the chance that a value will fall within a certain class. The probability function describes an essentially similar behavior.

For discrete random variables, the probability mass function, $\mathrm{P}(a)$, of a random variable X is

$$\mathrm{P}(a) = p(X = a). \qquad (2.50)$$

For a discrete random variable, X can take finite number of values $x_i, i = 1, n$. Therefore,

$$\sum_{i=1}^{n} \mathrm{P}(x_i) = 1. \qquad (2.51)$$

Numerical Example 2.11. Define the probability mass function for the rolling-a-die experiment. Show that Eq. 2.51 is satisfied for this function.

Solution. Knowing the random experiment, we can write, for example,

$$\mathrm{P}(1) = p(X = 1) = \frac{1}{6}.$$

Similarly, we can write

$$\mathrm{P}(2) = \mathrm{P}(3) = \mathrm{P}(4) = \mathrm{P}(5) = \mathrm{P}(6) = \frac{1}{6}.$$

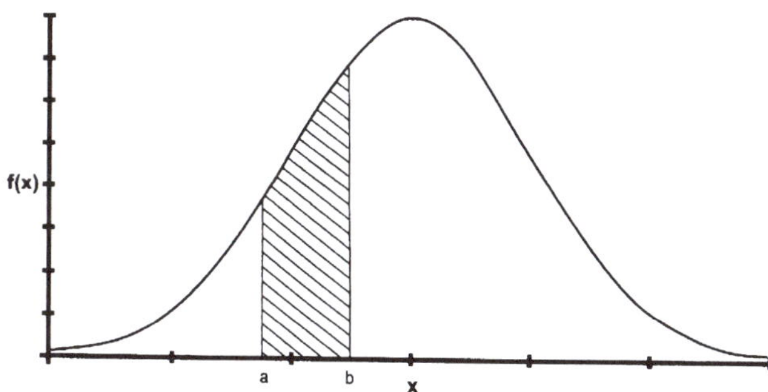

Fig. 2.28—Probability of value falling within an interval.

Applying Eq. 2.51 gives

$$\sum_{i=1}^{6} \mathrm{P}(x_i) = \frac{1}{6} + \frac{1}{6} + \frac{1}{6} + \frac{1}{6} + \frac{1}{6} + \frac{1}{6} = 1.$$

Therefore, Eq. 2.51 is satisfied.

For a continuous random variable, the probability density function, $\mathrm{f}(x)$, describes the behavior of a random variable. **Fig. 2.27** shows a representative probability density function. One requirement of the probability density function is that the area under the curve be equal to one. Mathematically,

$$\int_{-\infty}^{\infty} \mathrm{f}(x)\mathrm{d}x = 1. \qquad (2.52)$$

Recall that addition of relative-frequency distributions also adds to one. (Fig. 2.3 provides a good understanding of the definition of the probability density function.) Once the probability density function is described, the probability that the value of a random variable will fall within a certain interval can be calculated easily. For example,

$$p(a < X \leqq b) = \int_{a}^{b} \mathrm{f}(x)\mathrm{d}x. \qquad (2.53)$$

Schematically, the probability that a value will fall within a certain interval is represented by the area under the curve within that interval (**Fig. 2.28**). Again, it is important to have an intuitive feel for this equation (see Fig. 2.4 and the related discussion to understand the probability density function and its close relationship to the relative-frequency distribution).

Numerical Example 2.12. Pay-zone thickness in a reservoir is described by the following probability density function.

$$\mathrm{f}(x) = 0 \text{ for } x \leqq 20 \text{ ft}$$

$$= \frac{1}{50} \text{ for } 20 < x \leqq 70 \text{ ft}$$

$$= 0 \text{ for } > 70 \text{ ft}.$$

Show that this density function satisfies Eq. 2.52. Further, calculate the probability that the thickness at a particular location will fall between 30 and 50 ft thick.

Solution. Applying Eq. 2.52 to the probability density function gives

$$\int_{-\infty}^{20} 0\mathrm{d}x + \int_{20}^{70} \frac{1}{50}\mathrm{d}x + \int_{70}^{\infty} 0\mathrm{d}x = 0 + \frac{1}{50}(x)_{20}^{70} + 0 = 1;$$

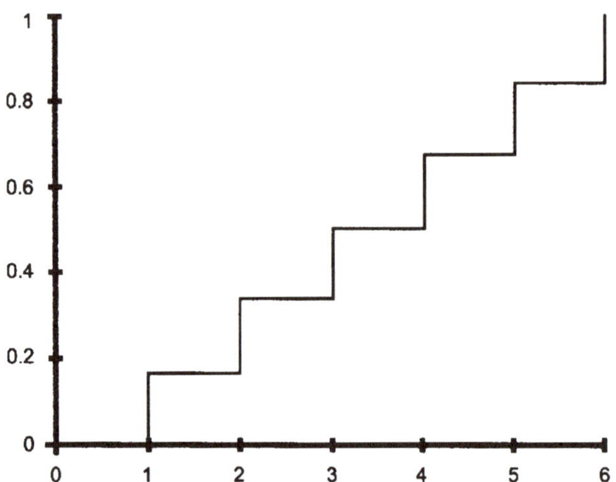

Fig. 2.29—Cumulative distribution function for rolling-a-die experiment.

therefore the density function satisfies Eq. 2.52.

Use Eq. 2.43 to calculate the probability that a value will fall within 30 and 50 ft.

$$p(30 \leqq x \leqq 50) = \int_{30}^{50} \frac{1}{50} dx = \frac{20}{50} = 0.4.$$

Therefore, there is a 40% probability that the pay zone will fall within the 30- to 50-ft interval.

Cumulative-Distribution Function. The cumulative-distribution function, F(x), is defined as

$$F(x) = p(X \leqq x). \quad \text{......................} (2.54)$$

It is the probability that a random variable X will be less than a particular value x. Knowing the definition of the cumulative-distribution function, we can use Eq. 2.53 to calculate the probability that a random variable will fall within a certain interval. For example, to compute $p(a < X \leqq b)$, we can write $p(X \leqq b)$ as comprising two mutually exclusive events or

$$p(X \leqq b) = p(X \leqq a) + p(a < X \leqq b). \quad \text{................} (2.55)$$

Therefore,

$$p(a < X \leqq b) = p(X \leqq b) - p(X \leqq a)$$
$$= F(b) - F(a). \quad \text{............} (2.56)$$

Eq. 2.56 calculates the probability that a random variable will fall within a certain interval.

For a discrete random variable, cumulative-distribution function can be calculated as

$$F(a) = \sum_{x_i \leqq a} P(x_i), \quad \text{........................} (2.57)$$

where $P(x_i)$ = the probability mass function of a random variable.

For a continuous random variable, the cumulative-distribution function can be calculated as

$$F(a) = \int_{-\infty}^{a} f(x)dx, \quad \text{......................} (2.58)$$

where f(x) = the probability density function of a random variable.

It is important to understand that cumulative-distribution function is closely related to cumulative-relative-frequency distribution. Like cumulative-relative-frequency distribution, cumulative-distribution function has a minimum value of zero and a maximum value of one. It is also a nondecreasing function, which means that it can stay constant over a certain interval but does not decrease. The following numerical examples illustrate the usefulness of the cumulative-distribution function.

Numerical Example 2.13. Calculate the cumulative-distribution function for the rolling-a-die experiment. What is the probability that the outcome will fall between two and five, $p(2 < X \leqq 5)$?

Solution. Using Eq. 2.57 gives

$$F(1) = \sum_{i=1}^{1} P(x_i) = \frac{1}{6}$$

$$F(2) = \sum_{i=1}^{2} P(x_i) = \frac{1}{6} + \frac{1}{6} = \frac{1}{3}.$$

We repeat this for the other four outcomes. **Fig. 2.29** shows the plot of the cumulative-distribution function. It starts with a value of zero and reaches a value of 1 at a value of the variable equal to six. Use Eq. 2.56 to calculate the probability that a value will fall between 2 and 5.

$$p(a < X \leqq b) = F(b) - F(a);$$

therefore,

$$p(2 < X \leqq 5) = F(5) - F(2) = \frac{5}{6} - \frac{1}{3} = \frac{1}{2}.$$

Numerical Example 2.14. In Numerical Example 2.12, we defined the probability density function as

$$f(x) = 0 \text{ for } x \leqq 20$$
$$= \frac{1}{50} \text{ for } 20 < x \leqq 70$$
$$= 0 \text{ for } x > 70.$$

Define the cumulative-distribution function for this function. Confirm that the probability that a thickness will fall between 30 and 50 ft is 0.4.

Solution. With Eq. 2.58, for a value of x between 20 and 70,

$$F(a) = \int_{-\infty}^{a} f(x)dx = \int_{-\infty}^{20} 0dx + \int_{20}^{a} \frac{1}{50} dx = \frac{(a-20)}{50}.$$

In general,

$$F(x) = 0 \text{ for } x \leqq 20$$
$$= \frac{x-20}{50} \text{ for } 20 < x \leqq 70$$
$$= 1 \text{ for } x > 70.$$

To calculate $p(30 \leqq X \leqq 50)$, we can write

$$p(30 < X \leqq 50) = F(50) - F(30)$$

$$= \left(\frac{50 - 20}{50}\right) - \left(\frac{30 - 20}{50}\right) = 0.4,$$

which confirms the previous answer.

Bivariate Functions. Here, we extend our definitions of probability functions to two variables. For discrete random variables, the probability mass function is

$$p(x_i, y_i) = p(X = x_i, Y = y_j). \quad \text{(2.59)}$$

For example, if we toss two coins simultaneously, the probability mass function of both coins sharing heads is

$$P(H, H) = p(X = H, Y = H) = \frac{1}{4}$$

because out of the four possible outcomes, both heads is one of the possible outcomes. Similarly,

$$P(H, T) = \frac{1}{4},$$

$$P(T, H) = \frac{1}{4},$$

and $P(T, T) = \frac{1}{4}$.

Adding these four values results in one. Similar to one variable, we can write

$$\sum_{i=1}^{n_X} \sum_{j=1}^{n_Y} P(x_i, y_j) = 1 \quad \text{(2.60)}$$

where n_X = number of possible outcomes of a random variable X and n_Y = number of possible outcomes of a random variable Y.

Considering the distribution of only one variable while ignoring the effect of other variable is called a marginal distribution. For example, the marginal distribution for Random Variable X, $P_X(x_i)$, can be written as

$$P_X(x_i) = \sum_{j=1}^{n_Y} P(x_i, y_j), \quad \text{(2.61)}$$

where all the possible values of Y are summed while the value of a random variable X is kept the same. Similarly, we can write the marginal distribution of the variable Y, $P_Y(y_j)$, as

$$P_Y(y_j) = \sum_{i=1}^{n_X} P(x_i, y_j). \quad \text{(2.62)}$$

Numerical Example 2.15 illustrates these concepts.

Numerical Example 2.15. Consider the rolling of two dice. Define Random Variable X as the addition of two outcomes from the two dice and Random Variable Y as the outcome of the first die. Obviously, X can take values between 2 and 12 and Y can take values between 1 and 6. **Table 2.10** shows the probability mass function of all possible outcomes.

For example, for an outcome of $Y = 2$, X can take values between 3 and 8. This is because the smallest outcome for the other die is 1 and the largest outcome is 6. Therefore, addition of the outcomes varies between 3 and 8. Note that for true dice, all the outcomes are equally likely. The missing values represent zero probabilities. For example, we can write

$$P(X = 5, Y = 2) = \frac{1}{36}$$

or $P(X = 12, Y = 6) = \frac{1}{36}$.

To calculate the marginal probabilities for X, we add all the probability mass functions over the entire range of Y. For example, marginal distribution for $X = 7$ can be calculated as

$$P_X(7) = \sum_{j=1}^{6} P(7, y_j) = \frac{1}{36} + \frac{1}{36} + \frac{1}{36} + \frac{1}{36} + \frac{1}{36} + \frac{1}{36}$$

$$= \frac{6}{36} = \frac{1}{6}.$$

We can calculate the marginal distribution for other values of X as well. To calculate the marginal distribution for Y, we add all the probability mass functions over the entire range of X

TABLE 2.10—PROBABILITY MASS FUNCTION OF ALL POSSIBLE OUTCOMES FOR NUMERICAL EXAMPLE 2.15

	Outcome of *Y*						
Outcome of *X*	1	2	3	4	5	6	Marginal Distribution of *X*
2	1/36						1/36
3	1/36	1/36					2/36
4	1/36	1/36	1/36				3/36
5	1/36	1/36	1/36	1/36			4/36
6	1/36	1/36	1/36	1/36	1/36		5/36
7	1/36	1/36	1/36	1/36	1/36	1/36	6/36
8		1/36	1/36	1/36	1/36	1/36	5/36
9			1/36	1/36	1/36	1/36	4/36
10				1/36	1/36	1/36	3/36
11					1/36	1/36	2/36
12						1/36	1/36
Marginal distribution of *Y*	1/6	1/6	1/6	1/6	1/6	1/6	1

for a given value of Y. For example, marginal distribution for $Y = 3$ can be calculated as

$$P_Y(3) = \sum_{i=1}^{11} P(x_i, 3) = 0 + 0 + \frac{1}{36} + \frac{1}{36} + \frac{1}{36} + \frac{1}{36} + \frac{1}{36} + \frac{1}{36} + 0 + 0 + 0 = \frac{1}{6}.$$

Other values are calculated similarly.

Note that marginal distributions are simply probability mass functions if we consider the distribution of a single variable. For example, if the outcome of a die is considered a random variable, the probability mass function is 1/6 for each outcome. Similarly, if the addition of a pair of dice is considered a random variable, the probability mass function of Outcome 7 is 1/6.

For continuous random variables, we can describe the probability density function, f(x,y), so that

$$\int_{-\infty}^{\infty}\int_{-\infty}^{\infty} f(x, y)dxdy = 1, \quad \text{.................. (2.63)}$$

where the volume under the surface equals one. Similar to a single-variable distribution, the probability that the values of random variables will fall within certain intervals can be calculated with the probability density function. For example,

$$p[a < X \leqq b, c < Y \leqq d] = \int_a^b\int_c^d f(x, y)dxdy. \quad \text{.................. (2.64)}$$

Like discrete distributions, marginal distributions for the two variables can be calculated as

$$f_X(x) = \int_{-\infty}^{\infty}\int_{-\infty}^{\infty} f(x, y)dy \quad \text{................. (2.65a)}$$

and $$f_Y(y) = \int_{-\infty}^{\infty} f(x, y)dx. \quad \text{.................. (2.65b)}$$

Numerical Example 2.16. A bivariate probability density function is given by $f(x, y) = e^{-x}e^{-y}$. Show that Eq. 2.64 is valid. Calculate $p[0 < X \leqq 1, 1 < Y < \infty]$. Calculate the marginal distributions for Random Variables X and Y.

Solution.

To validate Eq. 2.64,

$$\int_{-\infty}^{\infty}\int_{-\infty}^{\infty} 0dxdy + \int_0^{\infty}\int_0^{\infty} e^{-x}e^{-y}dxdy$$

$$= \int_0^{\infty}\left[-e^{-y}|_0^{\infty}\right]e^{-x}dx = \int_0^{\infty} e^{-x}dx = 1.$$

To calculate $p[0 < X \leqq 1, 1 < Y < \infty]$,

$$p[0 < X \leqq 1, 1 < Y < \infty] = \int_0^1\int_1^{\infty} e^{-x}e^{-y}dxdy$$

$$= \int_0^1 e^{-x}[-e^{-y}]_1^{\infty}dx = e^{-1}\int_0^1 e^{-x}dx$$

$$= e^{-1}[1 - e^{-1}].$$

Using Eq. 2.65a gives

$$f_X(x) = \int_0^{\infty} e^{-x}e^{-y}dy = e^{-x}.$$

Similarly, with Eq. 2.65b,

$$f_Y(y) = \int_0^{\infty} e^{-x}e^{-y}dx = e^{-y}.$$

The cumulative-distribution function for a bivariate distribution can be written like that for a single variable.

$$F(x, y) = p[X \leqq x, Y \leqq y]. \quad \text{............... (2.66)}$$

For discrete variables, we can write

$$F(a, b) = \sum_{x_i \leqq a}\sum_{y_j \leqq b} P[x_i, y_j]. \quad \text{................. (2.67)}$$

For continuous variables,

$$F(a, b) = \int_{-\infty}^{b}\int_{-\infty}^{a} f(x, y)dxdy. \quad \text{............... (2.68)}$$

Application of Eqs. 2.66 through 2.68 is straightforward.

Recall independent variables. Remember that, if two events are independent,

$$p(A \cap B) = p(A)p(B). \quad \text{.................... (2.40)}$$

Similarly,

$$p[X \leqq a, Y \leqq b] = p[X \leqq a]p[Y \leqq b]. \quad \text{.... (2.69)}$$

For discrete variables, for independence to be valid,

$$P[x_i, y_j] = P_X[x_i]P_Y[y_j] \quad \text{.................... (2.70)}$$

for all x_i and y_j.

For continuous variables,

$$f(x, y) = f_X(x)f_Y(y). \quad \text{..................... (2.71)}$$

Use of Eq. 2.71 easily shows that, for the probability density function provided in Numerical Example 2.16, the variables x and y are independent.

Conditional distribution, a distribution of one variable that is conditional on the other variable taking a certain value (discussed earlier), can also be defined. For a discrete variable,

$$P_{X|Y}(x|y) = P_{X|Y}(X = x|Y = y)$$

$$= \frac{P(X = x|Y = y)}{P(Y = y)} = \frac{P(x, y)}{P_Y(y)}, \quad \text{...... (2.72)}$$

where $P_{X|Y}(x|y)$ = the conditional distribution of Random Variable X taking a value of x given that Random Variable Y has taken the value of y. For continuous random variables, the conditional distribution is

$$f_{X|Y}(x|y) = \frac{f(x, y)}{f_Y(y)}. \quad \text{(2.73)}$$

Numerical Example 2.17. Numerical Example 2.15 considered two random variables, X and Y, where X = the addition of the outcomes from a pair of dice and Y = the outcome of the first die. The probability mass function for these two variables and the marginal-distribution function were calculated in Numerical Example 2.15. Calculate the probability that $Y = 3$ given that $X = 8$.

Solution. Using Eq. 2.72 gives

$$P_{Y|X}(y|x) = \frac{P(x, y)}{P_X(x)}.$$

The marginal distribution for $P_X(8) = 5/36$. The value of $P(8, 3) = 1/36$. Substituting gives

$$P_{Y|X}(y|x) = \frac{1/36}{5/36} = \frac{1}{5}.$$

That is, the conditional probability that $Y = 3$ is 20% given that $X = 8$.

The answer can be easily confirmed because the addition of eight can be achieved five different ways. The realizations have to be (2,6), (3,5), (4,4), (5,3), or (6,2). Of these five possibilities, the first die showing a 3 is possible in only one of these possibilities. This equates to 20% of the total possibilities, which is what the answer indicates.

Before concluding this section, we briefly review the concept of the random variable and some of its functions. It is not necessary to understand all the mathematical details covered in this section; however, understanding the concepts is critical for understanding future chapters.

A random variable is a numerically valued function whose values can be determined by some probabilistic function. The realizations of the random variable are the outcomes of a random experiment. A discrete random variable can take only a finite number of values, whereas a continuous random variable can take a large number of values. The probability mass or density function allows calculation of the probability that a random variable takes a certain value or a range of values. It is closely related to the relative-frequency-distribution function. Similar to relative frequencies, it adds up to one. The cumulative-distribution function allows determination of the probability that a random variable will be less than a certain value. It is closely related to cumulative-relative-frequency distribution and can take a value between 0 and 1.

These functions can be defined to describe the behavior of two variables to extend the analysis to two variables. Addition of another variable requires the description of two additional distribution functions. The marginal-distribution function describes the distribution of one of the variables by ignoring the variations in the other variable. The conditional distribution calculates the probability of one variable that depends on the other variable taking a certain value.

2.3.5 Mathematical Expectation. Mathematical expectation or expected value is defined as some weighted average outcome of a random experiment if the experiment is conducted a large number of times. For a discrete random variable X, the expected value of X is defined as

$$E[X] = \sum_{i=1}^{o} x_i P[X = x_i], \quad \text{(2.74)}$$

where $E[X]$ = expected value; x_i = outcome of the random variable, $P[X = x_i]$ = probability mass density function for the ith outcome, and o = number of possible outcomes. Numerical Examples 2.18 and 2.19 illustrate the concept.

Numerical Example 2.18. A wheel of fortune in a casino contains 80 slots. Each time you bet, you have to place \$1. Out of the 80 slots, if you hit one of seven special slots, you win \$10. If you hit one of the remaining 73 slots, you lose \$1. What is the expected value of this experiment?

Solution. Betting on a wheel can be considered a random experiment with two possible outcomes, success, S, or failure, F. The random variable can take two values: \$10 or \$1. S = a \$10 win and F = a \$1 loss. The probability mass functions for the two outcomes are

$$P[X = S] = \frac{7}{80}$$

and $P[X = F] = \dfrac{73}{80}$.

Using Eq. 2.74 gives

$$E[X] = (\$10)P[X = S] + (-\$1)P[X = F]$$

$$= (\$10)\left(\frac{7}{80}\right) + (-\$1)\left(\frac{73}{80}\right)$$

$$= -\$0.0375.$$

For a given bet, we either win \$10 or lose \$1. The expected value represents the average outcome of this random experiment if it is played a large number of times. That is, if we bet 1,000 times, we lose an average of \$0.0375 or 3.75¢/bet (or, on an average, approximately $\$0.0375 \times 1,000 = \$37.50$). Obviously, expected value achieves meaning only if a game is played a large number of times.

Numerical Example 2.19. What is the expected value of a rolling-a-die experiment?

Solution. We know that the random variable can take any of the six values, each having a probability of 1/6. Using Eq. 2.74 gives

$$E[X] = \sum_{i=1}^{6} x_i P[X = x_i]$$

$$= \left[1\left(\frac{1}{6}\right) + 2\left(\frac{1}{6}\right) + 3\left(\frac{1}{6}\right) + 4\left(\frac{1}{6}\right)\right.$$

$$\left. + 5\left(\frac{1}{6}\right) + 6\left(\frac{1}{6}\right)\right] = \frac{7}{2} = 3.5.$$

The expected value is 3.5 for the rolling-a-die experiment. As in the previous example, an outcome of 3.5 cannot be realized after a single roll of a die; it represents the average of all the outcomes if the experiment is repeated a large number of times. That is, if we roll a die a large number of times and note the outcome each time, the arithmetic average of all those realizations is close to 3.5.

The definition of expected value can be generalized for any real valued function u(x) of the variable X.

$$E[u(X)] = \sum_{i=1}^{o} u(x_i)P[X = x_i] \quad \text{.............. (2.75)}$$

for a discrete variable, and

$$E[u(X)] = \int_{-\infty}^{\infty} u(x)f(x)dx \quad \text{................. (2.76)}$$

for a continuous variable.

Eqs. 2.77 through 2.79 give some important characteristics of expected value.

$$E[K] = K; \quad \text{............................. (2.77)}$$

that is, the expected value of a constant is a constant.

$$E[Ku(X)] = KE[u(X)]; \quad \text{................... (2.78)}$$

that is, the expected value of a constant times a function is equal to the constant times the expected value of the function.

$$E[u_1(X) + u_2(X)] = E[u_1(X)] + E[u_2(X)];$$

$$\text{................... (2.79)}$$

that is, the expected value of a sum of two functions is equal to the sum of the expected values of the two functions. These properties are very useful in applying the expected-value equation.

The two most important expected values for a single variable are the arithmetic mean and the variance. The arithmetic mean, μ, is defined as

$$\mu = E[X]. \quad \text{............................ (2.80)}$$

μ is the expected value of the variable itself. The arithmetic mean is the population mean, not the sample mean; therefore, we differentiate it from the sample mean (denoted by $\bar{x}$) by using thc symbol μ to represent it.

Variance, σ^2, is defined as

$$\sigma^2 = E\left[(X - \mu)^2\right] = V[X]. \quad \text{............... (2.81)}$$

The notation σ^2 is different from s^2, which represents sample variance. The square root of σ^2, σ is the standard deviation. With the three characteristics of expected values, we can show that

$$\sigma^2 = E[X^2] - \{E[X]\}^2 \quad \text{.................. (2.82a)}$$

$$\text{or } \sigma^2 = E[X^2] - \mu^2. \quad \text{...................... (2.82b)}$$

Some important characteristics of the variance are

$$V[K] = 0, \quad \text{............................ (2.83a)}$$

where K is a constant;

$$V[KX] = K^2V[X], \quad \text{..................... (2.83b)}$$

where K is a constant; and

$$V[KX + b] = K^2[X], \quad \text{................... (2.83c)}$$

where K and b are constants.

Numerical Example 2.20. Calculate the variance for the rolling-a-die experiment.

Solution. We estimated the arithmetic mean in Numerical Example 2.19 and observed that

$$E[X] = \mu.$$

To calculate the variance, we need $E[X^2]$. With Eq. 2.75,

$$E[X^2] = \sum_{i=1}^{o} x_i^2P[X = x_i].$$

For six outcomes, we can write

$$E[X^2] = \left[1^2\left(\frac{1}{6}\right) + 2^2\left(\frac{1}{6}\right) + 3^2\left(\frac{1}{6}\right) + 4^2\left(\frac{1}{6}\right) + 5^2\left(\frac{1}{6}\right) + 6^2\left(\frac{1}{6}\right)\right] = 15.17.$$

Using Eq. 2.82b gives

$$\sigma^2 = E[X^2] - \mu^2 = 15.17 - (3.5)^2$$

$$= 15.17 - (3.5)^2 = 2.917$$

and $\sigma = 1.708$.

The variance of the random experiment is 2.917, and the standard deviation is 1.708.

In Sec. 2.2.2, we discussed calculation of the mean and the variance of a sample. The equations used (Eqs. 2.4 through 2.10) are closely related to the expected-value formulas presented in this section.

The expected-value relationship is easily extended to bivariate distributions. For a discrete variable,

$$E[u(X, Y)] = \sum_{i=1}^{n_X}\sum_{j=1}^{n_Y} u(x_i, y_j)P(x_i, y_j), \quad \text{......... (2.84)}$$

where $E[u(X, Y)]$ = expected value of the function $u(x_i, y_j)$, $P(x_i, y_j)$ = the probability mass function of the two variables X and Y, n_X = number of possible outcomes of Random Variable X and n_Y = is the number of possible outcomes of Random Variable Y.

For a continuous variable,

$$E[u(X, Y)] = \int_{-\infty}^{\infty}\int_{-\infty}^{\infty} u(x, y)f(x, y)dxdy, \quad \text{....... (2.85)}$$

where $f(x, y)$ = the probability density function of the two variables X and Y.

Covariance is one of the most important expected values for a bivariate distribution; it is defined as

$$C[X, Y] = E\left[(X - \mu_x)(Y - \mu_y)\right]. \quad \text{............ (2.86)}$$

Using the properties of expected values, we can show that

$$C[X, Y] = E[XY] - E[X]E[Y]. \quad \text{............. (2.87)}$$

Covariance is an indicator of the relationship between two variables. The stronger the relationship is, the higher the covariance value. The covariance is equal to zero for independent variables; that is,

$$C[X, Y] = 0 \quad \text{........................... (2.88)}$$

if X and Y are independent variables. Note that, if $X = Y$, the equation for covariance reduces to one for variance.

Some useful properties of covariance are

$$C[X + Z, Y] = C[X, Y] + C[Z, Y]. \quad \text{.........} \; (2.89)$$

where X, Y, and Z = random variables. In general, for multiple variables,

$$C\left[\sum_{i=1}^{n_X} X_i, Y\right] = \sum_{i=1}^{n_X} C[X_i, Y], \quad \text{..............} \; (2.90)$$

where X_i and Y = random variables. Eq. 2.90 indicates that the covariance of a summation of random variables equals a summation of the covariances. Eq. 2.90 can be generalized even further to state that

$$C\left[\sum_{i=1}^{n_X} X_i, \sum_{j=1}^{n_Y} Y_j\right] = \sum_{i=1}^{n_X}\sum_{j=1}^{n_Y} C[X_i, Y_j], \quad \text{........} \; (2.91)$$

Eq. 2.91 is extremely useful in geostatistical analysis. Chap. 4 discusses its usefulness.

Using the definition of covariance, we can also define the correlation coefficient.

$$\rho[X, Y] = \frac{C[X, Y]}{\sigma_X \sigma_Y}, \quad \text{......................} \; (2.92)$$

where $\rho[X, Y]$ = the correlation coefficient. Although similar to covariance, $\rho[X, Y]$ is a dimensionless quantity and can only take values between -1 and $+1$. $\rho[X, Y]$ also quantifies the relationship between the two variables.

Sec. 2.2.4 discussed the procedure for calculating the covariance and the correlation coefficient for sample data. The equations used to calculate the sample values are based on the definitions of the expected value with different notation. The population correlation coefficient is defined by $\rho[X, Y]$, whereas the sample correlation coefficient is defined by $r[X, Y]$. $C[X, Y]$ indicates the covariance for the population, and $c[X, Y]$ indicates the covariance for the sample. In practice, we can calculate only the sample values. Estimating values for the population requires that some assumptions be made. Sec. 2.3.7, which describes inference techniques, discusses the assumptions. Appendix B provides derivations of many equations presented.

2.3.6 Important Distribution Functions. In statistics, there are several important distribution functions with known characteristics. This section discusses three distribution functions for continuous random variables that play important roles in applying geostatistical techniques: uniform distribution, normal or Gaussian distribution, and log-normal distribution. Again, Appendix B provides detailed derivations of many equations.

Uniform Distribution. Uniform distribution is one of the simplest distributions of a continuous random variable. To define the uniform distribution properly, minimum and maximum values of the population are required. As the name indicates, uniform distribution assumes that any value between the minimum and the maximum has an equal probability of being selected. The probability density function for the uniform distribution is

$$f(x) = \frac{1}{b - a} \quad \text{.........................} \; (2.93a)$$

for $a \leqq x \leqq b$ and

$$f(x) = 0 \quad \text{............................} \; (2.93b)$$

otherwise.

The cumulative-distribution function is

$$F(x) = 0 \quad \text{.........................} \; (2.94a)$$

for $x < a$,

$$F(x) = \frac{x - a}{b - a} \quad \text{......................} \; (2.94b)$$

for $a \leqq x \leqq b$, and

$$F(x) = 1 \quad \text{.........................} \; (2.94c)$$

for xb. **Fig. 2.30** shows both the probability density function and the cumulative-distribution function.

With the definitions provided in Sec. 2.3.5, the mean of the uniform distribution is calculated as

$$\mu = \frac{a + b}{2} \quad \text{............................} \; (2.95)$$

and the variance is calculated as

$$\sigma^2 = \frac{(b - a)^2}{12}. \quad \text{........................} \; (2.96)$$

The most important application of uniform distribution is creating random-number realizations. A standard random-number generator creates a set of random numbers that are sampled from a uniform distribution. Typically, the random number is generated with a minimum of zero and a maximum of one. That is, any number between zero and one has

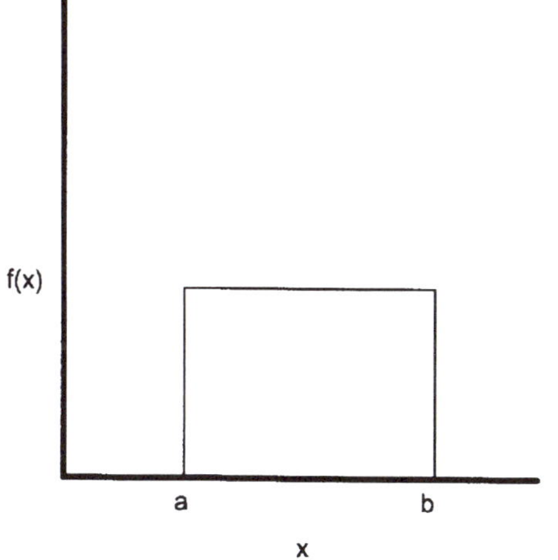

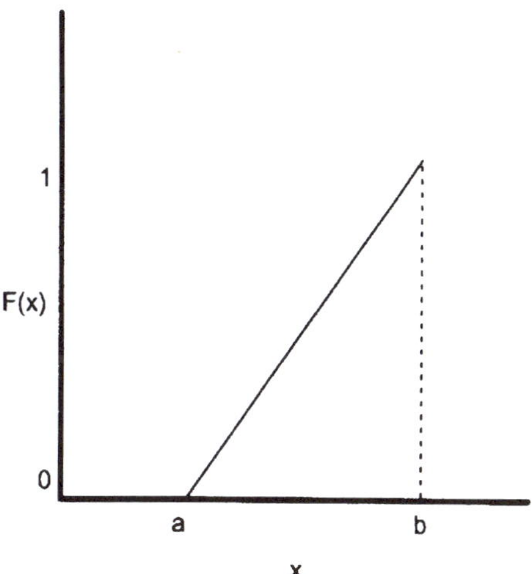

Fig. 2.30—Uniform distribution function.

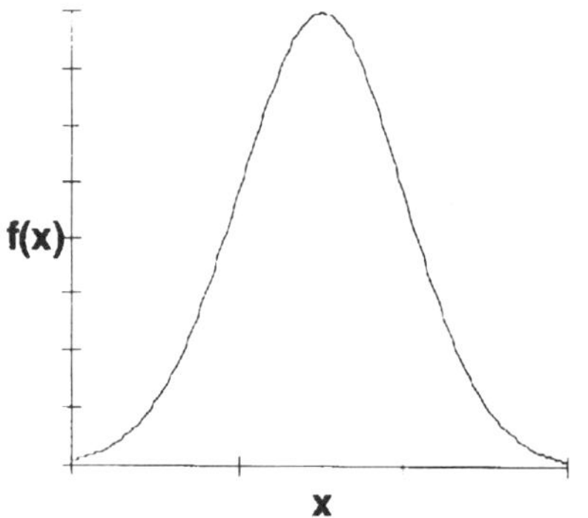

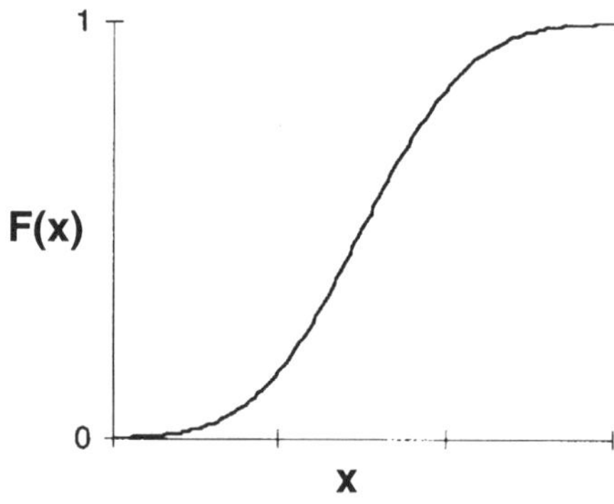

Fig. 2.31—Normal distribution function.

an equal likelihood of being selected. If the random-number generator needs to pick a value within a different range, we can easily convert a standard random-number generator to a new value by

$$N_{R,\text{modified}} = a + N_{R,\text{std}}(b - a), \quad \text{.............} \quad (2.97)$$

where $N_{R,\text{std}}$ = a standard random number between 0 and 1; $N_{R,\text{modified}}$ = a modified random number that falls in the range a to b; and a and b = minimum and maximum values of the uniform random number, respectively. In many of the geostatistics techniques, use of a uniform-random-number generator is quite common. Many algorithms have been proposed in the literature that generate "truly" uniform random numbers. Most of these algorithms can also generate an entirely new sequence of random numbers by changing what is called the "seed value." The ability of random-number generators to generate multiple sequences is useful in conditional-simulation techniques to generate multiple reservoir descriptions. Chaps. 5 through 7 provide a detailed discussion about conditional-simulation techniques.

Normal (Gaussian) Distribution. Normal distribution undoubtedly is the most famous distribution in the field of statistics. Its probability density function has a bell-shaped curve, which is well known to almost everyone, even those unfamiliar with statistical principles. The density function is given by

$$f(x) = \frac{1}{\sigma\sqrt{2\Pi}}\exp\left[-\frac{1}{2}\left(\frac{x-\mu}{\sigma}\right)^2\right] \text{for } -\infty < x < \infty. \quad \text{.................} \quad (2.98)$$

This distribution function has a mean of μ and a variance of σ^2. The maximum value of the density function is $0.4/\sigma$, which is reached at $x = \mu$. It is a symmetric function.

To use the normal-distribution function, it is much more convenient to define a standardized normal distribution. If we define a new variable,

$$z = \frac{x-\mu}{\sigma}, \quad \text{............................} \quad (2.99)$$

the probability density function is

$$f(z) = \frac{1}{\sqrt{2\Pi}}\exp\left(-\frac{z^2}{2}\right). \quad \text{..................} \quad (2.100)$$

This distribution has a mean of zero and a variance of one. With Eq. 2.100, the cumulative-distribution function is

$$F(z) = \frac{1}{\sqrt{2\Pi}}\int_{-\infty}^{z} e^{-t^2/2}dt. \quad \text{..................} \quad (2.101)$$

Eq. 2.101 has no analytical solution and has to be integrated numerically. The right side of Eq. 2.101 represents the error function, which is a standard mathematical function. **Fig. 2.31** shows the probability density function and the cumulative-distribution function for the normal distribution. Fortunately, for all practical purposes, $F(-3) \approx 0$ and $F(+3) \approx 1$. Standard tables, such as **Table 2.11**, are available in most statistics books. Table 2.11 provides the values of $F(z)$ for a range of z between -3 and $+3$ and can be used for a normal distribution with any mean and variance, as Numerical Example 2.21 example shows.

Numerical Example 2.21.

1. The porosity in the reservoir is estimated to have a mean of 0.2 and a variance of 0.0004. If the porosity is believed to be normally distributed, what is the probability that the porosity value will be between 0.18 and 0.22?

2. If rock with a porosity of less than 15% is believed to be nonreservoir rock, what is the probability that the rock at a given location will have a porosity of less than 15%?

Solution.

1. For the distribution, $\mu = 0.2$, $\sigma^2 = 0.0004$ or $\sigma = 0.02$. To calculate thc $p[0.18 \leqq \phi \leqq 0.22]$, we first need to standardize the two values 0.18 and 0.22. If $\phi_1 = 0.18$ and $\phi_2 = 0.22$,

$$z_1 = \frac{\phi_1 - \mu}{\sigma} = \frac{0.18 - 0.2}{0.02} = -1$$

$$\text{and } z_2 = \frac{\phi_2 - \mu}{\sigma} = \frac{0.22 - 0.2}{0.02} = 1,$$

respectively.

Once the values are standardized, we can look up the value of $F(z_1)$ and $F(z_2)$ in Table 2.11: $F(z_1) = 0.15866$ and $F(z_2) = 0.84134$.

Recall that

$$p[a < X \leqq b] = F(b) - F(a). \quad \text{............} \quad (2.56)$$

Therefore,

$$p\left[z_1 < \frac{\phi - \mu}{\sigma} \leqq z_2\right] = 0.84134 - 0.15866 = 0.6827.$$

That is, a 68% probability exists that a porosity value will fall between 0.18 and 0.22.

2. We first need to calculate $p[\phi \leqq 0.15]$. We can standardize the value with

$$z_3 = \frac{x_3 - \mu}{\sigma} = \frac{0.15 - 0.2}{0.02} = -2.5.$$

TABLE 2.11—z VALUES FOR NORMAL DISTRIBUTION (from Ref. 1)

Standard Deviation From Mean	Cumulative Probability	Standard Deviation From Mean	Cumulative Probability
−3.0	0.0014	+0.0	0.5000
−3.0	0.0014	+0.0	0.5000
−2.9	0.0019	+0.1	0.5398
−2.8	0.0026	+0.2	0.5793
−2.7	0.0035	+0.3	0.6179
−2.6	0.0047	+0.4	0.6554
−2.5	0.0062	+0.5	0.6915
−2.4	0.0082	+0.6	0.7257
−2.3	0.0107	+0.7	0.7580
−2.2	0.0139	+0.8	0.7881
−2.1	0.0170	+0.9	0.8159
−2.0	0.0228	+1.0	0.8413
−1.9	0.0287	+1.1	0.8643
−1.8	0.0359	+1.2	0.8849
−1.7	0.0446	+1.3	0.9032
−1.6	0.0548	+1.4	0.9192
−1.5	0.0668	+1.5	0.9332
−1.4	0.0808	+1.6	0.9452
−1.3	0.0968	+1.7	0.9554
−1.2	0.1151	+1.8	0.9641
−1.1	0.1357	+1.9	0.9713
−1.0	0.1587	+2.0	0.9773
−0.9	0.1841	+2.1	0.9821
−0.8	0.2119	+2.2	0.9861
−0.7	0.2420	+2.3	0.9893
−0.6	0.2743	+2.4	0.9918
−0.5	0.3085	+2.5	0.9938
−0.4	0.3346	+2.6	0.9953
−0.3	0.3821	+2.7	0.9965
−0.2	0.4207	+2.8	0.9974
−0.1	0.5602	+2.9	0.9981
−0.0	0.5000	+3.0	0.9987

From Table 2.11, $F(z_3) = 0.00621$. From Eq. 2.66, $F(z) = p[Z \leqq z]$. Therefore,

$$p\left[\frac{\phi - \mu}{\sigma} \leqq z_3\right] = 0.00621.$$

That is, there is a 0.62% probability that the porosity will be less than 15%.

One reason for the popularity of the normal-distribution function is the central-limit theorem, which states that the "sum of large number of independent random variables tends to be normally distributed." A good example of the application of this theorem is the measurement error in conducting an experiment. Typically, an error is a result of several possible independent sources that tend to be additive; therefore, the measurement error tends to be normally distributed for many experimental procedures. Later chapters show that many geostatistical techniques also assume that estimation error at unsampled locations is normally distributed.

A reservoir property that is commonly believed to be normally distributed is reservoir porosity. A quick way to check whether such an assumption can be justified is to plot the data on a probability plot (a special graph paper available in most engineering supply stores). Plotting the sample values in an ascending order on one scale and the probability on the other scale shows whether the data form a straight line on the graph paper, which indicates a normal distribution. Field Example 2.9 illustrates application of this procedure.

Field Example 2.9. This example uses porosity data from Well 34-29 and Flow Unit 3. **Fig. 2.32** shows the probability plots for both. The figure shows that the data for Well 34-29 were collected from different geological units. As a result, several inflection points are evident on the probability plots, indicating the possibility that multiple populations mixed together. The collected data for Flow Unit 3 show a smooth trend. Although the behavior is approximately linear, it is difficult to conclude that the distribution is normal.

Log-Normal Distribution. Log-normal distribution is closely related to normal distribution. If the logarithm of a variable is normally distributed, then the variable itself is log-normally distributed. **Fig. 2.33** shows this transformation schematically. In this figure, the log-normal distribution is skewed with a long tail on the right side. After transforming the data by taking the log of the variable, however, the distribution becomes symmetric and normal.

If we consider X to be a log-normally distributed variable, we can define $Y = \ln X$, where Y = the value of the natural logarithm of the random variable X. If the mean of the variable Y is α and the variance is β^2, we can write the probability density function for the variable X as

$$f(x) = \frac{1}{x\beta\sqrt{2\Pi}}\exp\left[-\frac{1}{2}\left(\frac{x-\alpha}{\beta}\right)^2\right] \text{ for } x > 0. \quad \text{(2.102)}$$

We can show that the mean and the variance of Random Variable X is related to the mean and variance of the transformed variable Y.

$$\mu = \exp\left[\alpha + \frac{\beta^2}{2}\right] \quad \text{(2.103)}$$

$$\text{and } \sigma^2 = \mu^2\left[e^{\beta^2} - 1\right], \quad \text{(2.104)}$$

where μ = the mean of Variable X and σ^2 = the variance of Variable Y. With Eqs. 2.103 and 2.104, we can still use Table 2.11 to determine the probability values for the log-normal distribution.

Numerical Example 2.22. Permeability values in a reservoir are expected to be log-normally distributed with a mean of 20 md and a variance of 2,000 md². What is the probability that the value of permeability in the reservoir at a given location will exceed 200 md?

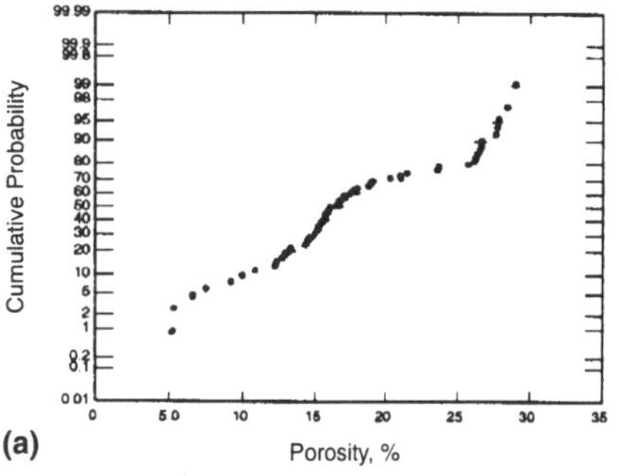

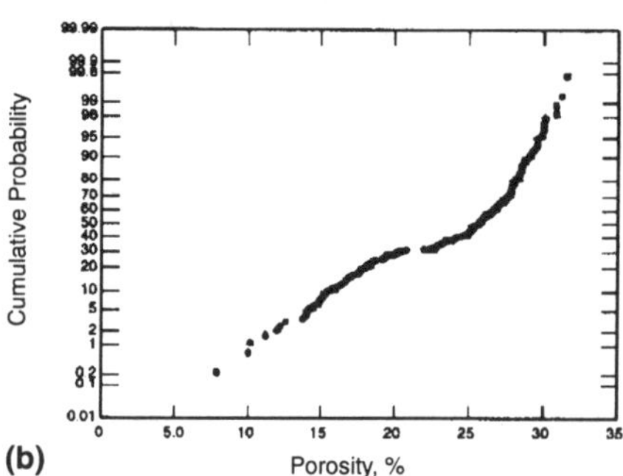

Fig. 2.32—Probability plots for porosity data: (a) Well 34-29 and (b) Flow Unit 3.

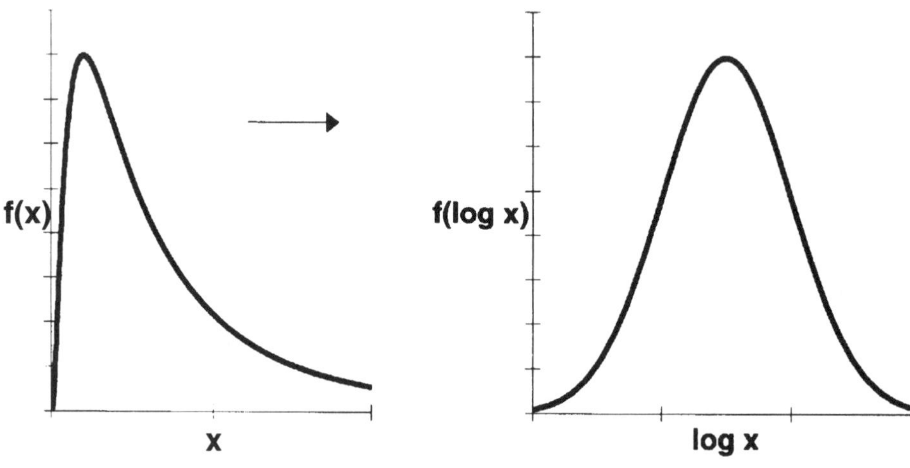

Fig. 2.33—Transformation for log-normal to normal distribution.

Solution. In this example, $\mu = 20$ md and $\sigma^2 = 2,000$ md^2. Rearranging Eqs. 2.104 and 2.103 gives, respectively,

$$\beta^2 = \ln\left[1 + \frac{\sigma^2}{\mu^2}\right] \quad \text{......................} \quad (2.105)$$

$$\text{and } \alpha = \ln\mu - \frac{\beta^2}{2}. \quad \text{......................} \quad (2.106)$$

Note that Table 2.11 is applicable to the normal distribution. Therefore, we need to consider the mean and the variance of the transformed variable before we can standardize it. Substituting into Eqs. 2.105 and 2.106 gives, respectively,

$$\beta^2 = \ln\left[1 + \frac{2,000}{400}\right] = 1.792$$

$$\text{and } \alpha = \ln(20) - \frac{1.792}{2} = 2.1$$

Standardizing gives

$$z = \frac{\ln x - \alpha}{\beta} = \frac{\ln(200) - 2.1}{\sqrt{1.792}} = 2.39.$$

From Table 2.11, F(2.39) = 0.992. That is, the probability that permeability will be less than 200 md is 99.2%. Or, in other words, the probability that the permeability will be greater than 200 md is (1 − 0.992 = 0.008) 0.8%.

Similar to normal distribution, the assumption of log-normal distribution can be validated by plotting the data on log probability paper. If one observes a straight line, the distribution is log-normal. Reservoir permeability is the property most commonly assumed to have log-normal distribution. The famous Dykstra-Parsons coefficient, V_{DP}, which describes reservoir heterogeneity for the well data, assumes a log-normal distribution of permeability and is defined as

$$V_{DP} = \frac{k_{50} - k_{15.9}}{k_{50}}, \quad \text{......................} \quad (2.107)$$

where k_{50} = the 50th percentile permeability value and $k_{15.9}$ = the 15.9th percentile permeability value when the data are ordered in an ascending fashion.

Field Example 2.10. With the permeability data from Well 34-29 and Flow Unit 3. **Fig. 2.34** shows the log probability plots for both the data sets. As Field Example 2.10 shows, the data collected from Well 34-29 are from several geological units. As a result, the data show several inflection points. The flow-unit data show a smoother trend. Although the data deviate at permeability values greater than 700 md, they show a fairly linear trend at less than 700 md. The Dykstra-Parsons coefficient for the flow-unit data is approximately 0.95, which indicates a highly heterogeneous reservoir.

Several advantages are associated with the assumption of one of the distributions describing reservoir properties. The most important is the ability to capture the characteristics of the distribution with a limited number of parameters. All three distributions described require only two parameters to describe the distributions completely. Uniform distribution requires the minimum and the maximum, and normal and log-normal distributions require the mean and the variance. Once those parameters are known, the entire distribution is known. However, as Field Example 2.9 illustrated, if the property cannot be described with a particular distribution function,

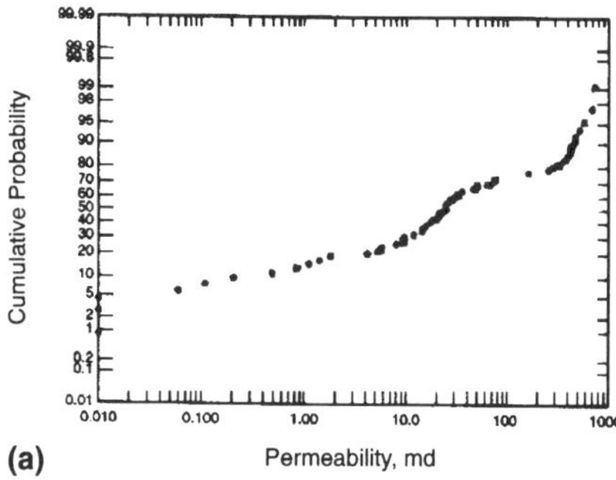

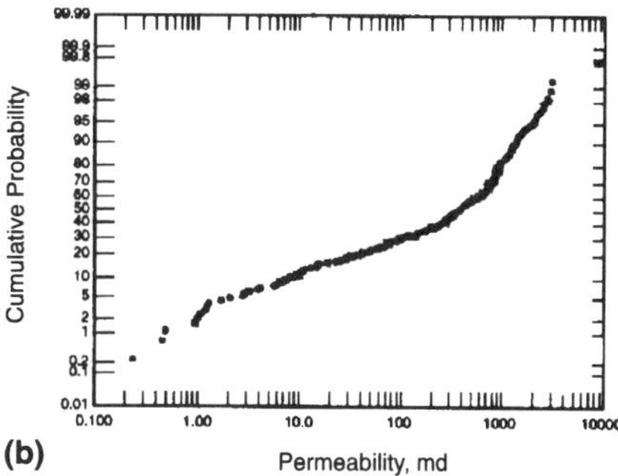

Fig. 2.34—Log probability for permeability data: (a) Well 34-29 and (b) Flow Unit 3.

the frequency distribution is the only way to capture the details of the distribution. Instead of requiring only two parameters, the frequency distribution requires storage of a lot more information to characterize the distribution.

Another advantage of assuming either a normal or log-normal distribution is the simplicity of the mathematical treatment. The normal distribution is especially amenable to many analytical modifications that allow much more simplified estimation procedures in geostatistical techniques. As a result, the assumption of a normal distribution makes the computational procedures much more efficient. However, these distribution functions should not be forced on any variable for the sake of computational speed or mathematical convenience because, unfortunately, these assumptions may sometimes result in erroneous results. We should take advantage of the distribution functions only if we are convinced that they do represent the reality of variations in the properties.

2.3.7 Inference of Parameters. Inference, as could be expected, is one of the important steps in inferential statistics. Inference allows estimation of population parameters from sample statistics. For example, the sample statistics may include determination of the sample mean, $\bar{x}$, and sample variance, s^2. From these statistics, we are interested in inferring the population parameters, such as the population mean, μ, and the population variance, σ^2. Because we do not know the true population, we can only estimate the parameters. We distinguish the symbols for estimated and true parameters by adding a hat to the symbol for estimated parameters (e.g., estimated population mean is $\hat{\mu}$ and true population mean is μ). Because we are estimating the parameter, some uncertainty is associated with it. The estimate can be considered to be a random variable with some distribution associated with it, and the distribution may indicate our confidence in the estimate. Because of these characteristics, two desirable characteristics in the estimate are unbiasedness and mimimum variance.

Unbiasedness. One desirable characteristic is that the estimate be close to the true value of the parameter. **Fig. 2.35** shows the estimated parameter $\hat{\theta}$ with its associated distribution. The true parameter θ ideally "falls" somewhere close to the middle of the distribution. This desirable property can be defined by stating that the expected value of the estimated parameter is the same as the true parameter.

$$E\left[\hat{\theta}\right] = \theta. \qquad (2.108)$$

In other words, "on average," the value of the estimate is equal to the true parameter.

Minimum Variance. **Fig. 2.36** shows three estimates of an unknown parameter. All three estimates are unbiased; however, Estimate 1 is more desirable than the other two because it shows the smallest range of uncertainty. A value estimated with less uncertainty can be predicted with more confidence. Variance indicates the spread of the estimate; therefore, minimizing the variance results in a desirable estimate.

Combining these two desirable properties, unbiased estimate and minimum variance, we can use the minimum-variance unbiased-estimate (MVUE) technique. This technique used to estimate values at unsampled locations in geostatistical methods. Appendix B provides a simple application of the MVUE technique for a linear-regression problem.

MVUE is not the only technique with which to estimate parameters. Other techniques exist in the literature for estimation of parameters (e.g, the method of moments; the method of maximum likelihood; and more sophisticated techniques, such as penalty-function methods). These techniques are

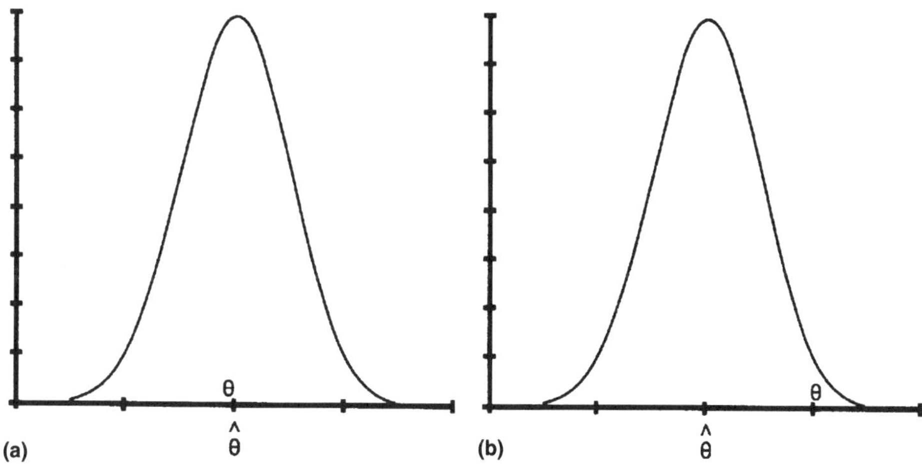

Fig. 2.35—(a) Desirable and (b) undesirable unbiased condition.

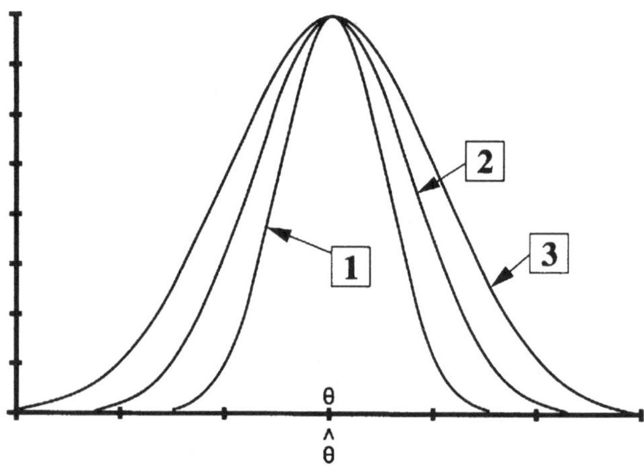

Fig. 2.36—Minimum variance condition.

based on heuristic principles in terms of what is considered a desirable characteristic of the estimate. For example, a biased estimate with a smaller variance may be preferable to an unbiased estimate with a larger variance. If the true population parameter is not known, the techniques cannot be compared to examine which is the best one. In practice, however, many of these techniques have worked well for estimating unknown parameters. For our purposes, we assume that the MVUE technique provides the best estimate.

Summary

This chapter covers a lot of material on understanding statistics principles. Many standard statistics books devote many chapters to the topics covered in this single chapter. Our goal, however, is not to make the reader expert in these techniques, but to familiarize the reader with the terminology of statistical principles and to expose the reader to the basic concepts of statistics. Appendix B covers the mathematical details of the equations used in the chapter to explain the concepts.

The chapter is divided into two main sections: principles of descriptive statistics and principles of inferential statistics. The descriptive-statistics section discusses several useful techniques that can be used to understand the characteristics of the sample data sets. Special emphasis is placed on the spatial data sets that are collected in analyzing reservoir characteristics. Field examples illustrate many techniques.

The section on inferential statistics introduces the concept of probability and its properties and applications. The random variable and its functions are illustrated, followed by the principle of expected value. Three distribution functions that have applications in geostatistical methods are described in additional detail, and the desirable characteristics of the estimator technique are presented.

Starting with Chap. 3, how these principles can be applied to quantify and to analyze spatial characteristics of the reservoir data is illustrated.

Nomenclature

a, b = minimum and maximum of uniform distribution, respectively
A_{sub} = subarea size, L^2
A_t = total area, L^2
A, B, C = event of a random experiment
b = intercept
c, d = constant used to define bivariate distribution
$c(x, y)$ = sample covariance between x and y variables
$C(X, Y)$ = population covariance between random variables X and Y
C_V = coefficient of variance.
E = event
$E[X]$ = expected value of random variable X
f_i = class frequency of Class i
f_{Ri} = relative class frequency of Class i
$\mathrm{f}(x)$ = probability density function for continuous random variable X
$\mathrm{f}(x, y)$ = probability density function for a bivariate distribution of continuous random variables X and Y
$f_X(x)$ = marginal distribution of continuous random variable X
$f_Y(y)$ = marginal distribution of continuous random variable Y
$f_{X|Y}(x|y)$ = conditional distribution of continuous random variable X given a particular value of continuous random variable Y
F = failure
F_j = cumulative relative frequency of Class j, defined in Eq. 2.3
$\mathrm{F}(x)$ = cumulative-distribution function of random variable X
$\mathrm{F}(x, y)$ = cumulative-bivariate-distribution function of random variables X and Y
k = permeability, L^2, md
K = constant
L = lag distance between two samples, L, ft
m = slope of best-fit line
n = total number of samples, n
n_c = number of times an experiment is conducted under controlled conditions, n
n_e = number of mutually exclusive events, n
n_X = number of possible outcomes of a random variable X, n
n_Y = number of possible outcomes of a discrete random variable Y, n
N = total number of classes, n
N_R = random-number realization
o = number of possible outcomes, n
$p(A)$ = probability of Event A
$p(A|B)$ = conditional probability of Event A given that Event B has occurred
$\mathrm{P}(x)$ = probability mass function for a discrete random variable
P_X, P_Y = marginal distributions of discrete random variables X and Y, respectively
$P_{X|Y}(x|y)$ = conditional distribution of a discrete random variable X given value of a discrete random variable Y
Q = quantile
$r(x, y)$ = sample correlation coefficient
R = sample range
R_x, R_y = rank of sample for variables x and y, respectively, when arranged in descending or ascending order
$R_{\log k}$ = rank of log k
R_ϕ = rank of ϕ
s = sample standard deviation

$s_{\log k}$ = standard deviation of log k
s_ϕ = standard deviation of ϕ
s^2 = sample variance
S = sample space
t = dummy variable
u = spatial location
$u(x)$ = function of variable x
$u(X)$ = function of random variable X
$u(x, y)$ = bivariate function of variables x and y
V_{DP} = Dysktra-Parsons coefficient
$V(X)$ = variance of random variable X
w_i = weight assigned to Sample i; Eq. 2.10
x_i = value of Sample i for Variable x
x_{max} = maximum value in the sample
x_{min} = minimum value in the sample
$\bar{x}$ = sample mean
$\tilde{x}$ = sample median
x^p = p percentile value of Variable X
$x(u)$ = variable x at location u
X, Y = random variables
y_i = value of Sample i for variable y
z = standard normal variable
α = arithmetic mean of the log of a log-normally distributed variable
β^2 = variance of the log of a log-normally distributed variable
θ = population parameter
μ = population mean
$\rho[X, Y]$ = population correlation coefficient
σ = population standard deviation
σ^2 = population variance
ϕ = porosity

Superscript

$\hat{}$ = estimated

References

1. Tukey, J.W.: *Exploration Data Analysis,* Addison-Wesley Publishing Co., Reading, Massachusetts (1977).
2. Mandenhall, W.: *Introduction to Probability and Statistics,* seventh edition, PWS Publishers, Boston, Massachusetts (1987).
3. Ross, S.M.: *Introduction to Probability and Statistics for Engineers and Scientists,* John Wiley & Sons, New York City (1987).
4. Hines, W.W. and Montgomery, D.C.: *Probability and Statistics in Engineering and Management Science,* third edition, John Wiley & Sons, New York City (1990).
5. Isaaks, E.K. and Srivastava, R.M.: *Applied Geostatistics,* Oxford U. Press, New York City (1989).
6. Journel, A.G.: *Fundamentals of Geostatistics in Five Lessons,* American Geophysical Union, Washington, DC (1989) **8.**
7. Journel, A.G.: "Non-Parametric Estimation of Spatial Distributions," *Math Geology* (1983) **15,** 445.
8. Deutsch, C.V. and Journel, A.G.: *GSLIB: Geostatistical Software Library and User's Guide,* Oxford U. Press, New York City (1992).
9. Davis, J.C.: *Statistics and Data Analysis in Geology,* John Wiley & Sons, New York City (1986).

SI Metric Conversion Factors

ft × 3.048* E − 01 = m
md × 9.869 233 E − 04 = μm^2

*Conversion factor is exact.

Chapter 3
Spatial Relationships: Estimation and Modeling

3.1 Introduction

In this chapter, we discuss various techniques of describing the spatial relationships for geoscience data. As Chaps. 1 and 2 explained, geoscience data sets are distinguished from other types of samples in one important aspect, they exhibit spatial relationship. In simple terms, neighboring values are related to each other. This relationship becomes stronger as the distance between neighboring values decreases and, in most cases, becomes uncorrelated beyond a certain distance. This type of qualitative information must be defined in a form suitable to be used to estimate values at unsampled locations. This chapter discusses various methods for estimating spatial relationships. It also describes the procedures used to model spatial relationships so that they can be incorporated into techniques that estimate values at unsampled locations.

The chapter is divided into several sections. The first section describes the importance of the assumption of stationarity before estimating spatial relationships. These spatial relationships are based on observed sample data; however, we use them to estimate values at locations where we have no samples. Understanding the assumptions that define the regions where estimated relationships can be applied is important.

In the second section, we present conventional summary statistics used to describe spatial relationships. Chap. 2 already described the covariance and the correlogram. The section also introduces the variogram, the most commonly used statistic to describe spatial relationships. As in other chapters, numerical and field examples illustrate the techniques.

Although the variogram is the most commonly used statistic, in practice, estimation of variograms requires great care and caution. In the next section, we discuss several potential problems in estimating variograms and the practical solutions that can be implemented to minimize their effects. These problems include lack of sufficient pairs at a given lag distance, instability in estimated values, undue influence of outlier data, and biased sampling. We propose solutions to all these problems.

Once the variogram is estimated, the next step is modeling the variogram to present it in a convenient format. Certain restrictions exist in modeling the estimated variogram. The next section discusses these restrictions and the type of models used. The section is divided in terms of models with a sill and models without a sill. The distinction is made by the type of variogram observed. Models with a sill are used when the variogram reaches a constant value after a certain lag distance, and those without a sill are used when the variogram does not reach a constant value over the region of interest. The section on models without a sill also includes discussions of fractal models and of models that capture the "hole effect" (an effect evident in the vertical direction where geological environments are repeated through depositional cycles).

The next section deals with constructing spatial relationships between two different variables situated at different locations in contrast with the previous sections, which dealt with the spatial relationship for one variable situated at different locations. Both estimation and modeling procedures for these cross relationships are presented.

The following section addresses alternative methods for describing spatial relationships. The recent literature has presented several new techniques that allow generation of reservoir descriptions in a more flexible format. Some of these methods are not restricted by conventional geostatistical paradigms and therefore allow more flexibility in describing the spatial relationships. We discuss some of these relationships.

In the last section, we summarize the major conclusions of this chapter. As in previous chapters, only necessary mathematical details are included within the text of the chapter. Appendix C provides additional details and derivations.

3.2 Random-Function Model

Chap. 2 introduced the concept of a random experiment: an experiment that may result in several possible outcomes, none of which can be predicted with certainty. The chapter also discusses reasons for considering the drilling, as well as a random experiment. Although drilling of a well may result in only one outcome, it is treated as a random experiment because of lack of knowledge about the outcome.

We can extend this concept to estimating values at unsampled locations. The goal of geostatistics is to estimate values at locations where there is no information. We use available sample data sets to develop certain models and use those models to predict the values at unsampled locations. If we can develop a completely deterministic model based on the evolution of the reservoir, we will be able to predict every reservoir property at every location with certainty. However, we do not have this knowledge; therefore, our approach has to be empirical and associated with uncertainty. Although we use all the available sample data in the best way possible, estimates at unsampled locations have an associated uncertainty. To reflect this uncertainty, we treat our estimates as random variables (denoted by uppercase letters). For example, if an unsampled location is u_O, the estimated value is denoted as $X(u_O)$, where X is the variable. In addition to the estimated values, we treat the sampled values as random variables because, as indicated earlier, we do not know in a deterministic way that we arrived at those sample values. Lack of full knowledge regarding the presence of a particular value of a variable at a particular location justifies treating the sampled locations as random variables. The actual samples are simply realizations of those random variables.

In describing both the sampled data and variable values at unsampled locations in terms of random variables, we are using what is called a random-function model. In reality, we may have only one value at each sampled location. However, in a deterministic sense, we do not know, for example, why a porosity of 15% is observed at a particular location. To account for this lack of knowledge, we treat both sampled and unsampled locations with a random-function model.

3.2.1 Requirement of Stationarity. In addition to assuming that all locations are described by random variables, we also have to consider the restrictions associated with using available sample data to predict values at unsampled locations. Chap. 1 briefly discussed the assumption of stationarity. This assumption, in qualitative terms, requires that the proposed model based on our sampled data can adequately describe the behavior of the population. We want to infer the population on the basis of the sample data. As in the case of any other statistical-inference technique, we cannot prove or disprove this assumption. We simply have to make an informed decision regarding what information we can use to describe the region of interest.

In practice, this decision obviously is based on all available information. There is, however, a danger in overcategorizing information in defining small regions of stationarity where a particular data set may be applicable. The region of stationarity also cannot be overly broad. If very distinct geological vertically stacked units are present, it may be inappropriate to combine them into a single region of stationarity. At the same time, if regions that are too small are specified, we may have very limited data for each region. Although it is difficult to state a general rule about how the region of stationarity should be defined, it suffices to state that the largest amount of relevant information should be collected to make reasonable predictions about the populations. The decision is necessarily subjective and varies from field to field. Also, during early stages of development, a large region of stationarity may have to be used because of limited data. As more information is gathered, we may be able to define the region. Note that geostatistical techniques are not the only techniques where this assumption has to be made. Every interpolation technique makes this implicit assumption before sample data are used.

In making the assumption of stationarity over the region of interest, we try to make it the least restrictive possible. For geostatistical purposes, the first and second orders of stationarities must be satisfied.

Mathematically, the first order of stationarity can be written as

$$f[X(\vec{u})] = f\left[X\left(\vec{u} + \vec{L}\right)\right], \qquad (3.1)$$

where f[ ] = any function of a random variable and $(\vec{u})$ and $(\vec{u} + \vec{L})$ define the two locations of the random variable. The most commonly used function is the expected value. Therefore, we can write

$$E[X(\vec{u})] = E\left[X\left(\vec{u} + \vec{L}\right)\right]. \qquad (3.2)$$

That is, the expected value of a random variable at $\vec{u}$ is the same as the expected value of a random variable $\vec{L}$ lag distance away. The value of $\vec{L}$ can vary from zero to the maximum distance between variables within the region of interest.

What does this definition mean in practice? The expected value of the variable itself is an arithmetic mean. Therefore, the definition means that arithmetic means of random variables across the region are the same. In practice, this translates into local means being approximately constant. If we divide the region of interest into small subregions and calculate the means of samples within these subregions (assuming that adequate numbers of samples are present within each subregion), those means should remain fairly close to each other. If the means vary significantly, the assumption of stationarity may not hold. Also, if the sampled data have a strong trend, the first order of stationarity may not hold.

In practice, local means frequently differ significantly. An example of this is Field Example 2.3 in Chap. 2, which presented the local means of Flow Unit 3 porosity data from the field data set. One option is to reduce the region where the means are approximately constant. Another option is to use modifications in describing our spatial relationships. This chapter discusses some of the modified techniques used to overcome local variations in sampled data sets. We can use the first option if adequate data exist within each of the smaller regions to pursue our studies. In practice, this may not be possible; therefore, the second option is used more often.

The second order of stationarity can be mathematically defined as

$$f\left[X(\vec{u}_1), X\left(\vec{u}_1 + \vec{L}\right)\right] = f\left[X(\vec{u}_2), X\left(\vec{u}_2 + \vec{L}\right)\right]. \qquad (3.3)$$

This relationship indicates that any function of two random variables located $\vec{L}$ distance apart is independent of the location and is a function of only the distance and the direction between the two locations. The arrows over the u and L indicate that locations can be treated in terms of vectors rather than distances.

In practice, we can use covariance as one of the functions that relates two variables located a certain distance and direction apart. In other words,

$$C\left[X(\vec{u}_1), X\left(\vec{u}_1 + \vec{L}\right)\right] = C\left[X(\vec{u}_2), X\left(\vec{u}_2 + \vec{L}\right)\right]. \qquad (3.4)$$

Depth (ft)	Porosity (%)
2,040	8.25
2,041	9.00
2,042	6.25
2,043	5.00
2,044	5.30
2,045	4.75
2,046	5.00

TABLE 3.1—POROSITY DATA FOR NUMERICAL EXAMPLE 3.1

That is, the covariance within the region of stationarity is a function of only the vector $\vec{L}$, not of the variable itself. This is an important assumption. It means that, as long as we know the distance and direction between any two points, we can estimate the covariance between the random variables at these two points; we do not need the actual random variables at those locations. Eq. 3.4 can therefore be written as

$$C\left[X(\vec{u}), X\left(\vec{u} + \vec{L}\right)\right] = C\left(\vec{L}\right). \quad \text{(3.5)}$$

With the definition of covariance,

$$C\left[X(\vec{u}), X\left(\vec{u} + \vec{L}\right)\right] = E\left[X(\vec{u}), X\left(\vec{u} + \vec{L}\right)\right] - E[X(\vec{u})] \cdot E\left[X\left(\vec{u} + \vec{L}\right)\right]. \quad \text{(3.6)}$$

However, Eq. 3.6 can be simplified because of Eq. 3.2: first-order stationarity assumption. We can write,

$$C\left[X(\vec{u}), X\left(\vec{u} + \vec{L}\right)\right] = E\left[X(\vec{u}), X\left(\vec{u} + \vec{L}\right)\right] - \{E[X(\vec{u})]\}^2. \quad \text{(3.7)}$$

We can define another function that describes the relationship between two variables located a certain distance apart: the variogram; this is discussed in the next section.

3.3 Spatial Relationship

In this section, we discuss the spatial relationships most commonly used to describe how neighboring values are related. Chap. 2 discussed two of the three relationships: covariance and correlation coefficient. We simply recall their introduction under assumptions of stationarity. We also introduce the most commonly used spatial relationship in geostatistics, the variogram.

3.3.1 Covariance. Chap. 2 introduced covariance as a way to capture the spatial relationship. Eq. 3.7 defines covariance under the first- and the second-order stationarity assumptions. With that definition, in practice, estimated covariance can be calculated as

$$c\left(\vec{L}\right) = \frac{1}{n\left(\vec{L}\right)} \sum_{i=1}^{n(\vec{L})} x(\vec{u}_i) x\left(\vec{u}_i + \vec{L}\right) - \left[\frac{1}{n} \sum_{i=1}^{n(\vec{L})} x(\vec{u}_i)\right]^2, \quad \text{(3.8)}$$

where $n(\vec{L})$ = number of pairs at vector distance $\vec{L}$; $x(\vec{u}_i)$, and $x(\vec{u}_i + \vec{L})$ = values of the variable at locations $\vec{u}_i$ and $\vec{u}_i + \vec{L}$, respectively; and n = total number of sample points. The second term on the right side of Eq. 3.8 represents the arithmetic mean of all the data points. $c(\vec{L})$ = the estimated value based on the sampled data. We can rewrite Eq. 3.8 as

$$c\left(\vec{L}\right) = \frac{1}{n\left(\vec{L}\right)} \sum_{i=1}^{n(\vec{L})} x(\vec{u}_i) x\left(\vec{u}_i + \vec{L}\right) - \bar{x}^2, \quad \text{(3.9)}$$

where $\bar{x}$ = the arithmetic mean of the sample data.

Numerical Example 3.1. This example uses the sample data used in Numerical Example 2.6. **Table 3.1** gives the porosity data as a function of depth in a vertical well. For convenience, we have used only seven porosity values; in practice, however, we will have a lot more values. Estimate the covariance as a function of lag distance.

Solution. To use Eq. 3.9, we first need to calculate the arithmetic mean.

$$\bar{x} = \frac{1}{n} \sum_{i=1}^{n} x(\vec{u}_i)$$

$$= \frac{1}{7}(8.25 + 9.0 + 6.25 + 5.00 + 5.3 + 4.75 + 5.0)$$

$$= 6.22.$$

Because the data are collected in only one direction (vertically), we do not need to worry about the direction. For a lag distance of 1.0 ft, we have six pairs. With Eq. 3.9,

$$c(1) = \frac{1}{6}(8.25 \times 9.0 + 9 \times 6.25 + 6.25 \times 5.0 + \ldots + 4.75 \times 5.0) - (6.22)^2 = 0.8408.$$

Similarly, at a lag distance of 2.0 ft, we have five pairs.

$$c(2) = \frac{1}{5}\left(8.25 \times 6.25 + 9.0 \times 5.0 + \ldots + 5.3 \times 5.0\right) - (6.22)^2 = -2.7009.$$

Using similar calculations, we can calculate the value of covariance at a lag distance of 3 ft as

$$c(3) = \frac{1}{4}(8.25 \times 5.0 + 9.0 \times 5.3 + 6.25 \times 4.75 + 5.0 \times 5.0) - (6.22)^2 = -2.7790.$$

We calculate only to this point because the number of pairs becomes smaller at additional lag distances. As a special case, the definition at $\vec{L} = 0$ is

$$c(0) = \frac{1}{n} \sum_{i=1}^{n} x(\vec{u}_i) x(\vec{u}_i) - \bar{x}^2.$$

This is the definition of variance. For the present data,

$$c(0) = \frac{1}{7}(8.25^2 + 9.0^2 + \ldots + 5^2) - (6.22)^2$$

$$= 2.565.$$

We repeat these calculations although we have already done them in Numerical Example 2.6. Re-examination of Numerical Example 2.6 shows that the answers differ slightly. For example, at $\vec{L} = 1$ ft, the estimated covariance in Numerical Example 2.6 was 0.43, while it is 0.823 in this case. Similar observations can be made at other lag distances. This difference is caused by the additional assumption of the first order of stationarity made in this example. The difference between the two answers is significant. The next section discusses which is more appropriate.

3.3.2 Correlation Coefficient. The correlation coefficient can also be used to describe the spatial relationship. Recall the definition of correlation coefficient for describing the spatial relationship.

$$\rho(\vec{L}) = \frac{C(\vec{L})}{\sigma_{\vec{u}}\sigma_{\vec{u}+\vec{L}}}, \quad (3.10)$$

where $\rho(\vec{L})$ = the correlation coefficient at lag distance $\vec{L}$; $C(\vec{L})$ = covariance; and $\sigma_{\vec{u}}$ and $\sigma_{\vec{u}+\vec{L}}$ = standard deviations for the data at locations $\vec{u}$ and $\vec{u}+\vec{L}$, respectively. However, if we make the assumption of the second order of stationarity, we can state that

$$V[X(\vec{u})] = V\left[X\left(\vec{u}+\vec{L}\right)\right] = C(0). \quad (3.11)$$

Therefore, we can also state that

$$\sigma_{\vec{u}} = \sigma_{\vec{u}+\vec{L}} = \sqrt{C(0)}. \quad (3.12)$$

Substituting Eq. 3.12 into Eq. 3.10 gives

$$\rho(\vec{L}) = \frac{C(\vec{L})}{C(0)}. \quad (3.13)$$

Using the sample data, we can write the estimated value of the correlation coefficient as

$$r(\vec{L}) = \frac{c(\vec{L})}{c(0)}, \quad (3.14)$$

where $c(0)$ = sample variance.

Numerical Example 3.2. Using the data from Numerical Example 3.1, calculate the correlation coefficient as a function of lag distance.

Solution. The values of covariance were computed earlier. Therefore, the correlation coefficient at $\vec{L} = 1$ ft is

$$r(1) = \frac{c(1)}{c(0)} = \frac{0.8408}{2.57} = 0.327.$$

Similarly,

$$r(2) = \frac{-2.7009}{2.57} = -1.05,$$

$$r(3) = \frac{-2.7790}{2.57} = -1.08,$$

and $r(0) = \dfrac{c(0)}{c(0)} = 1.0.$

These values are different from the ones calculated in Numerical Example 2.6 because of the assumption of second-order stationarity in this example. The next section discusses the appropriateness of the relevant equations for examining the field data.

3.3.3 Variogram. The variogram is the most commonly used geostatistical technique for describing the spatial relationship. Mathematically, it is defined as

$$\gamma(\vec{L}) = \frac{1}{2}V\left[X(\vec{u}) - X\left(\vec{u}+\vec{L}\right)\right]. \quad (3.15)$$

It is half of the variance of the difference between the two values located $\vec{L}$ distance apart. Before we expand this term, we need to understand the behavior of the variogram qualitatively. As the definition indicates, the value of the variogram is zero at $\vec{L} = 0$ because

$$\gamma(0) = \frac{1}{2}V[X(\vec{u}) - X(\vec{u})] = 0. \quad (3.16)$$

Because of our assumption for the geoscience data, the difference between the two values increases as the distance increases. That is, variance increases as lag distance increases, which is exactly opposite of covariance. Unlike covariance, the variogram starts with a zero value and increases as the lag distance between the two values increases.

With the definition of variance, we can expand Eq. 3.15 as

$$\gamma(\vec{L}) = \frac{1}{2}\left(E\left\{\left[X(\vec{u}) - X\left(\vec{u}+\vec{L}\right)\right]^2\right\} - E\left\{\left[X(\vec{u}) - X\left(\vec{u}+\vec{L}\right)\right]\right\}^2\right). \quad (3.17)$$

However, the assumption of first-order stationarity requires that

$$E[X(\vec{u})] = E\left[X\left(\vec{u}+\vec{L}\right)\right]. \quad (3.2)$$

Therefore, the second term on the right side of Eq. 3.17 is equal to zero. Simplifying,

$$\gamma(\vec{L}) = \frac{1}{2}E\left\{\left[X(\vec{u}) - X\left(\vec{u}+\vec{L}\right)\right]^2\right\}. \quad (3.18)$$

With the definition of covariance, Eq. 3.18 becomes

$$\gamma(\vec{L}) = C(0) - C(\vec{L}). \quad (3.19)$$

Appendix C gives the details of the derivation of Eq. 3.19.

As expected, the variogram is closely related to the covariance if the assumptions of the first- and second-order stationarities hold. The variogram increases as the covariance decreases. Both the variogram and covariance capture the spatial relationship. The variogram increases as the lag distance between the values increases (the relationship gets weaker); the covariance decreases as the lag distance between the two values increases.

In practice, Eq. 3.18 can be written as

$$\hat{\gamma}(\vec{L}) = \frac{1}{2n(\vec{L})}\sum_{i=1}^{n(\vec{L})}\left[x(\vec{u}_i) - x\left(\vec{u}_i+\vec{L}\right)\right]^2 \quad (3.20)$$

where $n(\vec{L})$ = number of pairs at lag distance $\vec{L}$; $x(\vec{u}_i)$ and $x(\vec{u}_i+\vec{L})$ = data values for the ith pair located $\vec{L}$ lag distance apart. The accent (hat) over γ indicates that it is an estimated value based on the sample data.

Numerical Example 3.3. Use the data from Numerical Example 3.1 to estimate the variogram as a function of lag distance.

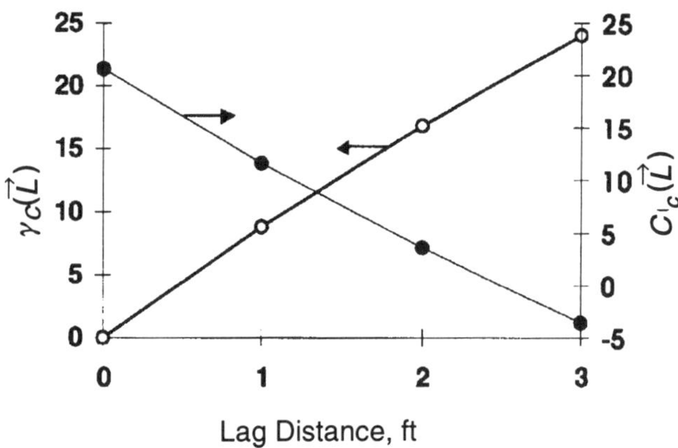

Fig. 3.1—Variogram and covariance for Numerical Example 3.3.

Solution. As in the case of covariance, first gather all the pairs at a given lag distance. For $\vec{L} = 1$ ft, there are six pairs. Therefore, using Eq. 3.20 gives

$$\hat{\gamma}(1) = \frac{1}{2 \times 6}\Big[(8.25 - 9.0)^2 + (9.0 - 6.25)^2 + \ldots + (4.75 - 5.0)^2\Big] = 0.845.$$

Similarly, for $\vec{L} = 2$ ft, we have five pairs;

$$\hat{\gamma}(2) = \frac{1}{2 \times 5}\Big[(8.25 - 6.25)^2 + (9.0 - 5.0)^2 + \ldots + (5.30 - 5.0)^2\Big] = 2.11.$$

For $\vec{L} = 3$ ft, we have four pairs;

$$\hat{\gamma}(3) = \frac{1}{2 \times 4}\Big[(8.25 - 5.0)^2 + (9.0 - 5.3)^2 + (6.25 - 4.75)^2 + (5.0 - 5.0)^2\Big] = 3.31.$$

We already established that $\hat{\gamma}(0) = 0$.

Fig. 3.1 shows a plot of the variogram as a function of lag distance. For comparison purposes, the covariance values are also shown. As discussed earlier, these show opposite trends as lag distance increases.

Both the variogram and covariance can capture the spatial relationship adequately. Why use the variogram instead of the covariance? One reason is tradition. By convention, the variogram has been used in geostatistics to describe spatial relationships. Most geostatistics literature also uses the variogram; therefore, an understanding of the variogram concept makes it much easier to follow the literature. The second reason is that the variogram can capture the spatial relationship under certain conditions where covariance may not. The variogram calculation requires only an assumption that variance of the difference between two values be finite (Eq. 3.15), while calculation of covariance requires that the variance of the data, $C(0)$, be finite. In some instances where data variability increases as distance increases within a region of interest, estimation of the value of $C(0)$ may not be possible but estimation of the variogram might. In other words, estimation of the variogram requires less restrictive assumptions (only that the variance of the increment be finite, which is also called the intrinsic hypothesis[1]) than the covariance. In most practical situations, this difference may not be important.[2-4] However, because of the popularity of the variogram, most of the spatial relationships in geostatistics still are described by variogram rather than covariance.

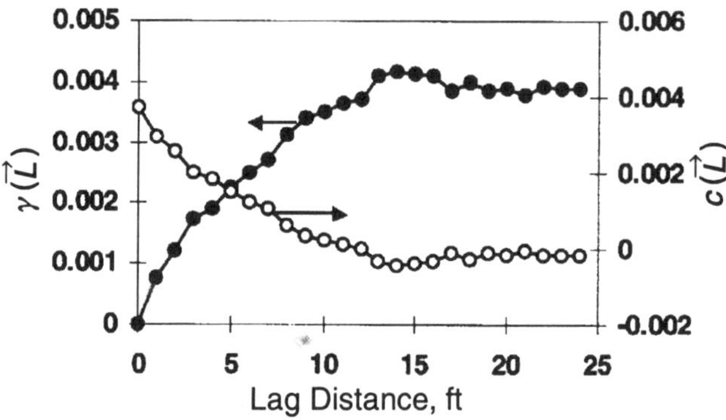

Fig. 3.2—Estimated variogram and covariance for Field Example 3.1.

Field Example 3.1. Generate the variogram and covariance for the well porosity data for Well Nos. 34–29.

Solution. We generated the covariance and correlogram figures (Fig. 2.23) for the well data in Field Example 2.8. Here, we use the definitions of the variogram and covariance, which are based on the assumption of second-order stationarity, to generate the variogram and covariance.

Fig. 3.2 shows a plot of the variogram as a function of lag distance for well-porosity data. As expected, the variogram starts with a value of zero and increases slowly. At approximately $(\vec{L}) = 13$ ft, the variogram reaches a constant value. In geostatistical terminology, the distance at which the variogram reaches a constant value is called the "range" of the variogram. The range represents the distance to which neighboring values are related to each other. The relationship gets weaker as lag distance increases, starting with a distance of zero. Beyond the lag distance equal to the range of the variogram ($\approx$13 ft in Fig. 3.2), the neighboring values are uncorrelated with each other.

The covariance plot in Fig. 3.2 is essentially a mirror image of the variogram plot. It starts with a value corresponding to $C(0)$ and reaches a constant value at $(\vec{L}) = 13$ ft. The covariance at that distance is approximately zero, indicating the uncorrelated nature of the data beyond that distance. As explained earlier, the distance at which the covariance reaches a constant value is called the range of the covariance.

With Eq. 3.19,

$$\gamma(\vec{L}) = C(0) - c(\vec{L});$$

therefore, at a lag distance of zero, $\gamma(\vec{L}) = 0$. Similarly, beyond the lag distance equal to the range, $C(\vec{L}) \approx 0$ (see Fig. 3.2). Substituting this value into Eq. 3.19 gives

$$\gamma(\vec{L}) \approx C(0) \quad \ldots\ldots\ldots\ldots\ldots\ldots\ldots\ldots\ldots (3.21)$$

for $\vec{L} \geqq a$, where a =range. That is, the value of the sill is approximately equal to the variance of the data. In practice, the sill may not be exactly equal to the variance of the sample. This may be because of the data configuration or sampling scheme. However, in many instances, the sill is very close to the variance of the data.

3.4 Estimation of Variogram

Sec. 3.3 discussed the basic definition of the variogram and estimation of the value of the variogram with an appropriate equation. In practice, several difficulties are encountered in estimating the variogram. Because the goal is to capture the spatial relationship in the best way possible, the original variogram equation is modified to capture the spatial relationship. This section discusses common problems encountered in variogram estimation and the proposed solutions implemented in practice. While many of these proposed solutions may not have a strong theoretical foundation, they have been proved to work in practice.

3.4.1 Lack of Sufficient Pairs. As defined, the variogram is a statistical property. We calculate the "average" squared difference between two values located a certain lag distance apart. For this average to be truly representative of a given lag distance, sufficient data pairs are needed for that lag distance. For the vertical data set examined in Field Example 3.1, the sample data were collected at uniform intervals. There were sufficient numbers of pairs for all lag distances in this case. Although defining a precise number of pairs as sufficient is difficult, at least seven to ten pairs are needed for a reliable estimate of the variogram for a given lag distance. Obviously, the more pairs we have for a given lag distance, the more precise the estimate of the variogram. Also, it is important to remember that the minimum number of pairs is only relevant to the extent that a reliable, interpretable experimental variogram can be observed. For example, if ten pairs do not provide an interpretable structure, alternative means of obtaining a more reliable variogram need to be explored.

Number of Pairs. A commonly used method to ensure sufficient pairs for a given lag distance is to restrict the maximum distance at which the variogram is computed. For a given region of stationarity, the possible number of pairs decreases as lag distance increases. Numerical Examples 3.1 and 3.2, where the number of pairs decreased as lag distance increased, illustrated this. Eventually, as the lag distance corresponds to the maximum possible distance between any two sample points, only one data pair may be available at that lag distance. Obviously, we may not be able to obtain a reliable estimate of the variogram at that lag distance. To avoid this problem, a rule of thumb used in the geostatistical literature is to use half the maximum possible distance within a region of interest as the maximum lag distance at which the variogram is calculated. For example, if 10,000 ft is the maximum distance between any two sample points within the region of interest, the variogram estimation is restricted to a maximum lag distance of 5,000 ft.

This rule of half the maximum distance also serves another important purpose. Recall the definition of the estimated variogram (Eq. 3.20).

$$\hat{\gamma}(\vec{L}) = \frac{1}{2n(\vec{L})}\sum_{i=1}^{n(\vec{L})}\left[x(\vec{u}_i) - x(\vec{u}_i + \vec{L})\right]^2. \quad \text{.......} (3.20)$$

Because the variogram is computed as the average of the squared difference, it is symmetric. That is, whether we take the difference as $[x(\vec{u}) - x(\vec{u} + \vec{L})]$ or $[x(\vec{u} + \vec{L}) - x(\vec{u})]$, the answer does not change. The variogram provides the same estimate by adding 180° to the given direction. Using the half-the-maximum-distance rule effectively ensures that representative pairs are collected on both sides of a given location.

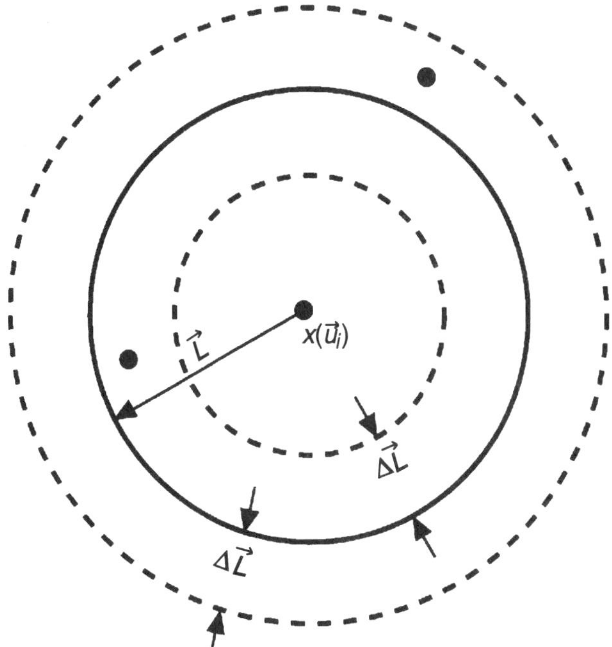

Fig. 3.3—Tolerance with respect to lag distance.

Lag Tolerance. The second alternative to ensure sufficient pairs for a given lag distance is to define the tolerance with respect to the distance as well as with respect to the direction. This is especially useful for areal data sets. Typically, examination of areally distributed data shows that the samples are not distributed at uniform intervals; therefore, a sufficient number of pairs for a precise lag distance cannot be obtained. For example, to estimate the variogram at a lag distance of 330 ft, we may have one data pair at precisely 330 ft. However, we may have three additional pairs between 310 and 330 ft and five additional pairs between 330 and 350 ft. Obviously, one pair, precisely at 330 ft, does not provide a reasonable estimate of the variogram. However, we can capture additional pairs for a better estimate of the variogram by defining a tolerance with respect to the lag distance. The variogram definition is modified as

$$\hat{\gamma}(\vec{L} \pm \Delta\vec{L}) = \frac{1}{2n(\vec{L} \pm \Delta\vec{L})}\sum_{i=1}^{n(\vec{L})}\left[x(\vec{u}_i) - x(\vec{u}_i + \vec{L})\right]^2, \quad \text{..................} (3.22)$$

where $\Delta\vec{L}$ = tolerance with respect to lag distance. To estimate the variogram at a lag distance of $\vec{L}$, we collect all the pairs within $\vec{L} \pm \Delta\vec{L}$ lag distance. **Fig. 3.3** illustrates tolerance with respect to distance for an isotropic search, which is concerned only with distance for all the angles. When searching for sample points away from $x(\vec{u})$ at a given distance L, we can search for all the data points within $L \pm \Delta L$ distance. In Fig. 3.3, no sample point is precisely at a distance L; however, by using a tolerance of ΔL, we can obtain two pairs at that lag distance.

In a similar fashion, **Fig. 3.4a** shows a tolerance with respect to distance and direction in a 2D domain. A tolerance of $L \pm \Delta L$ is defined with respect to distance, and $\theta \pm \Delta\theta$ is defined with respect to direction.

In addition to tolerance with respect to direction, another restriction can be placed on picking appropriate pairs. We can define a bandwidth, which is a distance perpendicular to the

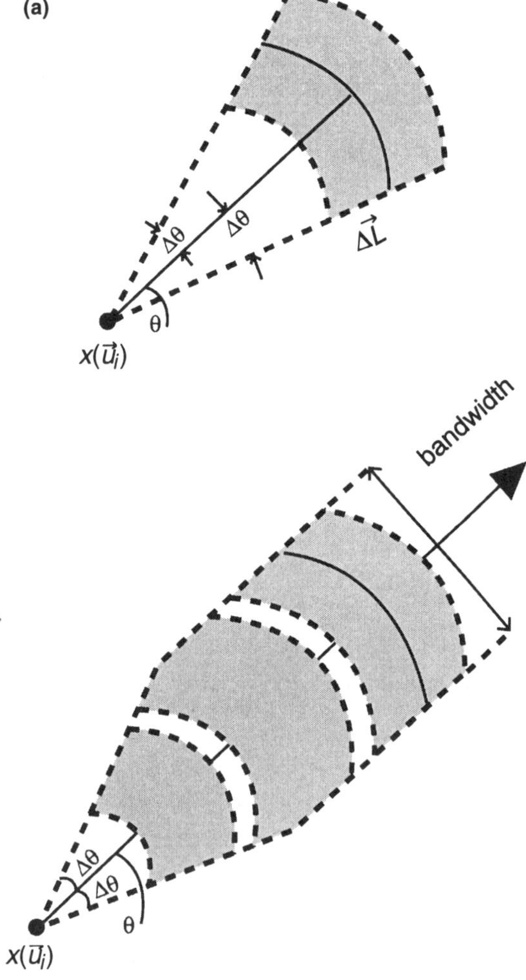

Fig. 3.4—(a) Tolerance with respect to distance and direction for variogram estimation. (b) Use of bandwidth to restrict directional tolerance.

direction in which we are computing the variogram. This distance defines the maximum width of the area to be used to select the data pairs. As **Fig. 3.4b** shows, at higher lag distances, once the area reaches the maximum width, the same width is used to select appropriate pairs for a given lag distance and to select the direction. In the absence of the bandwidth restriction, the width of the areal coverage increases as the lag distance increases. Bandwidth allows better control of the directionality of the estimated variogram.

In 3D data sets, if we define tolerances with respect to distance and two directions, the tolerance volume looks like a flattened cone. The bandwidth has to be defined for both directions.

The tolerance that is appropriate with respect to distance and direction is a subjective decision. The guiding principle should be to use the tolerance that is appropriate to achieve a clearly interpretable variogram structure. As Deutsch and Journel[5] state, in general, the spatial structure cannot be created by manipulating the tolerance and direction if it does not exist within the data. However, a poor choice of tolerance may mask the existing spatial structure. Therefore, a tolerance should be chosen so that the most interpretable variogram is obtained.

In practice, estimation of the variogram should begin with a small tolerance. If the variogram does not exhibit a clearly defined structure or shows too many fluctuations, the tolerance should be increased. The smallest possible tolerance at which an interpretable structure can be observed should be used to estimate the variogram. This technique applies whether the tolerance is defined with respect to distance or direction. If the tolerance is too large, the estimated variogram may not be as "crisp"; that is, some details in structures may be lost that may be important from a spatial relationship point of view. Specifically, if the spatial structure indicates anisotropy (different spatial relationships in different directions), a large tolerance with respect to direction may camouflage the anisotropic structure. For example, if the ratio of the range in the variogram in the x direction to the range of the variogram in the y direction (which is perpendicular to the x direction) is 5, using a tolerance of $\pm 45°$ with respect to the angle, may reduce the ratio to 2.4.[5] It is, therefore, important to use the smallest possible tolerance to capture both the true structure and the anisotropy, if present.

From an estimation point of view, it is always preferable to start with an isotropic variogram structure before investigating the presence of anisotropy. The isotropic variogram structure assumes that the variogram structure is a function only of distance, not of direction (see Fig. 3.3). Estimation of the isotropic variogram allows adjustment of some of the parameters (such as tolerance with respect to distance) before proceeding with estimation of the anisotropic variogram. Also, in almost all instances, some type of spatial structure should be evident in the isotropic variogram. If no apparent structure is observed in an estimated isotropic behavior, it is difficult to imagine that some type of structure can be captured by investigating the anisotropic behavior. It is important to remember that the isotropic variogram, by considering all the pairs at a given lag distance independent of direction, uses a lot more pairs at a given lag distance compared with the anisotropic estimated variograms, which are restricted by direction as well as distance. Therefore, the isotropic variogram should provide a more stable structure than any of the anisotropic variograms. If interpretable structure cannot be captured with an isotropic variogram, it is almost impossible to observe an interpretable structure in anisotropic variograms.

Field Example 3.2. Investigate the effects of distance and directional tolerances on the estimated variogram for the porosity data for Flow Unit 3.

Solution. This particular data set represents the arithmetically averaged log porosity data at each well location within Flow Unit 3.

As the preceding section discussed, we start with an isotropic variogram. **Fig. 3.5** shows the isotropic estimated variogram with different lag intervals with a fixed tolerance of 250 ft. That is, 800 ft of lag interval represents lag distances of 800; 1,600; 2,400, and so forth, with the tolerance defining a spread of ± 250 ft from each lag distance. For a fixed tolerance, the estimated variograms for different lag intervals are very similar. The 800-ft-interval variogram shows more fluctuations than those for the 1,000- or 1,400-ft lag distances because it is estimated at more lag distances, reflecting more fluctuations. Overall, however, the results are very similar, indicating that the choice of lag interval does not have a significant effect on the variogram computation.

Fig. 3.6 shows the estimated variogram plot for different lag tolerances for a lag interval of 1,400 ft for tolerances of 250, 500, and 700 ft. Although the estimated variograms for different tolerances look similar, closer examination reveals

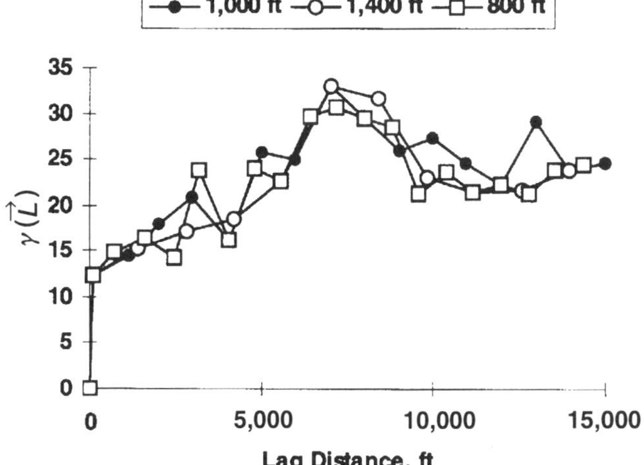

Fig. 3.5—Effect of lag intervals on estimated variogram.

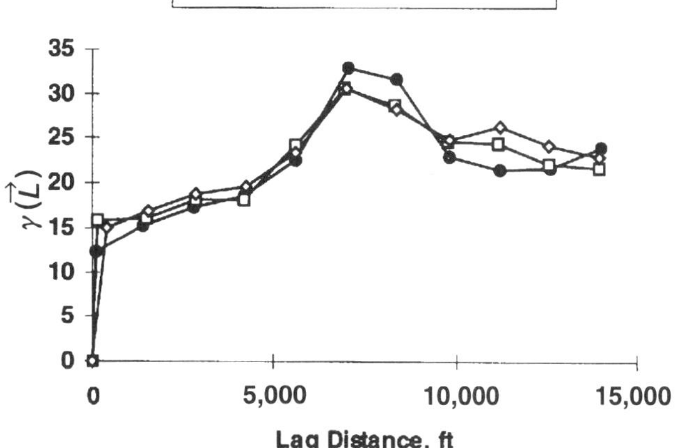

Fig. 3.6—Effect of tolerance on estimated variogram.

that the observed fluctuations become smaller as the tolerance increases. **Table 3.2** gives the values of the variogram, average lag distance, and number of pairs for different tolerances. As expected, the possible number of pairs for a given lag distance increases as tolerance increases. This also minimizes the fluctuations and "stabilizes" the estimated variogram. The differences in the average lag distances for different tolerances arise because the average of all the distances for the pairs within a given tolerance is considered as the average lag distance. For example, all the pairs within 1,400 ± 500 ft are gathered for a tolerance of 500 ft for the second lag. The individual distances for all pairs fall within that range. A total of 252 pairs fit within this tolerance. The average lag distance of all these pairs is 1,496.13 ft. This is the value reported in the table. Similarly, all the pairs within 1,400 ± 700 ft are gathered for the second lag for a tolerance of 700 ft. The average lag distance of these 330 pairs is 1,568.53 ft (Table 3.2).

On the basis of Figs. 3.5 and 3.6, we use a lag interval of 1,400 ft with a tolerance of 700 ft for further investigation. Although the decision to choose these parameters is subjective, they appear to provide the most stable variogram. Also, the maximum distance between any two points was observed to be 31,000 ft; therefore, all the variogram computations are restricted to a maximum distance of 15,000 ft.

Figs. 3.7 and 3.8 further examine the effect of tolerance with respect to direction. Fig. 3.7 presents the estimated variogram in four directions, 0, 45, 90, and 135°. For the purposes of orientation, 0° corresponds to the east direction and 90° corresponds to the north direction. The directional tolerance used is ± 20°. The estimated variograms show significant fluctuations. This is because of the lack of an adequate number of pairs for a given lag distance. For example, at 45°, the variogram value for the second lag is 54.6. Close examination, however, reveals only two pairs at that lag distance; therefore, that particular value is not representative of that lag distance. Although difficult to see, the variogram shows a lot more continuity (gradual change) at 0° than at 90°. Fig. 3.8, where the directional tolerance is ± 40°, clarifies this further. Compared with those for a 20° tolerance, the estimated variograms are smoother and show less variability. **Table 3.3** shows the numbers of pairs for each direction. For most of the lag distances, we have sufficient pairs for a given lag distance. Obviously, the number of pairs at any given angle is less than the number of pairs corresponding to the isotropic variogram. Overall, we can state that we have sufficient pairs to obtain representative variograms in different directions. Although hard to observe, the variogram in the 0° direction exhibits more continuity than the one in the 90° direction.

The apparent anisotropy in the variogram can be checked further by calculating the lag distance at which the estimated variogram in each direction reaches the variance of the sample. Recall that the sill of the variogram is approximately

TABLE 3.2—EFFECT OF TOLERANCE ON VARIOGRAM

Tolerance = 250 ft			Tolerance = 500 ft			Tolerance = 700 ft		
Average *L* (ft)	Variogram	Pairs	Average *L* (ft)	Variogram	Pairs	Average *L* (ft)	Variogram	Pairs
93.23	12.3	18	162.29	15.91	22	388.66	15.06	42
1,379.77	15.16	148	1,496.13	16.11	252	1,568.53	16.77	330
2,812.29	17.29	270	2,818.80	18.03	416	2,828.80	18.68	524
4,174.04	18.59	248	4,171.38	18.08	436	4,211.03	19.56	596
5,565.14	22.62	200	5,589.50	24.26	376	5,592.38	23.48	476
7,020.50	32.98	122	6,949.53	30.47	318	6,955.44	30.51	432
8,395.42	31.66	138	8,339.69	28.64	274	8,369.49	28.15	364
9,823.15	23.05	128	9,783.31	24.63	250	9,798.59	24.78	384
11,197.12	21.46	96	11,161.19	24.36	192	11,189.22	26.41	272
12,589.89	21.61	94	12,571.33	22.23	178	12,565.65	24.34	252
13,987.26	23.95	66	13,966.57	21.81	146	13,971.90	23.09	202

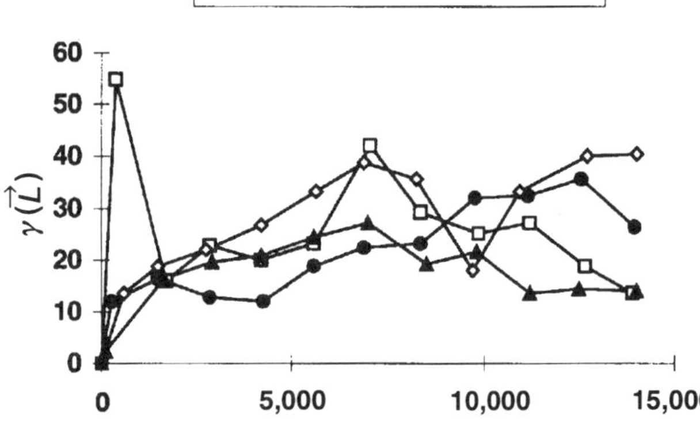

Fig. 3.7—Effect of direction on anisotropic variogram calculation; tolerance = ±20°.

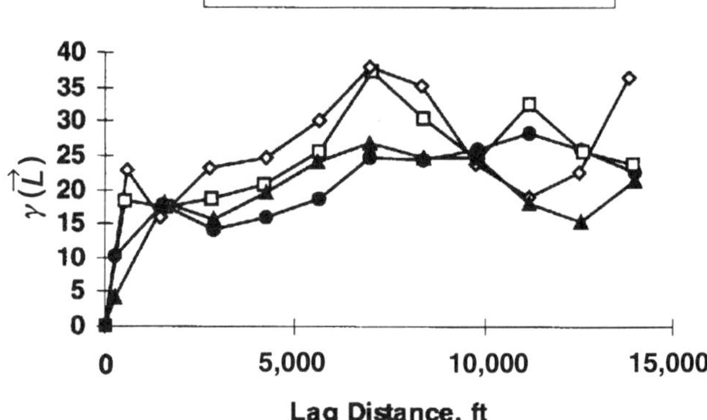

Fig. 3.8—Effect of directional tolerance on anisotropic variogram; tolerance = 40°.

equal to the variance of the sample. For our data set, the variance is equal to 23.1. Therefore, by drawing a horizontal line at the variogram value of 23.1, we can calculate the lag distance where the variogram in a given direction reaches a value of 23.1. We also estimated the variograms at 22.5, 67.5, 112.5, and 157.5° but do not show the results here. **Table 3.4** presents the calculated lag distance where the variogram reaches the variance value for each of the eight variograms. Because the variogram exhibits symmetry after every 180°, the variogram at 0° is the same as that at 180°. Obviously, the lag distance corresponding to the variance at 157.5° is the largest, and the lag distance at 90° is the smallest; that is, the porosity data show maximum continuity at 157.5° and minimum continuity at 90°.

For modeling convenience, we always assume that the directions of maximum and minimum continuities are perpendicular to each other. **Fig. 3.9** plots the lag distance in Table 3.4 at appropriate directions. Obviously, at 157.5° the lag distance is maximum. Also note the symmetry after 180°. The distances are repeated after every 180°. An ellipse is superimposed on top of these lag distances. Although, in reality, the lengths of the lag distances in various directions do not fall precisely on the perimeter of the ellipse, for modeling purposes, we assume that the ellipse represents the range of the variograms in different directions. At any angle, the distance between the center of the ellipse and the perimeter of the ellipse corresponds to the range in that direction. The modeling of that variogram is discussed later, and the reasons for such an approximation become clearer.

Suffice it to state that the exploratory procedure of estimating the variograms in different directions and plotting the directional lag distances corresponding to the variance of the sample may provide information about the directions of maximum and minimum continuities. Although we can fine tune the precise directions of maximum and minimum continuities even further, we assume that 157.5° represents the direction of the maximum continuity, and 67.5° represents the direction of minimum continuity.

3.4.2 Instability. The estimated variogram represents the arithmetic average of the squared differences of variable pair values at a particular lag distance. Because it uses the square of the difference, any large difference between a given pair is magnified. If pairs exhibit a large difference, the squared difference may have a significant impact on the arithmetically averaged variogram value. This effect may change the variogram value disproportionately at a particular lag distance, resulting in instability of the estimated variogram. This instability may prevent capturing the underlying variogram structure that may be present and also causes fluctuations in the estimated variogram as lag distance increases. The instability must be minimized to model the variogram.

TABLE 3.3—ANISOTROPIC VARIOGRAM

Direction = 0°		Direction = 45°		Direction = 90°		Direction = 135°	
Lag (ft)	Number of Pairs	Lag (ft)	Number of Pairs	Lag (ft)	Number of Pairs	Lag (ft)	Number of Pairs
0	68	0	68	0	68	0	68
241	11	549	7	604	9	290	5
1,511	68	1,666	51	1,479	56	1,593	62
2,872	130	2,796	101	2,772	115	2,874	113
4,257	138	4,187	119	4,231	115	4,256	136
5,613	120	5,635	87	5,599	89	5,589	119
6,946	116	6,996	83	6,938	86	6,957	103
8,376	98	8,378	86	8,319	63	8,380	69
9,827	104	9,821	92	9,757	71	9,802	72
11,209	91	11,204	64	11,169	26	11,161	59
12,578	99	12,599	53	12,511	15	12,563	57
13,976	88	13,935	37	13,850	6	13,974	51

TABLE 3.4—LAG DISTANCE CORRESPONDING TO VARIOGRAM EQUAL TO VARIANCE	
Direction (degrees)	Lag Distance (ft)
0	6,300
22.5	6,000
45.0	5,000
67.5	2,800
90.0	2,700
112.5	4,300
135	5,000
157.5	8,400
180	6,300

The two methods commonly used to minimize fluctuations are to increase the possible number of pairs for a given lag distance or to remove certain pairs for a given lag distance. The previous section discussed the first possibility: increasing the possible number of pairs for a given lag distance by use of appropriate tolerance values with respect to the distance and the direction. That discussion showed that increasing the number of pairs for a given lag distance does improve stability of the variogram.

An alternative for improving the stability of the estimated variogram is to examine the possible pairs used for estimation of the variogram for a given lag distance. The difference between the two point values in a pair is what affects the variogram. If the difference is very large, the squared difference can have a significant impact on the estimated variogram. If we can eliminate certain "extreme" pairs that have a significant impact on the variogram computations, we may be able to obtain a better estimate of the variogram that is less affected by these extreme pairs.

Scatter plots are one way to examine these extreme pairs.[6] Plotting one data point of a pair vs. the other data point from the same pair may reveal the differences between the two data points. If the match between the two points is exact, the point falls on a 45° line. On the basis of the scatter plots, certain pairs can be removed, and the variogram can be recomputed for a given lag distance. Alternatively, a certain percentage of the pairs showing the maximum deviation can be removed to create more uniformity in the analysis and to eliminate subjectivity in deciding which pairs should be removed. For example, for every lag distance, 10% of all pairs in the order of showing the maximum deviations can be removed. Under these circumstances, the variogram represents a truncated mean of the differences squared for a particular lag distance. Such truncated means are often used in statistics to reduce the adverse effects of erratic values (e.g., in figure skating in the Olympic Games, the two extreme scores are removed from the final tally). This procedure also has the advantage of being objective.

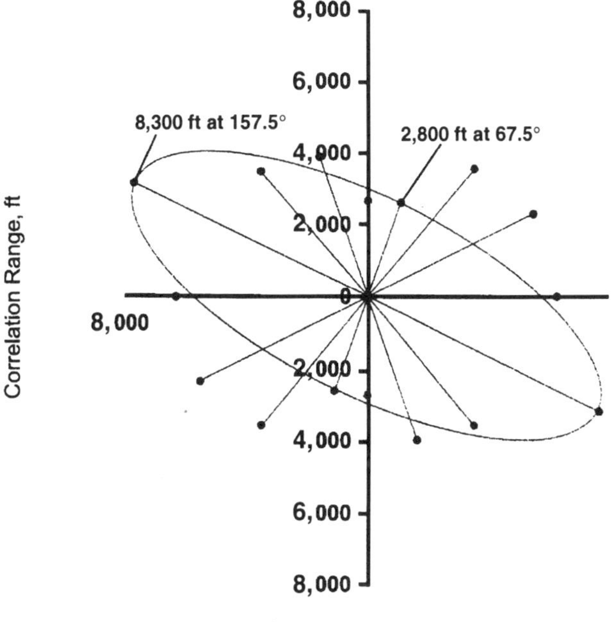

Fig. 3.9—Rose diagram of lag distances in different directions.

Field Example 3.3. Analyze the Flow Unit 3 porosity data and re-estimate the variogram by removing the effect of extreme pairs.

Solution. For this exercise, we assume that the average lag interval is still maintained at 1,400 ft, with a tolerance of 700 ft. **Fig. 3.10** shows the scatter plots for two different lag intervals

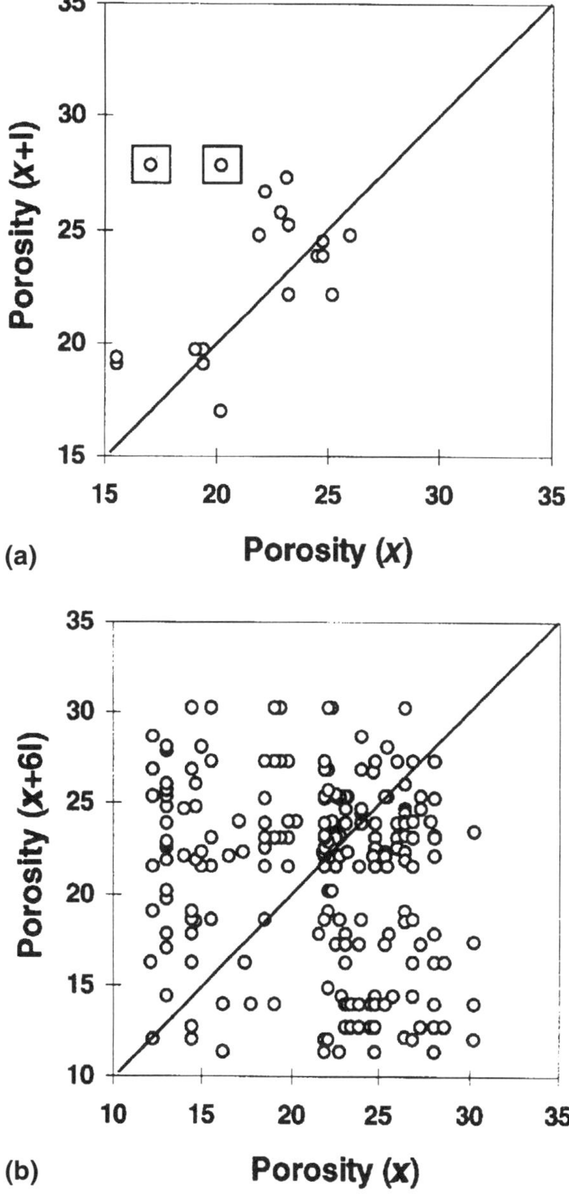

Fig. 3.10—Scatter plot at (a) a first lag interval and (b) a sixth lag interval.

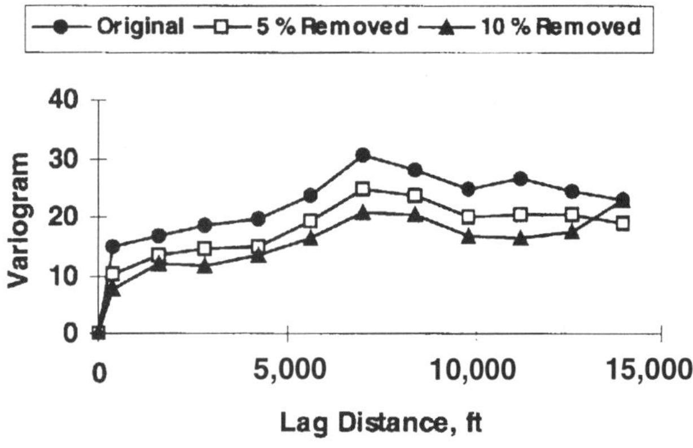

Fig. 3.11—Effect of pair removal on estimated variogram.

vals; the first and the sixth. For the first lag intervals, we have 42 pairs (see Table 3.2). However, we show only 21 pairs here. The other 21 pairs are symmetrical with the first 21 pairs, and do not change the variogram estimation [i.e., $x(u_1)$ vs. $x(u_2)$ is symmetrical to $x(u_2)$ vs. $x(u_1)$ and use of both pairs does not add any additional information because the difference squared would be the same for both the pairs]. In the scatter plot in Fig. 3.10a, we can remove two pairs that can be considered as extreme. The points surrounded by open boxes show these pairs. The choice of what data pairs to consider as extreme is arbitrary. In Fig. 3.10b, which shows 216 pairs, the decision becomes more difficult. The figure shows five pairs that are considered extreme. However, with the large number of pairs in the figure, it is hard to decide what pairs to remove as extreme. It is easy to come up with 8 or 15 pairs that could be considered extreme. This process can become extremely difficult and cumbersome. As a result of the extremely subjective nature of removing individual pairs on the basis of the scatter plot (unless it is so obvious that anyone would remove it), we have not attempted to estimate the variogram after subjective removal of certain pairs for each lag interval.

Instead, we adopted the approach of removing a certain percentage of pairs from the total number of pairs. **Fig. 3.11** shows an estimated variogram plot after removal of 5 and 10% of the extreme pairs. Note that extreme refers to the pairs showing the largest differences: pairs that show the smallest differences are not considered to be extreme. This is because the pairs that show a large difference affect the average much more than the pairs that show a small difference.

In Fig. 3.11, the overall sill (maximum variogram value) decreases as the extreme pairs are removed. This is to be expected because the average of the squared differences is smaller after removal of pairs showing the largest differences. The effect is more pronounced after removal of 10% of the extreme pairs than after removal of 5% of the extreme pairs. Unfortunately, the overall structure of the variograms (including the fluctuations) is largely unaffected by this removal. Because the goal is to capture the spatial structure and not necessarily the exact sill value, removal of extreme pairs has not added any new information to our understanding of spatial relationships in this case.

Note that, although we did not achieve the desired smoothness in this case by removing a certain number of pairs, this technique may be able to be applied to other data sets. It is important to remember that the overall objective is to capture the most interpretable structure. Therefore, it is important to try different modifications and techniques to capture that structure. The modification that gives the most interpretable structure should be used for further analysis.

3.4.3 Influence of Outliers. Outliers are hard to define. In a conventional sense, outlier data are data points that fall outside the "norm." For a normal distribution, a data point falling outside the mean plus or minus three standard deviations can be considered an outlier. However, for distributions that cannot be described by parametric distribution functions, it is hard to define precisely what constitutes outlier data. Specifically, if data exhibit several-orders-of-magnitude variations, it is difficult to define the value beyond which data can be considered as outliers. For example, permeability data at well locations typically exhibit several-orders-of-magnitude variations. This becomes evident when the coefficient of variation (the ratio of standard deviation to mean) is >2. For permeability data, a typical value of coefficient of variation is in the range of two to five. Under such a large variation, it is very difficult (and subjective) to define "anomalous" data that can be considered as outlier data.

Outlier data can significantly affect the variogram estimation. As previously explained, use of an extreme value in variogram estimation can amplify the effect because the squared difference between a data pair is used. If the difference between a given pair is several orders of magnitude, the squared difference is large enough to influence the estimated variogram at a particular lag distance. Remember that the variogram is an arithmetic average of squared differences; therefore, one large squared difference can significantly alter the variogram value. This may create instability in the variogram estimation and also may prevent us from clearly identifying the spatial structure for a particular variable.

The simplest way to deal with the outlier information that causes this instability is to remove the data point from the estimation process. If sufficient physical reason exists for removal, we can simply remove the data point or points and reestimate the variogram. In the absence of a satisfactory reason, it is hard to justify removal of a particular data point or points for mere mathematical convenience. If a particular data point is eliminated, valuable information could be lost that might be hard to find otherwise. Specifically, when the sample data set shows several-orders-of-magnitude variations, it is hard to eliminate only certain data points.

A better way to deal with these variations in the sample data is to use some type of nonlinear transformation to minimize the variation. In this section, we discuss many of the commonly used transforms to minimize the effect of outliers or extreme values. Note, however, that use of nonlinear transforms may create additional difficulties during the estimation process. Chap. 4 discusses these difficulties.

Log Transform. The most commonly used transform is to use the logarithm of the sample value. By taking either natural (base e) or base 10 logs, the order-of-magnitude variations are translated into variations in the integer part of the log of the variable. This should minimize the effect of extreme and order-of-magnitude variations within the data points.

Field Example 3.4. This field example examines the effect of a log transform on the variogram estimation for two variables. The porosity for Flow Unit 3 is not used in this example because the porosity data do not show significant enough variations to necessitate use of a nonlinear transformation. Instead, we use initial-potential (IP) data, collected from several wells within the field, as the variable. We also use net kh as another variable. Net-kh value in each well is determined by adding core permeabilities (collected at 1-ft intervals) over the entire

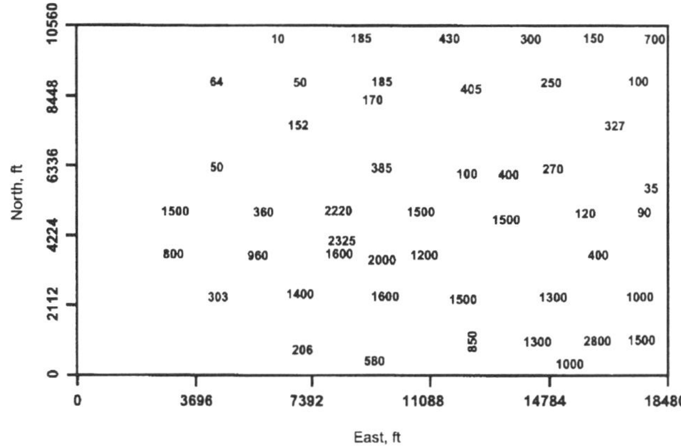

Fig. 3.12—IP data for 48 cored wells in barrels per day.

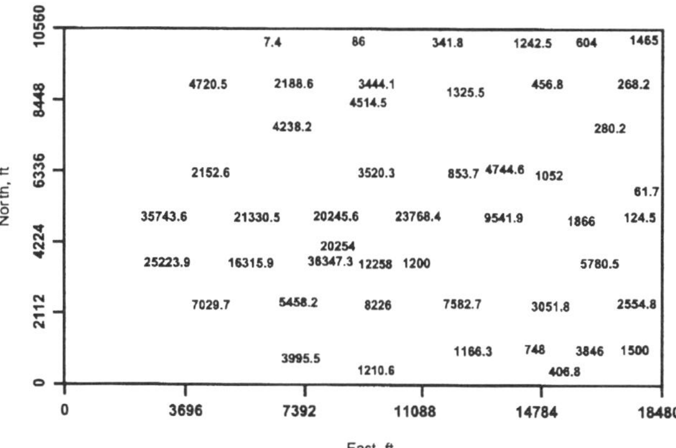

Fig. 3.13—*kh* data for 48 cored wells.

pay-zone interval. This value reflects the contribution from all the flow units at a particular well location.

Fig. 3.12 shows the spatial locations for IP data and the associated IP values at each location. **Fig. 3.13** shows net-*kh* values at the same locations. We have a total of 48 values of both IP and net-*kh* data. **Fig. 3.14** shows the histograms for the IP and net-*kh* data. Both variables show a significant number of values at a lower range and a long tailing. The variation in the values is over several orders of magnitude. The IP data range from 10 B/D to as high as 2,800 B/D, and the net-*kh* data range from 7.0 to 36,347.0 md-ft. The coefficient of variation for the IP data is 0.93, and the coefficient of variation for the net-*kh* data is 1.33. If we had considered individual foot-by-foot permeability data instead of net-*kh* data, we would have observed a much higher coefficient of variation. However, "averaging" over the entire pay zone interval reduces the coefficient of variation substantially.

This field example illustrates the application of a log transform for both IP and net-*kh* data, with attention to only isotropic variograms. Similar results can be obtained for anisotropic variograms as well. An average lag interval of 1,400 ft with a distance tolerance of 700 ft is assumed. These values are the same as those used in Field Examples 3.2 and 3.3. Because the data are collected from the same field with approximately the same density, we can assume that the lag interval and the lag tolerance do not change significantly.

Fig. 3.15 compares the conventional variogram and the variogram of the log-transformed data. The conventional variogram, especially at large lag distances, shows significant variations. In contrast, the variations exhibited by the transformed data are small. The only exception is at the largest lag distance. The changes in the variogram of the transformed variable are much more gradual than in the conventional variogram. Clearly, the log transform has minimized the fluctuations in the estimated variogram values.

This effect is even more pronounced for the net-*kh* data, which exhibit a higher coefficient of variation than the IP data. As **Fig. 3.16** shows, the conventional variogram hardly shows any discernible spatial structure, fluctuations dominate the variations, and it is hard to capture any gradual trend in the data. In contrast, the log-transformed variable shows a nicely developing spatial structure. Starting with a very small value, the estimated variogram increases and reaches a sill value at approximately 7,000 ft. Beyond that, the variogram is fairly constant. As before, the only exception to this gradual trend is the estimated value at a lag distance of 14,000 ft. Otherwise, the log transform clearly has helped to identify the variogram structure for the net-*kh* data.

Overall, for both the IP and net-*kh* data, the log transform has resulted in better identification of the spatial structure. It is safe to state that the higher the variations are in the original data set, the greater the impact of the log transform on the estimated variogram. As stated earlier, if the goal is to capture the spatial structure that is exhibited by the sample data, the log transform may be a useful tool, especially for the data showing order-of-magnitude variations.

Power Transform. Use of power transform is similar to log transformation. Instead of using the sample data itself, take a power transform by

$$x_T(u) = [x(u)]^p, \quad \ldots\ldots\ldots\ldots\ldots\ldots\ldots\ldots (3.23)$$

where $x_T(u)$ = transformed variable, $x(u)$ = original variable, and p = power. The value of p has to be less than one. The smaller the power we use, the smaller the variation of the

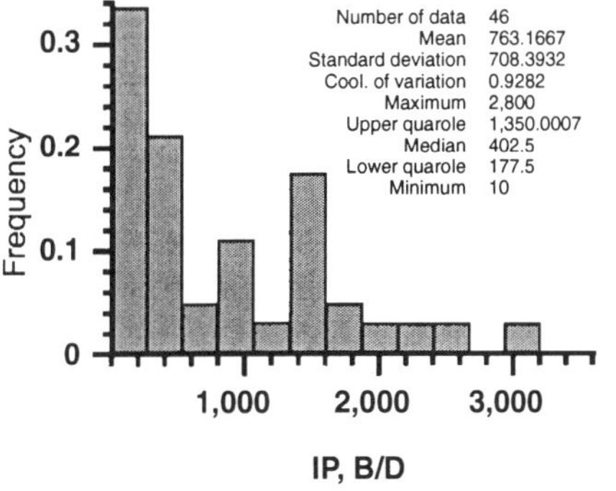

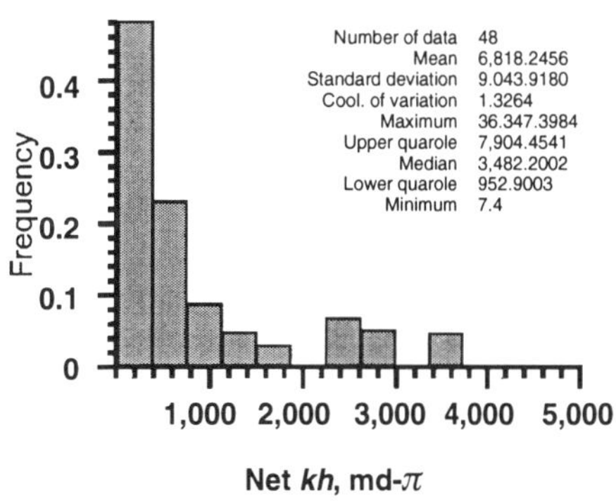

Fig. 3.14—Histogram for IP and net-*kh* data.

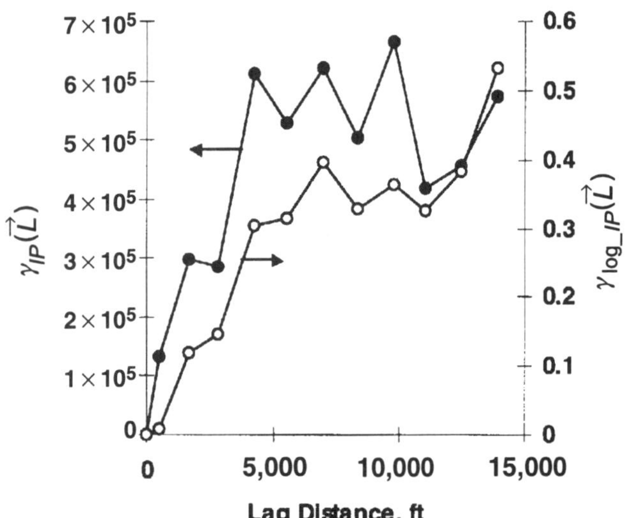

Fig. 3.15—Effect of log transform on estimated IP variogram.

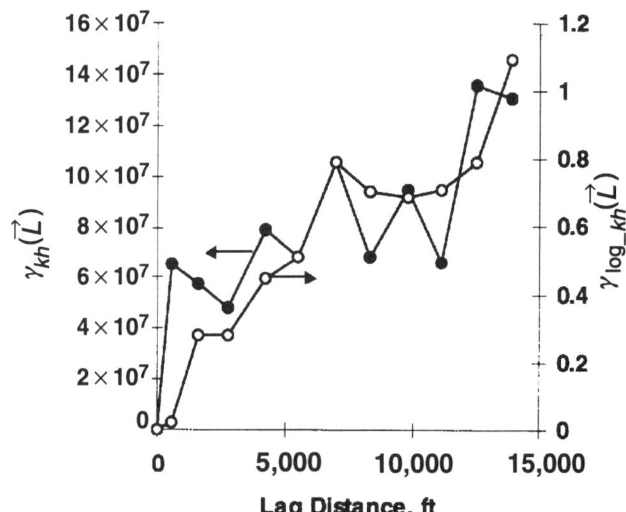

Fig. 3.16—Effect of log transform on net-*kh* variogram.

data values. The most commonly used value of p is 0.5, which represents the square root of the sample data value.

Field Example 3.5. This example uses the same data as used in Field Example 3.4 with the addition of use of the square root of the sample data as a new variable in estimating the variogram.

Fig. 3.17 compares the conventional variogram with that of the transformed values for the IP data. The square-root transform does not show any significant improvement over the conventional data. The magnitude of the fluctuations for both the transformed data and the original data is similar. **Fig. 3.18** shows the results of square-root transformation on the net-*kh* variograms. The results are similar to the IP data results. Although the variogram of the transformed data shows a slightly better structure at smaller lag distances, the fluctuations in the estimated variograms have not been minimized.

Overall, the square-root transform improves the estimated variogram structure slightly compared with the conventional variogram. Obviously, with an even lower value of the power, slightly more improvement might be observed. However, instead of using a trial-and-error procedure of different power values, it might be easier to use the log transform discussed in the previous section.

To summarize, the power transform does not appear to be as effective in transforming the data as the log transform does. If the goal is to reduce the variability of the data through nonlinear transformation, log transform may achieve that objective better than the power transform.

Rank Transform. Chap. 2 briefly discussed the ranking of a given sample data set. The same principle can be used to estimate the variogram of the rank-transformed data. To calculate the ranking of a given sample value, all the sample data are arranged in an ascending order, with the lowest value first and the highest value last. The rank of the ith value in the sequence is calculated by

$$R_i = \frac{i}{n+1}, \qquad (3.24)$$

where n = total number of sample points and R_i = rank of the ith value in the sequence. The ranking of the highest value is slightly less than one to ensure that only a small possibility exists that an estimated value at an unsampled location will be less than the smallest value or greater than the largest value.

Using the rank transform ensures that no significant difference exists between sample values because all the sample values fall between zero and one. Rank transforms, however, provide information only about the relative ordering of

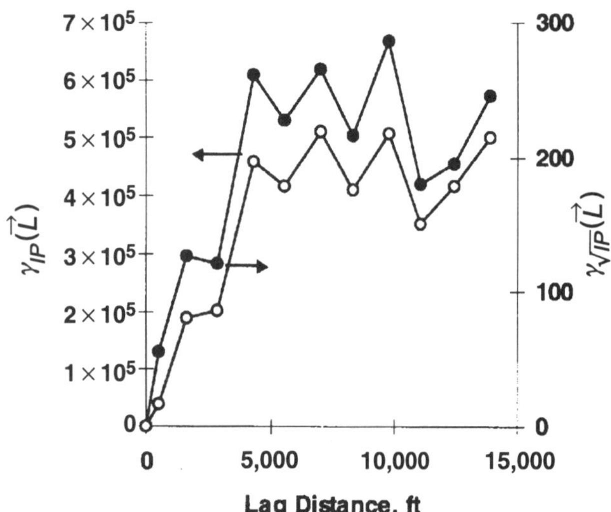

Fig. 3.17—Effect of square-root transform on IP variogram.

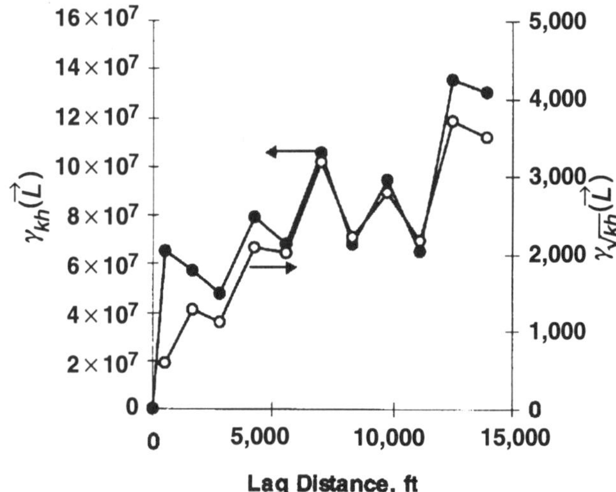

Fig. 3.18—Effect of square-root transform on net-*kh* variogram.

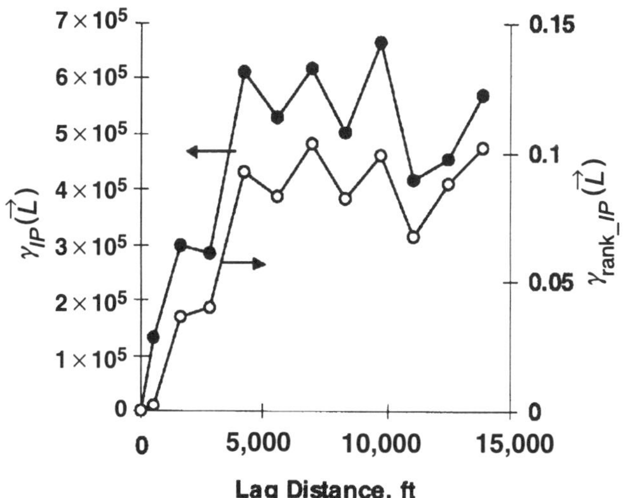

Fig. 3.19—Effect of rank transform on IP variogram.

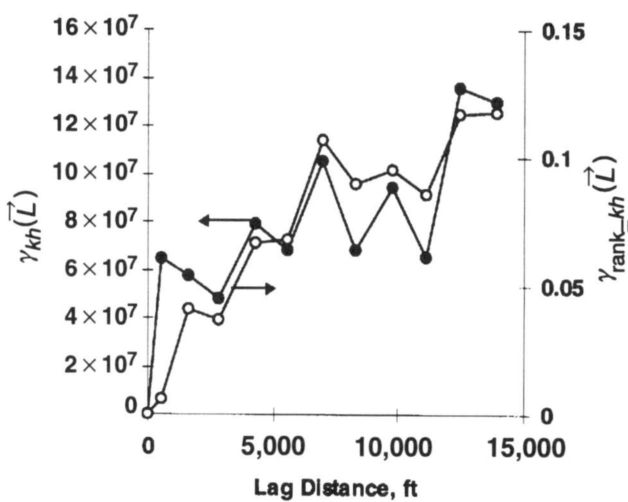

Fig. 3.20—Effect of rank transform on net-*kh* variogram.

sampled data. They do not convey any information about the magnitude of the sample. If we are interested only in the relative ranking of a particular variable at unsampled locations, a rank transform can capture that information.

Field Example 3.6. This example uses the same data as those used in Field Examples 3.4 and 3.5. **Fig. 3.19** compares the conventional variogram with the variogram of the rank-transformed variable for the IP data. For this data set, we see no significant improvement in terms of discerning the variogram structure. In **Fig. 3.20**, which shows the same plot for the net-*kh* data, the rank transform improved the overall variogram structure. The variogram of the transformed data does not show as many fluctuations as the conventional variogram. Further, at smaller lag distances, the change is much more gradual than that shown by the conventional variogram.

Overall, the rank transform may help to define the variogram structure better. The rank transform does not retain any information about the absolute magnitude of the sampled data. However, restricting the values between zero and one does eliminate the influence of outlier data, irrespective of the actual range of the sample data. In addition, the rank transformation has been observed to be useful in certain estimation procedures that attempt to define the uncertainties at unsampled locations in terms of a probability distribution.[7] Chap. 4 discusses the particular application.

Indicator Transform. The indicator transform allows transformation of a continuous variable into a discrete variable. Many reservoir variables, such as permeability and porosity, are continuous variables. These continuous variables can be represented in terms of discrete variables with an indicator transform, which is defined as[8]

$$I(\vec{u}, x_t) = 1, \qquad (3.25a)$$

if $X(\vec{u}) \leqq x_t$ and

$$I(\vec{u}, x_t) = 0, \qquad (3.25b)$$

if $X(\vec{u}) > x_t$, where $I(\vec{u}, x_t)$ = the indicator variable at location $(\vec{u})$, x_t = a threshold value, and $X(\vec{u})$ = the original variable that is being transformed. In practice, each sample value can be transformed into an indicator value by[8]

$$i(\vec{u}_j, x_t) = 1, \qquad (3.26a)$$

if $x(\vec{u}_j) \leqq x_t$, and

$$i(\vec{u}_j, x_t) = 0, \qquad (3.26b)$$

if $x(\vec{u}_j) > x_t$, where $i(\vec{u}_j, x_t)$ = the indicator value at threshold x_t, and $x(\vec{u}_j)$ = the value of the variable at location $(\vec{u}_j)$.

The indicator variable can take only two values: one or zero. If the value of the sample is less than a given threshold, the indicator takes a value of one. If the value of the sample is greater than a given threshold, the indicator takes a value of zero. By specifying multiple threshold values, we can define multiple indicator values at each threshold. Appendix C provides some additional properties of the indicator function.

Numerical Example 3.4. Table 3.5 provides the potential data at various locations. Using threshold values of 178, 403, and 1,350 B/D, transform the data into indicator values.

Solution. This example uses part of the sample IP data used in the previous field examples. The selection of threshold values is subjective. For this example, we selected the values corresponding to 25th, 50th (median), and 75th percentiles.

To calculate the corresponding indicator value at each threshold for each sample, compare the sample data with the threshold. For example, for the first sample, at threshold one (178 B/D), $850 > 178$, $i = 0$; at threshold two (403 B/D), $850 > 403$, $i = 0$; and at threshold three (1,350 B/D), $850 \leqq 1,350$, $i = 1$. Compare the sample value with the threshold value. If the sample value is less than or equal to the threshold value, assign an indicator value of one. If the sample value is greater than the threshold value, assign an indicator value of zero. Repeat this procedure for every sample.

TABLE 3.5—POTENTIAL DATA FOR NUMERICAL EXAMPLE 3.4

x Location (ft)	*y* Location (ft)	IP B/D
12,560	890	850
15,125	6,170	270
8,485	4,860	2,220
17,915	4,860	90
9,770	6,170	385
10,985	3,477	1,200

TABLE 3.6—RESULTS FOR NUMERICAL EXAMPLE 3.4

x Location (ft)	y Location (ft)	IP (B/D)	i_1	i_2	i_2
12,560	890	850	0	0	1
15,125	6,170	270	0	1	1
8,485	4,860	2,220	0	0	0
17,915	4,860	90	1	1	1
9,770	6,170	385	0	1	1
10,985	3,477	1,200	0	0	1

Table 3.6 shows the results. In Table 3.6, i_1 = the indicator value corresponding to the first threshold, i_2 = the indicator value corresponding to the second threshold, and i_3 = the indicator value corresponding to the third threshold. Four possible combinations of indicator values exist, depending on whether the value is less than the first threshold (1,1,1), between the first and the second threshold (0,1,1), between the second and third threshold (0,0,1), or greater than the third threshold (0,0,0). Use of additional thresholds allows definition of additional indicator values at each sample point.

Defining each sample point in terms of either zero or one eliminates the effect of outlier data. The difference squared between the two sample indicator values can be either zero or one. The extreme values, therefore, cannot adversely impact the variogram estimates. One disadvantage of defining the indicator values is that the subtle differences between the data values in a particular class (between two successive threshold values) are lost. For example, in Numerical Example 3.4, we can no longer distinguish between 270 and 385 B/D because both are described by the same set of indicator values. To some extent, defining additional thresholds can eliminate this drawback. However, there is a limit to the number of thresholds that can be defined. Later in the chapter, we discuss the difficulty of modeling indicator variograms if too many thresholds are defined. Also, the estimation procedure becomes computationally more demanding (Chap. 4 discusses this).

In addition to eliminating the impact of outlier data, describing the sample values in terms of indicator values has two additional advantages. First, by appropriately defining the threshold values and estimating the indicator variograms at each threshold, we can examine how the sample values are connected at different thresholds. For example, we may observe that relatively small values exhibit better continuity (smooth variogram change over a bigger range) than high values, which exhibit less continuity. Such types of behavior may not be observed through a conventional variogram of sample values.[9,10]

Second, the indicator definition may allow qualitative information to be translated into a unique quantitative form. For example, a geological facies description may be translated into an indicator function. If three facies are observed at a particular well location, Facies 1 may be described by (1,0,0) indicator values, Facies 2 by (0,1,0) indicator values, and Facies 3 by (0,0,1) indicator values. Once numbers are assigned to each location corresponding to each facies, the spatial continuity of geological facies can be evaluated quantitatively. This is an application of the indicator transform for discrete variables.[11]

Field Example 3.7. This example uses the same data used in Field Examples 3.4 through 3.6: IP and net-*kh* values. To estimate the indicator variogram, we transformed both the IP and the net-*kh* values by a median value. For IP, the median is 403 B/D; for net-*kh*, the median is 3,482 md-ft. Depending on whether a value falls below or above the median, an appropriate indicator value is assigned at each location.

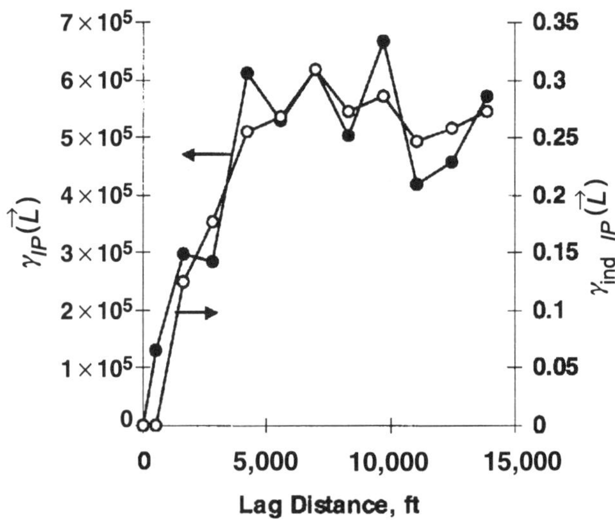

Fig. 3.21—Effect of median indicator transform on IP variogram.

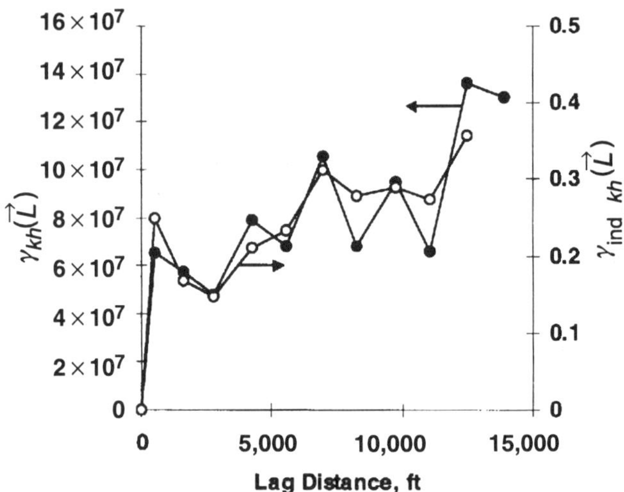

Fig. 3.22—Effect of median indicator transform on net-*kh* variogram.

Fig. 3.21 compares the conventional variogram and the variogram of the median indicator value for the IP data. The indicator variogram is smoother than the conventional variogram and characterizes the spatial structure much better. As **Fig. 3.22** indicates, the improvement with the indicator function is significant for net-*kh* data. Although the first value (corresponding to the shortest lag distance) is much higher than expected, it can be ignored because it contains only two pairs. After ignoring this point, all the other points clearly show an identifiable spatial structure.

Overall, the indicator function has the ability of eliminating the effect of outlier data through the indicator transformation. Using the indicator transform minimizes the fluctuations observed in conventional variograms.

Normal-Score Transform. The normal-score transform allows transformation of sample data into equivalent data that follow a normal (Gaussian) distribution. **Fig. 3.23** shows this

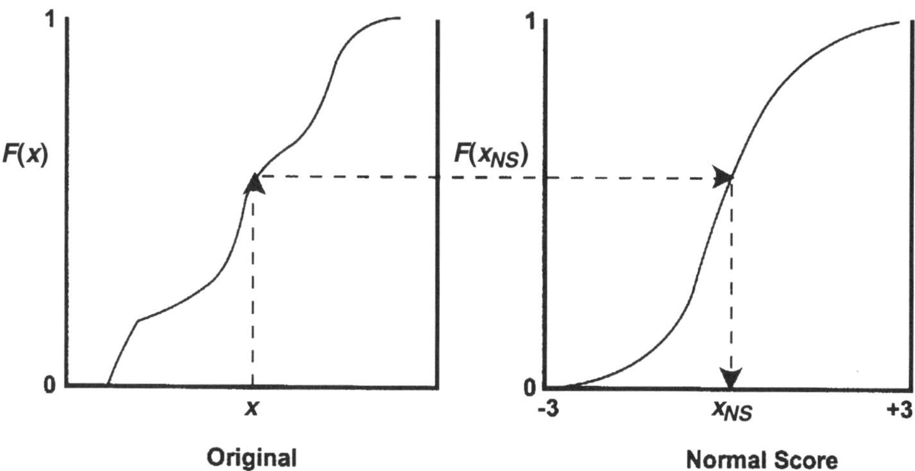

Fig. 3.23—Transform of original data to normal score.

transformation schematically. A cumulative distribution function can be constructed for any sample data. The cumulative distribution function has a minimum value of zero and a maximum value of one. For the normal distribution function, the cumulative distribution function has the same range. Each sample value is associated with a corresponding cumulative distribution function. For the same value of the cumulative distribution function, we can define an equivalent normal-score-transformed value. Typically, the normal-score-transformed values have a mean of zero and a variance of one; therefore, they range from -3 to $+3$. Using such a transform for every sample value allows transformation of all the sample values to a corresponding normal score. There is a one-to-one correspondence between the sample value and the transformed value. That is, if we know the transformed value, we can backcalculate the corresponding sample value.

With transformation of the sample data to a normal score, the variability of the data set is restricted to -3 to $+3$. Effectively, there are no order-of-magnitude variations within the transformed data set. The effect of extreme data on the variogram should be minimized with the transformation. Another advantage of the normal-score transform is that certain estimation techniques, discussed in later chapters, work better with normal-score-transformed data. After obtaining the estimation with the transformed data, we can backtransform the data to original variable values.

Field Example 3.8. This example illustrates the application of the normal-score transform on the variogram estimation for the IP and the net-*kh* data. **Fig. 3.24** shows the effect of transformation for the IP data, and **Fig. 3.25** shows the same effect for the net-*kh* data. No significant improvement can be noted for the IP data with the transform; however, for the net-*kh* data, the estimated variogram with the transformed data shows much better definition of the spatial relationship than the variogram with the raw data. It seems that the greater the variation within the data set, the more the normal-score transform improves the estimated variogram. Note that the sills of the variograms are close to one because the variance of normal scores is equal to one.

To summarize the material covered in this section, nonlinear transforms are useful for minimizing the impact of outlier data on the estimated variogram function. Five transforms were examined: log, power, rank, indicator, and normal score. Of the five, the log and indicator transforms have been the most successful in reducing the fluctuations of the estimated variogram. The log transform, although simple to use, may create difficulties in estimating values at unsampled locations. The indicator transform loses some of the subtle variations in the sample values; however, it has some additional ad-

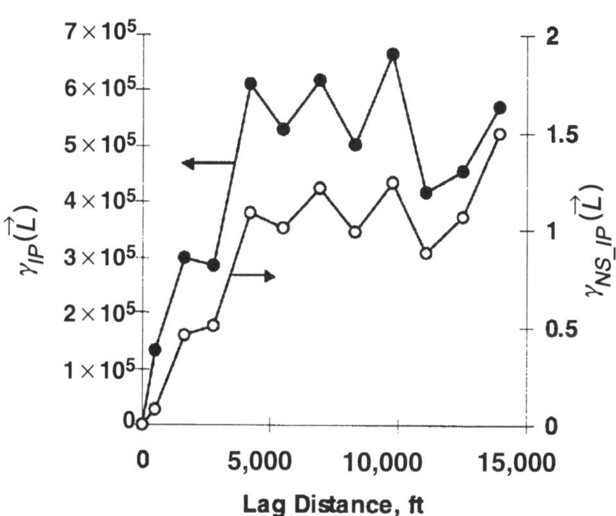

Fig. 3.24—Effect of normal-score transform on IP variogram.

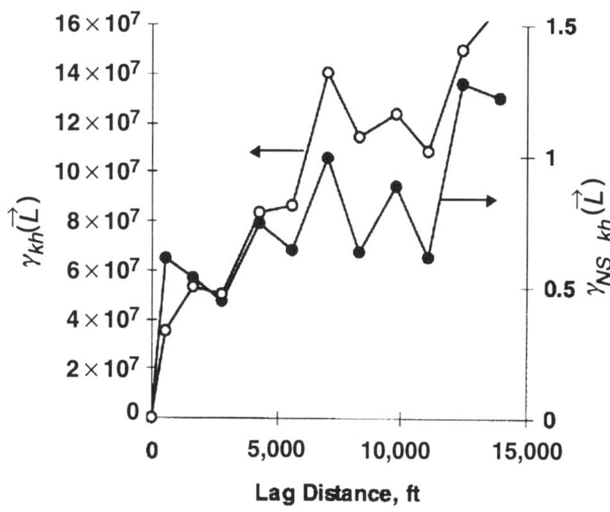

Fig. 3.25—Effect of normal-score transform on net-*kh* variogram.

vantages other than simply eliminating the effect of the outlier data. The power transform does not appear to be very successful in eliminating variogram fluctuations. This transform, therefore, is rarely used. The rank transform is useful in defining the continuity of sample data if we are interested only in relative ordering of the data. It also can be used in certain estimation techniques. The normal-score transform is reasonably successful in minimizing fluctuations in the estimated variogram and has the additional advantage of being useful in some estimation procedures.

In addition to these transformations, there may be other nonlinear transforms that can be used to reduce fluctuations in the estimated variogram. The key issue to consider in use of a nonlinear transformation is that the variogram is eventually going to be used to estimate values at unsampled locations. If we use the transformed variable in generating the spatial relationship, we can estimate only the transformed variable at the unsampled locations. An effective method is needed for converting the transformed data back to the original variable. As long as an effective method can be devised for such an inverse transformation, such transformations are effective in capturing the spatial relationship, which may otherwise be hard to discern with the original data.

3.4.4 Biased Sampling. Chap. 2 briefly discussed the impact of biased sampling on univariate statistics and also presented the results of declustering that type of data set on the sample mean and the sample variance. In this section, we discuss the techniques used for removing the impact of biased sampling on the estimation of a variogram.

Briefly, biased sampling results from the way wells are drilled in an oil field. The first few exploration wells may be drilled on the basis of almost the same information. However, as information is gathered from these wells, subsequent wells are drilled on the basis of the additional information. Obviously, the new wells are drilled in areas where the maximum potential for additional oil recovery exists. To use a geological analogy, if the reservoir includes channel sand and a mud plane, an effort would be made to drill wells in the channel sand rather than in the mud plane.

When we analyze collected samples to describe the spatial relationship, we must understand that the well density may be higher in regions where better prospects exist and lower in regions where the potential is low. This type of biased sampling may impact our ability to describe the spatial relationship. For example, if more wells are drilled in high-porosity regions, the estimated variogram at small lag distances uses information only in a high-porosity region. At a lag distance of 600 ft, with all the pairs at that lag distance in high-porosity regions, only those pairs at 600 ft are used for the variogram estimation. This means that the pairs used are selectively taken from high-potential regions (i.e., those regions representing high porosity, high permeability, high pay-zone thickness, or a combination of these variables). Under these circumstances, the arithmetic means of the data values used for a particular lag distance may not be equal to the sample mean.

If we define

$$\bar{x}_{-\vec{L}} = \frac{1}{n(\vec{L})}\sum_{i=1}^{n(\vec{L})} x(\vec{u}_i) \quad \text{(3.27a)}$$

and $$\bar{x}_{+\vec{L}} = \frac{1}{n(\vec{L})}\sum_{i=1}^{n(\vec{L})} x(\vec{u}_i + \vec{L}), \quad \text{(3.27b)}$$

where $\bar{x}_{-\vec{L}}$ = the arithmetic mean of the first data points within data pairs and $\bar{x}_{+\vec{L}}$ = the arithmetic mean of all the second data points within data pairs for a particular lag distance, then the average of the two,

$$\bar{x}_{\vec{L}} = \frac{\bar{x}_{-\vec{L}} + \bar{x}_{+\vec{L}}}{2}, \quad \text{(3.28)}$$

which represents the arithmetic average of all the data points used for estimating the variogram at a particular lag distance.

If we define the arithmetic mean of all the samples as

$$\bar{x} = \frac{1}{n}\sum_{i=1}^{n} x(\vec{u}_i), \quad \text{(3.29)}$$

where n = total number of samples, ideally, $\bar{x}_{\vec{L}}$ and $\bar{x}$ should be very similar to each other. If we have sufficient pairs for each lag distance and all the pairs gathered represent the entire region in a representative fashion, the lag mean, $\bar{x}_{\vec{L}}$, and the sample mean, $\bar{x}$, should be similar. However, if for a particular lag distance, we are using pairs from a particularly small region because of biased sampling, these two means may not be equal to each other. A similar argument can be advanced for the differences between the sample variance and the lag variance. The sample variance is defined as

$$s^2 = \frac{1}{n}\sum_{i=1}^{n} x^2(\vec{u}_i)\left[\frac{1}{n}\sum_{i=1}^{n} x(\vec{u}_i)\right]^2, \quad \text{(3.30)}$$

and the lag variances are defined as

$$s^2_{-\vec{L}} = \frac{1}{n(\vec{L})}\sum_{i=1}^{n(\vec{L})} x^2(\vec{u}_i) - \left[\frac{1}{n(\vec{L})}\sum_{i=1}^{n(\vec{L})} x(\vec{u}_i)\right]^2 \quad \text{(3.31)}$$

and $$s^2_{+\vec{L}} = \frac{1}{n(\vec{L})}\sum_{i=1}^{n(\vec{L})} x^2(\vec{u}_i + \vec{L}) - \left[\frac{1}{n(\vec{L})}\sum_{i=1}^{n(\vec{L})} x(\vec{u}_i + \vec{L})\right]^2. \quad \text{(3.32)}$$

For selective sampling in estimating the variograms, the lag variances and the sample variance may be different from each other. This type of variation, in the strictest sense, violates the requirement of first- and second-order stationarities. Recall that under first-order stationarity, the mean does not change, and, under second-order stationarity, the variogram is a function of only the lag distance and not of the lag mean or variance. This difficulty must be overcome to estimate the spatial relationships for the sample data correctly.

For univariate statistics, we considered a simple procedure of declustering to estimate unbiased sample mean and sample variance. Unfortunately, a simple procedure does not exist for estimation of the variogram. Several procedures have been proposed in the literature to account for biased sampling. The following discusses some of these techniques.

General-Relative Variogram. The general-relative variogram is defined as[6]

$$\gamma_{GR}(\vec{L}) = \frac{\gamma(\vec{L})}{\left(\bar{x}_{\vec{L}}\right)^2}, \quad \text{(3.33)}$$

where $\gamma_{GR}(\vec{L})$ = the general-relative variogram at a lag distance $\vec{L}$, $\gamma(\vec{L})$ = the conventional variogram, and $\bar{x}_{\vec{L}}$ = the lag mean for a particular distance. Normalizing the variogram with a local mean eliminates the influence of the variations in lag mean. The denominator uses the square of a lag mean instead of a lag mean because the units of the square of the lag mean and $\gamma(\vec{L})$ are the same. As a result, $\gamma_{GR}(\vec{L})$ is a dimensionless or normalized variogram.

Pairwise-Relative Variogram. The pairwise-relative variogram is defined as[6]

$$\gamma_{PR}(\vec{L}) = \frac{1}{2n(\vec{L})} \sum_{i=1}^{n(\vec{L})} \frac{\left[x(\vec{u}_i) - x(\vec{u}_i + \vec{L})\right]^2}{\left[\frac{x(\vec{u}_i) + x(\vec{u}_i + \vec{L})}{2}\right]^2}, \quad \ldots\ldots (3.34)$$

where $\gamma_{PR}(\vec{L})$ = the pairwise-relative variogram and the other symbols have been defined previously. The pairwise-relative variogram also normalizes the conventional variogram with a squared mean. The only difference between the general-relative and the pairwise-relative variograms is that, in the pairwise-relative variogram, each pair difference squared is normalized with respect to the square of the pair mean rather than to the square of the lag mean, which is done in the general-relative variogram.

Neither the general-relative variogram nor the pairwise-relative variogram has the same theoretical foundation as the conventional variogram. However, they have been shown to provide better results in practical applications.[12]

Nonergodic Variogram. Isaaks and Srivastava[13] first introduced the concept of nonergodic variograms. Before we discuss the nonergodic variogram, we must define ergodicity. According to Christoakos,[14] a random function is called ergodic if "its mean and/or covariance (variogram) coincide with the corresponding spatial averages calculated over the single available realization." In practice, we have only one possible realization of the true distribution: the sample value at each location. We use these sample points to estimate the mean, variograms, and covariances. We assume that each variogram or covariance estimation is based on the sample points distributed over the entire region of interest. No matter at what lag distance we are computing the variogram, the ergodicity assumption requires that samples be collected from the entire region so that the spatial averages of the sample coincide with the true parameters. However, as discussed earlier, the sample sets used for estimating the variogram at each lag distance are different. For every sample point within the region of interest, we do not have a corresponding data point(s) that can form a pair with that sample point at every lag distance. Depending on the availability of the samples, only selected data points are used to estimate the variogram at a particular lag distance. We use a different realization of sample points for estimating the variogram at each lag distance. This represents nonergodic behavior.

To account for this nonergodic behavior, Isaaks and Srivastava[13] defined a nonergodic covariance as

$$c_{NE}(\vec{L}) = \frac{1}{n(\vec{L})} \sum_{i=1}^{n(\vec{L})} x(\vec{u}_i) + x(\vec{u}_i + \vec{L})$$
$$- \frac{1}{n(\vec{L})} \sum_{i=1}^{n(\vec{L})} x(\vec{u}_i) \frac{1}{n(\vec{L})} \sum_{i=1}^{n(\vec{L})} x(\vec{u}_i + \vec{L}). \quad \ldots\ldots (3.35)$$

Substituting Eqs. 3.26 and 3.27, we can write Eq. 3.35 as

$$c_{NE}(\vec{L}) = \frac{1}{n(\vec{L})} \sum_{i=1}^{n(\vec{L})} x(\vec{u}_i) x(\vec{u}_i + \vec{L}) - \bar{x}_{-\vec{L}} \bar{x}_{+\vec{L}}, \quad \ldots\ldots (3.36)$$

where $\bar{x}_{-\vec{L}} \bar{x}_{+\vec{L}}$ = the lag means. This nonergodic covariance differs from the conventional covariance (defined in Eq. 3.9). In Eq. 3.9, we defined covariance as

$$c(\vec{L}) = \frac{1}{n(\vec{L})} \sum_{i=1}^{n(\vec{L})} x(\vec{u}_i) x(\vec{u}_i + \vec{L}) - \bar{x}^2. \quad \ldots\ldots (3.9)$$

Comparison of Eqs. 3.9 and 3.36 shows that the difference in the two equations is in the use of the means on the right side of the equations. The nonergodic covariance accounts for the variations in the lag means, while the conventional definition uses the sample mean. Obviously, with a large number of samples so that the lag mean approximately coincides with the sample mean for every lag distance, Eqs. 3.9 and 3.36 do not differ. However, if the sampling is biased so that we selectively choose pairs from different regions at different lag distances, the nonergodic covariance differs from the conventional definition of covariance.

The nonergodic covariance can be used as the measure of spatial continuity, but conventional geostatistics use the variogram. We can define the nonergodic variogram as

$$\hat{\gamma}_{NE}(\vec{L}) = c(0) - c_{NE}(\vec{L}), \quad \ldots\ldots (3.37)$$

where $\hat{\gamma}_{NE}(\vec{L})$ = the nonergodic variogram and $c(0)$ = sample variance. Eq. 3.37 is the same as Eq. 3.19. Appendix C provides additional details.

Nonergodic Correlogram. The concept of a nonergodic variogram can be taken one step further by defining a normalized covariance, which accounts for the variations in both the lag means as well as the lag variances. We can define the nonergodic correlogram as[15]

$$r_{NE}(\vec{L}) = \frac{c_{NE}(\vec{L})}{s_{-\vec{L}} s_{+\vec{L}}}, \quad \ldots\ldots (3.38)$$

where $c_{NE}(\vec{L})$ = a nonergodic covariance defined in Eq. 3.36; the lag variances are defined in Eqs. 3.31 and 3.32, respectively. This definition differs from that used in Eq. 3.14.

$$r(\vec{L}) = \frac{c(\vec{L})}{c(0)}. \quad \ldots\ldots (3.14)$$

The denominators of Eqs. 3.14 and 3.38 are identical only if the lag variances are the same as the sample variance. If the lag variances reflect the variations in samples used for a particular lag distance, the nonergodic correlogram will differ from the conventional definition of the correlogram. As before, to follow the convention used in geostatistics, we use $[1 - r_{NE}(\vec{L})]$ as our measure of spatial continuity because it behaves in the same way as the variogram.

All four measures discussed in this section improve the interpretation of spatial continuity if there is a significant difference between lag properties and sample properties. Field example 3.9 illustrates this.

Field Example 3.9. This field example uses the three data sets used in the previous field examples: porosity data for Flow

TABLE 3.7—COMPARISON OF LAG MEANS AND VARIANCES, FIELD EXAMPLE 3.9

Porosity for Flow Unit 3			IP, B/D			Net kh, md-ft		
Lag (ft)	Lag Mean	Lag Variance	Lag (ft)	Lag Mean	Lag Variance	Lag (ft)	Lag Mean	Lag Variance
0.00	21.62	23.11	0.00	763.17	5.02×10^5	0.00	6,818.24	8.18×10^7
388.66	22.31	15.43	526.27	1,070.00	1.66×10^6	526.27	16,140.00	3.24×10^8
1,568.53	22.52	18.36	1,623.52	970.97	6.06×10^5	1,623.52	7,926.29	9.18×10^7
2,828.80	22.23	19.34	2,811.62	794.98	4.66×10^5	2,811.62	7,674.42	8.36×10^7
4,211.03	22.40	20.55	4,262.66	779.79	5.49×10^5	4,262.66	6,877.47	7.16×10^7
5,592.38	21.57	23.28	5,546.14	764.16	4.95×10^5	5,546.14	7,026.20	7.63×10^7
6,955.44	21.27	24.50	6,962.50	818.51	5.14×10^5	6,962.50	7,279.39	7.42×10^7
8,369.49	20.61	24.32	8,307.26	738.36	4.87×10^5	8,307.26	5,770.67	5.78×10^7
9,798.59	20.51	23.87	9,749.58	764.31	5.39×10^5	9,749.58	5,838.94	6.28×10^7
11,189.22	20.82	23.79	11,087.19	620.15	4.00×10^5	11,087.19	5,237.65	4.91×10^7
12,565.65	21.38	19.95	12,453.84	613.57	3.87×10^5	12,453.84	7,097.74	7.57×10^7
13,971.90	22.28	19.06	13,895.77	634.09	3.30×10^5	13,895.77	6,817.49	8.45×10^7

Unit 3, IP data, and net-kh data. **Table 3.7** shows how the lag means and lag variances change at different lag distances for these three data sets. The lag variance is the variance calculated on the basis of all the data points used for estimating the variogram at a particular lag distance. By definition, it is calculated as

$$s_{\vec{L}}^2 = \frac{1}{2n(\vec{L})}\sum_{i=1}^{n(\vec{L})}\left[x^2(\vec{u}_i)x^2(\vec{u}_i + \vec{L})\right] - \left[\frac{1}{2n(\vec{L})}\sum_{i=1}^{n(\vec{L})}\bar{x}_h(\vec{u}_i) + \bar{x}_h(\vec{u}_i + \vec{L})\right]^2. \quad \ldots\ldots. (3.39)$$

The lag mean, $\bar{x}_{\vec{L}}$, is calculated with Eq. 3.28.

For all three variables, the sample mean and the sample variance are shown at a lag distance of 0 ft. For porosity data, for example, the sample mean is 21.62 and the sample variance is 23.11.

The porosity data do not exhibit significant spatial variability. This is also reflected in the lag means and lag variances. The lag mean is fairly constant for all lag distances; the variation within the values is small. The lag variance shows some variations at smaller lag distances; however, the value is stabilized at large lag distances. In contrast, for both the IP and net-kh data, the lag means are higher than the sample mean at smaller lag distances. As the lag distance increases, the lag means are either equal to or slightly less than the sample mean. This type of behavior is to be expected because more wells are drilled in regions with a high potential, which also are those regions with higher net-kh values. Therefore, at shorter distances, the pairs used for variogram estimation reflect relatively high values. The lag variances are also higher at smaller lag distances; however, these values decrease or stabilize as lag distance increases. This may be an indication of what the mining literature calls the "proportional effect," an effect that reflects an increase in the local variance as the local mean increases. In other words, high values tend to exhibit more variability than low values.

Knowing the behavior of lag means and lag variances, we should expect that the modifications suggested in this section may not significantly impact the porosity variogram; whereas, for the IP and net-kh data, we should observe some improvement in estimation of the variogram.

We first examine the effect of the modifications on the porosity variogram. **Fig. 3.26** shows the use of a general-relative variogram for isotropic data; the conventional variogram is included for comparison. As expected, no significant improvement is observed in the variogram with the general-relative variogram. Similar results can be noted for the pairwise-

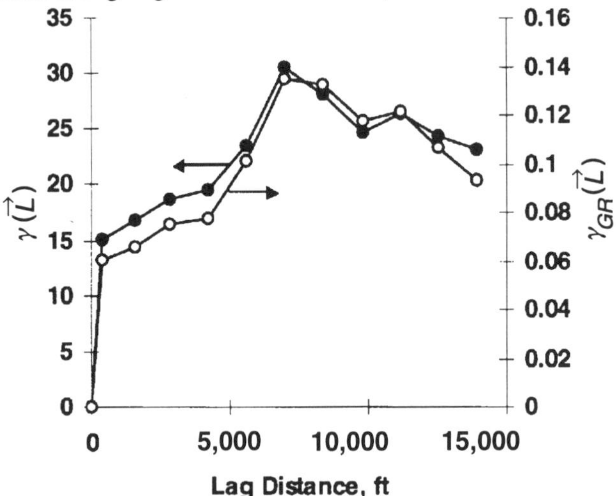

Fig. 3.26—General-relative variogram for porosity data.

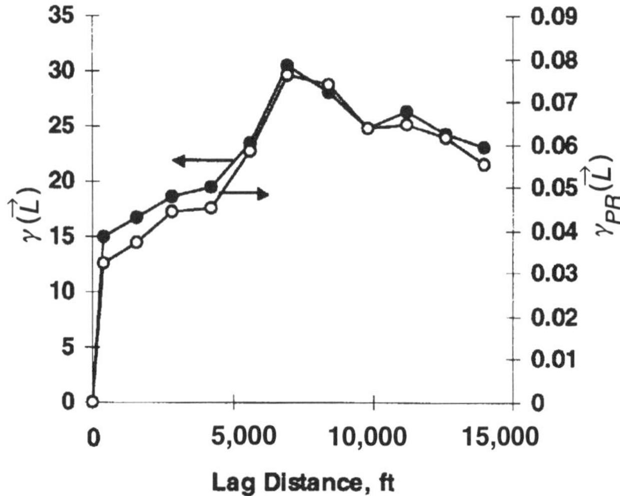

Fig. 3.27—Pairwise-relative variogram for porosity data.

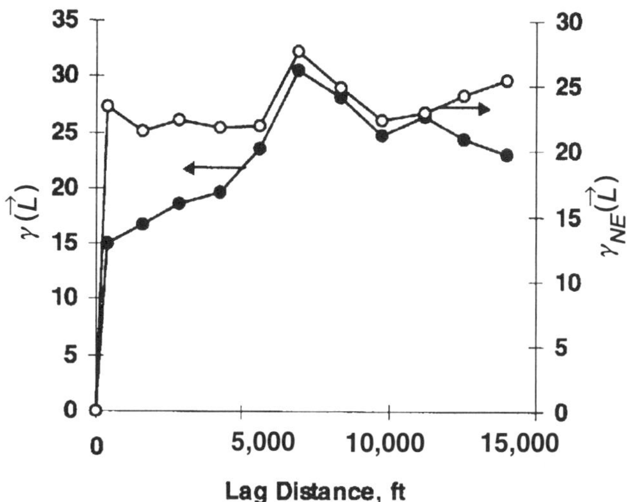

Fig. 3.28—Nonergodic variogram for porosity data.

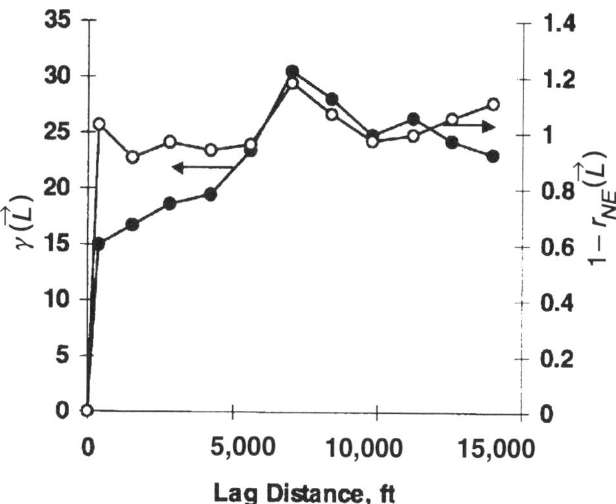

Fig. 3.29—Nonergodic correlogram for porosity data.

relative variogram in **Fig. 3.27**. **Figs. 3.28 and 3.29** show the nonergodic variogram and correlogram plots, respectively. The variograms estimated with both these modifications show essentially a total lack of spatial relationship. For both these modifications, the variogram rises quickly to a sill value, indicating a total lack of spatial continuity. Because the estimated variogram that provides the best interpretable structure should be used, we should use the conventional variogram for these data.

Fig. 3.30 shows the effect of the modified variograms for the IP data. To eliminate clutter, the generalized-relative variogram is not shown, but its behavior is very similar to that of the pairwise-relative variogram. The conventional variogram exhibits the most fluctuation. At large distances, instead of reaching a stable value, there is considerable fluctuation in the variogram estimation. The pairwise-relative variogram shows fairly smooth variations in the estimated variogram, which increases gradually and reaches essentially a constant value at large distances. The nonergodic variogram and nonergodic correlogram also show less fluctuation and better continuity in the estimated variogram. Overall, all four modifications clearly improve the variogram estimation to define the spatial structure better. Comparison of the different modifications shows that the pairwise-relative variogram provides the clearest interpretable structure.

A similar conclusion can be reached for the net-*kh* data as well. The conventional variogram fails to exhibit any spatial structure (**Fig. 3.31**). Any structure is masked by the fluctuations in the estimated variogram values. In contrast, the pairwise-relative variogram and the nonergodic correlogram exhibit a smooth change with a clearly interpretable structure. The nonergodic variogram shows some improvement over the conventional variogram; however, it also shows some fluctuation in the estimated values.

Overall, the modifications suggested in this section could improve our ability to describe the spatial relationship. This is especially true where the sampling of the data is biased, which results in variations in lag means and lag variances. In studying the effect of these various modifications, the modification that provides the most interpretable structure should be favored.

3.5 Modeling Variograms

Sec. 3.4 considered various methods and modifications used to estimate variograms. This section discusses methods used to model these variograms. Estimation of variograms allows us to know variogram values only at the lag distances at which we computed the variogram. The primary purpose, however, in estimating the variogram is to use this information to esti-

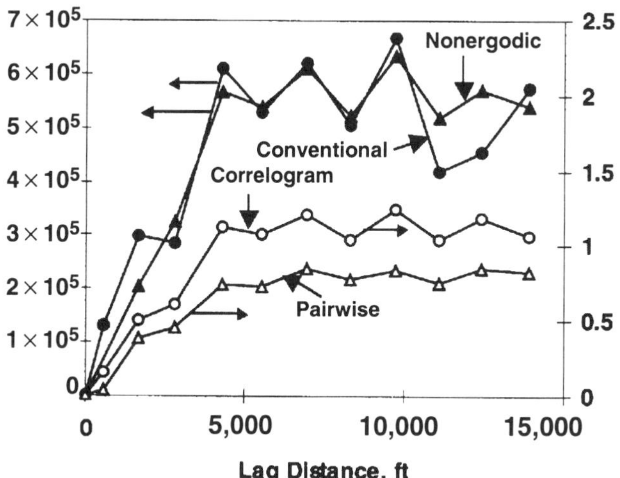

Fig. 3.30—Effect of modified methods on estimated IP variogram.

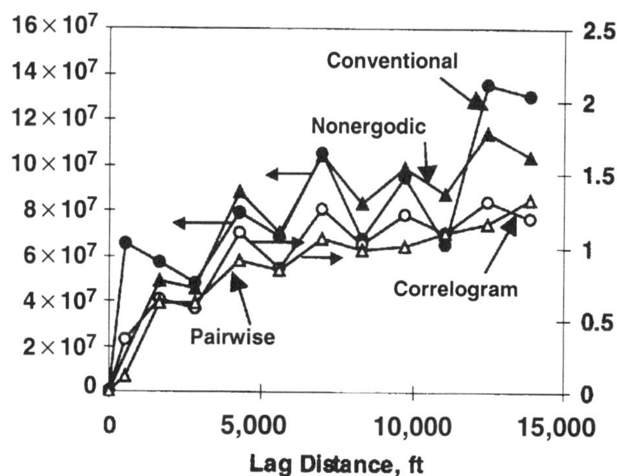

Fig. 3.31—Effect of modified methods on estimated net-*kh* variogram.

mate variable values at unsampled locations. The information about the spatial relationship is used to estimate those values. When estimating, it is not necessary that the distances between the samples and the unsampled locations coincide with the distances at which we estimated the variogram. In fact, the distances rarely coincide with the distances of the estimated variogram. That means we will have to determine the value of the variograms at distances between the lag distances at which we have estimated the variogram. One option is simply to interpolate between the estimated values so that we know the variogram value at any lag distance. Unfortunately, such simple approximation may create problems in the estimation procedure. Therefore, we have to restrict ourselves to certain types of models that can be used to model the variogram.

This section is divided into five subsections. In Sec. 3.5.1, we discuss the requirements that have to be honored in modeling variograms. Sec. 3.5.2 covers the types of models that can be used to model variograms that reach a constant value (models with a sill). For these models to fit, the estimated variogram should increase with distance at smaller lag distances and reach a constant value at a certain distance called the range. We examine the applications of these models through field examples.

In Sec. 3.5.3, we consider the modeling variograms that do not reach a constant value within a region of interest. These variograms show increases in value as lag distance increases. Certain types of models are adequately suited to capture these variograms. We discuss the applications of these models through examples.

Sec. 3.5.4 addresses modeling variograms with "hole effects," a common effect observed when modeling variograms in the vertical direction. These variograms show periodic behavior that represents cyclical geological processes. We study two types of models that allow modeling of hole effects.

In Sec. 3.5.5, we examine modeling of anisotropic variograms. Certain additional constraints are imposed in modeling the anisotropic variogram, including the requirement that the directions of maximum and minimum continuities be perpendicular to each other. We explain the procedure used in modeling anisotropic variograms and provide examples.

In studying this section, it should be understood that variogram modeling is a time-consuming exercise. The job can be made easier through interactive software, which allows changing the parameters of the model with ease. Modeling variograms correctly requires practice. We highly recommend the use of interactive software for a better modeling exercise. One such public-domain software, GeoEAS,[16] allows interactive modeling of variograms in two dimensions.

3.5.1 Modeling Requirements. In modeling the estimated variogram, we must consider two requirements. The first is the use of a minimum number of parameters and models to model the variogram. We need to capture all the essential features observed in an estimated variogram, some of which have physical relevance and need to be honored. However, the model does not have to pass through every estimated variogram value. We need to decide which spatial features of the estimated variogram are important, then try to capture those features with as few parameters as possible. For example, if an estimated variogram can be reasonably honored with either one type of model or a linear combination of two models, modeling with only one type is preferred over modeling with the combination. Using as few parameters as possible makes it easier to model the anisotropic variograms, as well as cross variograms (discussed in Sec. 3.4.2). Furthermore, the estimation process requires less computational effort with a simpler variogram model.

The second requirement is the condition of positive definiteness.[17] Any model used to model the variogram (or covariance) should satisfy the requirement of the condition of positive definiteness, which ensures that a unique solution is obtained during the estimation process. Recall from Chap. 2 that, in the estimation process, we assume that the variable at the unsampled location is linearly related to the surrounding samples. Mathematically, we can write

$$x^*(\vec{u}_o) = \sum_{i=1}^{n} \lambda_i x(\vec{u}_i), \quad \text{.......... (3.40)}$$

where $x^*(\vec{u}_o)$ = the estimated value at the unsampled location, $\vec{u}_o$, $x(\vec{u}_o)$ = the sample value at location $\vec{u}_i$, and λ_i = the weight assigned to sample $x(\vec{u}_i)$. In other words, we assume that the value at the unsampled location is a weighted linear combination of the nearby n samples. Because the estimated value is a random variable with associated uncertainties, we can also estimate the variance associated with it. Taking the variance on both sides of Eq. 3.40, we can write

$$V[x^*(\vec{u}_o)] = V\left[\sum_{i=1}^{n} \lambda_i x(\vec{u}_i)\right]. \quad \text{.......... (3.41)}$$

Expanding the right side of Eq. 3.41 gives

$$V\left[\sum_{i=1}^{n} \lambda_i x(\vec{u}_i)\right] = \sum_{j=1}^{n}\sum_{i=1}^{n} \lambda_i \lambda_j C(\vec{u}_i, \vec{u}_j), \quad \text{.......... (3.42)}$$

where $C(\vec{u}_i, \vec{u}_j)$ = the covariance between the two sample points located at $\vec{u}_i$ and $\vec{u}_j$. Appendix B provides the details of the expansion. We know that variance must always be positive; i.e., the left and right sides of Eq. 3.42 must be positive.

$$\sum_{i=1}^{n}\sum_{j=1}^{n} \lambda_i \lambda_j C(\vec{u}_i, \vec{u}_j) \geqq 0. \quad \text{.......... (3.43)}$$

The left side of Eq. 3.43 is zero only if all the weights, λ_i, are zero; otherwise, it should always be positive. This is the condition of positive definiteness. The model should be chosen so that Eq. 3.42 is always satisfied.

In cases where we need to express the equation in terms of a variogram model, we can write

$$-\sum_{i=1}^{n}\sum_{j=1}^{n} \lambda_i \lambda_j \gamma(\vec{u}_i, \vec{u}_j) \geqq 0, \quad \text{.......... (3.44)}$$

where all the weights must add to zero, or

$$\sum_{i=1}^{n} \lambda_i = 0. \quad \text{.......... (3.45)}$$

Eq. 3.44 requires that the variogram model be conditionally positive definite (i.e., conditional to the weights summing to zero). In contrast, Eq. 3.43 must be satisfied for any weights. Restricting the type of weights that can be used to satisfy Eq. 3.44 imposes fewer restrictions on variogram models compared with covariance models; therefore, more allowable variogram models are available.

To prove that a particular type of model satisfies the positive definiteness condition is tedious and requires the use of Fourier transforms. Christoakos[14,18] provides different criteria of permissibility to satisfy the requirement of positive definiteness.

Fortunately, several models have already been tested for the condition of positive definiteness. In addition, if certain models satisfy the requirement of the condition of positive definiteness, any linear combination of these models can also be used as an acceptable model. For example, if we write

$$C(\vec{L}) = \sum_{i=1}^{N} a_i C_i(\vec{L}) \quad \text{.................... (3.46a)}$$

$$\text{or } \gamma(\vec{L}) = \sum_{i=1}^{N} a_i \gamma_i(\vec{L}), \quad \text{.................... (3.46b)}$$

$C(\vec{L})$ and $\gamma(\vec{L})$ are acceptable models if $C_i(\vec{L})$ and $\gamma_i(\vec{L})$ satisfy the requirement of the condition of positive definiteness and a_i are positive. Furthermore, any product of acceptable covariance models is also an acceptable covariance model. That is,

$$C(L) = \prod_{i=1}^{N} a_i C_i(\vec{L}) \quad \text{..................... (3.47)}$$

is an acceptable covariance model, where $\prod$ = the product, a_i = the coefficient of Model C, and $C_i(\vec{L})$ = a covariance model that must satisfy the condition of positive definiteness. As subsequent sections discuss, we can model essentially any observed variogram with available covariance and variogram models using these types of combinations. In practice, we do not need to invent a new model to fit an estimated variogram.

In modeling the variogram, one additional restriction must be kept in mind. The condition of positive definiteness can be satisfied in different dimensions. Certain models satisfy the requirement in only one dimension, while other models satisfy the requirement in three dimensions. A model that satisfies the requirement in higher dimensions also satisfies the positive-definiteness requirement in a lower dimension. That is, a model that satisfies the requirement in three dimensions can be used to model in one or two dimensions. The reverse is, however, not true. Most problems that are of interest in reservoir characterization require a model that can be applied in three dimensions. Therefore, in subsequent sections we present models that satisfy the condition of positive definiteness in three dimensions. If a particular model does not satisfy the requirement in three dimensions, we explicitly state it. Otherwise, it can be assumed that the models presented can be applied in three dimensions.

3.5.2 Models With Sill. This section covers four models that are most commonly used in modeling the estimated variograms. All four are applicable in three dimensions.[17]

Nugget-Effect Model. The simplest model is the nugget-effect model. In practice, it can be written as

$$\gamma(\vec{L}) = 0, \quad \text{............................ (3.48a)}$$

if $L = 0$,

$$\text{and } \gamma(\vec{L}) = C_0, \quad \text{......................... (3.48b)}$$

if $L > 0$, where C_0 = the sill value. The variogram by definition has to be zero at a lag distance equal to zero. The nugget effect assumes that the variogram abruptly increases to a value of C_0 for any lag distance greater than zero. Beyond that point, it remains constant. Knowing the relationship between the variogram and the covariance,

$$\gamma(\vec{L}) = C(0) - C(\vec{L}), \quad \text{.................... (3.19)}$$

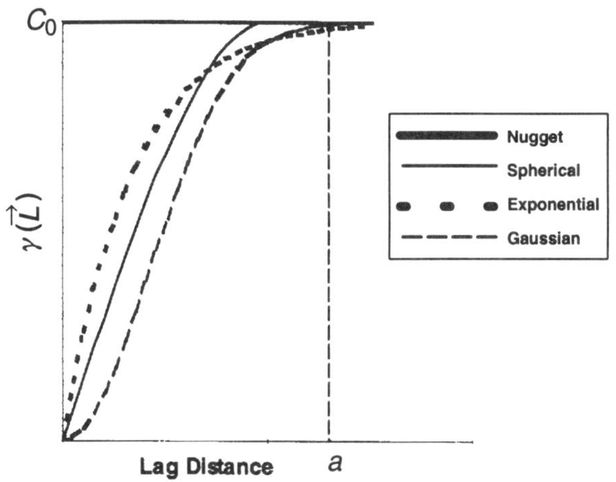

Fig. 3.32—Variogram models with sill.

we can write the corresponding covariance equation as

$$C(\vec{L}) = C_0, \quad \text{........................... (3.49a)}$$

if $L = 0$,

$$\text{and } C(\vec{L}) = 0, \quad \text{......................... (3.49b)}$$

if $L > 0$. Note that the sill value is equivalent to $C(0)$.

The nugget effect indicates a total lack of information with respect to spatial relationship. If a particular estimated variogram is observed to present a pure-nugget effect, it is an indication that no quantitative information is available about the spatial relationship about that variable. There may be two reasons for this. First, the shortest distance at which we have sample pairs may be greater than the range of the variogram. Therefore, we do not have data pairs that exhibit some type of correlation. Second, the measurement error can add uncertainty in the estimation, which is reflected in a nugget value. In practice, it may be hard to determine which reason is responsible for the nugget effect. What is important is the fact that a high value of nugget indicates a lack of information about the spatial relationship. **Fig. 3.32** shows the nugget-effect model and other sill models. As explained, the nugget effect is a horizontal line with a constant sill value of C_0. **Fig. 3.33** shows an equivalent covariance plot for the nugget-effect model and the other models. Zero value or a lack of correlation indicates the nugget effect.

Spherical Model. The spherical model is probably the most commonly used model to demonstrate variograms with a sill. The equation for the spherical variogram can be written as

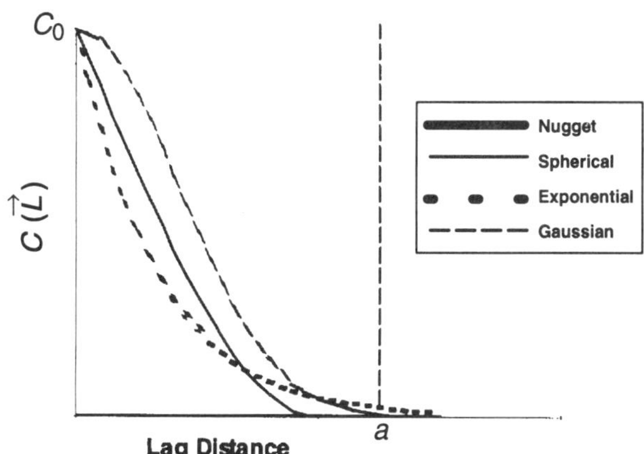

Fig. 3.33—Covariance models with sill.

$$M_{S_a}(L) = \gamma(L) = C_0\left[\frac{3}{2}\left(\frac{L}{a}\right) - \frac{1}{2}\left(\frac{L}{a}\right)^3\right], \quad \ldots\ldots (3.50a)$$

if $L \leqq a$,

and $M_{S_a}(L) = C_0$, $\quad\ldots\ldots\ldots\ldots\ldots\ldots\ldots\ldots$ (3.50b)

if $L \geqq a$, where $M_{S_a}(L)$ = a spherical model with a range a. An equivalent covariance model can be written as

$$C(L) = C_0\left[1 - \frac{3}{2}\left(\frac{L}{a}\right) + \frac{1}{2}\left(\frac{L}{a}\right)^3\right],$$

if $L \leqq a$,

and $C(L) = 0$, $\quad\ldots\ldots\ldots\ldots\ldots\ldots\ldots\ldots\ldots$ (3.51)

if $L \geqq a$.

The spherical model is defined with a range of a and a sill value of C_0. The variogram reaches a sill value at a lag distance of a. Figs. 3.32 and 3.33, respectively, show the spherical variogram and the associated covariance model. Note that, of all the models, the spherical model has the highest slope at the origin. Therefore, if an estimated variogram rises rapidly for a given range, the spherical model is a good option to try to model the variogram.

Exponential Model. The equation for the exponential model can be written as

$$M_{E_a}(L) = \gamma(L) = C_0\left[1 - \exp\left(\frac{-3L}{a}\right)\right], \quad \ldots\ldots (3.52)$$

if $L \geqq 0$, where a = effective range of the variogram and $M_{E_a}(L)$ = exponential model with a range a. The corresponding covariance model can be written as

$$C(L) = C_0 \exp\left(\frac{-3L}{a}\right), \quad \ldots\ldots\ldots\ldots\ldots (3.53)$$

if $L \geqq 0$. For the exponential model, the variogram reaches the sill value only asymptotically. Therefore, for practical purposes, a range is defined as the lag distance where the variogram reaches 95% of the sill value. Substituting $\vec{L} = a$ into Eq. 3.52 gives

$$\gamma(L) = C_0(1 - e^{-3}) = 0.95C_0 \approx C_0,$$

and we obtain $\gamma(L)$ equal to 95% of the sill value. Figs. 3.32 and 3.33, respectively, show the variogram and corresponding covariance models. As can be seen, for the same range, the slope at the origin for the exponential model is smaller than that of the spherical model, indicating a much more gradual change in the variogram value.

Gaussian Model. The variogram equation for the Gaussian model can be written as

$$M_{G_a}(L) = \gamma(L) = C_0\left[1 - \exp\left(-3\frac{L^2}{a^2}\right)\right], \quad \ldots\ldots (3.54)$$

if $L \geqq 0$, where $M_{G_a}(L)$ = a Gaussian model with a range a. The corresponding covariance model can be written as

$$C(L) = C_0 \exp\left(-3\frac{L^2}{a^2}\right), \quad \ldots\ldots\ldots\ldots\ldots (3.55)$$

if $L \geqq 0$. For the Gaussian model, the slope at the origin is zero, indicating an extremely smooth variation in properties as a function of distance. The effective range for the Gaussian model is a, which represents the lag distance at which the variogram reaches 95% of the sill value. Figs. 3.32 and 3.33, respectively, show the Gaussian variogram model and the associated covariance model. The figures show that the variogram changes very gradually at the origin and exhibits an S-shaped curve before reaching a sill value.

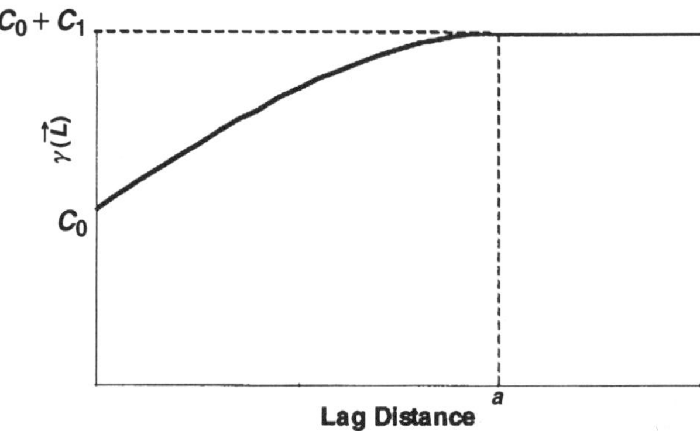

Fig. 3.34—Linear combination of nugget and spherical model.

Combination Models. Using any of the four models described, we can create multiple linear combinations of these models to model a given estimated variogram. Most commonly, the nugget-effect model becomes an inherent part of the variogram model. For example, we can write a variogram model as

$$\gamma(L) = C_0 + C_1 M_{S_a}(L), \quad \ldots\ldots\ldots\ldots\ldots (3.56)$$

where C_0 = the sill value of the nugget effect, C_1 = the sill value of a spherical model, and a = the range of a spherical model. **Fig. 3.34** shows this model schematically. The variogram starts at a value of C_0, representing a certain lack of information about the spatial relationship. The variogram finally rises to a value of $(C_0 + C_1)$, which is a total sill of the variogram. C_1 is the contribution of the spherical variogram to the total sill.

Similar to Eq. 3.56, we can describe an estimated variogram with any linear combination. For example, we can write

$$\gamma(L) = C_0 + C_1 M_{E_{a_1}}(L) + C_2 M_{G_{a_2}}(L), \quad \ldots\ldots (3.57)$$

where the variogram is represented by a linear combination of three models: nugget, exponential, and Gaussian. The total sill of the variogram is $C_0 + C_1 + C_2$, and the range of the exponential model is a_1. In general, we can write

$$\gamma(L) = \sum_{i=1}^{n} C_i \gamma_{i_{a_i}}(L), \quad \ldots\ldots\ldots\ldots\ldots (3.58)$$

where $\gamma_{i_{a_i}}(L)$ = a variogram model with a range a_i, and C_i = the associated sill value. This type of model is also called a nested model, which represents a composite structure that combines multiple models. In practice, more than four models are rarely used in a linear combination.

Field Example 3.10. This example illustrates modeling of the estimated variograms for the porosity data of Flow Unit 3, the IP data, and the net-*kh* data. For simplicity, we model only isotropic or omnidirectional variograms. Modeling of anisotropic variograms is discussed later.

Fig. 3.35 compares the estimated variogram for Flow Unit 3 porosity data and the model. The equation for the model is

$$\gamma(L) = 15 + 10 M_{G_{8,000}}(L).$$

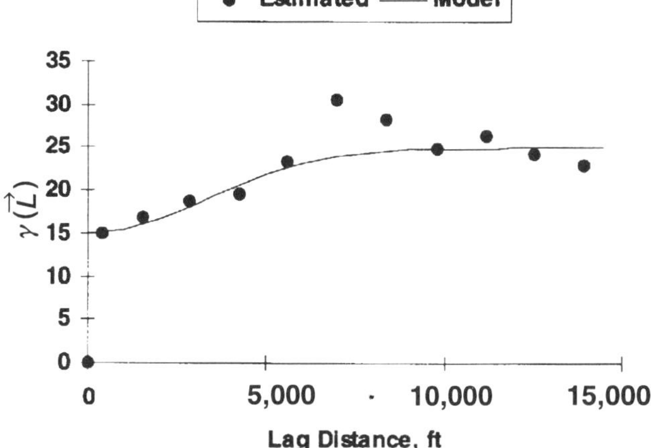

Fig. 3.35—Estimated and model variograms for Flow Unit 3 porosity data.

The estimated variogram is a conventional variogram without modifications. Recall that most of the modifications discussed in the previous section did not significantly improve the estimated variogram for porosity. The match between the variogram and the model is reasonable. Obviously, all the observations do not fall on the model variogram. However, the basic trend is captured.

Fig. 3.36 compares the estimated variogram for the log of IP and two possible models. Recall that the log of IP did result in a better interpretable variogram structure; therefore, we used the log of IP. As an alternative, we could have used the normal-score transform or pairwise-relative variogram for modeling purposes. Both these modifications resulted in improvements of the structure. We illustrate only the application of the log of IP data. Similar methods can be used for other modifications as well. The two models in Fig. 3.36 are

$$\gamma(L) = 0.06 + 0.3M_{G_{8,000}}(L)$$

and $\gamma(L) = 0.36M_{S_{7,400}}(L)$.

Both are equally applicable. The first estimated value at a lag distance of 389 ft can be ignored because it contains only two pairs. The last estimated value is at a lag distance beyond half the maximum distance; therefore, we can also ignore that value. Both models adequately capture the overall spatial trend.

Having alternative models that can capture the spatial trend raises the question of which to use in the estimation process. The answer is whichever is more "robust."[19] Robustness should be defined in terms of the model's relative insensitivity to perturbations with respect to sample data as well as the model's underlying assumptions.

As noted previously, both spherical and Gaussian models satisfy the requirement of the condition of positive definiteness; however, they are not equally robust. It has been shown that the Gaussian model, especially with a zero-nugget effect, is much more sensitive to data variations than either the spherical or exponential model.[20] While the Gaussian model can be made more robust by adding the nugget effect to it, it still is less robust than the spherical or exponential model. As a result, in the absence of compelling external physical reasons to use the Gaussian model (i.e., the presence of extreme continuity at short distances), it is better to choose an alternative model. If the two alternative models are either spherical or exponential, the choice of one over the other is not that critical.[21]

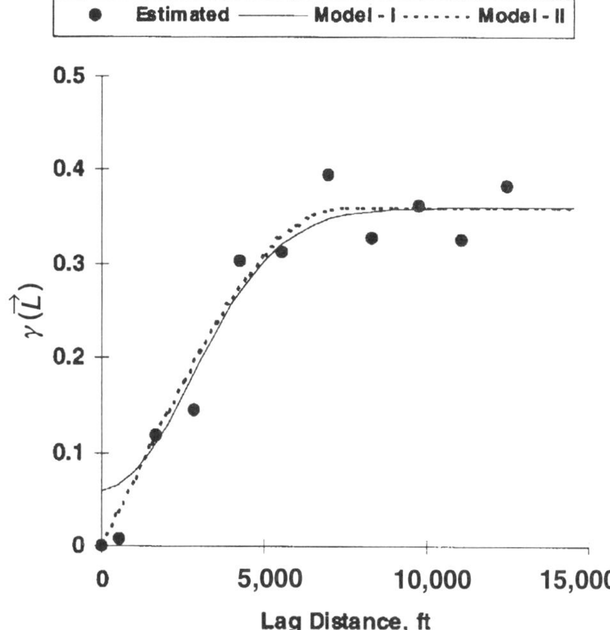

Fig. 3.36—Estimated and model variograms for log IP data.

Fig. 3.37 illustrates the modeling of the log of net *kh* with two alternative models. The log of net *kh* was used for better interpretability of the structure. As stated before, we can also use the normal-score transform or the pairwise-relative variogram to capture the overall spatial relationship. The models used are

$$\gamma(L) = 0.15 + 0.58M_{G_{7,400}}(L)$$

and $\gamma(L) = 0.72M_{S_{8,000}}(L)$.

As in the case of IP data, the estimated variogram at the lag distance of 389 ft can be ignored because there are only two pairs at that lag distance. Both models adequately capture the spatial trend observed by the estimated variogram. For the

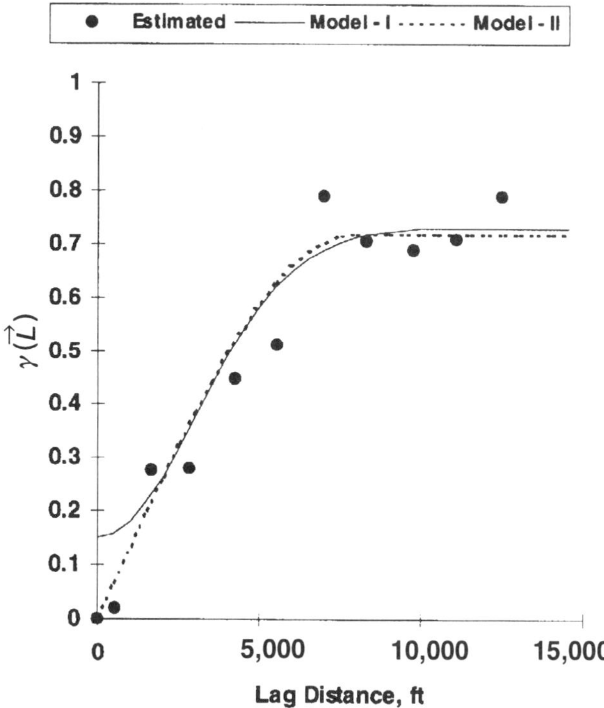

Fig. 3.37—Estimated and model variograms for log net-*kh* data.

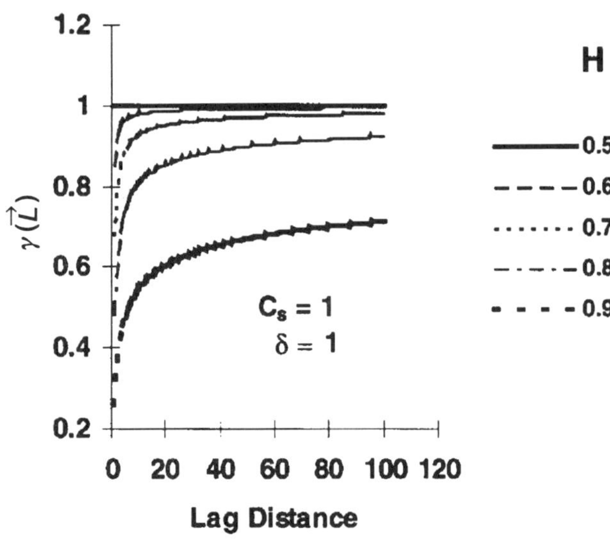

Fig. 3.38—Effect of intermittency exponent on f_{Gn} model.

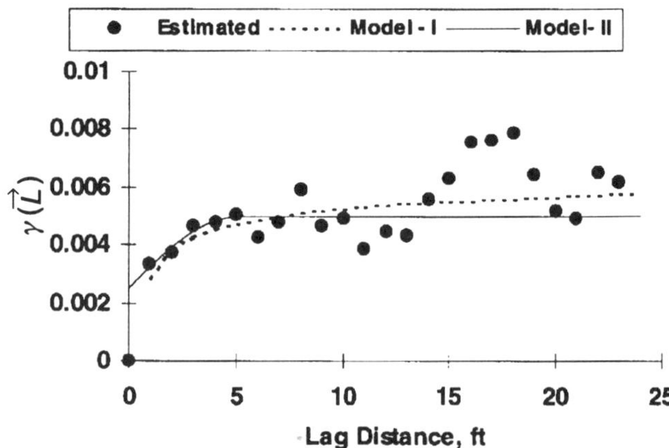

Fig. 3.39—Estimated and model variograms for porosity data from Well 36-W27.

reasons given earlier, the spherical model is preferred over the Gaussian model.

3.5.3 Models Without a Sill. Models without a sill are used to model variograms that continually increase as lag distance increases. These variograms may eventually reach a sill value; but, within the region of interest, they exhibit a continuing increase in value as the lag distance is increased. Three types of models are used most commonly to model these types of variograms. The first two, which have recently become popular,[22] are fractal models. The third, the logarithmic model, is popular in the mining industry but has rarely been used to describe the variogram of reservoir properties.

Fractional Gaussian Noise, f_{Gn}, Model. The f_{Gn} model, M_{fGn}, was first made popular by Hewett,[22] who demonstrated that it could represent well-log data. The f_{Gn} model is a fractal type of model, and the variogram equation can be written as

$$\gamma(L) = \frac{1}{2}C_s\delta^{2H-2}$$
$$\times \left[2 - \left(\frac{|L|}{\delta} + 1\right)^{2H} + 2\left|\frac{L}{\delta}\right|^{2H} - \left(\frac{|L|}{\delta} - 1\right)^{2H}\right], \quad \text{(3.59)}$$

where C_s = a scaling parameter that is different from the sill value, H = the intermittency exponent, which can take a value between zero and one excluding the limits (to ensure that the requirement of the condition of positive definiteness is satisfied), and δ = an averaging parameter that represents the "support" of a measurement. For example, $\delta = 3$ ft, if log data represent a resolution of 3 ft, and $\delta = 1$ ft, if core data are assumed to represent a value covering a 1-ft interval.

When an f_{Gn} model is used, sample data must be spaced uniformly. The three parameters, C_s, δ and H, need to be estimated for proper modeling. δ can be reasonably guessed. One option for estimating these three parameters is to use an interactive trial-and-error procedure to obtain the values as we did for the models with a sill. Another option is to use specialized procedures to estimate the value of H separately; these include R/S analysis,[22] spectral methods,[22] and a box-counting method.[23] All these methods require uniform spacing of the data. The box-counting method also requires that a sequence of data be graded before its use. Refs. 22 and 23 provide detailed discussions of these methods. Once H and δ (resolution of the sample) are known, the remaining unknown, C_s, can be estimated easily.

The intermittency exponent, H, represents the degree of continuity between neighboring sample values. For $H < 0.5$, the behavior is called antipersistent; i.e., a high value is more likely to be followed by a low value and vice versa. For $H > 0.5$, the behavior is called persistent; i.e., a high value is more likely to be followed by a high value. In analyzing geoscience data, neighboring values are expected to have a greater likelihood of being similar to each other than dissimilar. Therefore, H is expected to be >0.5. Experimental observations match this expectation. The analysis of the well-log data reveals that H is always >0.5, with the most likely values falling between 0.7 and 0.9. **Fig. 3.38** shows sample variograms for the f_{Gn} model with different values of H. The values of δ and C_s are fixed at 1.0. The figure shows that, for $H = 0.5$, the variogram is a pure nugget. This is equivalent to "white noise." As the value of H increases, the correlation between the neighboring values increases. At $H = 0.9$, the change in the variogram value is much more gradual.

Examination of these variograms raises another issue. When an estimated variogram is modeled, several options are available. The f_{Gn} model does have a sill value. Asymptotically, at a very large lag distance, the variogram reaches a value of $C_s\delta^{2H-2}$. However, within the region of interest, the variogram may show an increase in value for all the lag distances. When modeling a variogram, it could be modeled either with an f_{Gn} model or, possibly, with the models with sill. Both models may fit the estimated variogram equally; therefore, it is unnecessary to prefer one type of model over the other. As long as we can capture the spatial characteristics adequately, we can model it with any appropriate model. The use of these alternative models should not significantly affect estimations at unsampled locations. The following example illustrates alternative modeling.

Field Example 3.11. This field example analyzes the vertical data for Well 36-W27. Core porosity data are collected at uniform 1-ft intervals. We assume that $\delta = 1.0$ ft for the f_{Gn} model. **Fig. 3.39** shows the estimated variogram fits for the two models. Model 1 represents the f_{Gn} model with $H = 0.85$, $\delta = 1$ ft, and $C_s = 0.0075$. The value of H was obtained with the box-counting method. The fit is reasonable.

H=0.1

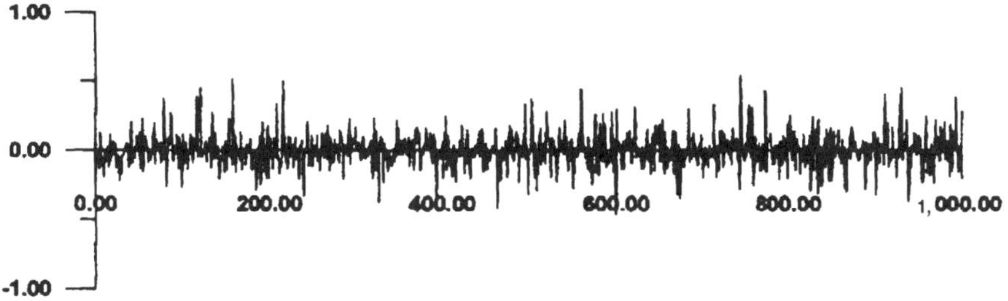

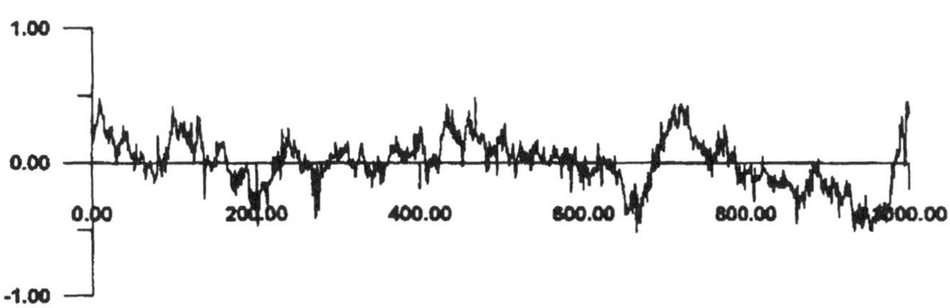

H=0.9

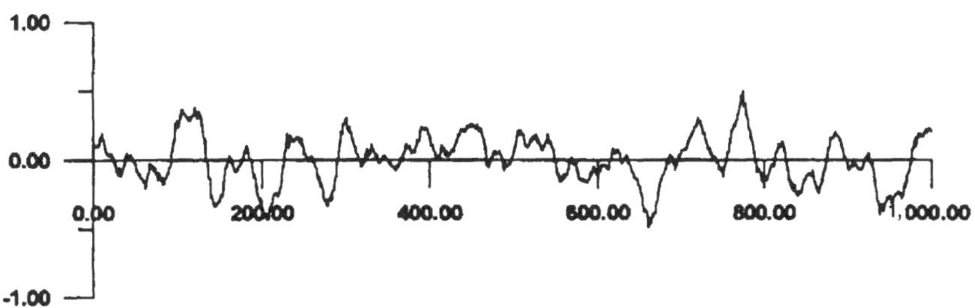

Fig. 3.40—Effect of intermittency exponent on f_{Bm} trace.

The other model, Model 2, is a spherical plus nugget model that can be written as

$$\gamma(L) = 0.0025 + 0.0025M_{S_5}(L).$$

The fit with the early lag variogram values is good. The spherical model does not capture the slightly increasing trend in the estimated variogram; however, this may not be significant.

When examining the two alternative models, it is difficult to choose one over the other. Both models can be considered to describe the spatial trend adequately. The difference between the absolute values between the two models at a given lag distance is fairly small. As a result, the use of either of the two models (one with and one without sill) should not significantly affect the estimation at the unsampled locations.

Fractional Brownian Motion, f_{Bm}, Model. The f_{Bm} model, $M_{f_{Bm}}$, has been observed to capture features of several naturally occurring objects.[24] The variogram equation for the f_{Bm} model can be written as

$$\gamma(L) = C_s L^{2H}, \qquad (3.60)$$

where C_s = a scaling parameter, L = a lag distance, and H = an intermittency exponent. H can take values between zero and one, excluding these two limits. f_{Gn} and f_{Bm} are related to each other. For a given H, f_{Bm} can be written as

$$f_{Bm} = \int f_{Gn}. \qquad (3.61)$$

That is, f_{Bm} is an integral of f_{Gn}. Because of this integration, the f_{Bm} model shows much smoother behavior than the f_{Gn} model. As in the case of the f_{Gn} model, the behavior is called antipersistent for $H < 0.5$ and persistent for $H > 0.5$. **Fig. 3.40** shows typical f_{Bm} model traces for values of H ranging from 0.1 to 0.9. The figure shows that the behavior becomes smoother as H increases. For small values of H, significant

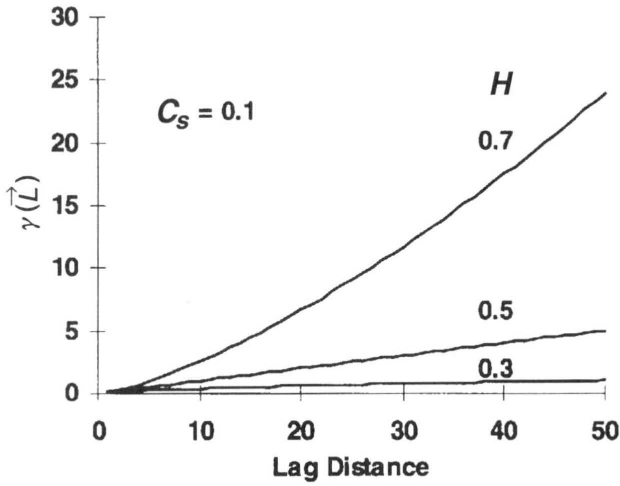

Fig. 3.41—Effect of intermittency exponent on f_{Bm} variogram.

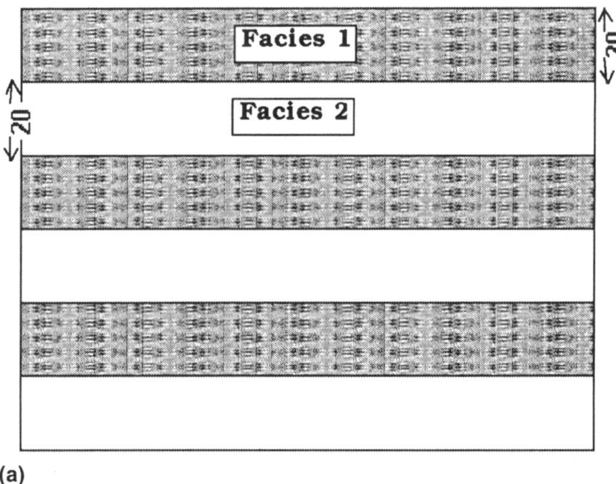

(a)

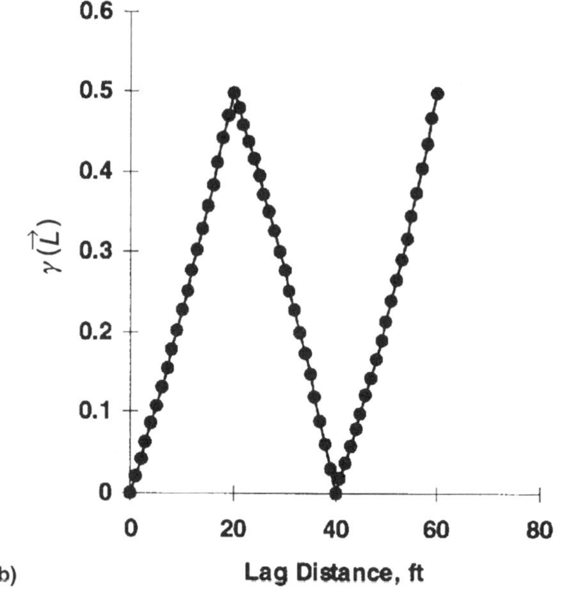

(b)

Fig. 3.42—(a) Stacked channels. (b) Estimated variogram of cyclic data.

fluctuation exists. **Fig. 3.41** shows the variogram plots for three values of H. Depending on whether H is less than or greater than 0.5, the variogram behavior is either parabolic or hyperbolic. For $H = 0.5$, we obtain a perfect straight line.

In fitting an estimated variogram with an f_{Bm} model, the parameters for the model can be obtained through a trial-and-error procedure. To obtain the value of H, we can use the R/S analysis method,[22] the spectral analysis method,[22] or the box-counting method.[23] Or the slope can be calculated on the basis of the best-fit line by plotting $\gamma(L)$ vs. h on a log-log scale. The slope is equal to $2H$; therefore, the value of H can be calculated from the slope.

In a reservoir-description process, Emanual *et al.*,[25] Hewett and Behrens,[26] and Tang *et al.*[27] were the first to assume that reservoir petrophysical properties in the horizontal direction can be described by the f_{Bm} model. This has not been proved with actual data. Analysis of horizontal well logs by some investigators revealed that the variogram in the horizontal direction can also be described by an f_{Gn} model rather than by an f_{Bm} model.[28-30] In short, whether an f_{Bm} model is appropriate for describing petrophysical properties in the horizontal direction is a subject of debate. Because we have insufficient data in the horizontal direction at uniform intervals in the Burbank field, we cannot provide a field example to demonstrate modeling of the variogram with an f_{Bm} model.

Logarithmic Model. The variogram equation for the logarithmic model can be written as

$$\gamma(L) = C_0 \log L \quad \text{(3.62)}$$

for $L > 0$. The logarithmic model is one of the first models used in geostatistical applications. It is still a popular model in mining engineering; however, its popularity has decreased somewhat because of the advances in interactive modeling of the variogram through computers. This allows the use of nested structures to model estimated variograms easily. Otherwise, the logarithmic model has the advantage of obtaining the model parameters by simply plotting the estimated variogram vs. log of the lag distance. The logarithmic model has rarely been used to describe spatial continuity of reservoir properties, and we do not see any application of this model in the foreseeable future.

3.5.4 Hole-Effect Models. Hole-effect models capture the cyclical behavior of variograms. When the estimated variogram shows cyclical behavior instead of a monotonic increase, hole-effect models are used. This type of behavior is more common in describing the spatial relationship in the vertical direction because geological processes often repeat in cycles. The vertical stacking of the geological environment is reflected in cyclical variograms. For example, if several channels are superimposed on top of each other with each channel consisting of upper, middle and lower subfacies, the vertical variogram reflects cyclical behavior.

As a simple example, **Fig. 3.42a** shows a three-layered (stacked channels) structure, with each layer having a thickness of 40 ft. Each channel is divided into two 20-ft-thick subfacies, upper (labeled 1) and lower (labeled 2). **Fig. 3.42b** shows the generated variogram in the vertical direction. Although this is an exaggerated and oversimplified description, the cyclical behavior is evident in the variogram. The wavelength reflects the length of each cycle, and the maximum value reflects the transition from one subfacies to another. In most instances, such transitions may not be clear-cut; however, the average length of a geological cycle can be obtained through variogram analysis and can be corroborated with an independent geological interpretation. Assuming that reproduction of such cyclical behavior is important, two potential models are described next.

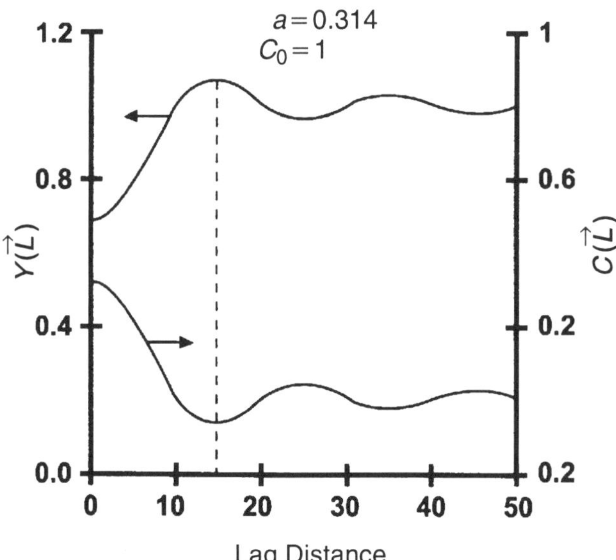

Fig. 3.43—Sine models.

Sine Model. The sine model can be written as

$$\gamma(L) = C_0\left[1 - \frac{\sin(aL)}{L}\right], \quad \ldots\ldots\ldots\ldots\ldots\ldots\ldots (3.63)$$

and the corresponding covariance model can be written as

$$C(L) = C_0\frac{\sin(aL)}{L}, \quad \ldots\ldots\ldots\ldots\ldots\ldots\ldots\ldots (3.64)$$

where the product (aL) is defined in radians. Where (aL) reaches a value of $3\mu/2$, the variogram reaches a maximum value. At a large value of L, $\gamma(L)$ converges to C_0 or the sill value. **Fig. 3.43** shows a typical sine variogram model and corresponding covariance model. Sine models can be applied in three dimensions. One restriction of the sine model is its relative amplitude. Relative amplitude is defined as

$$a = \frac{\max\gamma(L) - C_0}{C_0}, \quad \ldots\ldots\ldots\ldots\ldots\ldots\ldots\ldots (3.65)$$

where $\gamma(L)$ = the maximum value of the variogram and C_0 = the sill value. If the estimated variogram exhibits a value of $a > 0.212$, the sine model may not be applicable.

Cosine Model. The cosine variogram model can be written as

$$\gamma(L) = C_0[1 - \cos(aL)], \quad \ldots\ldots\ldots\ldots\ldots\ldots\ldots (3.66)$$

and the corresponding covariance model can be written as

$$C(L) = C_0\cos(aL). \quad \ldots\ldots\ldots\ldots\ldots\ldots\ldots\ldots (3.67)$$

The cosine model is applicable in only one dimension; however, the relative amplitude can be as high as one. This allows for a much bigger range of oscillations in the variogram than possible with the sine model. The cosine model is a periodic model with no dampening effect. However, the model can be dampened if used in conjunction with other nested models.

Field Example 3.12. This example examines the vertical variogram of porosity data for Well 35-W21. **Fig. 3.44** shows the estimated variogram and three possible models. The first model, a combination of the nugget and exponential models, can be written as

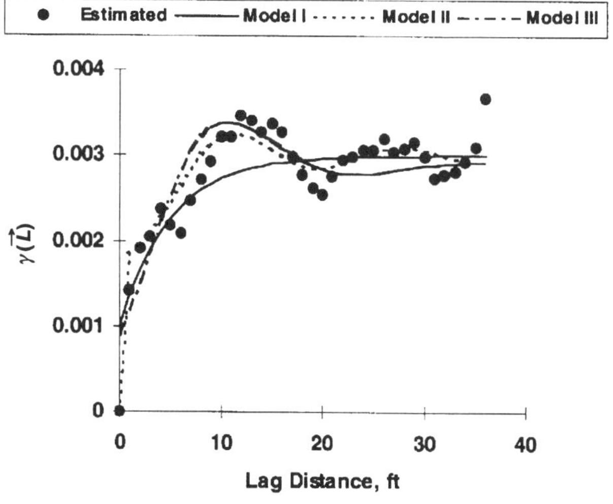

Fig. 3.44—Estimated and model variograms for porosity data from Well 35-W21.

$$\gamma(L) = 0.001 + 0.002\left[1 - \exp\left(-\frac{3L}{25}\right)\right].$$

The fit between the model and the experimental variogram is reasonable. However, we ignored the oscillations observed in the estimated variogram. If we believe that the oscillations are not important enough to honor strictly, we may ignore them and capture the average variogram trend with the model.

The second model is a cosine model. The relative amplitude, a, can be calculated as

$$a = \frac{0.0034 - 0.003}{0.003} = 0.133.$$

Because $a < 0.212$, we can apply the sine model, which can be written as

$$\gamma(L) = 0.003\left[1 - \frac{\sin(0.3926L)}{L}\right].$$

The model matches the estimated variogram much better, because it captures the oscillations observed in the variogram.

The third model is a combination of the exponential and cosine models and can be written as

$$\gamma(L) = 0.0009 + 0.002\left[1 - \exp\left(-\frac{3L}{25}\right)\cos(0.25L)\right].$$

Recall that Eq. 3.47, which allows a covariance model, is a product of two individual covariance models. The fit is not as good as with a sine model. Therefore, this model is a valid model. Because we used the cosine model, this model is applicable in only one dimension. In most reservoir studies, we are interested in generating a description of the reservoir properties in three, or at the least, two dimensions. Therefore, application of a variogram model, which is applicable in only one dimension, is doubtful in practice. However, in defining anisotropic structures, if we restrict the cosine model to only one direction, we may still be able to use it. We discuss such a modeling exercise in the next section.

In summary, we examined the two important rules in modeling of isotropic variograms: (1) the minimum number of parameters should be used to model the basic characteristics of the structure and (2) the condition of positive definiteness should be observed. We explained three types of models: with

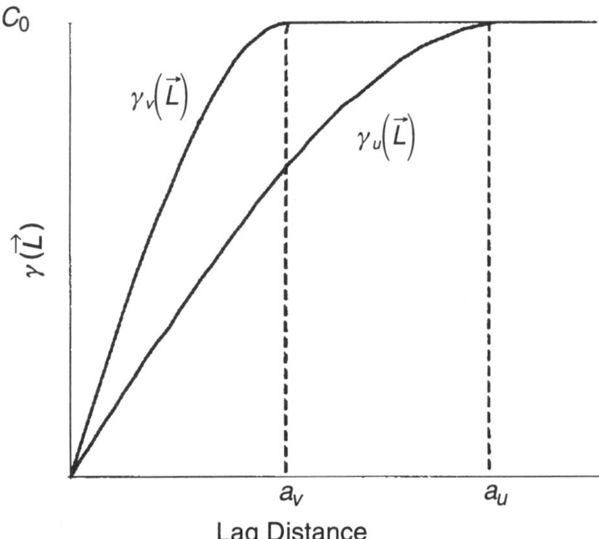

Fig. 3.45—Geometric anisotropy.

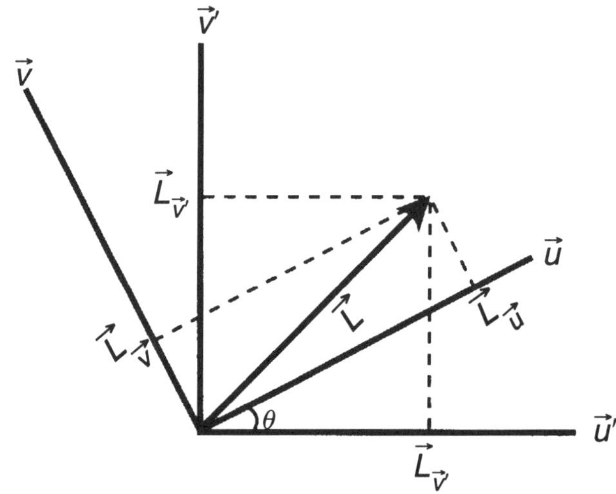

Fig. 3.46—Rotational transformation.

a sill, without a sill, and with hole effects. Except for the cosine model, which describes the hole effect in only one dimension, all the other models are applicable in three dimensions. Also, using linear combinations of these models, we can create multiple possibilities that, in practice, allow us to model any estimated variogram. More than one way may exist to model an observed variogram. If the alternative models capture the essential features of the variogram, subsequent estimations at unsampled locations are not affected by their use.

3.5.5 Anisotropic Models. Anisotropic models are needed when the estimated variograms show different spatial continuities in different directions. For example, in a channel sand, the variograms of porosity may show better continuity in the direction of the channel than across the channel. Anisotropic models need to be used to capture these types of continuities.

The same two rules apply in modeling anisotropic variograms as apply to modeling isotropic variograms: (1) use the minimum number of parameters needed to model the variograms and (2) satisfy the condition of positive definiteness.

These restrictions require careful selection of variogram models to describe the anisotropy. For modeling convenience, we further assume that directions of maximum and minimum continuity are perpendicular to each other. In practice, the two most commonly observed types of anisotropy are geometric and zonal anisotropies.

Geometric Anisotropy. Geometric anisotropy typically is observed when the variograms in the directions of maximum and minimum continuity show a similar shape and sill but different ranges (**Fig. 3.45**). In Fig. 3.45, the range in the direction of maximum continuity, $\vec{u}$, is a_u, while the range in the direction of minimum continuity is a_v. We assume that the two directions are perpendicular to each other.

To model the two variograms with the same sill, we have to use the same combination of linear models in both directions except with different ranges. For example, with a linear combination of nugget and spherical models, the model in the $\vec{u}$ direction is

$$\gamma_u(L) = C_0 + C_1 M_{S_{a_u}}(L), \quad \text{(3.68)}$$

and the model in the $\vec{v}$ direction is

$$\gamma_v(L) = C_0 + C_1 M_{S_{a_v}}(L). \quad \text{(3.69)}$$

The models in both directions contain the same linear combination; however, the range in each direction is different. It is not necessary to restrict the model to a combination of only two models; we can use as many as we want as long as the sills of the models in both directions are the same and the types of models used are the same. For example, we can model the variogram in the $\vec{u}$ direction as

$$\gamma_u(L) = C_0 + C_1 M_{E_{a_{u_1}}}(L) + C_2 M_{G_{a_{u_2}}}(L) \quad \text{(3.70)}$$

and the variogram in the $\vec{v}$ direction as

$$\gamma_v(L) = C_0 + C_1 M_{G_{a_{v_1}}}(L) + C_2 M_{E_{a_{v_2}}}(L). \quad \text{(3.71)}$$

Note that we have the same linear combination in both directions except that the ranges differ in the two directions.

This type of modeling satisfies the requirement for a condition of positive definiteness. Further, by defining the model in only two directions, we can estimate the variogram in any other direction. Before we examine the method of estimating the variogram in any direction, we must understand that any variogram model with a sill can be normalized to convert to an equivalent model with a range equal to one. For example, a spherical model can be written as

$$M_{S_a}(L) = \gamma(L) = C_0\left[\frac{3}{2}\left(\frac{L}{a}\right) - \frac{1}{2}\left(\frac{L}{a}\right)^3\right] \quad \text{(3.50a)}$$

for $L \leqq a$, and

$$M_{S_a}(L) = \gamma(L) = C_0 \quad \text{(3.50b)}$$

for $L \geqq a$.

This model has a range of a. If we define a variable

$$L_D = \frac{L}{a}, \quad \text{(3.72)}$$

we can write Eq. 3.50 as

$$\gamma(L_D) = C_0\left(\frac{3}{2}L_D - \frac{1}{2}L_D^3\right) \quad \text{(3.73a)}$$

for $L \leqq a$, and

$$\gamma(L_D) = C_0 \quad \text{(3.73b)}$$

for $L \geqq a$, where Eqs. 3.73a and 3.73b represent a variogram model with a unit range. Similar transformations also can be used for exponential and Gaussian models. By appropriately dividing the lag distance by the range in a given direction, we can normalize the variogram equation.

In **Fig. 3.46,** the directions of maximum and minimum continuities are shown as $\vec{u}$ and $\vec{v}$, respectively. These directions need not coincide with the original axes, $\vec{u}'$ and $\vec{v}'$, which are based on the sampled information. If the direction of maximum continuity ($\vec{u}$) is θ, then the direction of minimum continuity ($\vec{v}$) is $(\theta+90°)$. To calculate the variogram for vector $\vec{L}$ defined in an arbitrary direction, we can first calculate the components $\vec{L}_{u'}$ and $\vec{L}_{v'}$ based on the original axes $\vec{u}'$ and $\vec{v}'$ through a simple coordinate transformation. We can write

$$\vec{L}_u = \vec{L}_{u'}\cos\theta + \vec{L}_{v'}\sin\theta, \quad \ldots\ldots\ldots\ldots\ldots\ldots (3.74)$$

$$\text{and } \vec{L}_v = -\vec{L}_{u'}\sin\theta + \vec{L}_{v'}\cos\theta. \quad \ldots\ldots\ldots\ldots (3.75)$$

We can then normalize the vector $\vec{L}$ with the transformation

$$L_D = \sqrt{\left(\frac{\vec{L}_u}{a_u}\right)^2 + \left(\frac{\vec{L}_v}{a_v}\right)^2}, \quad \ldots\ldots\ldots\ldots\ldots (3.76)$$

where a_u and a_v = ranges in the $\vec{u}$ and $\vec{v}$ directions, respectively. We can then calculate the variogram at dimensionless lag distance L_D with the appropriate equation.

This type of transformation satisfies the limiting cases. For example, when calculating the variogram in the $\vec{u}$ direction, $\vec{L}_v = 0$ and $L_D = \vec{L}_u/a_u$. Substituting $\vec{L}_D$ in the variogram equation with a unit range results in a variogram in the $\vec{u}$ direction. Similarly, in calculating the variogram in the $\vec{v}$ direction, $\vec{L}_u = 0$ and $L_D = \vec{L}_v/a_v$. Substituting in a variogram equation with a unit range results in calculating the variogram in the $\vec{v}$ direction.

To consider this transform in a different way, normalizing the lag distance in different directions appropriately ensures that the variogram can be defined by an isotropic structure with a unit range in every direction. The following example explains this further.

Numerical Example 3.5. From the estimated variograms in different directions, we observe that the direction of maximum continuity is N30°E. The variogram model in that direction is defined as

$$\gamma(L) = 0.2 + 0.4M_{S_{8,000}}(L).$$

The variogram in the direction of minimum continuity (N30°W) is defined as

$$\gamma(L) = 0.2 + 0.4M_{S_{5,000}}(L).$$

Estimate the variogram value between two points located at (1,500, 4,000) and (4,000, 6,000).

Solution. Typically, in practice, for a 2D system, we need to estimate a variogram between two points whose coordinates based on the original axes are known. In this case, the (u',v') coordinates for the two points are given. Therefore, we can estimate the value of $\vec{L}_{u'}$ and $\vec{L}_{v'}$, respectively, by

$$\vec{L}_{u'} = u_2' - u_1' = 4,000 - 1,500 = 2,500,$$

$$\text{and } \vec{L}_{v'} = v_2' - v_1' = 6,000 - 4,000 = 2,000.$$

These coordinates need to be transformed in the directions of maximum and minimum continuities, respectively, before we estimate the variogram. With Eq. 3.74, knowing that $\theta = 30°$,

$$\begin{aligned}\vec{L}_u &= \vec{L}_{u'}\cos\theta + \vec{L}_{v'}\sin\theta \\ &= 2,500(0.867) + 2,000(0.5) \\ &= 3,167.5,\end{aligned}$$

$$\begin{aligned}\text{and } \vec{L}_v &= -\vec{L}_{u'}\sin\theta + \vec{L}_{v'}\cos\theta \\ &= -2,500(0.5) + 2,000(0.867) \\ &= 484.\end{aligned}$$

With Eq. 3.76, knowing that the range in the $\vec{u}$ direction is 8,000 ft and the range in the $\vec{v}$ direction is 5,000 ft,

$$\begin{aligned}L_D &= \sqrt{\left(\frac{\vec{L}_u}{a_u}\right)^2 + \left(\frac{\vec{L}_v}{a_v}\right)^2} \\ &= \sqrt{\left(\frac{3,167.5}{8000}\right)^2 + \left(\frac{484}{5000}\right)^2} \\ &= 0.4076.\end{aligned}$$

For $L \leqq 1$, substituting this value into an equation with a unit range gives

$$\begin{aligned}\gamma(L_D) &= 0.2 + 0.4M_{S_1}(L_D) \\ &= 0.2 + 0.4\left[1.5(L_D) - 0.5(L_D^3)\right] \\ &= 0.2 + 0.4\left[1.5(0.408) - 0.5(0.408)^3\right] \\ &= 0.4312.\end{aligned}$$

The variogram in any given direction and at any given distance can be calculated in a similar manner. As stated previously, we are not restricted to a particular set of models in applying this technique. For example, if our variogram model in N30°E is defined as

$$\gamma(L) = 0.1 + 0.2M_{S_{8,000}}(L) + 0.3M_{E_{5,000}}(L),$$

and the variogram model in N30°W is defined as

$$\gamma(L) = 0.1 + 0.2M_{S_{5,000}}(L) + 0.3M_{E_{7,000}}(L),$$

we can apply the same principles.

In this case, we have to calculate two values of L_D corresponding to two different models. For the spherical model,

$$L_{D_1} = \sqrt{\left(\frac{3,167.5}{8,000}\right)^2 + \left(\frac{484}{7,000}\right)^2} = 0.408,$$

and for the exponential model,

$$L_{D_2} = \sqrt{\left(\frac{3,167.5}{5,000}\right)^2 + \left(\frac{484}{7,000}\right)^2} = 0.637.$$

The variogram equation with a unit range can be written as

$$\gamma(L) = 0.1 + 0.2M_{S_1}\left(L_{D_1}\right) + 0.3M_{E_1}\left(L_{D_2}\right)$$

for $L_{D_1} \leqq 1, L_{D_2} \leqq 1$, and as

$$\gamma(L) = 0.6$$

for $L_{D_1} \geqq 1, L_{D_2} \geqq 1$.

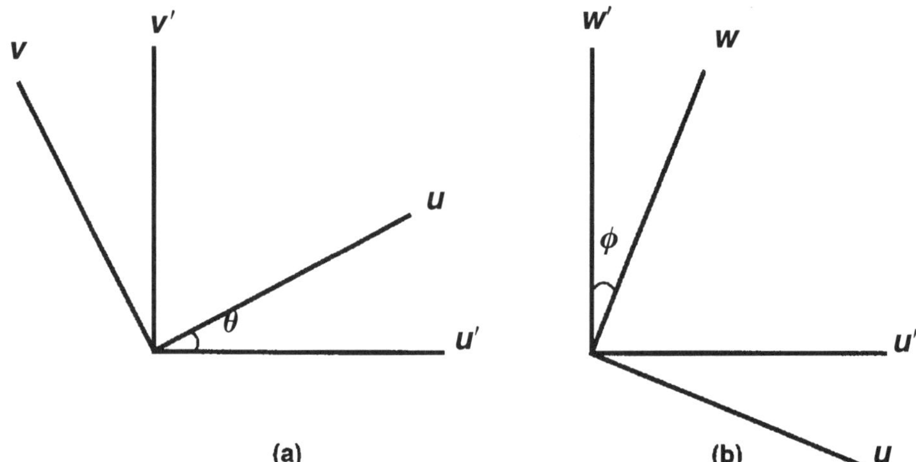

Fig. 3.47—Rotation in three dimensions.

Depending on the values of L_{D_1} and L_{D_2}, the appropriate value of the variogram can be calculated at any given lag distance and direction. Substituting,

$$\gamma(L) = 0.1 + 0.2\left[1.5(0.408) - 0.5(0.408)^3\right]$$
$$+ 0.3[1 - \exp(-3 \times 0.637)]$$
$$= 0.4712.$$

It often is convenient to use a matrix notation to describe the transformation process. Recall that, as a first step, we use the appropriate notation to transform the original axes to the directions of the maximum and the minimum continuity respectively. If we define the rotation matrix, **[R]**, as

$$[\mathbf{R}] = \begin{bmatrix} \cos\theta & \sin\theta \\ -\sin\theta & \cos\theta \end{bmatrix}, \quad \ldots\ldots\ldots\ldots\ldots\ldots\ldots (3.77)$$

we can write

$$\begin{bmatrix} \vec{L}_u \\ \vec{L}_v \end{bmatrix} = \begin{bmatrix} \cos\theta & \sin\theta \\ -\sin\theta & \cos\theta \end{bmatrix} \begin{bmatrix} \vec{L}_{u'} \\ \vec{L}_{v'} \end{bmatrix}. \quad \ldots\ldots\ldots\ldots (3.78)$$

Eq. 3.78 is the same as Eq. 3.74 except written in a matrix format. Alternatively, we can write Eq. 3.78 as

$$[\mathbf{L}] = [\mathbf{R}][\mathbf{L'}]. \quad \ldots\ldots\ldots\ldots\ldots\ldots\ldots\ldots\ldots (3.79)$$

Once we estimate **[L],** we multiply it by the appropriate scaling factor to normalize it to L_D (see Eq. 3.76).

This type of transformation can be translated easily to three dimensions. Consider coordinate systems with u', v', and w' coordinates. **Fig. 3.47a** shows the clockwise rotation defined with respect to the w' axis with angle θ. Note that the w' axis is perpendicular to the plane pointing away from the reader. Angle ϕ represents a clockwise rotation with respect to the v axis, which is perpendicular to the plane pointing away from the reader in **Fig. 3.47b**.

The rotation matrix, **[R],** can be written as

$$[\mathbf{R}] = \begin{bmatrix} \cos(\theta)\cos(\phi) & \sin(\theta)\cos(\phi) & \sin(\phi) \\ -\sin(\theta) & \cos(\theta) & 0 \\ -\cos(\theta)\sin(\phi) & -\sin(\theta)\sin(\phi) & \cos(\phi) \end{bmatrix}. \quad \ldots\ldots\ldots\ldots\ldots\ldots (3.80)$$

We can then write the transformed coordinates with Eq. 3.79 as

$$[\mathbf{L}] = [\mathbf{R}][\mathbf{L'}]. \quad \ldots\ldots\ldots\ldots\ldots\ldots\ldots\ldots\ldots (3.79)$$

where,

$$[\mathbf{L}] = \begin{bmatrix} \vec{L}_u \\ \vec{L}_v \\ \vec{L}_w \end{bmatrix}. \quad \ldots\ldots\ldots\ldots\ldots\ldots\ldots\ldots\ldots (3.81)$$

Once **[L]** is calculated, the normalized vector magnitude can be calculated as

$$L_D = \sqrt{\left(\frac{\vec{L}_u}{a_u}\right)^2 + \left(\frac{\vec{L}_v}{a_v}\right)^2 + \left(\frac{\vec{L}_w}{a_w}\right)^2}, \quad \ldots\ldots (3.82)$$

where a_u, a_v, and a_w = ranges in the $\vec{u}$, $\vec{v}$, and $\vec{w}$ directions, respectively. The normalized value, L_D, can be substituted into an appropriate variogram equation with a unit range to estimate the variogram.

Field Example 3.13. This example uses the variograms constructed in Field Example 3.2 for the porosity data for Flow Unit 3. Recall that the porosity data showed maximum continuity at 157.5° and minimum continuity at 90° (Fig. 3.9). As discussed previously, for modeling purposes, we assume that the directions of maximum and minimum continuity are perpendicular to each other. After superimposing an ellipse on a rose diagram (a diagram that represents the approximate ranges in various directions), we determined that we could assume that the direction of minimum continuity was 67.5°.

We model the variograms in the directions of 157.5° and 67.5°, ensuring that the same linear combinations of models are used. **Fig. 3.48** shows the comparison between the model and the estimated variograms for these two directions.

The model in the direction of 157.5° is

$$\gamma(L) = 9 + 15M_{S_{8,500}}(L),$$

and the model in the direction of 67.5° is

$$\gamma(L) = 9 + 15M_{S_{4,000}}(L).$$

The only difference between the two models is the range.

The appropriateness of this anisotropic modeling can be validated by calculating the variograms in other directions at

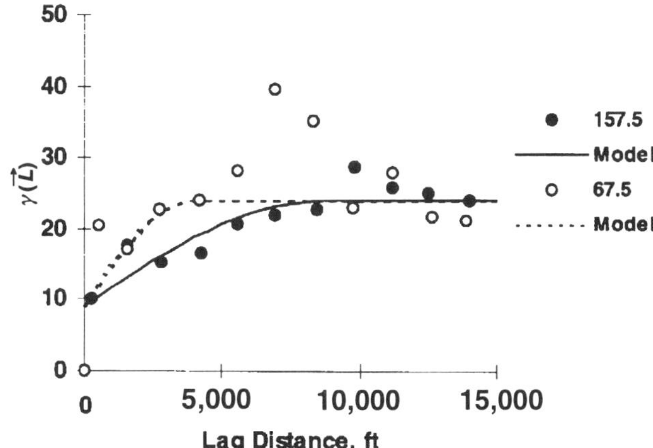

Fig. 3.48—Estimated and model anisotropic variograms for Flow Unit 3 porosity data.

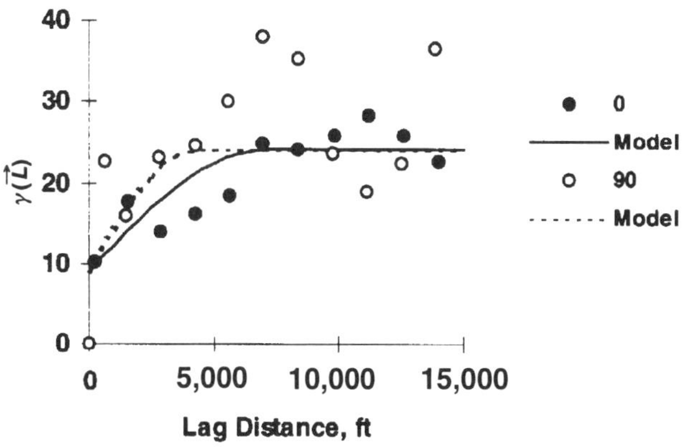

Fig. 3.49—Estimated and model variograms at 0° and 90° for Flow Unit 3 porosity data.

various distances with the procedure described in Numerical Example 3.5. The calculated variogram is then compared with the estimated variogram based on the sample data. **Fig. 3.49** shows the comparison between the calculated and estimated variograms in two directions, 0 and 90°. Again, the calculated variograms are based on the models in the directions of maximum and minimum continuities. Although the estimated variograms exhibit significant fluctuation, the comparison is reasonable, indicating that the proposed model is representative of spatial relationships in different directions.

Before an anisotropic model is used to represent spatial relationships, such cross validation must be carried out. If the calculated variograms do not reasonably match the estimated variograms in directions other than the directions of maximum and minimum continuities, the appropriateness of the directions of maximum and minimum continuities may have to be examined in addition to the ranges and models in those two directions.

Zonal Anisotropy. Zonal anisotropy is observed where the variogram in the directions of maximum and minimum continuity show different sills. Typical zonal anisotropy is observed when vertical vs. horizontal variograms are compared. Changes in the vertical direction are more rapid, representing higher variance or sill value compared with relatively smooth changes in the horizontal direction, representing a smaller sill value (**Fig. 3.50**). Zonal anisotropy is more generalized than geometric anisotropy because it allows for different sill values in two directions as well as different ranges. Zonal anisotropy, however, does have some restrictions. It requires that the directions of maximum and minimum continuity be perpendicular to each other and that linear combinations of models used for modeling the two directions be the same.

For example, if we use a linear combination of a nugget and a spherical model in the $\vec{u}$ direction as

$$\gamma_u(L) = C_0 + C_1 M_{S_{a_u}}(L), \quad \text{.......... (3.83)}$$

we can write a model in the direction $\vec{w}$ as

$$\gamma_w(L) = C_0 + (C_1 + C_1')M_{S_{a_w}}(L). \quad \text{.......... (3.84)}$$

Note that both the range and the sill values differ in the $\vec{u}$ and $\vec{w}$ directions; however, the same model is used in both directions. As before, we are not restricted to a combination of only two linear models; we can use more than two models to describe the variograms in two different directions. For example, we can write the variogram in the $\vec{u}$ direction as

$$\gamma_u(L) = C_0 + C_1 M_{S_{a_{u_1}}}(L) + C_2 M_{E_{a_{u_2}}}(L) \quad \text{...... (3.85)}$$

and in the $\vec{w}$ direction as

$$\gamma_w(L) = C_0 + (C_1 + C_1')M_{S_{a_{w_1}}}(L) + (C_2 + C_2')M_{S_{a_{w_2}}}(L). \quad \text{.......... (3.86)}$$

The sills and ranges for individual models may differ.

This type of modeling satisfies the condition of positive definiteness. By defining the variograms in only two directions, we can estimate the variogram in any other direction using the same principle as in the case of geometric anisotropy. That is, we convert the variograms into equivalent normalized variograms with a unit range. After using the appropriate rotation transform, we define the magnitude of the normalized vectors as

$$L_D = \sqrt{\left(\frac{\vec{L}_u}{a_u}\right)^2 + \left(\frac{\vec{L}_w}{a_w}\right)^2} \quad \text{.......... (3.87)}$$

$$\text{and } L_D' = \sqrt{\left(\frac{\vec{L}_w}{a_w}\right)^2} = \frac{\vec{L}_w}{a_w}, \quad \text{.......... (3.88)}$$

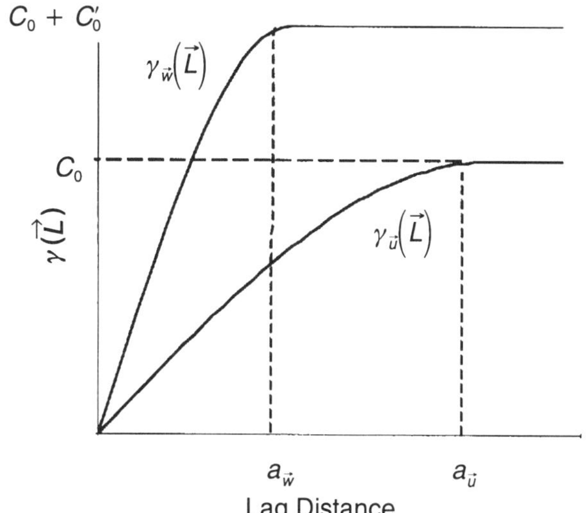

Fig. 3.50—Zonal anisotropy.

where a_u and a_w = ranges in the $\vec{u}$ and $\vec{w}$ directions, respectively. We then substitute the values of L_D and L'_D into a normalized variogram equation. For example, the normalized variogram equation is

$$\gamma(L_D) = C_0 + C_1 M_{S_1}(L) \quad \text{(3.89)}$$

and $\gamma(L'_D) = C'_1 M_{S_1}(L'_D)$, (3.90)

and the resulting variogram is

$$\gamma(L) = \gamma(L_D) + \gamma(L'_D). \quad \text{(3.91)}$$

Note that the two components are divided so that one component represents the variogram that is common to both directions, and the other component represents the variogram that is valid only in the $\vec{w}$ direction. This type of separation ensures that the variogram values in the limiting directions are satisfied. For example, to calculate the variogram only in the $\vec{u}$ direction,

$$L_D = \frac{L_u}{a_u} \quad \text{(3.92a)}$$

and $L'_D = 0$. (3.92b)

Substituting in Eq. 3.89 results in

$$\gamma(L_D) = C_0 + C_1 M_{S_1}\left(\frac{L_u}{a_u}\right) \quad \text{(3.93a)}$$

and $\gamma(L'_D) = 0$, (3.93b)

which is an equation for the $\vec{u}$ direction only. To calculate the variogram in only the $\vec{w}$ direction,

$$L_D = \frac{L_w}{a_w} \quad \text{(3.94a)}$$

and $L'_D = \frac{L_w}{a_w}$. (3.94b)

Substituting these values into Eqs. 3.89 and 3.90 gives

$$\gamma(L_D) = C_0 + C_1 M_{S_1}\left(\frac{L_w}{a_w}\right) \quad \text{(3.95a)}$$

and $\gamma(L'_D) = C'_1 M_{S_1}\left(\frac{L_w}{a_w}\right)$. (3.95b)

Substituting into Eq. 3.91 gives

$$\gamma(L) = C_0 + (C_1 + C'_1) M_{S_1}\left(\frac{L_w}{a_w}\right), \quad \text{(3.96)}$$

which is the variogram equation for the $\vec{w}$ direction. The following numerical example illustrates these principles.

Numerical Example 3.6. The variogram in the $\vec{u}$ direction is given as

$$\gamma_u(L) = 5 + 20M_{E_{300}}(L) + 30M_{S_{100}}(L),$$

and the variogram in the $\vec{w}$ direction is given as

$$\gamma_w(L) = 5 + 30M_{E_{50}}(L) + 40M_{S_{30}}(L).$$

If the $\vec{u}$ direction represents a horizontal plane and the $\vec{w}$ direction represents a vertical plane, estimate the variogram between two points located at (50,100) and (13,060).

Solution. Because the major and minor axes coincide with the original axes, no rotation transformation is needed. We can calculate

$$\vec{L}_u = 130 - 50 = 70$$

and $\vec{L}_w = 60 - 100 = -40$.

From the variogram equations, we can write

$$C_0 = 5,$$
$$C_1 = 20,$$
$$C'_1 = 30 - 20 = 10,$$
$$C_2 = 30,$$

and $C'_2 = 40 - 30 = 10$.

For the exponential variogram,

$$L_{D_1} = \sqrt{\left(\frac{\vec{L}_u}{a_u}\right)^2 + \left(\frac{\vec{L}_w}{a_w}\right)^2}$$

$$= \sqrt{\left(\frac{70}{300}\right)^2 + \left(\frac{-40}{50}\right)^2}$$

$$= 0.833$$

and $$L'_{D_1} = \sqrt{\left(\frac{\vec{L}_w}{a_w}\right)^2}$$

$$= \frac{40}{50}$$

$$= 0.8.$$

For the spherical variogram,

$$L_{D_2} = \sqrt{\left(\frac{70}{100}\right)^2 + \left(\frac{-40}{30}\right)^2} = 1.503$$

and $L'_{D_2} = \frac{40}{30} = 1.33$.

Using a normalized variogram equation gives

$$\gamma(L_D) = C_0 + C_1 M_{E_1}(L_{D_1}) + C_2 M_{S_1}(L_{D_2})$$

$$= 5 + 20[1 - \exp(-3 \times 0.833)] + 30$$

$$= 53.36.$$

Note that because $h_{D_2} > 1$, the spherical component contribution is equal to C_2.

Similarly,

$$\gamma(L'_D) = C'_1 M_{E_1}(L'_{D_1}) + C'_2 M_{S_1}(L'_{D_2})$$

$$= 10[1 - \exp(-3 \times 0.8)] + 10$$

$$= 19.1.$$

Again, $L'_{D_2} > 1$, which makes the spherical component contribution equal to C'_2. Therefore, the resulting variogram is

$$= \gamma(L_D) + \gamma(L'_D)$$

$$= 53.36 + 19.1.$$

$$= 72.46.$$

Similar calculations can be repeated at any other direction and for any other distance.

As in the case of geometric anisotropy, this analysis can be extended to three dimensions. In practice, the most common types of anisotropies observed are geometric anisotropy in the horizontal plane and zonal anisotropy in the vertical direction. If $\vec{u}$ and $\vec{v}$ represent the directions in the horizontal plane and $\vec{w}$ represents the vertical direction, after proper coordinate transformations, we get Eqs. 3.82 and 3.88

$$L_D = \sqrt{\left(\frac{\vec{L}_u}{a_u}\right)^2 + \left(\frac{\vec{L}_v}{a_v}\right)^2 + \left(\frac{\vec{L}_w}{a_w}\right)^2} \quad \ldots\ldots (3.82)$$

$$\text{and } L'_D = \sqrt{\left(\frac{\vec{L}_w}{a_w}\right)^2}, \quad \ldots\ldots (3.88)$$

These values then are substituted into an appropriate variogram equation.

Note that using a 3D variogram requires the use of a variogram model that satisfies the condition of positive definiteness in three dimensions. A special exception is defining the zonal anisotropy. If the hole effect in the vertical direction needs to be represented by a cosine model (a 1D model), this type of modeling is permissible as long as the variograms in the horizontal directions do not include a cosine model. This is because the cosine-model component appears only in the vertical direction. For example, we can define the variogram models in the $\vec{u}$, $\vec{v}$, and $\vec{w}$ directions, respectively, as

$$\gamma_u(L) = C_0 + C_1 M_{S_{a_u}}(L), \quad \ldots\ldots (3.97)$$

$$\gamma_v(L) = C_0 + C_1 M_{S_{a_v}}(L), \quad \ldots\ldots (3.98)$$

$$\text{and } \gamma_w(L) = C_0 + (C_1 + C'_1) M_{S_{a_w}}(L) + C'_2[1 - \cos(a'_w L)], \quad \ldots\ldots (3.99)$$

where there is geometric anisotropy in the horizontal plane and a zonal anisotropy with a hole effect in the vertical direction. Although the cosine model is a 1D model, this set of equations satisfies the requirement of a condition of positive definiteness.

In all these equations, we considered variogram models with a sill. Obviously, the concept of anisotropy is much easier to define when the spatial continuity is defined in terms of how quickly the variogram reaches a sill value. For the f_{Bm} model, which is a model without a sill, we can also write an anisotropic equation as

$$\gamma(L) = C_s(\theta) L^{2H}, \quad \ldots\ldots (3.100)$$

where C_s, the scaling coefficient, is defined as a function of the direction θ. Eq. 3.100 satisfies the condition of positive definiteness. If C_s is constant in all the directions, we have an isotropic variogram model. If C_s is smaller in a given direction, the rise in the variogram is slower, representing better spatial continuity compared with a direction where C_s is large. Similar to geometric anisotropy, we can also define an ellipse that represents C_s values and can be defined, where C_s is smallest in the direction of maximum continuity and largest in the direction of minimum continuity. The value of C_s in any other direction can be calculated with the equation of an ellipse if it is a 2D variogram and with the equation for an ellipsoid if it is a 3D variogram.

To summarize, when modeling anisotropic variograms, the same two conditions need to be satisfied (i.e., using a minimum number of modeling parameters and satisfying the condition of positive definiteness). For convenience, the directions of maximum and minimum continuity are assumed to be perpendicular to each other. The two types of anisotropy commonly observed are geometric and zonal. In geometric anisotropy, the sill values in different directions must be the same but the ranges can be different. In zonal anisotropy, both the sill and the range values can be different in different directions. In modeling both types of anisotropy, the same combination of models must be used in both the directions of maximum and minimum continuities.

Modeling anisotropic variograms is a trial-and-error procedure. It is more important to use a linear combination of model fit reasonably well in both directions of maximum and minimum continuity rather than modeling one direction well and leaving the other direction with a poor fit. In addition, after the modeling is completed, comparison between the calculated and estimated variograms is essential to ensure that the established model does an adequate job of capturing the spatial relationships in all directions.

3.6 Cross Variograms

Cross variograms (and the corresponding cross covariance) represent the spatial relationship between two variables located a certain lag distance apart. For example, we may want to determine whether a permeability value at a given location is spatially related to a porosity value a certain lag distance apart. If such a relationship exists, we may be able to use it to improve the estimate of permeability values at unsampled locations.

The cross variogram is especially useful in cases with two variables where one is measured more extensively than the other. One example is having only one cored well with measured permeability values and several logged wells with log porosity values. If we can establish a spatial relationship between log porosity and core permeability at the cored-well location, we may be able to use that relationship to estimate permeability values at wells where we do not have cored data. Another example would be the use of 3D-seismic data that are collected at a high density. During the exploration phase, we may have drilled only a limited number of wells and collected a limited amount of information on petrophysical properties. If we can establish a spatial relationship between seismic data and a petrophysical property (such as porosity), we may be able to estimate porosity values at unsampled locations where we have not yet drilled the wells.

3.6.1 Estimation of Cross Variograms. The equation for the cross variogram is

$$\gamma_c(\vec{L}) = \frac{1}{2} E\left\{\left[X(\vec{u}) - X(\vec{u} + \vec{L})\right]\left[Y(\vec{u}) - Y(\vec{u} + \vec{L})\right]\right\}. \quad \ldots\ldots (3.101)$$

In practice, we can estimate the cross variogram between the variables x and y as

$$\gamma_c(L) = \frac{1}{2n(L)} \sum_{i=1}^{n(L)} \left[x(\vec{u}_i) - x(\vec{u}_i + \vec{L})\right] \times \left[y(\vec{u}_i) - y(\vec{u}_i + \vec{L})\right]. \quad \ldots\ldots (3.102)$$

Obviously, the estimation requires that both variable values be available at locations $\vec{u}_i$ and $\vec{u}_i + \vec{L}$. Unlike the standard variogram, the cross variogram does not need to be strictly positive. If the two variables are negatively correlated with each other (as x increases, y decreases), the cross variogram is also negative.

The cross-covariance equation can be written as

$$C_c(\vec{L}) = E\left[X(\vec{u})Y(\vec{u}+\vec{L})\right] - E[X(\vec{u})]E\left[Y(\vec{u}+\vec{L})\right]. \quad (3.103)$$

In practice, we can write Eq. 3.103 as

$$c_c(\vec{L}) = \frac{1}{n(L)}\sum_{i=1}^{n(L)} x(\vec{u}_i)y(\vec{u}_i+\vec{L}) - \frac{1}{n(L)}\sum_{i=1}^{n(L)} x(\vec{u}_i) \times \frac{1}{n(L)}\sum_{i=1}^{n(L)} y(\vec{u}_i+\vec{L}). \quad (3.104)$$

Unlike the standard covariance equation, the cross covariance is not symmetrical. That is, $C_c(\vec{L})$ is not necessarily equal to $C_c(-\vec{L})$; however, in most cases, we assume that they are. The cross variogram, in contrast, is symmetric. That is, we can show that $\gamma_c(\vec{L}) = C_c(-\vec{L})$. If we assume that $C_c(-\vec{L}) = C_c(\vec{L})$,

$$\gamma_c(\vec{L}) = C_c(0) - C_c(\vec{L}). \quad (3.105)$$

Eq. 3.105 is similar to the equation we write for the standard variogram, where $C_c(0)$ is the cross covariance at a lag distance of zero. This is calculated as

$$c_c(0) = \frac{1}{n}\sum_{i=1}^{n} x(\vec{u}_i)y(\vec{u}_i) - \frac{1}{n}\sum_{i=1}^{n} x(\vec{u}_i)\frac{1}{n}\sum_{i=1}^{n} y(\vec{u}_i), \quad (3.106)$$

where n = total number of locations where both x and y variables are measured. Appendix C provides additional details.

Numerical Example 3.7. Table 3.8 shows data collected from a well. Estimate the cross variogram between the permeability and the porosity data. Also estimate the corresponding cross covariance between the same two variables.

Solution. Table 3.8 provides the same data as used in Numerical Example 3.1 except that permeability data have been added. Because of the limited number of pairs, we use only three lag distances. With Eq. 3.102,

$$\gamma_c(L) = \frac{1}{2n(L)}\sum_{i=1}^{n(L)}\left[x(\vec{u}_i) - x(\vec{u}_i+\vec{L})\right] \times \left[y(\vec{u}_i) - y(\vec{u}_i+\vec{L})\right]. \quad (3.102)$$

TABLE 3.8—PERMEABILITY AND POROSITY DATA FOR NUMERICAL EXAMPLE 3.7

Depth (ft)	ϕ	k (md)
2,040	8.25	35
2,041	9.0	50
2,042	6.25	21
2,043	5.0	16
2,044	5.3	20
2,045	4.75	8
2,046	5.0	14

At L = 1 ft, there are six pairs and we can write

$$\hat{\gamma}_c(1) = \frac{1}{2\times 6}\left[(8.25-9.0)(35-50) + (9.0-6.25) \times (50-21) + (6.25-5.0)(21-16) + (5.0-5.3)(16-20) + (5.3-4.75) \times (20-8) + (4.75-5.0)(8-14)\right] = 8.88.$$

At L = 2 ft, there are five pairs and we can write

$$\hat{\gamma}_c(2) = \frac{1}{2\times 5}(8.25-6.25)(35-21) + (9.0-5.0) \times (50-16) + (6.25-5.3)(21-20) + (5.0-4.75)(16-8) + (5.3-5.0) \times (20-14)] = 16.89.$$

Similarly, at L = 3 ft, we have four pairs and can calculate

$$\hat{\gamma}_c(3) = 24.03.$$

To estimate the cross covariance, we first need to estimate $c_C(0)$. Eq. 3.106 states that

$$c_c(0) = \frac{1}{n}\sum_{i=1}^{n} x(\vec{u}_i)y(\vec{u}_i) - \frac{1}{n}\sum_{i=1}^{n} x(\vec{u}_i)\frac{1}{n}\sum_{i=1}^{n} y(\vec{u}_i) \quad (3.106)$$

$$= \frac{1}{7}[(8.25)(35) + (9)(50) + \ldots + (5)(14)] - \frac{1}{7}[8.25 + \ldots + 5]\frac{1}{7}[35 + \ldots + 14] = 20.53.$$

With the traditional relationship between the variogram and the covariance as stated by Eq. 3.105,

$$\gamma_c(\vec{L}) = C_c(0) - C_c(\vec{L}), \quad (3.105)$$

for L = 1 ft,

$$c_c(1) = c_c(0) - \hat{\gamma}_c(1) = 20.53 - 8.88 = 11.65.$$

Similarly, for L = 2 and 3 ft, respectively,

$$c_C(2) = 20.53 - 16.89 = 3.64$$

and $c_C(3) = 20.53 - 24.03 = -3.50$.

Fig. 3.51 shows a plot of the cross variogram and the cross covariance as functions of lag distance. The behavior is similar to the standard variogram and covariance, respectively. This is possible when the two variables are positively related to each other.

It is important to note that Eq. 3.104, which is a standard covariance equation, gives a different result for the cross covariance from this result. With Eq. 3.104,

$$c_c(\vec{L}) = \frac{1}{n(L)}\sum_{i=1}^{n(L)} x(\vec{u}_i)y(\vec{u}_i+\vec{L}) - \frac{1}{n(L)}\sum_{i=1}^{n(L)} x(\vec{u}_i) \times \frac{1}{n(L)}\sum_{i=1}^{n(L)} y(\vec{u}_i+\vec{L}). \quad (3.104)$$

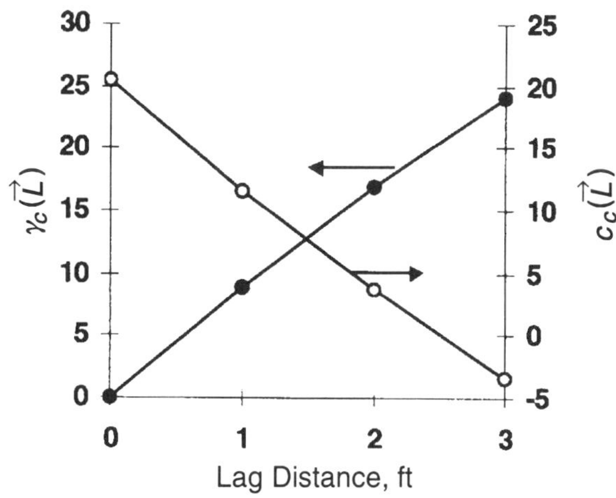

Fig. 3.51—Cross variogram and cross covariance for data in Numerical Example 3.7.

For L = 1 ft, we have six pairs and can write

$$c_c(1) = \frac{1}{6}[(8.25)(50) + (9.0)(21) + \ldots + (4.75)(5.0)]$$

$$- \frac{1}{6}[8.25 + \ldots + 4.75]\frac{1}{6}[50 + \ldots + 14]$$

$$= 13.60.$$

Similarly, for L = 2 and 3 ft, respectively,

$$c_c(2) = 4.48$$

and $c_c(3) = 4.69$.

As discussed in the section on nonergodic variograms, difference is caused by differences in lag means compared with sampled means. In Eq. 3.104, local lag means are used on the right side of the equation. If these means differ from sample means of both variables, the answer for $c_c(L)$ will differ with the two approaches. Appendix C provides additional details.

Similar to the nonergodic variogram, we can estimate the nonergodic cross variogram with Eq. 3.104. Then with Eq. 3.105,

$$\hat{\gamma}_{c_{NE}}(L) = c_c(0) - c_c(L).$$

In this example, for L = 1, 2 and 3 ft, respectively,

$$\hat{\gamma}_{c_{NE}}(1) = 20.53 - 13.60 = 6.93,$$

$$\hat{\gamma}_{c_{NE}}(2) = 20.53 - 4.48 = 16.05,$$

and $\hat{\gamma}_{c_{NE}}(3) = 20.53 - 4.69 = 15.84$.

As in the case of variograms, when estimating the cross variogram, we should try to find the most interpretable structure. Both the conventional cross variogram and the nonergodic cross variogram should be used to determine which type provides a better defined structure and that structure should be used for modeling purposes.

Field Example 3.14. This example uses core permeability and porosity data from Well 34-29. Estimate the cross variogram and cross covariance between the porosity and the log permeability data. Compare them with their nonergodic counterparts.

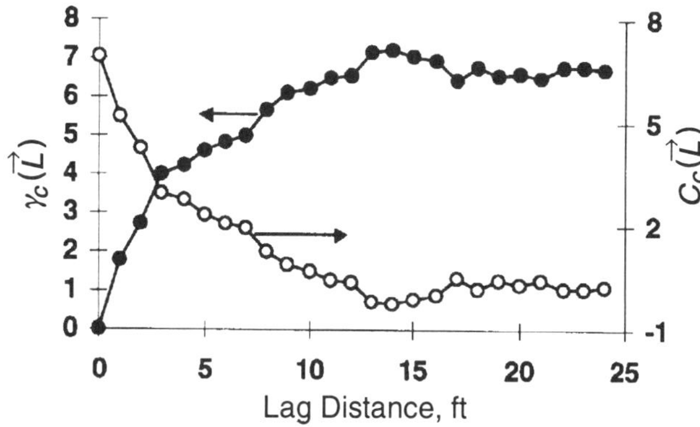

Fig. 3.52—Cross variogram and cross covariance for data from Well 34-29.

Solution. **Fig. 3.52** shows the cross variogram and cross covariance for the well. There is a well-defined structure, indicating a strong spatial relationship between the porosity and the log of permeability. **Fig. 3.53** shows the nonergodic counterparts for the same data. In this particular example, both the conventional and the nonergodic spatial functions are very similar to each other, an indication that the local and sample properties do not differ significantly.

Field Example 3.15. Using data from Well 34-29, estimate the cross variogram between the log of IP and the log of net-*kh* data. Intuitively, these two variables should be related to each other. Recall that the estimated variograms of the log of IP showed a reasonably interpretable structure. Also, use of the log of net *kh* instead of simply net *kh* significantly improved the structure. **Fig. 3.54** shows the cross variogram and the nonergodic cross variogram between the two variables. As expected, the spatial structure indicates a spatial relationship between the two variables. The difference between the cross variogram and the nonergodic cross variogram is small, indicating that either one can be used for modeling.

3.6.2 Modeling of Cross Variograms. Cross variograms can be modeled with the same models used to model variograms. As before, the rule of the use of a minimum number of parameters to model the variogram applies, and the rule of the condition of positive definiteness should be satisfied.

To satisfy the condition of positive definiteness, certain additional restrictions are imposed when modeling the cross variogram.[17] If x and y are considered as two variables, the variograms for the two variables x and y are modeled, respectively, as

$$\gamma_x(L) = c_{0_x} + c_{1_x} M_{S_a}(L) \quad \ldots\ldots\ldots\ldots\ldots\ldots (3.107)$$

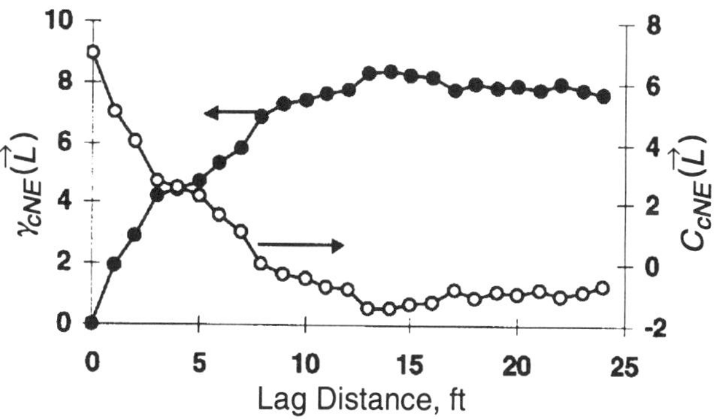

Fig. 3.53—Nonergodic cross variogram and cross covariance for data from Well 34-29.

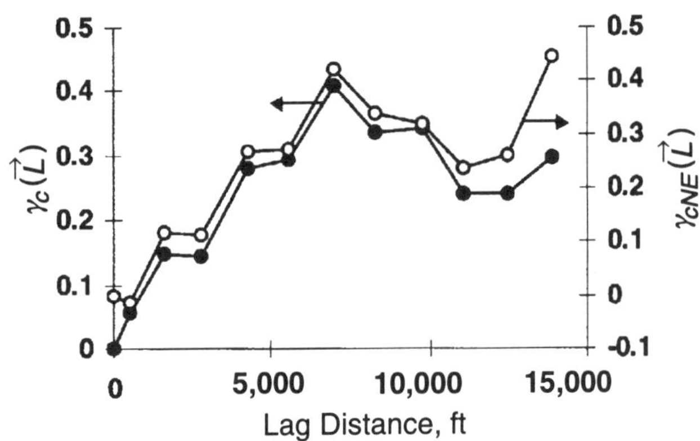

Fig. 3.54—Conventional and nonergodic cross-variograms for log IP and log net-*kh* data.

$$\text{and } \gamma_y(L) = c_{0_y} + c_{1_y} M_{S_a}(L). \quad (3.108)$$

Note that both variograms need to be modeled with the same linear combination of structures and that the range for both variograms must be the same. The only difference between the two structures are the sill values, which can be different. To model the cross variogram between the two variables, we can write

$$\gamma_{c_{xy}}(L) = c_{0_{xy}} + c_{1_{xy}} M_{S_a}(L). \quad (3.109)$$

As in modeling variograms, we are restricted by the same linear combinations of models in modeling the cross variogram, and the range must also be the same for the cross-variogram structure. In addition, the coefficients of the model should be defined so that the following two conditions are satisfied.

$$c_{0_x} c_{0_y} > c_{0_{xy}}^2 \quad (3.110)$$

$$\text{and } c_{1_x} c_{1_y} > c_{1_{xy}}^2. \quad (3.111)$$

That is, for a given model, the product of the coefficients of individual variable variogram models should be greater than the square of the coefficient of the cross variogram.

This requirement does not restrict the modeling to a combination of only two models. To generalize, if we model the variograms for x and y variables, respectively, using

$$\gamma_x(L) = \sum_{i=0}^{I} c_{i_x} \gamma_{a_i}^i(L) \quad (3.112)$$

$$\text{and } \gamma_y(L) = \sum_{i=0}^{I} c_{i_y} \gamma_{a_i}^i(L), \quad (3.113)$$

where c_i = coefficients for individual models, $\gamma_i(L)$ = the ith model that can be any acceptable model, and a_i = ranges for individual models. Then the cross variogram model is

$$\gamma_{xy}(L) = \sum_{i=0}^{I} c_{i_{xy}} \gamma_{a_i}^i(L). \quad (3.114)$$

To satisfy the condition of positive definiteness requires that

$$c_{i_x} c_{i_y} > \left(c_{i_{xy}}\right)^2 \quad (3.115)$$

for all i. All the components do not have to be present in modeling all three variograms (two variograms and one cross variogram). An additional modeling component may be present in one of the variograms, and it may be absent in the cross variogram as well as the other variogram. For example, if the variogram for the x variable is modeled as

$$\gamma_x(L) = c_{0_x} + c_{1_x} M_{E_a}(L) + c_{2_x} \frac{\sin(a_x L)}{L}, \quad (3.116)$$

the variogram for the y variable is modeled as

$$y_y(L) = c_{0_y} + c_{1_y} M_{E_a}(L), \quad (3.117)$$

and the cross variogram is modeled as

$$y_{xy}(L) = c_{0_{xy}} + c_{1_{xy}} M_{E_a}(L). \quad (3.118)$$

This satisfies the condition of positive definiteness, and the additional structure present for the x variable does not violate Eq. 3.115. However, an additional structure in the cross variogram model, without the corresponding structure in the variograms, does violate Eq. 3.115. That is, the models used in modeling the cross variogram must be present in modeling the variograms; however, extra model components can be used in modeling the variograms without including them in modeling the cross variogram.

The approach for modeling anisotropic cross variograms is essentially the same as that for modeling anisotropic variograms. For an anisotropic model, the directions of maximum and minimum continuity for the variograms and the cross variogram have to be the same. Otherwise, there are no additional restrictions on anisotropic modeling.

As can be seen from this discussion, modeling anisotropic variograms requires practice and some creativity. It is much more important to model all the variograms and cross variograms reasonably well rather than modeling one variogram well and sacrificing the others. If the variable values vary by several orders of magnitude, the sample values probably should be normalized by the individual sample means before the variogram is estimated. For example, for variable x, a normalized variable may be defined as

$$x_{ri} = \frac{x_i}{\bar{x}}, \quad (3.119)$$

where $\bar{x}$ = the sample mean and x_{ri} = a normalized sample value at location i. With normalizing, both variables have the same order of magnitude ranges, and modeling, as well as future estimation at unsampled locations, becomes relatively easy. Normalizing alters only the magnitude of the variogram, not the structure.

Field Example 3.16. Model the cross variogram between porosity and permeability for Well 34-29.

Solution. We generated a variogram for the porosity data in Field Example 3.1 and a cross variogram between the porosity and the log of permeability data in Field Example 3.14. Before modeling the cross variogram between the porosity and log of permeability data, we need to model the porosity and the log of permeability variograms. **Fig. 3.55** shows the comparison between the model and the estimated variograms for the porosity and the log of permeability data. The model for porosity data is

$$\gamma(L) = 5.0 + 35.0 M_{S_{13}}(L).$$

The model for the log of permeability data is

$$\gamma(L) = 0.35 + 1.02 M_{S_{13}}(L).$$

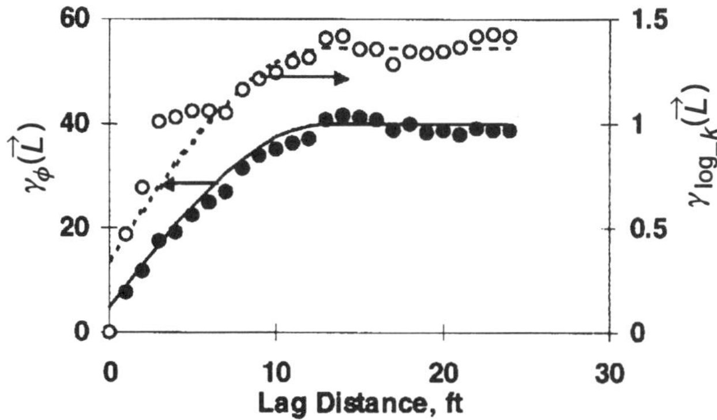

Fig. 3.55—Estimated and model variograms for Well 34-29.

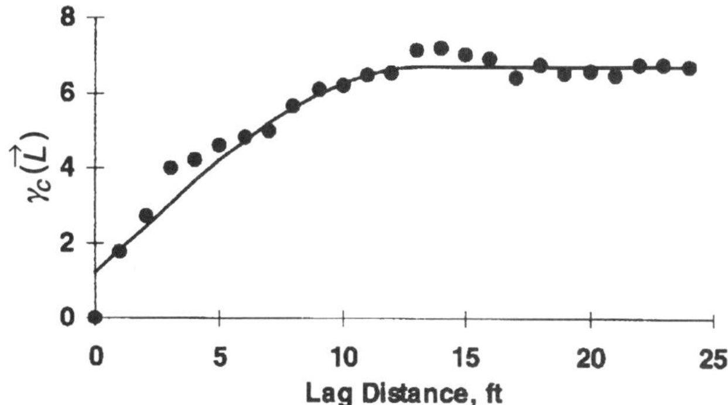

Fig. 3.56—Estimated and model variograms for Well 34-29 data.

Fig. 3.56 shows the comparison between the model and the estimated cross variogram. The model can be written as

$$\gamma_c(L) = 1.2 + 5.5M_{S_{13}}(L).$$

As previously discussed, in addition to using the same linear combination of models, we also must satisfy Eq. 3.115. To satisfy Eq. 3.115, we need to ensure that

$$(5.0)(0.35) > (1.2)^2 = 1.75 > 1.44$$

and $(35.0)(1.02) > (5.5)^2 = 35.7 > 30.25$.

As can be seen, Eq. 3.115 is valid, ensuring that the condition of positive definiteness is satisfied.

Field Example 3.17. Model the cross variogram between the log of IP and log-*kh* data for Well 34-29.

Solution. **Fig. 3.57** shows the comparison between the model and the estimated variograms for the log of IP and the log-*kh* data. The model for the log of IP data is

$$\gamma(L) = 0.06 + 0.3M_{G_{7,700}}(L),$$

and the model for log-*kh* data is

$$\gamma(L) = 0.15 + 0.58M_{G_{7,700}}(L).$$

These models are slightly different from the ones presented in Field Example 3.10 because of the requirement that both models be the same type and have the same range.

The model for the cross variogram is

$$\gamma_c(L) = 0.06 + 0.3M_{G_{7,700}}(L).$$

The comparison between the model and the cross variogram is shown in **Fig. 3.58**. Eq. 3.115 requires that

$$(0.06)(0.15) > (0.06)^2$$

and $(0.3)(0.58) > (0.3)^2$.

These equations are satisfied; therefore, the condition of positive definiteness is satisfied.

Although Field Examples 3.16 and 3.17 illustrate that modeling can be carried out so that the condition of positive definiteness is satisfied, modeling the cross variogram is tedious and requires several false starts before the correct type of modeling is used. Xu *et al.*[31] proposed one way to avoid modeling cross variograms. If we assume that the *x* and *y* variables are linearly related (i.e., log permeability vs. porosity or seismic impedance vs. porosity), we can model the variogram for the primary variable, *x*. The primary variable is the variable we are interested in estimating at unsampled locations (e.g., permeability in a logged well). By making the additional assumption that variable *y* at a given location is influenced only by variable *x* at the same location, we can write

$$C_{c_{xy}}(h) = C_{c_{xy}}(0) \times \frac{C_x(h)}{C_x(0)}, \qquad (3.120)$$

where $C_{c_{xy}}(L)$ = cross covariance at a lag distance L, $C_{c_{xy}}(0)$ = cross covariance at $L = 0$, $C_x(L)$ = covariance for variable x at a lag distance L, and $C_x(0)$ = variance for variable x. Once we model the variogram for variable x, we can estimate the covariance for any lag distance L. Then the cross covariance at $L = 0$ and the variance for variable x are easy to calculate. We can then use Eq. 3.120 to estimate the cross covariance at any lag distance without ever having to model it. This equation also satisfies the condition of positive definiteness. Appendix C provides details for Eq. 3.120.

In summary, the cross variogram provides a quantitative measure of the spatial relationship between the two variables.

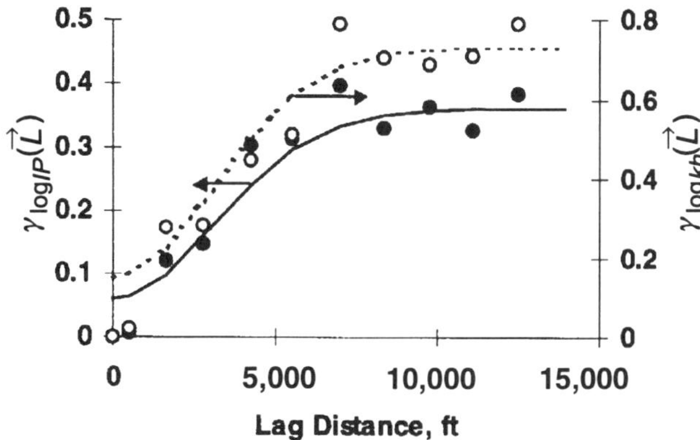

Fig. 3.57—Estimated and model variograms for log IP and log net-*kh* data.

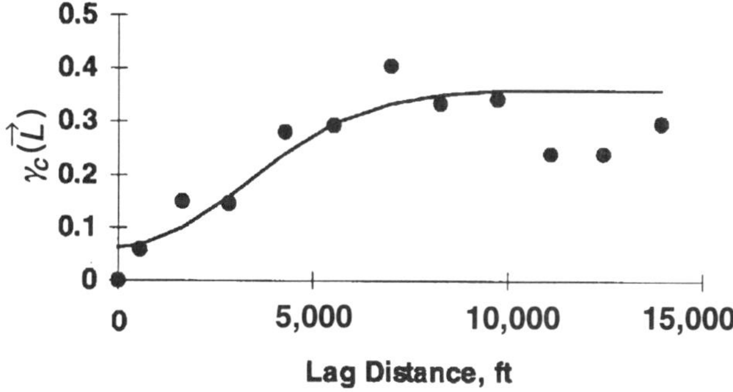

Fig. 3.58—Estimated and model cross variograms for log IP and log net-*kh* data.

Estimation of the cross variogram requires the use of a simple equation. Similar to the variogram, appropriate tolerances have to be defined with respect to the lag distance and the direction to ensure sufficient numbers of pairs. In addition, we can estimate the nonergodic cross variogram to examine the effect of variations in the lag means from the sample mean.

Once estimated, modeling the cross variogram can be done with the same set of models used to model conventional variograms. To satisfy the condition of positive definiteness, the same linear combination of models must be used to model the cross variogram as is used to model variograms for the original variables, and the coefficients of the models must satisfy Eq. 3.115. As a result, modeling cross variograms requires practice and ingenuity. Making some additional assumptions allows us to infer the cross variogram if we know the variogram of the primary variable. This avoids cumbersome modeling of the cross variogram as long as the assumptions used are valid.

3.7 Alternative Methods of Spatial Relationships

In the previous sections, we discussed estimation and modeling of variograms. The variogram is the most commonly used statistical measure to describe the spatial relationship. As Sec. 3.6 discussed, however, modeling the variogram is restricted by certain conditions. Further, the variogram can define only connectivity of two points: a variable at a given point and at another point located lag distance L away. It does not allow simultaneous definition of connectivity of multiple points, which may be important under certain conditions.

Some new estimation methods have appeared in recent literature that may relax the traditional restrictions on the spatial relationships. These methods, which are discussed in Chap. 6, are much more flexible and allow wide latitude in quantifying the spatial relationships. In this section, we present only some of the alternative techniques used to describe spatial relationships. The section is divided into two subsections: the first discusses some possible modifications of traditional variogram methods, and the second discusses multipoint histogram techniques that allow simultaneous definition of connectivity of more than two points.

3.7.1 Modified Variograms. The definition of the variogram is based on the variance of the difference in variable measured at two points located a certain distance apart. Recall that Eq. 3.15 states that

$$\gamma(\vec{L}) = \frac{1}{2}V\left[X(\vec{u}) - X\left(\vec{u} + \vec{L}\right)\right]. \quad \ldots\ldots\ldots\ldots (3.15)$$

On the basis of the first-order stationarity assumption, Eq. 3.15 reduces to

$$\gamma(\vec{L}) = \frac{1}{2}E\left\{\left[X(\vec{u}) - X\left(\vec{u} + \vec{L}\right)\right]^2\right\}. \quad \ldots\ldots\ldots (3.18)$$

This traditional variance-based definition is convenient because it allows us to define the uncertainty with respect to estimation at an unsampled location. As we discuss in Chap. 4, when we estimate a value at an unsampled location using the modeled variogram, we also estimate an error variance associated with the estimate. The error variance quantifies the uncertainty with respect to the estimate. Specifically, if we assume a Gaussian distribution with respect to the error, we can completely define our uncertainty by knowing the mean (the estimated value) and the error variance.

As an alternative, if we quantify the uncertainty using some other method, we may not need to use a variance-based spatial relationship, such as the variogram. These alternative methods currently are available. They are called conditional simulation methods, and they quantify uncertainties by generating multiple images of the unsampled space. Chaps. 6 and 7 discuss these methods in more detail.

Once the restriction of a variance-based spatial relationship is removed, the traditional variogram equation can be modified as[32]

$$\gamma(\vec{L}) = \frac{1}{2}E\left\{\left|X(\vec{u}) - X\left(\vec{u} + \vec{L}\right)\right|^w\right\} \text{ for } w > 0, \quad \ldots\ldots\ldots\ldots\ldots\ldots (3.121)$$

where w can take a value between zero and two, excluding zero. If we choose $w = 2$, Eq. 3.121 becomes a conventional variogram. If $w = 1$, it is called a madogram, which is a measure of the absolute deviation between the two values. If $w = 0.5$, it is called a rodogram, which is the square root of the absolute deviation. The smaller the value of w, the more resistant the spatial relationship is to outlier data. In practice, we can estimate the modified variogram by

$$\hat{\gamma}(\vec{L}) = \frac{1}{2n(\vec{L})}\sum_{i=1}^{n(\vec{L})}\left\{\left|x(\vec{u}_i) - x\left(\vec{u}_i + \vec{L}\right)\right|^w\right\}, \quad \ldots\ldots (3.122)$$

where $n(\vec{L})$ = the number of pairs at a lag distance $\vec{L}$ apart. Once we estimate the variogram, we can estimate the modified covariance by

$$C(\vec{L}) = \gamma(\infty) - \gamma(\vec{L}), \quad \ldots\ldots\ldots\ldots\ldots\ldots (3.123)$$

where $\gamma(\infty)$ = the modified variogram value at a large lag distance. It can be considered as a sill value if the variogram reaches a sill; otherwise, it has to be defined as the value at a distance that is greater than the maximum distance within a region of interest.

Although we modified the variogram equation that minimizes the influence of the outlier data, we still need to model the modified variogram using the same models that satisfy the condition of the positive definiteness.

Field Example 3.18. This example uses the net-*kh* data used in the previous field examples. Estimate the madogram and rodogram for these data and model them using conventional models.

Solution. **Figs. 3.59 and 3.60** compare the conventional and modified variograms. The modified variograms do show some improvement over conventional variograms in terms of better stability. However, the improvement is not as significant as with some other transforms presented previously. Fig. 3.59 also shows the comparison between the model and the estimated madogram. The model is written as

$$\gamma(L) = 2,500 + 3,000M_{E_{15,000}}(L).$$

Fig. 3.60 shows the comparison between the model and the estimated rodogram. The model is written as

$$\gamma(L) = 30 + 15M_{E_{15,000}}(L).$$

Both modified variograms allow a slightly better definition of the variogram. Another advantage of using these variograms is that they do not require nonlinear transformation of the sample data to achieve better stability with respect to the spatial relationship.

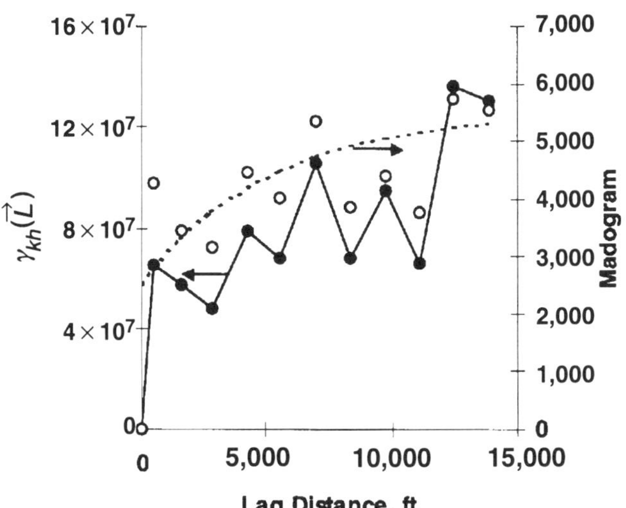

Fig. 3.59—Conventional variogram and madogram for net-*kh* data.

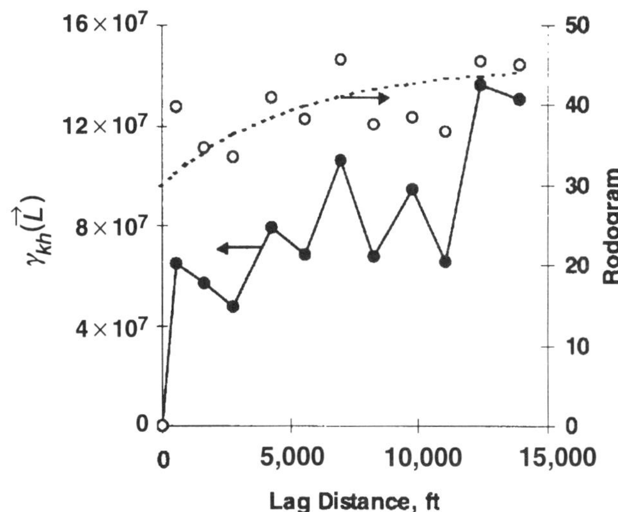

Fig. 3.60—Conventional variogram and rodogram for net-*kh* data.

3.7.2 Multipoint Histograms. Multipoint histograms allow definition of the connectivity of multiple points instead of only two points as the variogram does. Currently, application of multiple histograms is restricted to discrete variables rather than continuous variables.

To understand multipoint histograms, we first need to understand the Markov-chain principle, which is commonly applied in describing geological features.[33] A Markov chain is a method by which the ordering of geological observations can be quantified in terms of transition probabilities. Transition probabilities represent the probabilities that one feature will be followed by another feature. Numerical Example 3.8 illustrates the calculation of transition probabilities.

Numerical Example 3.8. In a vertical well, three prominent facies that are present have been identified at every 1-ft interval. **Table 3.9** shows the ordering of these three facies from top to bottom, with the facies defined as A, B, or C. Estimate the transition probabilities.

TABLE 3.9—FACIES FOR NUMERICAL EXAMPLE 3.8
Top
B
B
B
A
B
C
C
C
B
B
A
A
A
B
C
C
Bottom

Solution. We know the ordering of the facies at the well. If we assume that the probability that a given facies at a certain location is dependent only on the facies at the previous location, we can build a transition-frequency matrix as shown in **Fig. 3.61a.** This transition-frequency matrix shows the number of times a given facies succeeds the other facies. The first number in the matrix, 2, (top-left corner) represents the two times that Facies A follows Facies A. The number 0 in the top-right corner indicates that Facies C never follows Facies A. Other numbers are calculated similarly. The total number of transition frequencies is 16, which represents the total number of transitions because there is a total of 17 observations.

To define the frequencies in a probabilistic framework, we can divide the frequencies in each row by the row totals. We then obtain the matrix shown in **Fig. 3.61b.** This matrix is called a transition-probability matrix. Each number in this matrix represents a conditional probability. Probability $P(A|B)$ is the probability that Facies A occurs at the next location given that Facies B is observed at the present location. In the first row, if Facies A is observed at a given location, then the probability that Facies A is present below that location is 50% and that Facies B is present below that location is 50%. Facies C never follows Facies A.

The transition-probability matrix quantifies the spatial relationship for discrete variables by defining conditional probabilities for each facies at a given location, depending on what is present at the previous location. This is especially useful in describing geological variables, which are qualitative in nature but can be quantified with this method. Further, in geological environments, some transition probabilities need to be zero, and some need to be very high. Some geological

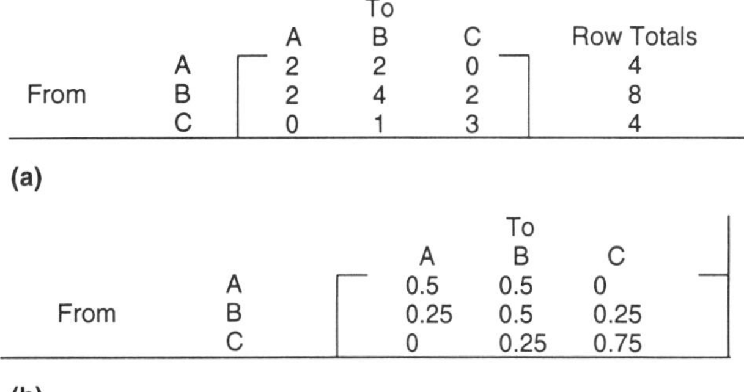

		To: A	To: B	To: C	Row Totals
From	A	2	2	0	4
	B	2	4	2	8
	C	0	1	3	4

(a)

		To: A	To: B	To: C
From	A	0.5	0.5	0
	B	0.25	0.5	0.25
	C	0	0.25	0.75

(b)

Fig. 3.61—Transition matrices for Numerical Example 3.8.

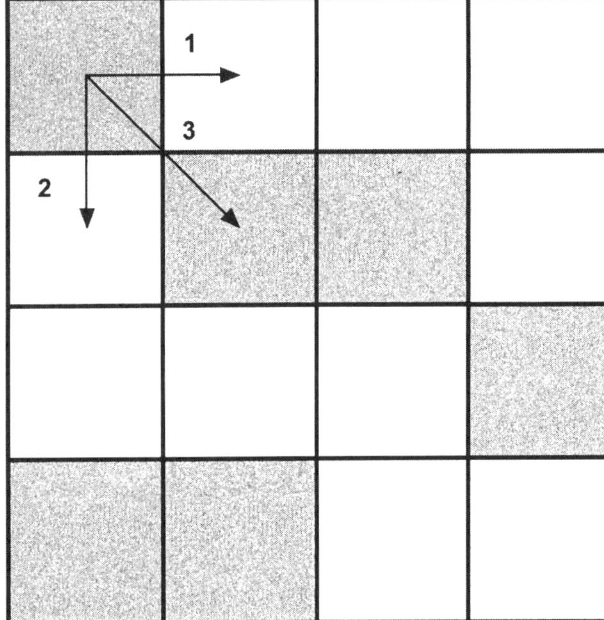

Fig. 3.62—Configuration for multipoint histogram example.

processes require that two facies never be close to each other, while other processes require that two facies always be close to each other. This type of information can be quantified through the transition-probability matrix.

While extremely useful, the Markov-chain analysis has one major disadvantage: it does not allow extension of spatial relationships beyond adjacent points. Although higher-order Markov chains can be defined that allow for relationships between two points that are more than one step apart, these relationships still are based on the underlying transition probabilities, which are described on the basis of one step distance. Further, like the variogram, Markov chains can define only the joint probability of two points being a certain combination of facies. It does not allow, for example, definition of the joint probability that A, B, and C are present one after the other.

To overcome these drawbacks of the Markov-chain principle, we can use multipoint histograms. Multipoint histograms are extremely flexible and allow definition of the connectivity of multiple points in a probabilistic framework.[34] This is illustrated best through an example. **Fig. 3.62** consists of 16 squares, either black or white. This configuration can be captured through various means. If we consider a single-point histogram, we can state that out of a total of 16 squares, six are black and ten are white. In a probabilistic framework, 0.625 ($= 10/16$) are white and 0.375 are black. This is the simplest information we can infer from the figure. Extending it to a two-point histogram (which is similar to the Markov-chain analysis), we can consider multiple directions. Fig. 3.62 shows three of those directions. For direction 1, there are four possible combinations for two-point histograms. **Fig. 3.63** shows that there could be two black squares, two white squares, a black square followed by a white square, or a white square followed by a black square. Examination of Fig. 3.62 allows calculation of how many of each of the configurations are present. We can estimate that combination 1 is present two times, combination 2 is present three times, combination 3 is present two times, and combination 4 is present five times. In terms of a probabilistic framework, the probability of combination 1 is $2/12 = 0.167$, the probability of combination 2 is $3/12 = 0.25$, the probability of combination 3 is

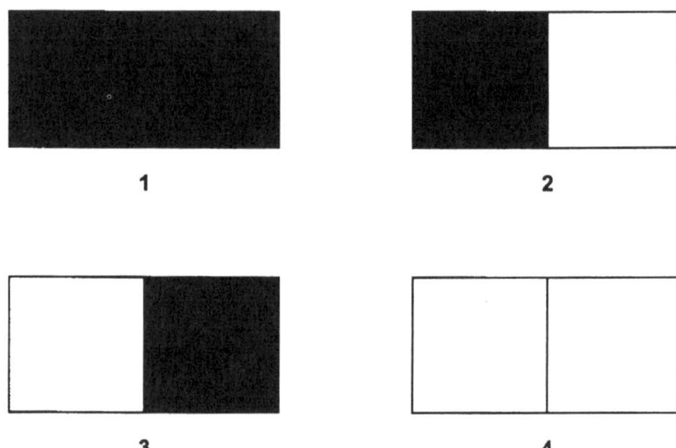

Fig. 3.63—Four possible combinations for Direction 1.

$2/12 = 0.167$, and the probability of combination 4 is $5/12 = 0.416$. Similar calculations can be carried out for other directions.

For example, **Fig. 3.64** shows that we also have four combinations in direction 3. Examining Fig. 3.62, we can state that the probability of combination 1 is $2/9 = 0.222$, the probability of combination 2 is $1/9 = 0.111$, the probability of combination 3 is $2/9 = 0.222$, and the probability of combination 4 is $4/9 = 0.445$. We can also calculate the probabilities of the four combinations for direction 2.

Obviously by collecting this additional information, we now know more about the figure. We can extend this even further by considering a four-point histogram (**Fig. 3.65**). For this histogram, 16 possible combinations exist. Fig. 3.65 shows some of these combinations. Examining Fig. 3.62, we can count the number of times each combination occurs. **Table 3.10** shows each combination and its associated probability. The probabilities are based on nine repetitions of this four-point histogram. This type of information captures more of the pattern and includes the one- and two-point histograms. That is, if we calculate four-point histograms, we do not need to account for the lower-point histograms separately.

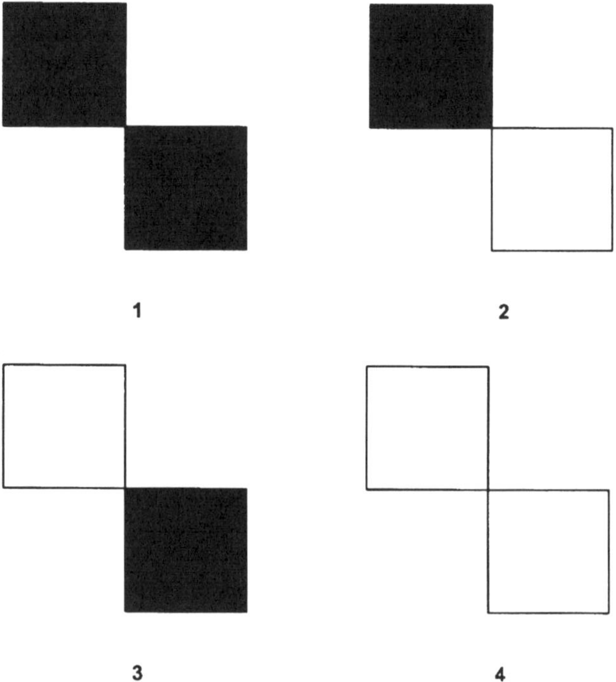

Fig. 3.64—Four possible combinations for direction 3.

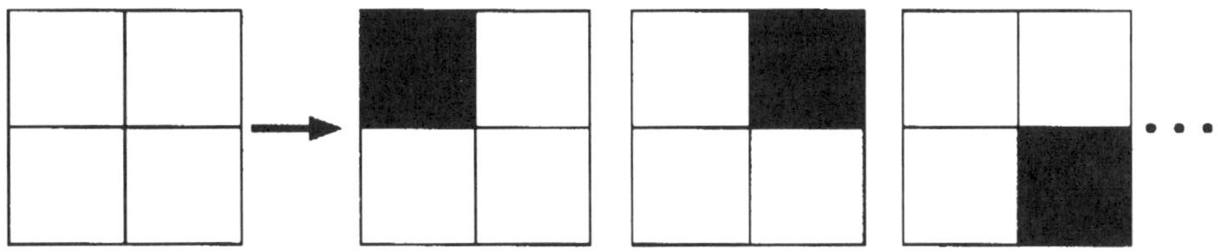

Fig. 3.65—Four-point histogram.

TABLE 3.10—COMBINATION PROBABILITIES FOR FOUR-POINT HISTOGRAM

Combination	Frequency	Probability
□□ □□	0	0
■□ □□	0	0
□■ □□	0	0.222
□□ □■	0	0
□□ ■□	0	0.222
■□ □■	0	0.222
□■ ■□	0	0
■■ □□	1	0.111
□■ □■	0	0
□□ ■■	2	0.222
■□ ■□	0	0
■■ □■	0	0
□■ ■■	0	0
■□ ■■	0	0
■■ ■□	0	0
■■ ■■	0	0

Fig. 3.66—Configurations of four-point histogram.

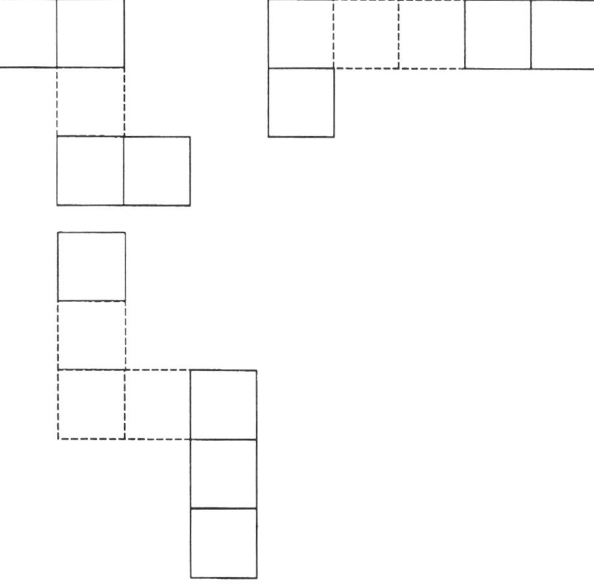

Fig. 3.67—Examples of noncontiguous four-point histogram.

This discussion shows that the multipoint histogram is a very flexible technique that can incorporate different patterns as well as different numbers of facies. The pattern has to be determined on the basis of the appropriate shape and size of the facies we need to capture. For example, a four-point histogram can take several alternative shapes (**Fig. 3.66**).

Which shape to choose depends on what we want to preserve. For example, to capture a channel sand, the shape on the extreme left may be better than the other shapes. Or, to capture thin layers of shale, the middle shape may be more appropriate. In addition, the multipoint histogram does not have to represent contiguous blocks. It may establish a relationship between blocks that are not attached to each, as **Fig. 3.67** shows for a four-point histogram.

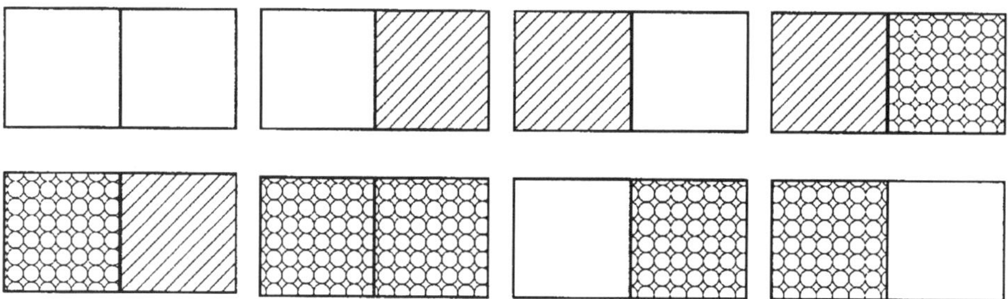

Fig. 3.68—Combinations for two-point histogram with three categories.

These examples concentrated on the presence of only two facies. The technique can be extended easily to more facies or categories. A simple equation allows calculation of how many possible combinations have to be dealt with for a particular multipoint histogram. For example, for F categories and an N-point histogram, the total number of combinations is F^N. For example, for three facies and a two-point histogram, there are nine possible combinations (**Fig. 3.68**). With three categories and a four-point histogram, the total number of combinations is $3^4 = 81$. Obviously, as the number of category increases and the number of points in the multipoint histogram increases, the number of combinations increases rapidly. Further, we can have several configurations within each multipoint type (Figs. 3.66 and 3.67). For each configuration, we have the same number of combinations. In principle, we can gather information on multipoint histograms for different configurations. However, collecting this information and eventually honoring it during the estimation process can be a computational burden. Applications of multipoint histograms to reservoir description are relatively new, and the optimal number of multipoint histograms needed for proper characterization is a guess that is affected by the type of geological process one needs to capture and the level of detail one wants to incorporate.

The multipoint-histogram technique can also be extended to continuous variables.[35] First, the continuous variables need to be discretized by use of an indicator transform. Recall that Eq. 3.26 states that

$$i(\vec{u}_j, x_t) = 1, \qquad \text{(3.26a)}$$

if $x(\vec{u}_j) \leqq x_t$, and

$$i(\vec{u}_j, x_t) = 0, \qquad \text{(3.26b)}$$

if $x(\vec{u}_j) > x_t$, where each variable value can be defined in terms of an indicator function that can take a value of either one or zero. Depending on the number of thresholds, we have an equivalent number of indicator values at each sampled location. We can define another type of indicator variable as

$$K(\vec{u}_j, x_t) = 0,$$

if $x(\vec{u}_j) \leqq x_t$, and

$$K(\vec{u}_j, x_t) = 0, \qquad \text{(3.124)}$$

if $x(\vec{u}_j) > x_t$, where K = a variable that is zero if the sample value is less than or equal to the threshold and one if the sample value is greater than the threshold. Obviously, $i + K$ always equals one. To define connectivity, we use an equation of the multipoint histogram.[33]

$$H_I(\vec{L}_1, \ldots, \vec{L}_N; x_{t_1}, \ldots, x_{t_N}) = E\left[\prod_{i=1}^{N} I(\vec{L} + \vec{L}_j; x_{t_j})\right]. \qquad \text{(3.125)}$$

This is an N-point histogram among locations placed at $\vec{h}_i, \ldots, \vec{h}_N$ distances, respectively, and the threshold values at each location are defined differently from x_{t_i} to x_{t_N}. A similar equation for the K variable is

$$H_K(\vec{L}_1, \ldots, \vec{L}_N; x_{t_1}, \ldots, x_{t_N}) = E\left[\prod_{j=1}^{N} K(\vec{L} + \vec{L}_j; x_{t_j})\right]. \qquad \text{(3.126)}$$

This represents a multipoint connectivity of K variables. Because, different thresholds rarely are defined in practice at each location, the histogram is estimated as

$$H_I(\vec{L}_1, \ldots, \vec{L}_N; x_t) = \frac{1}{n}\sum_{l=1}^{n}\left[\prod_{j=1}^{N} i(\vec{u}_l + \vec{L}_j; x_t)\right]. \qquad \text{(3.127)}$$

Similar equations can be written for the K variable. This multipoint histogram (Eq. 3.127) represents a joint probability that all values located from $\vec{h}_1$ to $\vec{h}_N$ will be less than the threshold x_t. Instead of defining only a relationship between two points, as the traditional variogram does, the multipoint histogram describes the joint relationship for multiple points. An equation like Eq. 3.127 written for the K variable also represents the probability that all values located from $\vec{h}_1$ to $\vec{h}_N$ will be greater than a certain threshold value. Such knowledge may be very important. For example, if we suspect a thief zone with a high-permeability value, it is very important that we know the probability that several values in a layer "simultaneously" will be greater than a given threshold. If the probability is high, it represents a high-permeability zone. The two-point variogram simply cannot capture this type of behavior.

As a special case, if we only consider $N = 2$ (a two-point histogram), Eq. 3.127 reduces to

$$H_I(\vec{L}_1, \vec{L}_2; x_t) = \frac{1}{n}\sum_{j=1}^{n} i(\vec{u}_l + \vec{L}_1; x_t)\, i(\vec{u}_j + \vec{L}_2; x_t). \qquad \text{(3.128)}$$

An equation for covariance can be written as

$$C_I(\vec{L}_1, \vec{L}_2; x_t) = \frac{1}{n}\sum_{j=1}^{n} i(\vec{u}_j + \vec{L}_1; x_t)\, i(\vec{u}_j + \vec{L}_2; x_t) - \frac{1}{n}\sum_{j=1}^{n} i(\vec{u}_j + \vec{L}_1; x_t)\frac{1}{n}\sum_{j=1}^{n} i(\vec{u}_j + \vec{L}_2; x_t). \qquad \text{(3.129)}$$

Note that the two-point histogram is the first part of the right side of a covariance equation; therefore, it is also called the noncentered covariance. Although Eq. 3.129 looks slightly

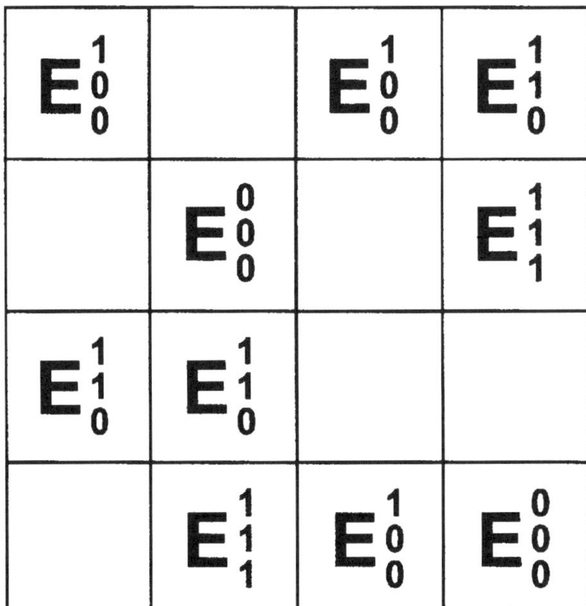

Fig. 3.69—Sample indicator values for Numerical Example 3.9.

different from the covariance equation we have seen in practice, substituting $\vec{L} = 0$ and $\vec{L}_2 = \vec{L}$ reduces Eq. 3.129 to

$$C_I(\vec{L};x_t) = \frac{1}{n}\sum_{j=1}^{n} i(\vec{u}_j;x_t)i(\vec{u}_j + \vec{L};x_t) - \frac{1}{n}\sum_{j=1}^{n} i(\vec{u}_j;x_t)\cdot\frac{1}{n}\sum_{j=1}^{n} i(\vec{u}_j + \vec{L};x_t), \quad (3.130)$$

which is a standard form of the covariance equation.

Numerical Example 3.9 illustrates calculation of the multipoint histogram or noncentered covariance.

Numerical Example 3.9. Fig. 3.69 presents the available sample data in terms of indicator variables. The sample data are already transformed into indicator variables by use of three threshold values. Estimate the following.

1. Estimate the two-point histogram at $\vec{L} = 1$ in an easterly direction for the configuration (Threshold 2) in **Fig. 3.70a.**

2. Estimate the two-point histogram at $\vec{L} = 1$ in a diagonal direction for the configuration (Threshold 1) in **Fig. 3.70b.**

3. Estimate the three-point histogram at $\vec{L} = 1$ for the configuration (Threshold 2) in **Fig. 3.70c.**

Solution. We do not have sample values at each grid location.

1. The two-point histogram at a lag interval for Threshold 2. For the configuration in **Fig. 3.71a**, we have four pairs. Using Eq. 3.128 gives

$$H_I(\vec{L}_1,\vec{L}_2;x_t) = \frac{1}{n}\sum_{j=1}^{n} i(\vec{u}_l + \vec{L}_1;x_t)i(\vec{u}_j + \vec{L}_2;x_t)$$

$$H_I(1,x_{t_2}) = \frac{1}{4}\left[(0 \times 1) + (1 \times 1) + (1 \times 0) + (0 \times 0)\right] = 0.25.$$

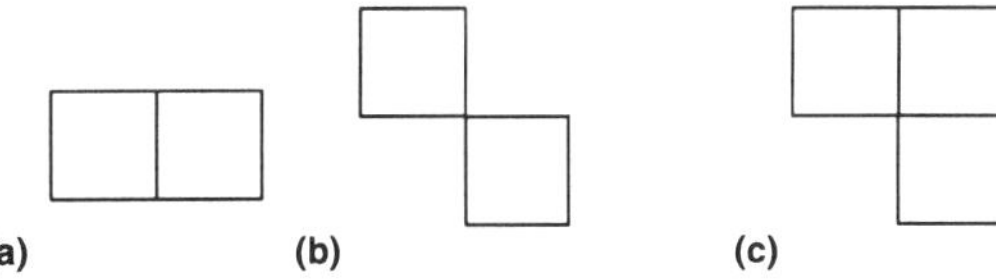

Fig. 3.70—Histograms for Numerical Example 3.9, Question 1.

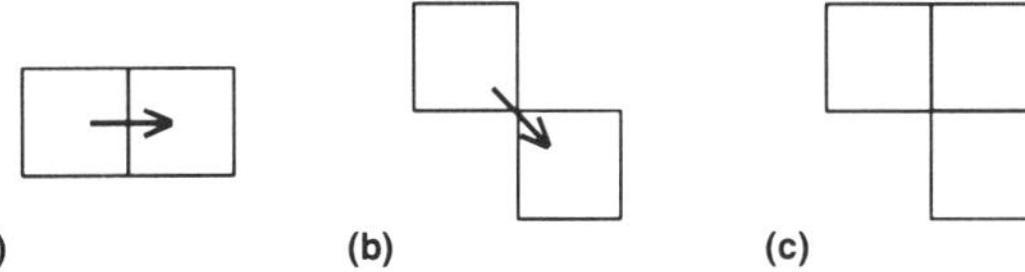

Fig. 3.71—Configurations for Numerical Example 3.9, Question 2.

That is, there is a 25% probability that two values located one lag distance apart are both less than the second threshold.

2. The two-point histogram at a lag interval for Threshold 1. For the configuration in **Fig. 3.71b,** we have four pairs. Using Eq. 3.128 gives

$$H_I(1,x_{t_1}) = \frac{1}{4}\left[(0 \times 0) + (0 \times 1) + (0 \times 1) + (0 \times 0)\right] = 0.$$

That is, there is a 0% probability that the two values located one lag distance apart are both less than the first threshold.

3. The three-point histogram at $\vec{L} = 1$ for Threshold 2. Two sample data sets satisfy the configuration in **Fig. 3.71c.** Using Eq. 3.127 gives

$$H_I\left(\vec{L}_1,\vec{L}_2,\vec{L}_3;x_t\right) = \frac{1}{n}\sum_{l=1}^{n}\left[\prod_{j=1}^{N} i(\vec{u}_l + \vec{L}_j;x_t)\right] = \frac{1}{2}\left[(0 \times 1 \times 1) + (1 \times 1 \times 1)\right] = 0.5.$$

That is, there is a 50% probability that all three values will be less than the second threshold. Similar configurations can be repeated for other thresholds and other configurations.

To summarize the multipoint histogram is a very powerful technique that allows definition of multipoint connectivity of various configurations. The technique has been widely applied in pattern-recognition applications[4]; however, its use in reservoir description processes is relatively new. In principle, many geological features as well as petrophysical characteristics can be captured better by the multipoint histogram. However, a balance must be maintained between the need to capture detailed information vs. the computational demands placed in storing, analyzing, and applying such information. Currently, no optimization technique or procedure exists that allows for use of the right combination of multipoint histograms to capture essential information about reservoir characteristics. Once such a procedure is established, the multipoint histogram will become widely accepted methods for characterizing spatial connectivity.

Summary

The basic distinction between geostatistical techniques and other interpolation techniques is the ability of geostatistical techniques to incorporate customized spatial relationships. Defining and modeling spatial relationships is probably the practitioner's most important task. All subsequent decisions will be consistent with the assumptions made with respect to the spatial relationship. It is, therefore, very important that all possible avenues be explored to define as representative a spatial relationship as possible. The process can be time-consuming, is often frustrating, and, in many instances, can result in dead ends. This chapter attempted to provide the reader with multiple avenues to explore with the spatial data to find the spatial relationship. If enough sample data are not available, examining reservoirs located in similar depositional environments and studying the outcrop data are worthwhile. Use of multipoint histograms can be greatly enhanced if the geologic features can be captured through the outcrop data. In certain cases, geological contour maps can be used to capture spatial relationships in the absence of adequate data. Any possible information, qualitative or quantitative, that can be used to define spatial characteristics better should be explored, and, if possible, incorporated in the spatial-relationship description. As subsequent chapters indicate, once the spatial relationship is identified, the estimation at the unsampled locations is relatively straightforward.

Nomenclature

a = range
C = sill value
C_s = scaling parameter for f_{Bm} and/or f_{Gn} models
$C(\vec{L})$ = covariance at lag distance $\vec{L}$
$C(X, Y)$ = covariance between variables X and Y
$c(\vec{L})$ = estimated covariance at lag distance $\vec{L}$
$E[\ ]$ = expected value of the bracketed term
f[ ] = function of the bracketed term
f_{Bm} = fractional Brownian motion
f_{Gn} = fractional Gaussian noise
h = thickness, L, ft
H = intermittency exponent
$H_I(h_1,\ldots,H_N;x_{t_1},\ldots,x_{t_N})$ = N-point indicator histogram with N different thresholds
$i(\vec{u}, x_t)$ = indicator value at Location $\vec{u}$ for threshold x_t
$I(\vec{u}, x_t)$ = indicator variable at Location $\vec{u}$ for threshold x_t
k = permeability
$K(u_j, x_t)$ = indicator value at Location u_j for threshold x_t (Eq. 3.123)
L = lag distance, L, ft
$\vec{L}$ = lag-distance vector
$\vec{L}_O$ = distance vector based on original axes (Eq. 3.26)
$\vec{L}_R$ = distance vector based on rotated axes
$\Delta\vec{L}$ = lag-distance-vector tolerance
$M_{E_a}(L)$ = exponential variogram model with a range a
$M_{f_{Bm}}$ = fractional Brownian motion model
$M_{f_{Gn}}$ = fractional Gaussian noise model
$M_{G_a}(L)$ = Gaussian variogram model with a range a
$M_{S_a}(L)$ = spherical variogram model with a range a
n = total number of samples, n
$n(\vec{L})$ = number of pairs at lag distance $\vec{L}$, n
N = number of points connected in a multipoint histogram, n
p = power in a power transform
R_i = rank of ranked value i among n samples
$[\mathbf{R}]$ = rotation matrix
$r(\vec{L})$ = correlation coefficient at lag distance $\vec{L}$
s^2 = variance of sample data
$\vec{u}$ = location $\vec{u}$ or axis direction
$\vec{v}$ = axis direction
$V[\ ]$ = variance of the bracketed term
$\vec{w}$ = axis direction
$\bar{x}$ = sample mean
x_t = threshold value for indicator transform
$x(\vec{u})$ = sample value of variable X at location $\vec{u}$
$x_T(\vec{u})$ = transformed sample value at location $\vec{u}$
$x^*(\vec{u}_o)$ = estimated value of variable x at location $\vec{u}_o$
X = variable X
$X(\vec{u})$ = variable X at location $\vec{u}$
$X(\vec{u} + \vec{L})$ = variable X at location $\vec{u} + \vec{L}$
$y(\vec{u})$ = sample value of variable Y at location $\vec{u}$
Y = variable Y
α = relative amplitude in sine model
$\gamma(\vec{L})$ = variogram at lag distance $\vec{L}$
$\hat{\gamma}(\vec{L})$ = estimated variogram at lag distance $\vec{L}$
δ = resolution of a measurement
θ = angle at which $u-v$ plane is rotated with respect to the w axis to coincide with the axis of anisotropy, degrees
$\Delta\theta$ = tolerance with respect to angle θ
λ_i = weight assigned to sample point i for estimation purposes
$\rho(\vec{L})$ = correlation coefficient at lag distance $\vec{L}$
σ = standard deviation
ϕ = angle through which $u-w$ plane is rotated with respect to the v axis, degrees
ω = exponent in a modified variogram equation (Eq. 3.120)

Subscripts

a = range
c = cross relationship—either variogram or covariance
D = dimensionless
GR = general relative
IP = initial potential
$\vec{L}$ = all samples corresponding to lag distance $\vec{L}$
$-\vec{L}$ = first data points in sample pairs located $\vec{L}$ lag distance apart
$+\vec{L}$ = second data points in sample pairs located $\vec{L}$ lag distance apart
N = number of points in a multipoint variogram
NE = nonergodic
PR = pairwise relative
r = normalized
$\vec{u}, \vec{v}, \vec{w}$ = directions of axes
$\vec{u}', \vec{v}', \vec{w}'$ = directions of original axes
x = variable X
y = variable Y
0, 1, 2 = sill values

References

1. Matheron, G.: "The Intrinsic Random Functions and Their Applications," *Advances in Applied Probability* (1973) **5,** 439.

2. Isaaks, E. and Srivastava, R.: "Spatial Continuity Measures for Probabilistic and Deterministic Geostatistics," *Math Geology* (1988) **20,** No. 4, 313.
3. Journel, A. and Rossi, M.: "When Do We Need a Trend Model in Kriging?" *Math Geology* (1989) **21,** No. 7, 715.
4. Ripley, B.: *Statistical Inference for Spatial Processes,* Cambridge U. Press, New York City (1988).
5. Deutsch, C.V. and Journel, A.G.: *Geostatistical Software Library and User's Guide,* Oxford U. Press, New York City (1992) Chap. 3.
6. Isaaks, E.H. and Srivastava, R.M.: *An Introduction to Applied Geostatistics,* Oxford U. Press, New York City (1989) Chap. 7.
7. Isaaks, E.: "Risk Qualified Mappings for Hazardous Waste Sites: A Case Study in Distribution-Free Geostatistics," MS thesis, Stanford U., Stanford, California (1984) 85.
8. Alabert, F.: "Stochastic Imaging of Spatial Distributions Using Hard and Soft Information," MS thesis, Stanford U., Stanford, California (1987) 185.
9. Journel, A.G. and Alabert, F.: "New Method for Reservoir Mapping," *JPT* (February 1990) 212.
10. Journel, A.G. and Alabert, F.: "Non-Gaussian Data Expansion in the Earth Sciences," *Terra Nova* (1989) **1,** 123.
11. Suro-Perez, V. and Journel, A.G.: "Indicator Principal Component Kriging," *Math Geology* (1991) **23,** No. 5, 759.
12. David, M.: *Geostatistical Ore Reserve Estimation,* Elsevier Publishing Co., Amsterdam (1977).
13. Isaaks, E.H. and Srivastava, R.M.: "Spatial Continuity Measures for Probabilistic and Deterministic Geostatistics," *Math Geology* (1988) **20,** No. 4, 313.
14. Christoakos, G.: *Random Field Models in Earth Sciences,* Academic Press, New York City (1992) Chap. 2.
15. Srivastava, R.M. and Parker, H.M.: "Robust Measures of Spatial Continuity," *Proc.*, Third Intl. Geostatistics Congress M. Armstrong (ed.), Lugar Academic Publishers (1989) 295.
16. *GeoEAS (Geostatistical Environmental Assessment Software): User's Guide,* Environmental Monitoring Systems Laboratory, U.S. Environmental Protection Agency (EPA), Las Vegas, Nevada (1988).
17. Journel, A.G.: *Geostatistics for the Environmental Sciences,* Environmental Monitoring Systems Laboratory, U.S. EPA, Las Vegas, Nevada (1987) Chap. 5.
18. Christoakos, G.: "On the Problem of Permissible Covariance and Variogram Models," *J. Water Resources Res.* (February 1984) **20,** No. 2, 251.
19. Diamond, P. and Armstrong, M.: "Robustness of Variograms and Conditioning of Kriging Matrices," *Math Geology* (1984) **16,** 809.
20. Posa, D.: "Conditioning of the Stationary Kriging Matrices for Some Well Known Covariance Models," *Math Geology* (1989) **21,** No. 4, 755.
21. Journel, A.G. and Huijbregts, C.J.: *Mining Geostatistics,* Academic Press, New York City (1989) Chap. 3.
22. Hewett, T.A.: "Fractal Distribution of Reservoir Heterogeneity and Their Influence on Fluid Transport," paper SPE 15386 presented at the 1986 SPE Annual Technical Conference and Exhibition, New Orleans, 5–8 October.
23. Feder, J.: *Fractals,* Plenum Press, New York City (1988) Chap. 10.
24. Voss, R.F.: "Fractals in Nature: From Characterization to Simulation," *Science of Fractal Images*, Peitgen and Saupe (eds.), Springer Verlag, New York City (1988) 21–69.
25. Emanual, A.S. *et al.*: "Reservoir Prediction Methods Based on Fractal Geostatistics," *SPERE* (August 1989) 311; *Trans.*, AIME, **287.**
26. Hewett, T.A. and Behrens, R.A.: "Conditional Simulation of Reservoir Heterogeneity With Fractals," *SPEFE* (September 1990) 217; *Trans.*, AIME, **289.**
27. Tang, R.W., Behrens, R.A., and Emanuel, A.S.: "Reservoir Studies Using Geostatistics To Forecast Performance," *SPERE* (May 1991) 253.
28. Crane, S.D. and Tubman, K.M.: "Reservoir Variability and Modeling with Fractals," paper SPE 20606 presented at the 1990 SPE Annual Technical Conference and Exhibition, New Orleans, 23–26 September.
29. Perez, G. and Kelkar, M.: "Assessing Distributions of Reservoir Properties Using Horizontal Well Data," *Reservoir Characterization III*, B. Linville (ed.), PennWell Publishing Co., Tulsa, Oklahoma (1993) 399–436.
30. Beier, R.A. and Hardy, H.H.: "Comparison of Different Horizontal Fractal Distributions in Reservoir Simulation," *Reservoir Characterization III*, B. Linville (ed.), PennWell Publishing Co., Tulsa, Oklahoma (1993) 271–296.
31. Xu, W. *et al.*: "Integrating Seismic Data in Reservoir Modeling: The Collocated Cokriging Alternative," paper SPE 24742 presented at the 1992 SPE Annual Technical Conference and Exhibition, Washington, D.C., 4–7 October.
32. Journel, A.G.: "New Distance Measures: The Route Toward Truly Non-Gaussian Geostatistics," *Math Geology* (1988) **20,** No. 4, 459.
33. Davis, J.C.: *Statistics and Data Analysis in Geology,* second edition, John Wiley and Sons, New York City (1986) Chap. 4.
34. Farmer, C.: "Numerical Rocks," *The Mathematical Generation of Reservoir Geology,* J. Fayers and P. King (eds.), Oxford U. Press, New York City (1991).
35. Deutsch, C.V. and Journel, A.G.: "Annealing Techniques Applied to the Integration of Geological and Engineering Data," fifth annual report, Stanford Center for Reservoir Forecasting, Stanford, California (1992) 120.

SI Metric Conversion Factor

ft × 3.048* E − 01 = m

*Conversion factor is exact.

Chapter 4
Conventional Estimation Techniques

Chap. 3 discussed various modeling techniques to capture and quantify spatial relationships of reservoir variables. Chap. 4 utilizes those models to estimate values at unsampled locations. Here, discussion focuses on conventional estimation techniques, commonly known as kriging techniques. Subsequent chapters discuss other types of estimation techniques, called conditional simulation techniques.

The term kriging comes from Danny Krige, a South African geoscientist, who first applied this technique to gold mines. Later, the mathematical validity and foundation was provided by Matheron.[1]

The kriging technique uses a linear estimation procedure to estimate a value at unsampled locations. In principle, the technique assumes that the value at the unsampled location is estimated by

$$X^*(\vec{u}_o) = \sum_{i=1}^{n} \lambda_i X(\vec{u}_i), \quad \text{(4.1)}$$

where $X^*(\vec{u}_o)$ = the estimated value at the unsampled location, $X(\vec{u}_i)$ = the value at the neighboring location, $\vec{u}_i$, and λ_i = the weight assigned to the neighboring value. In other words, the estimated value is a weighted average of the neighboring values. All of the kriging algorithms use this basic equation with minor variations, depending on the particular application.

The goal, in the estimation procedure, is to calculate the weights assigned to the individual neighboring points. These weights depend on the spatial relationship between the unsampled location and the neighboring values, as well as the relationship among the neighboring values. These relationships are obtained from the modeling of the variogram. The kriging algorithms use the minimum variance unbiased estimation (MVUE) technique to estimate the weights—the estimated value is unbiased and will result in minimum error variance. We observed the application of the MVUE technique for linear regression and will extend the application to kriging techniques.

We divided this chapter into several sections. Sec. 4.1, "Preliminary Considerations," discusses some of the common issues applicable to kriging algorithms. These issues include determining which neighboring values to use in an estimation process. This question is important because a large number of neighbors makes computations demanding. Further, the local variability of sample values may dictate the size of neighborhoods. This section also discusses the related topic, cross validation. Cross validation is tantamount to dress rehearsal. It allows us to test the procedure on a smaller scale and eliminates any major errors in the kriging process. Beforehand, one must estimate the values at the "sampled" locations and compare them to actual samples. This procedure helps to plan a strategy in the actual kriging process, in which we estimate the values at unsampled locations. The types of information to consider during the cross-validation process is discussed in this section, as well.

Sec. 4.2 discusses linear kriging procedures. Depending on the objective, various kriging procedures are proposed in this text. We begin with simple kriging—the simplest kriging procedure available. It requires a knowledge of population mean, which may not be known in practice without prior assumptions. Therefore, this type of kriging procedure is not very popular.

Ordinary kriging is the most popular technique, which eliminates the need for knowledge of the mean value. It also is easier to adopt to local variations. It is, without question, the most widely applied kriging technique.

If the objective is to use information from other variables to estimate a value of a particular variable, the cokriging technique can be used. If the data exhibit a trend in a particular direction, universal kriging is probably more appropriate.

All four of these techniques can be applied to estimate a point value (without any support) or a block value (with a known support or volume). We discuss the applications for both point and block estimates and state the advantages and disadvantages of these methods. We also elaborate on a particular set of considerations one must give when applying these techniques. In addition, we present some variations of these method that are used in practice.

Sec. 4.3 discusses the nonlinear kriging procedures. The term, nonlinear kriging, is a misnomer because we still use a linear estimation technique. The difference is that we apply the linear estimation technique to a transformed variable, rather than applying it to the original variable. This transformation is nonlinear; therefore, the procedure is called nonlinear kriging. Many of the procedures are applied for two reasons. One, the variable values exhibit too much variability, making it difficult to either model and estimate the spatial relationship or estimate the value of the variable. Two, in addition to estimating the value, kriging procedures allow the estimation of uncertainties. By using a certain type of nonlinear transformation, it is easier to quantify uncertainties with respect to estimates.

We briefly discuss three nonlinear kriging procedures. Lognormal kriging uses a log transform to convert original data. Multi-Gaussian kriging converts data into normal score transform. The indicator kriging procedure changes data into binary indicator values. We show examples and applications of these techniques and, in addition, discuss their advantages and disadvantages. Some related applications are also presented.

When we estimate values, it is beneficial to know our confidence level. Sec. 4.4 discusses techniques that determine uncertainty regarding the estimated values. One common technique, known as confidence interval, estimates values and identifies the boundary of uncertainty, as well as the probability of confidence. Defining the confidence interval is simple, if we are able to make certain assumptions with respect to the distribution of uncertainty. The most common assumption is that uncertainties are normally distributed. The first subsection, Sec. 4.4.1, shows the applications of this assumption for determining local uncertainties with respect to estimates. Sec. 4.4.2 shows the methods of quantifying the uncertainties in the absence of any underlying distribution. These techniques, called nonparametric methods, are much more flexible than the techniques that require an assumption of uncertainty distribution. At the same time, application of these techniques may require more extensive data, as well as additional modeling and computational efforts.

The last subsection shows the utility of loss functions in quantifying the uncertainties. In some applications, all the errors (i.e., positive vs. negative) may not be equally critical. By penalizing the critical errors more than the not-so-critical errors, much more realistic estimates of uncertainties can be obtained. We review some of the loss functions commonly used to assess uncertainties.

The final section summarizes the findings of this chapter. The overall advantages and disadvantages of kriging procedures are illustrated through several numerical and field examples. Only minimal mathematical details are provided in the main text. Additional details are found in Appendix D.

4.1 Preliminary Considerations

This section discusses two common considerations applicable to all kriging techniques—defining a search neighborhood and conducting a cross-validation exercise. These two considerations are recommended before initiating a kriging procedure to estimate values at unsampled locations.

4.1.1 Search Neighborhood. The search neighborhood defines the neighboring sample points used in estimating values at the unsampled location. As explained in Eq. 4.1, all kriging procedures use a linear estimation technique, which assumes that the unsampled value is linearly related to the neighboring values. In principle, we can use all available samples to estimate the values at unsampled locations. In practice, we restrict the number of samples to a smaller neighborhood for several reasons.[2,3]

- Kriging algorithms involve inverting a matrix to obtain the weights (λ_i) described in Eq. 4.1. The larger the number of data points selected, the larger is the matrix size. As the matrix size gets larger, both the memory and computation time requirements increase. For example, doubling the number of sample points requires a fourfold increase in memory requirements and an eightfold increase in central processing unit (CPU) time requirements. By restricting the number of sampling points to a reasonable size, we can restrict the memory and computation time requirements to a manageable level.
- As stated before, kriging algorithms involve the inversion of a matrix. If too many sample points are used, especially clustered sample points, there is a possibility that the matrix will become close to singular, making inversion difficult.
- As discussed in Chap. 3, we typically restrict variogram estimations to half the maximum distance. This ensures that we obtain a reliable estimate of the variogram. We rarely have estimated variogram values beyond that distance. If we include sample data points at larger distances, we must infer the spatial relationships between the unsampled location and the sample point, as well as between all sample points. This may only be available from a variogram model, which must be extrapolated beyond the estimated range of the variogram value. By restricting sample points to nearby values, we can only use variogram models with estimated distances for variogram values.
- When sample points are farther away from an unsampled location, their influence tends to be screened by closer sample points, which are between the unsampled location and the sample point farthest away. Therefore, additional sample points, farther apart in the estimation process, may not necessarily improve the estimate at the unsampled location.
- Kriging algorithms require the implicit assumption of first- and second-order stationarities. As noted in the previous chapters, this assumption is rarely satisfied in practice. Local properties can differ from sample properties. In estimating values at unsampled locations, it is critical to choose samples that reflect the behavior of the unsampled location. This is especially true when significant local variability exists. By restricting the search neighborhood to local neighborhoods, we ensure that the estimated value is much more representative.

In selecting an appropriate search neighborhood, the following factors must be considered: number of data points necessary for a representative sample, size of search neighborhood, and shape of search neighborhood. These factors determine the final shape and size of the search neighborhood. One should consider each factor separately.

Number of Samples. The number of samples dictates the size of the matrix and, hence, the computation time required to estimate the value. If very few samples are chosen, the estimation will not be representative; if too many samples are chosen, the estimation will be computationally intensive. A careful balance must be maintained so that sufficient samples are selected to obtain a representative estimate. At minimum, the search neighborhood should be large enough to select at least four of the closest samples. In general, we need 12 samples surrounding the unsampled location to obtain a reasonable estimate.[4]

There is no maximum number of samples one must select, as long as one uses an appropriately powerful computer. However, in practice, 32 sample points are selected as the

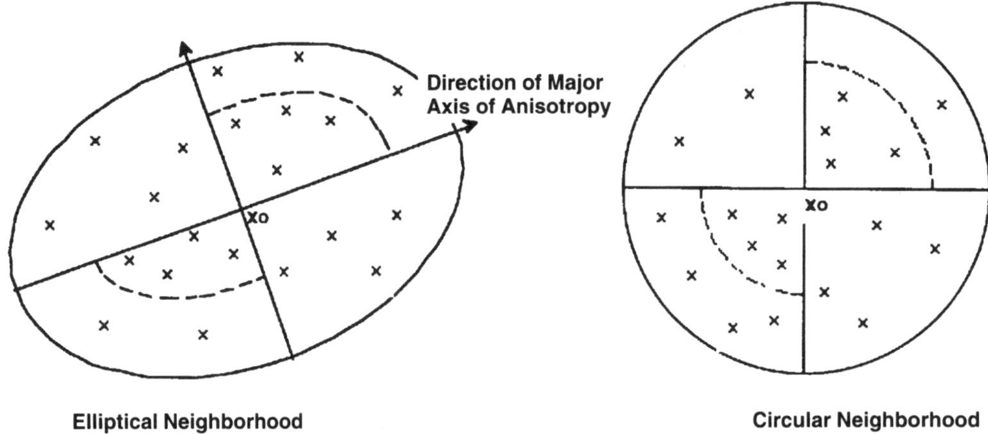

Fig. 4.1—Search neighborhood shapes.

maximum. As sample points get farther apart from an unsampled location, the weight assigned to the sample point becomes increasingly smaller. Adding more points, beyond the maximum of 32, will not significantly improve the estimate.

A related question to the selection of sample points is how these sample points should be selected. Should we select the closest sample points until we reach the desired maximum, or should we be more discerning in terms of selected sample points? One potential disadvantage of selecting the closest sample points is that if the sample points are clustered in one particular direction, all sample points will be selected from that cluster, if they are the closest. This sample set may not be representative of the neighborhood. As shown in **Fig. 4.1,** one way to minimize the clustering effect is to divide the search neighborhood into small sections and select a certain number of points within each section. A two-dimensional (2D) search neighborhood can be divided into quadrants, or a three-dimensional (3D) search neighborhood can be divided into octants. Once the sections are defined, a certain maximum number of sample points, within each section, can be selected.

Size of Neighborhood. The size of the neighborhood is relatively difficult to determine. If the size of the neighborhood is too small, we may not have sufficient samples within the neighborhood to estimate a representative value. If the search neighborhood is too large, we might select samples outside of the local stationarity region, which could bias our estimate.

One consideration, which will determine the minimum size, is the minimum number of samples needed to obtain a representative estimate. As discussed, we must have the size large enough to at least select four of the closest sample points.

The maximum size of the search neighborhood is more difficult to determine. One misconception is that the search neighborhood should be dictated by the range of the variogram model. Although it could provide a starting point, it does not necessarily provide the maximum limit. For sparse data sets, if the first-order stationarity assumption is reasonably valid over the region of interest, the sample data beyond the range of the variogram can still provide information about the mean and, hence, improve the local estimate.

The best way to determine the maximum size of a search neighborhood is through a cross-validation process, which is explained in the next section. Briefly, the maximum size of the search neighborhood should be determined so that it is large enough to include sufficient points; at the same time, it should be small enough so that the selected sample points are within the local stationarity region.

Shape of Neighborhood. The shape of the neighborhood is primarily dictated by variogram models. If, in two dimensions, an isotropic variogram model is used, the search neighborhood will be circular. If an anisotropic variogram model is used, the search neighborhood is elliptical with the ratio of major and minor axes of ellipse equal to the anisotropy ratio of a variogram mode. Examples of circular and elliptical search neighborhoods, with a maximum of four sample points in each quadrant, are shown in Fig. 4.1. In three dimensions, the search neighborhood will be spherical, if the spatial continuity is described by an isotropic variogram model. For an anisotropic variogram model, the search neighborhood will be an ellipsoid, where the major and minor axes dimensions will be related to the anisotropy ratios of a variogram model.

In addition to the shape, the orientation of the neighborhood is also important. A common practice uses the orientation of the major axes of the ellipse or ellipsoid to coincide with the major direction of spatial continuity. This ensures that we select the nearest sample points that have the strongest spatial relationship to the unsampled location.

It is not always necessary for the search neighborhood to coincide with the spatial continuity models. Sometimes, geological inferences about facies may dictate the shape to coincide with the geological model. Also, remember that, in a variogram estimation process, because of the use of tolerance with respect to direction, the anisotropy ratio (ratio of range in the direction of minimum continuity to the range in the direction of maximum continuity) is observed to be larger than what may be the real value. We can, therefore, define a search neighborhood with a different anisotropy ratio, if it improves the estimates during a cross-validation exercise. In general, using the search neighborhood, which mirrors the variogram model, will be adequate for estimations at an unsampled location.

Once the search neighborhood shape and size are determined, a relevant computational issue must be addressed—searching for neighboring points in an efficient manner. We can use several strategies, such as super block search, spiral search, and two-part search. For a detailed discussion and implementation of these strategies, please see Ref. 2.

4.1.2 Cross Validation. Cross validation involves the estimation of values at the sampled locations so that the estimation can be compared with the sampled values. Although several versions of cross validation are used in practice, the most common version of cross validation involves the "leaving one out" method.[5] In this method, one sample point at a time is removed from the sample data, and using the remaining sample points, the value of the variable at the now "unsampled" location is estimated. The estimated value is then compared with the observed value. The procedure is repeated at all sampled locations, so we can have an estimated value at every sampled location. In the "leaving one out" method, the spatial model used for the estimation process is based on all available data. Therefore, the estimation of the value at a sampled location is not strictly independent. Some information about the sample point is already used in the form of a spatial relationship.

A stricter test, called "jackknifing," can also be used to test the validity of an estimation process. In this technique, certain sample points are removed before the spatial relationship is estimated and modeled. Using the kriging parameters, the values are estimated at the sampled locations, which were removed at the beginning. Then, the observed values are compared with the estimated values. Unfortunately, in most reservoir applications, we do not have enough data to remove certain sample points at the beginning. Therefore, we generally use the conventional cross-validation method.

Cross validation is similar to a dress rehearsal[3] before an actual production. Therefore, it should be conducted under similar operating conditions to the actual estimation procedure. The same variogram model, search neighborhood, number of sample points, and any other relevant parameters should be used. For example, if sample data are concentrated along the wellbore, the estimation of values at the sampled location should only be done by removing all sampled data along that wellbore. This is because estimation at interwell locations does not involve use of nearby samples in a vertical direction. The configuration of available samples for the cross-validation exercise should be similar to the actual estimation exercise.

Similar to a dress rehearsal, cross validation can identify obvious glaring errors, including decisions about stationarity, appropriateness of the variogram model, and the search neighborhood. However, it does not guarantee a success during the actual production.

After cross validation is finished, we have two values at every sampled location—an observed, or true, value and an estimated value. By comparing the true value and the estimated value, we might be able to detect problems related to the estimation process. We first define

$$e(\vec{u}_i) = x(\vec{u}_i) - x^*(\vec{u}_i), \quad \ldots\ldots\ldots\ldots\ldots\ldots\ldots (4.2)$$

where $e(\vec{u}_i)$ = the estimation error at location $\vec{u}_i$, $x(\vec{u}_i)$ = the true value, and $x^*(\vec{u}_i)$ = the estimated value. The following observations, regarding $e(\vec{u}_i)$, $x(\vec{u}_i)$ and $x^*(\vec{u}_i)$, must be examined as part of the cross-validation process.

• Plot the true value vs. the estimated value. Some examples of this plot are shown in **Fig. 4.2.** In Fig. 4.2a, the spread of error is uniform around the 45° line. An approximately equal number of values are under- and overestimated over the entire region. This is an ideal situation. In contrast, in Fig. 4.2b, the spread of error (deviation from the straight line) increases as the true value increases. In other words, the error is dependent on the magnitude of the sample. This type of behavior reflects the heteroscedasticity of error variance. In Fig. 4.2c, up to a certain range, the sample value is overestimated; and over that range, the sample value is underestimated. This type of estimate is called a conditionally biased estimate.

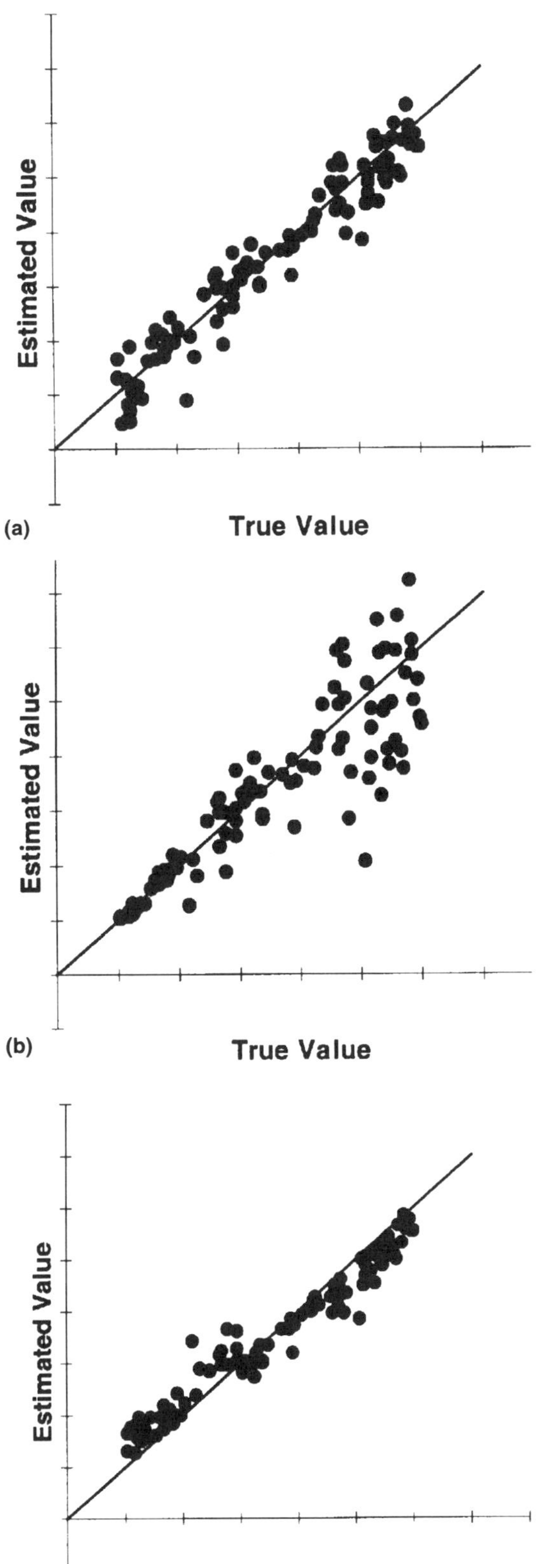

Fig. 4.2—True vs. estimated values during cross validation.

Figs. 4.2b and 4.2c exhibit undesirable characteristics. Fig. 4.2b could be a reflection of a search neighborhood being too large, resulting in an inclusion of data points that are not within the local stationarity region. Fig. 4.2c could be a reflection of mixing two regions of stationarity. Our assumption of stationarity must be re-examined to ensure that the proper sample set is included within the region.

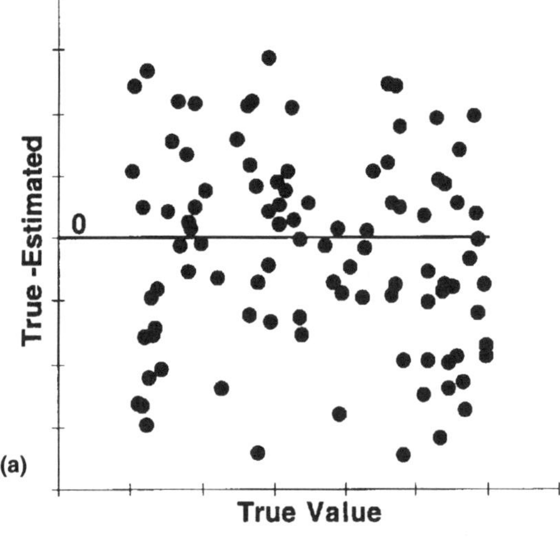

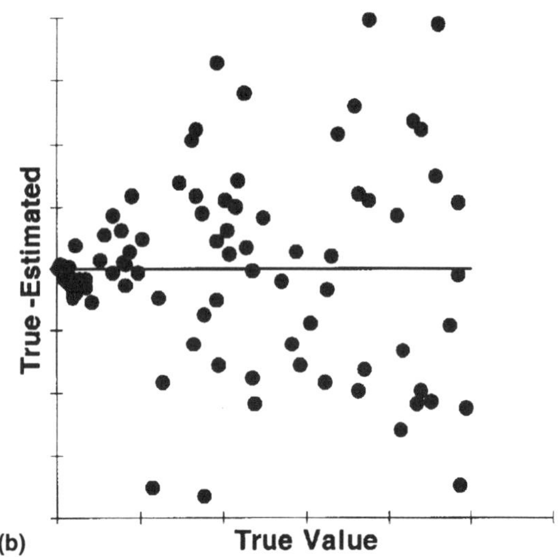

Fig. 4.3—Error (true-estimated) vs. true value during cross validation.

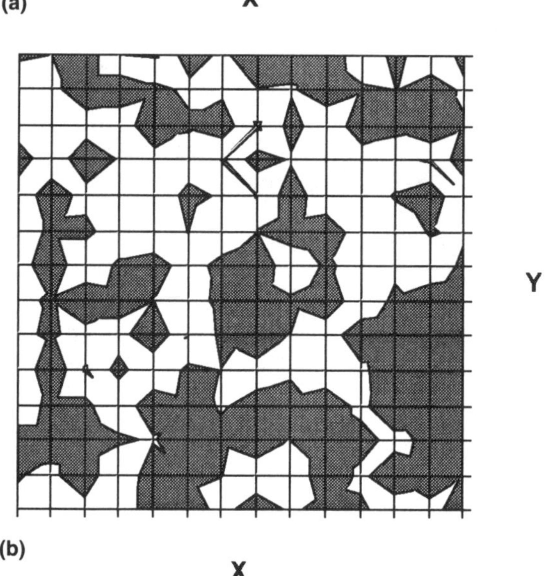

Fig. 4.4—Spatial correlation of errors.

• Plot the estimation error, $e(\vec{u})$, vs. the estimated value. **Fig. 4.3a** shows an example where the error is centered around a zero line with equal spread. Ideally, for n samples,

$$\frac{1}{n}\sum_{i=1}^{n} e(\vec{u}_i) \approx 0, \qquad (4.3)$$

indicating that estimation is conditionally unbiased. The equal spread around the zero line indicates the homoscedasticity of error variance. That is, the error is independent of the estimated value. In contrast, in Fig. 4.3b, although the estimation is conditionally unbiased, the magnitude of error increases as the sample value increases. This is a reflection of heteroscedastic behavior. As before, the size and slope of the neighborhood must be examined to ensure that the error variance has homoscedastic characteristics.

In practice, the condition of homoscedasticity is difficult to achieve; the error must be independent of the estimate. In terms of absolute magnitude, the error should not change. For example, if we estimate a value of 48 md, where the true value is 50 md, this estimate is worse than an estimated value of 0.1 md in place of a true value of 1.0 md. This is true because, in absolute terms, the former estimate represents an error of −2.0 md; whereas, in the later case, the error is only −0.9 md. Intuitively though, we should agree that the first estimate is better in terms of relative error than the second estimate. In practice, we are willing to accept a larger absolute error, as long as relative errors are comparable. This means that the estimates, which are heteroscedastic (errors increasing with the estimates), might be acceptable.

• Plot the contour maps of $e(\vec{u}_i)$ values. **Fig. 4.4a** shows a contour map of errors with several bull's-eyes indicating no spatial relationship of error values. In contrast, Fig. 4.4b indicates a trend or correlation among error values. Ideally, error estimates should show no spatial correlation because values should be independent of spatial location. Spatial correlation of error values indicates either a wrong assumption of stationarity or a lack of capturing a trend that is evident within the sample data. Both must be examined before proceeding further.

Obviously, cross validation is a trial-and-error procedure. It involves adjusting some of the operational parameters such as size and shape of search neighborhood, assumption of region of stationarity, or variogram model to capture the trend. These parameters are adjusted until the desired results (e.g., conditionally unbiased with respect to the estimation, homoscedasticity of error, and lack of spatial relationships among errors) are obtained as close to ideal conditions as possible. Out of these three desirable characteristics, the requirement of homoscedasticity of error can be relaxed in many situations because of the reasons stated previously. Fortunately, cross validation involves the estimation of values at relatively few locations. Therefore, this trial-and-error procedure is not computationally intensive. Within a few iterations, a desired result can be obtained.

As explained before, cross validation's use is mostly limited as a qualitative tool. In the past, cross validation has also been used as a quantitative tool[6] to select correct variogram models for estimation purposes. However, recent studies show that cross validation, as a quantitative tool, may be inappropriate and will not result in improvements in the estimation of unsampled values.[4,7] Therefore, in this chapter, we restrict the use of cross validation as a qualitative tool only.

4.2 Linear Kriging Procedures

This section discusses several kriging procedures and assumes that the estimated value of the variable is linearly related to the nearby samples. Depending on the specific application, different procedures have been used for the purpose of estimation. We begin with a simple kriging procedure—the simplest but not necessarily the most practical. Discussion follows with ordinary kriging, the most popular kriging procedure. Ordinary kriging is more flexible than simple kriging and allows for variations in local changes. We also discuss cokriging, which allows the estimation of one variable based on the spatial information of other related variables. This procedure is especially useful when there is one extensively sampled variable and one sparsely sampled variable, and they are spatially related. The last procedure we discuss is universal kriging, in which the sample data exhibit a trend, and the assumption of stationarity may not be valid. Within each of the four techniques, we also discuss the estimation at a location without support and block estimates involving the estimation of a location which includes support or associated volume. All techniques are explained with numerical and field examples.

4.2.1 Simple Kriging. Simple kriging starts with the assumption that a value at an unsampled location can be estimated by

$$X^*(\vec{u}_o) = \lambda_o + \sum_{i=1}^{n} \lambda_i X(\vec{u}_i), \quad \text{(4.4)}$$

where $X^*(\vec{u}_o)$ = the value to be estimated, $X(\vec{u}_i)$ = a nearby sample value at location $\vec{u}_i$, n = the total number of samples selected within a search neighborhood, λ_i = the weight assigned to each sample, and λ_o = a constant.

To estimate the value of λ_i, we use the MVUE technique. An unbiased condition requires that

$$E[X^*(\vec{u}_o) - X(\vec{u}_o)] = 0. \quad \text{(4.5)}$$

Substituting $X^*(\vec{u}_o)$ from Eq. 4.4,

$$\lambda_o + \sum_{i=1}^{n} \lambda_i E[X(\vec{u}_i)] = E[X(\vec{u}_o)]. \quad \text{(4.6)}$$

If we assume $E[X(\vec{u}_i)] = E[X(\vec{u}_o)]$, based on the first order stationarity, we can write

$$\lambda_o = m\left(1 - \sum_{i=1}^{n} \lambda_i\right). \quad \text{(4.7)}$$

In addition to the unbiased requirement, the condition of minimum variance must also be satisfied. Mathematically, weights should be chosen so

$$\sigma^2[X(\vec{u}_o) - X^*(\vec{u}_o)] \quad \text{(4.8)}$$

is minimized.

The result of this condition is equation

$$\sum_{j=1}^{n} \lambda_j C(\vec{u}_i, \vec{u}_j) = C(\vec{u}_i, \vec{u}_o) \text{ for } i = 1,\ldots,n, \quad \text{(4.9)}$$

where $C(\vec{u}_i,\vec{u}_j)$ = the covariance value between points located at $\vec{u}_i$ and $\vec{u}_j$, respectively, and $C(\vec{u}_i,\vec{u}_o)$ = the covariance between the sampled location, $\vec{u}_i$, and the unsampled location, $\vec{u}_o$. The covariance values are obtained based on the spatial model. For details regarding the derivation of Eq. 4.9, refer to Appendix D.

In matrix form, Eq. 4.9 can be written as

$$\begin{bmatrix} C(\vec{u}_1, \vec{u}_1) & \ldots & C(\vec{u}_1, \vec{u}_n) \\ \vdots & & \vdots \\ C(\vec{u}_n, \vec{u}_1) & \ldots & C(\vec{u}_n, \vec{u}_n) \end{bmatrix} \begin{bmatrix} \lambda_1 \\ \vdots \\ \lambda_n \end{bmatrix} = \begin{bmatrix} C(\vec{u}_1, \vec{u}_o) \\ \vdots \\ C(\vec{u}_n, \vec{u}_o) \end{bmatrix}, \quad \text{(4.10)}$$

or

$$|C||\Lambda| = |c|.$$

To calculate the value of $|\Lambda|$, we write

$$|\Lambda| = |C|^{-1}|c|, \quad \text{(4.11)}$$

where $|C|^{-1}$ = the inverse of matrix $|C|$.

It is important to examine Eq. 4.11, so we can understand how individual weights are assigned to sample points to estimate the value at the unsampled location. Eq. 4.11 states that the weights assigned to samples are directly proportional to $|c|$ and are inversely proportional to $|C|$, where $|c|$ represents the covariance between the sample point and the unsampled location. The stronger the spatial relationship between the sample point and the unsampled location, the larger is the value of $C(\vec{u}_i,\vec{u}_o)$. As a result, the weight assigned to the sample point at $\vec{u}_i$ is greater. This is intuitively correct because the sample point with a strong spatial relationship with the unsampled location should receive a bigger weight. On the other hand, $|C|$ represents the covariance among the sampled points. If a particular sample point at $\vec{u}_i$ is very close to the surrounding sample points, $\vec{u}_j$, $C(\vec{u}_i,\vec{u}_j)$ will be large. If these values are large, because the weight assigned to sample points is inversely proportional to $|C|$, the weight will be reduced, if $C(\vec{u}_i,\vec{u}_j)$ values are large. This observation also makes intuitive sense. If sample points are clustered together, they should not individually receive large weights because they do not provide independent information on an individual basis. Clustered samples, being spatially related, are similar to each other and assigned smaller weights. In contrast, if a sample point is isolated, and is away from other sample points, $C(\vec{u}_i,\vec{u}_j)$ will be smaller, and the weight assigned to that sample point will be proportionately bigger.

To summarize, the weight assigned to an individual sample is dependent on two factors. One is its spatial relationship to the unsampled location. The stronger the relationship, the larger is the assigned weight. Two is the sample point's spatial relationship to other sample points. The stronger the relationship, the less independent information that point can provide.

As a result, the weight assigned to that sampled point will be smaller.

Once the weights are estimated, Eq. 4.4 estimates the value of $X^*(\vec{u}_o)$. In addition to the estimation, it also estimates the error variance associated with the estimate. It is written as (see Appendix D for details)

$$\hat{\sigma}_E^2 = C(\vec{u}_o, \vec{u}_o) - \sum_{i=1}^{n} \lambda_i C(\vec{u}_i, \vec{u}_o) . \qquad (4.12)$$

By examining Eq. 4.12, we observe the maximum value of error variance, $C(\vec{u}_o,\vec{u}_o)$, which is the data variance. In the absence of spatial information, uncertainty with respect to estimation should be represented by the variance of the data. As spatial relationship information becomes available, error variance is reduced. Reduction is related to two parameters—weight assigned to sample points, and covariance between sample points and the unsampled location. Error variance is reduced when there is a larger $C(\vec{u}_i,\vec{u}_o)$ value and a larger weight is assigned. As previously discussed, a sample point, with a very strong relationship to the unsampled location, is isolated from other sample points and has a higher weight assignment. This, coupled with a large value of covariance, results in the largest error variance reduction. On the other hand, if a sample point has a strong spatial relationship with the unsampled location but is surrounded (clustered) by other samples, the weight assigned to that point will be smaller and, hence, will result in a smaller error variance reduction. The smallest reduction in error variance is achieved by a sample point with a weak spatial relationship to the unsampled point (small covariance value) and represents a point within a cluster of samples. This is logical because that sample point provides the least information about the unsampled location; therefore, it provides the least reduction in the uncertainty of the estimate.

Note, all previous equations are written in terms of covariance instead of a variogram. Although similar equations can be written in terms of a variogram,[8] it is much easier to invert a matrix when it is written in terms of covariance. Therefore, covariance values are preferred over variogram values.

Numerical Example 4.1. Consider the data configuration, $x(\vec{u}_1)=10, x(\vec{u}_2)=20$, and $m=18$ (**Fig. 4.5**). Further, assume that the isotopic variogram, representing the spatial relationship, is given as

$$\gamma(L) = 100\left[1 - \exp\left(\frac{-3L}{100}\right)\right] \text{ for } L \geq 0.$$

Estimate the value at location $\vec{u}_0$. Also, estimate the error variance. If the spatial model is given as

$$\gamma(L) = 200\left[1 - \exp\left(\frac{-3L}{100}\right)\right] \text{ for } L \geq 0,$$

how would the estimate and error variance change?

Solution—Part 1. In this part, we assume the variogram is represented by

$$\gamma(L) = 100\left[1 - \exp\left(\frac{-3L}{100}\right)\right].$$

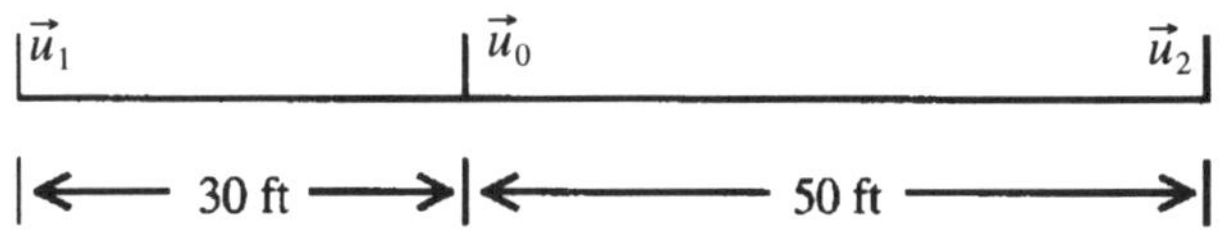

$\vec{u}_1$ $\vec{u}_0$ $\vec{u}_2$

|← 30 ft →|← 50 ft →|

Fig. 4.5—Data configuration in Numerical Example 4.1.

In this example, we have two sample points, or $n=2$. Therefore, the matrix equation can be written as

$$\begin{bmatrix} C(\vec{u}_1, \vec{u}_1) & C(\vec{u}_1, \vec{u}_2) \\ C(\vec{u}_2, \vec{u}_1) & C(\vec{u}_2, \vec{u}_2) \end{bmatrix} \begin{pmatrix} \lambda_1 \\ \lambda_2 \end{pmatrix} = \begin{bmatrix} C(\vec{u}_1, \vec{u}_0) \\ C(\vec{u}_2, \vec{u}_0) \end{bmatrix}.$$

We know the relationship between the covariance and the variogram,

$$C(L) = C(0) - \gamma(L).$$

Therefore, if we calculate $\gamma(L)$, we can calculate $C(L)$.

The second-order stationarity assumption requires a variogram or covariance to be only a function of distance and direction. Therefore, if we know the distance and the direction (only distance for an isotropic model), we can calculate the associated variogram value.

In the previous matrix, the only unknowns are λ_1 and λ_2.

$$C(\vec{u}_1, \vec{u}_1) = C(\vec{u}_2, \vec{u}_2) = C(0) = 100$$

because the distances between $\vec{u}_1$ and $\vec{u}_1$, and $\vec{u}_2$ and $\vec{u}_2=0$. $C(0)$ is the sill of the variogram.

$$C(\vec{u}_1, \vec{u}_2) = C(\vec{u}_2, \vec{u}_1) = C(0) - \gamma(\vec{u}_1, \vec{u}_2) = C(0) - \gamma(80)$$

$$= 100 - 100\left[1 - \exp\left(\frac{-3 \times 80}{100}\right)\right] = 9.07.$$

Because the covariance and the variogram are symmetric, the variogram between two given points is the same, irrespective of the directions.

$$C(\vec{u}_1, \vec{u}_0) = C(0) - \gamma(\vec{u}_1, \vec{u}_0)$$

$$= 100 - 100\left[1 - \exp\left(\frac{-3 \times 30}{100}\right)\right] = 40.66,$$

and $C(\vec{u}_2, \vec{u}_0) = C(0) - \gamma(\vec{u}_2, \vec{u}_0)$

$$= 100 - 100\left[1 - \exp\left(\frac{-3 \times 50}{100}\right)\right] = 22.31.$$

Note, the distance between $\vec{u}_1$ and $\vec{u}_0$ is 30 ft, and between $\vec{u}_2$ and $\vec{u}_0$, it is 50 ft.

Substituting all the values,

$$\begin{pmatrix} 100 & 9.07 \\ 9.07 & 100 \end{pmatrix} \begin{pmatrix} \lambda_1 \\ \lambda_2 \end{pmatrix} = \begin{pmatrix} 40.66 \\ 22.31 \end{pmatrix},$$

or

$$100\lambda_1 + 9.07\lambda_2 = 40.66,$$

and $9.07\lambda_1 + 100\lambda_2 = 22.31$.

Solving for λ_1 and λ_2, we obtain $\lambda_1=0.389$, and $\lambda_2=0.188$.

Using Eq. 4.7,

$$\lambda_o = m\left(1 - \sum_{i=1}^{n} \lambda_i\right) = 18(1 - 0.389 - 0.188)$$

$$= 7.614.$$

Using Eq. 4.4,

$$x^*(\vec{u}_0) = \lambda_0 + \sum_{i=1}^{2} \lambda_i x(\vec{u}_i),$$

and $x^*(\vec{u}_0) = 7.614 + 0.389 \times 10 + 0.188 \times 20$

$= 15.264.$

This is the estimated value at the unsampled location. The error variance is calculated with Eq. 4.12.

$$\hat{\sigma}_E^2 = C(\vec{u}_0, \vec{u}_0) - \sum_{i=1}^{2} \lambda_i C(\vec{u}_i, \vec{u}_0)$$

$$= 100 - 0.389 \times 40.66 - 0.188 \times 22.31$$

$$= 79.99.$$

The error variance is not an absolute measure of uncertainty in estimation; however, it does indicate how well the sample data are configured surrounding the unsampled location. This statement will become clear as we go through additional numerical examples, as well as Sec. 4.4.

Solution—Part 2. Here, we assume that the variogram is represented by

$$\gamma(L) = 200\left[1 - \exp\left(\frac{-3L}{100}\right)\right] \text{ for } L \geq 0.$$

Note, the only difference between Part 1 and Part 2 is the sill value of the variogram. The variogram model and the range are the same.

We have two sample points; therefore, using the same logic as Part 1, we state,

$$C(\vec{u}_1, \vec{u}_1) = C(\vec{u}_2, \vec{u}_2) = C(0) = 200,$$

$$C(\vec{u}_1, \vec{u}_2) = C(\vec{u}_2, \vec{u}_1)$$

$$= 200 - 200\left[1 - \exp\left(\frac{-3 \times 80}{100}\right)\right] = 18.14,$$

$$C(\vec{u}_1, \vec{u}_0) = 200 - 200\left[1 - \exp\left(\frac{-3 \times 30}{100}\right)\right] = 81.31,$$

and

$$C(\vec{u}_2, \vec{u}_0) = 200 - 200\left[1 - \exp\left(\frac{-3 \times 50}{100}\right)\right] = 44.63.$$

Substituting in the matrix form,

$$\begin{pmatrix} 200 & 18.14 \\ 18.14 & 200 \end{pmatrix}\begin{pmatrix} \lambda_1 \\ \lambda_2 \end{pmatrix} = \begin{pmatrix} 81.31 \\ 44.63 \end{pmatrix}.$$

Solving for λ_1 and λ_2, $\lambda_1 = 0.389$, and $\lambda_2 = 0.188$.

Therefore,

$$\lambda_o = 18(1 - 0.389 - 0.188) = 7.614,$$

and

$$x^*(\vec{u}_0) = 15.264.$$

The error variance can be estimated as

$$\hat{\sigma}_E^2 = 200 - 0.389 \times 81.31 - 0.188 \times 44.63$$

$$= 159.98.$$

By comparing answers from Parts 1 and 2, you will notice that the estimate is not affected by changing the sill value, whereas the error variance doubled as the sill value is doubled. In other words, the estimated value at the unsampled location is independent of the sill value. This important characteristic is not restricted to just simple kriging. It is common to all kriging procedures. This characteristic proves useful in quantifying uncertainties with respect to kriging estimates.

Field Example 4.1—Simple Kriging Cross Validation. This field example illustrates the cross-validation procedure for simple kriging with the gross thickness data for Flow Unit 5. Cross validation quantifies the errors of kriging estimates and test assumptions made about variogram models and other kriging parameters. There are two major steps in the cross-validation process. First, each data value is removed from the conditioning data set, and kriging is used to estimate it with the remaining data. This step is repeated for all data points. Second, estimated values are compared to actual data. Crossplots of actual vs. estimated values, histograms of the errors, and kriging variance are used to evaluate the validity of kriging and assumptions.

The following information is given:

• Conditioning data for gross thickness of Flow Unit 5. (See Appendix A.)

• Variogram models for gross thickness of Flow Unit 5 are:

$$\gamma(\vec{L}) = 5 + 16.25 M_{s_{1,400}}(\vec{L}) \text{ principal direction: east/west}$$

$$\gamma(\vec{L}) = 5 + 16.25 M_{s_{6,000}}(\vec{L}), \text{ minor direction: north/south}$$

Find:

• Calculate simple kriging cross-validation estimates for gross thickness data of Flow Unit 5.

• Evaluate the effect of search neighborhood and number of samples for kriging on cross-validation estimates.

• Compare statistics of data, cross-validation estimates and errors.

• Evaluate the kriging error variance.

Solution. **Fig. 4.6** shows the field location for the 103 wells, with gross thickness data for Flow Unit 5. The data show that this flow unit tends to thin from north to south. The variograms indicate that gross thickness varies the most in the north/south direction, and it is most continuous in the east/west direction. The histogram of the gross thickness data is shown in **Fig. 4.7.** Gross thickness ranges from 0 to 27 ft, and it has a mean of 5.05 ft. Only one well has a gross thickness greater than 20 ft, and 75% of the wells have a gross thickness less than 7 ft.

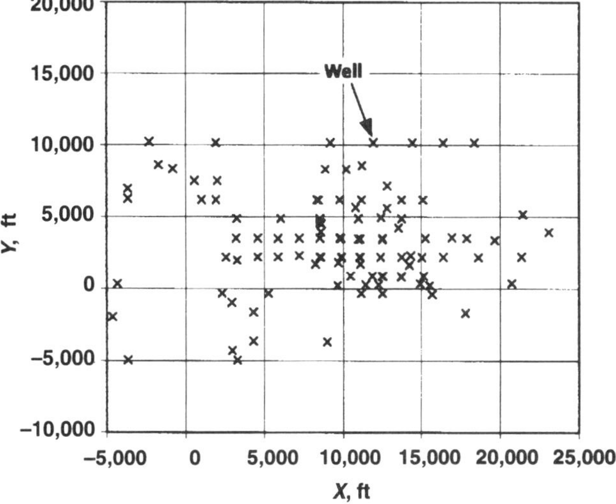

Fig. 4.6—Location of Flow Unit 5 gross thickness data.

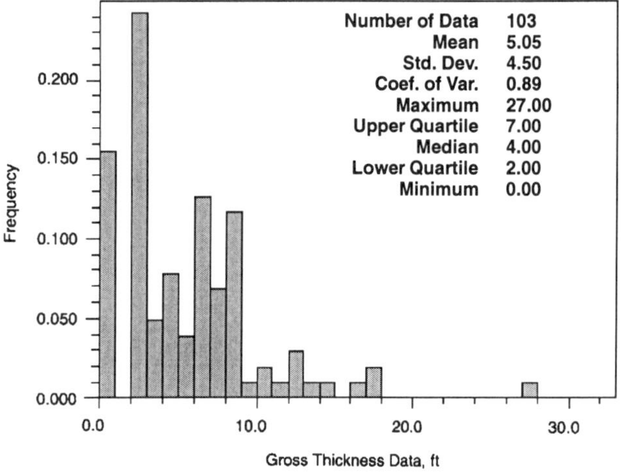

Fig. 4.7—Histogram of Flow Unit 5 gross thickness data.

One of the objectives of cross validation is to select optimum kriging parameters. Cross validations were generated for several values of the number of samples for kriging, with and without octant search and size of the search neighborhood. The simple kriging cross-validation estimates for an initial configuration of parameters are compared to the gross thickness data in **Fig. 4.8.** This initial configuration uses the four closest samples for the kriging system and omnidirectional search radius equal to 10,000 ft. Overall, simple kriging tends to underestimate gross thickness. The correlation coefficient between cross-validation estimates and the data is only 0.60 because it is heavily affected by data points greater than 7 ft. Note, simple kriging assumes that the mean is constant throughout the area of interest. Therefore, the largest estimation errors occur when data values are significantly different from the mean. For wells with data deviating significantly from the mean, simple kriging estimates break down. In this example, the simple kriging mean was assumed to be equal to the mean of the data (5.05 ft). If cross validation indicates that this assumption does not hold, another technique such as ordinary kriging, which accounts for changes in the mean, should be tested. Another configuration of kriging parameters, which yielded slightly better results, is shown in **Fig. 4.9.** This

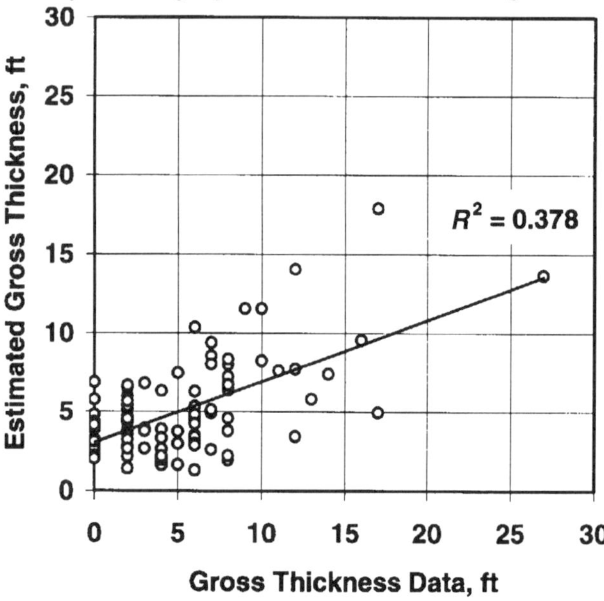

Fig. 4.9—Simple kriging cross validation of Flow Unit 5 gross thickness data using 16 samples for kriging, octant search and anisotropic search neighborhood.

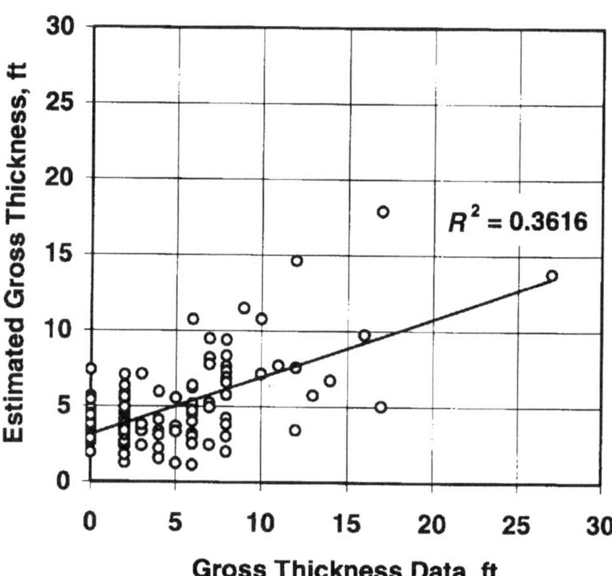

Fig. 4.8—Simple kriging cross validation of Flow Unit 5 gross thickness data using four samples for kriging and omni-directional search neighborhood.

configuration consists of 16 samples for the kriging system and an anisotropic search neighborhood. In addition, octant search was used to select the 16 samples. This means that a maximum of four samples is selected for each octant around the point being estimated. Octant search is useful to get a more uniform sampling when the data tend to be clustered because it prevents too many samples from being selected from only one area around the estimation location. The anisotropic search consisted of an ellipse with axes equal to the variogram correlation range and the same orientation as the variogram. This configuration yielded the highest correlation coefficient (0.62) between the data and kriging estimate and the lowest estimation error.

The histogram shown in **Fig. 4.10,** cross-validation estimates for optimum configuration, indicates that the mean, median, and mode are close to the statistics of the gross thickness data (Fig. 4.6). The standard deviation of the cross-validation estimates is 2.83 ft, and it is smaller than the standard deviation of the data (4.50 ft). In general, simple kriging does not closely reproduce the extreme values of a distribution. The maximum cross-validation value is 17.9 ft, while the

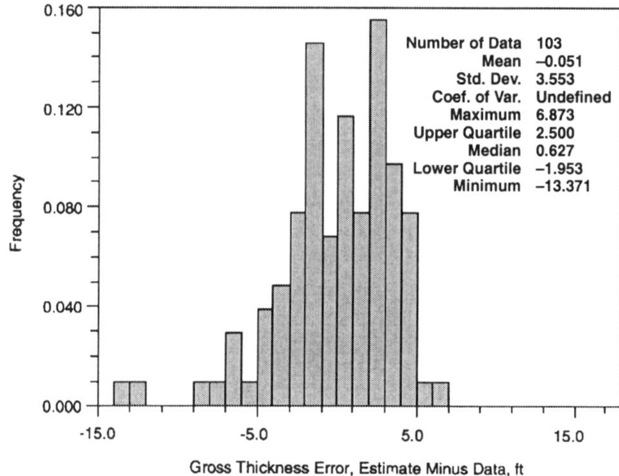

Fig. 4.10—Histogram of simple kriging cross validation errors for Flow Unit 5 gross thickness using optimum search neighborhood.

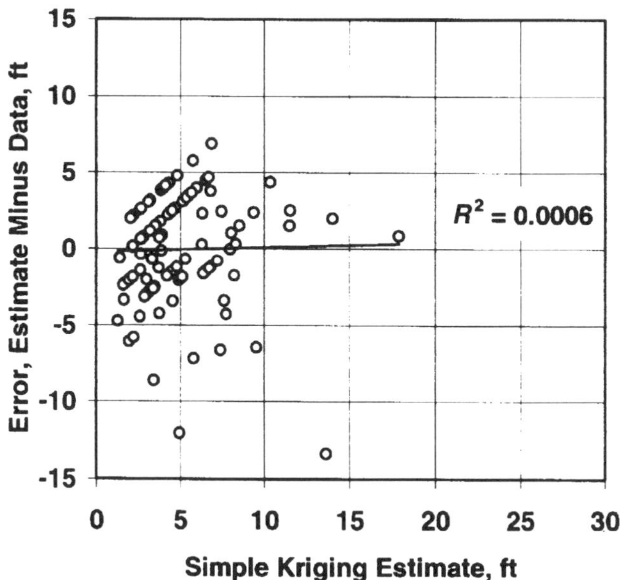

Fig. 4.11—Simple kriging cross validation errors for Flow Unit 5 gross thickness data using optimum search neighborhood.

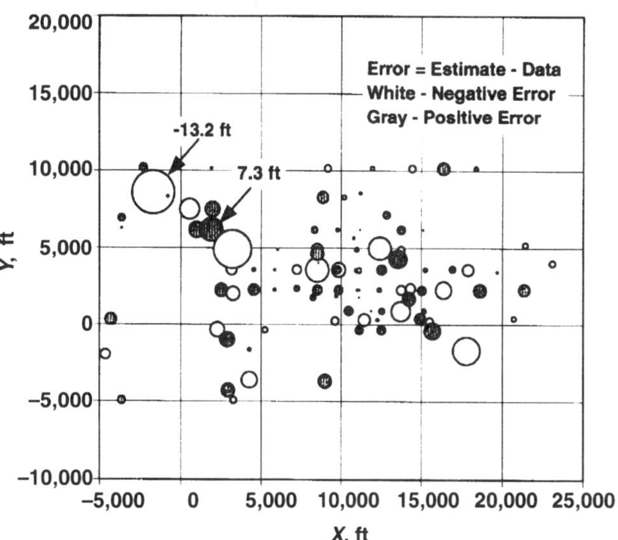

Fig. 4.12—Location of simple kriging cross validation errors for Flow Unit 5 gross thickness using optimum search neighborhood.

maximum for the data is 27 ft. Reproduction of extreme values of a histogram can be achieved with indicator kriging, discussed later in this chapter.

The quality of cross validation can be examined by analyzing the estimation errors. The uniform distribution of the error around the zero error line is a desirable feature in cross validation. **Fig. 4.11** shows that there is a small correlation between the estimated values and errors. Here, estimation error is defined as the simple kriging estimate minus the data value. The largest deviation occurs for the data with an error of –13.2 ft, which corresponds to the well with a maximum gross thickness of 27 ft and simple kriging estimate of 13.8 ft. The histogram of estimation errors (Fig. 4.9) shows that the mean error is close to zero, and the median is 0.427 ft. The errors for most data points, with a few exceptions for extreme values, fall between –5 and 5 ft. **Fig. 4.12** shows that the estimation errors do not exhibit any noticeable spatial trend. Examination of spatial trends in the errors is another way to diagnose deficiencies of the kriging assumptions.

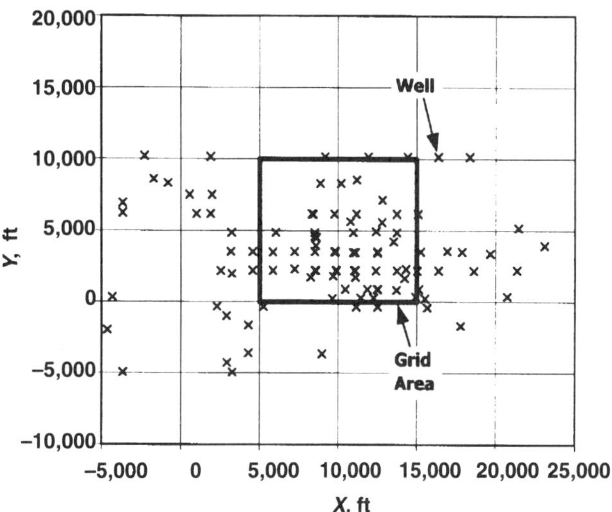

Fig. 4.13—Location of grid used for kriging.

Field Example 4.2—Simple Kriging To Generate Gross Thickness Maps. In this field example, simple kriging is used to generate a map of gross thickness for Flow Unit 5. Mapping in a regular grid is one of the most common reservoir description applications of kriging. Gross thickness maps are used to estimate reservoir bulk volume of rock and to describe the internal zonation.

The following information is given:

• Conditioning data and variograms are given in Field Example 4.1.

• The location of the grid is shown in **Fig. 4.13,** and it is described by the following parameters, as shown in **Table 4.1.**

Find the following:

• Generate a map of gross thickness for Flow Unit 5 with simple kriging.

• Compare statistics of conditioning data and map.

• Generate a map of kriging error variance.

Solution. Simple kriging was used to generate a gross thickness map in the area shown in Fig. 4.13. The grid covers an area of 10,000 by 10,000 ft in the field. Even though only 37 wells fall inside the grid, data outside the grid, but within the search neighborhood, were also included in the kriging calculations.

Fig. 4.14 shows the gross thickness map generated with simple kriging. Overall, the map has a smooth appearance that is typical of simple kriging. The good spatial continuity from east to west corresponds to the principal direction of the variogram. The north/south trend, observed in a small area in the northwest corner of the map, is a result from the configuration of the conditioning data and the search neighborhood. In the northwest corner, the conditioning data are undersampled with respect to the rest of the grid (Fig. 4.13). The optimum search neighborhood, found in Field Example 4.1, was used in the kriging calculations, and it prevented the use of data from the closest clustered points. The simple kriging variance map is shown in **Fig. 4.15.** The estimation variance

TABLE 4.1—GRID INFORMATION

	East	North
Origin, ft	5,000	0
Number of Blocks	50	50
Block Size, ft	200	200

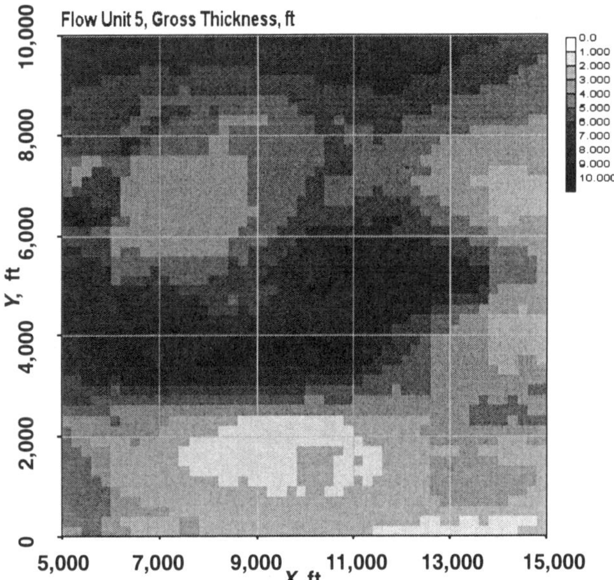

Fig. 4.14—Simple kriging map for Flow Unit 5 gross thickness using optimum search neighborhood.

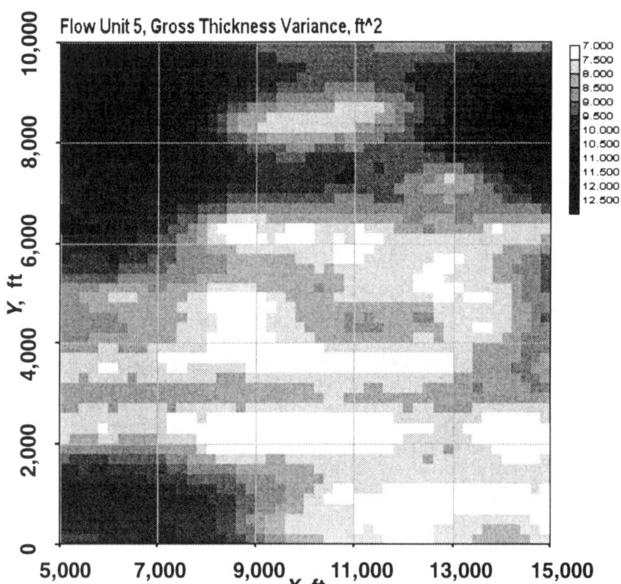

Fig. 4.15—Simple kriging variance map for Flow Unit 5 gross thickness.

is small in gridblocks close to the conditioning data, and it becomes large in areas far from the data. Near the conditioning data, the kriging variance becomes close to the nugget of the variogram (5).

The histogram, for simple kriging estimates. is shown in **Fig. 4.16.** The mean, median and mode of the kriging estimates are close to the conditioning data (Fig. 4.7). The histogram, for kriging estimates, is narrower than the data histogram. The standard deviation for kriging estimates is 1.82 ft, while for conditioning data, it is 2.86 ft. Simple kriging does not reproduce extreme values with a gross thickness greater than 10 ft, observed in the conditioning data.

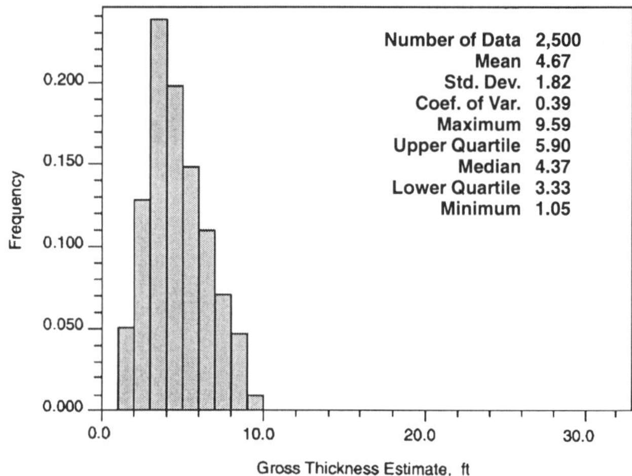

Fig. 4.16—Histogram of simple kriging estimates for Flow Unit 5 gross thickness.

Block Estimation. In the previous example, we considered the estimation of sample values at point locations. This section extends the estimation procedure for block values, using a simple kriging technique. The difference between the point estimation and the block estimation is the support of the location. Point estimation allows the estimation of the value at unsampled locations without any significant volume or area association. In contrast, block kriging allows the estimation of an unsampled location associated with a certain area (in two dimensions) or volume (in three dimensions). This is especially useful when we are interested in generating gridblock properties for a reservoir simulator or estimating in-place hydrocarbons, based on estimated reservoir properties.

In principle, block kriging calculates the arithmetic average of several points located within that support. If we estimate several point values within a block and calculate an arithmetic average, we get the same value as the block estimate. Obviously, we must ensure to choose a sufficient number of points within the block so we obtain a representative block estimate.

The need for block kriging might be questioned because we can always estimate the point values with point kriging and then estimate a block value by calculating the arithmetic average of point values. The following equations show it is more efficient to use block kriging to estimate block values, rather than to use point kriging to estimate the point values and then estimate the arithmetic average.

Another related question we must address is the applicability of block kriging to adequately describe block values of reservoir variables. Remember, block kriging only provides the arithmetic average of point values. If a reservoir variable can be adequately represented on a block level by calculating the arithmetic average of small scale values, block kriging is adequate. One example is porosity of a block. An arithmetic average of small scale porosity values provides an accurate block level porosity measurement. The same reason can be applied to estimate a block saturation value. On the other hand, the arithmetic average of small scale permeability values is not a good representation of a block permeability value. As Chap. 8 discusses, upscaling permeability is much more involved than simply calculating the arithmetic average. Block kriging, obviously, will not be very effective for estimating block values of permeability. To generalize, we observe that block kriging is useful for estimating block values of static properties that can be upscaled using the arithmetic average. Block kriging is not useful for estimating block values of dynamic properties that are difficult to upscale and also flow dependent. This important limitation must be understood before

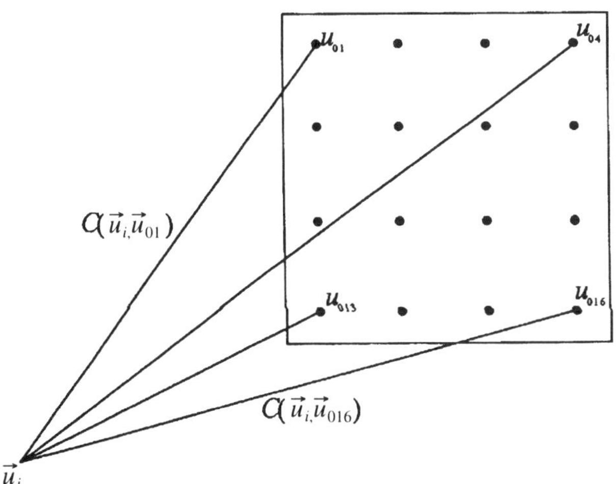

Fig. 4.17—Covariance between a point and a block.

proceeding with the necessary equations required for applying the block kriging technique.

In estimating block values, we use an equation similar to point kriging. The equation is written as

$$X_v^*(\vec{u}_0) = \lambda_0 + \sum_{i=1}^{n} \lambda_i X(\vec{u}_i), \quad \text{.................} \quad (4.13)$$

where $X_v^*(\vec{u}_0)$ = the estimate of a block value located at $\vec{u}_0$, and $X(\vec{u}_i)$ = the point values surrounding the unsampled block. As before, we apply the MVUE technique to estimate the weights assigned to each of the sampled locations within the search neighborhood.

By applying the unbiased condition, we obtain

$$\lambda_o = m\left(1 - \sum_{i=1}^{n} \lambda_i\right). \quad \text{.....................} \quad (4.7)$$

This is the same equation we obtained for simple point kriging. The second condition of minimum variance results in equation,

$$\sum_{j=1}^{n} \lambda_j C(\vec{u}_i, \vec{u}_j) = C_v(\vec{u}_i, \vec{u}_0) \text{ for } i = 1, \ldots, n. \quad \text{...} \quad (4.14)$$

This equation is very similar to Eq. 4.9, except for the right side of the equation. The right side of the equation represents a covariance between a sampled point location at $\vec{u}_i$ and a block located at $\vec{u}_0$. $C_v(\vec{u}_i, \vec{u}_0)$ = an arithmetic average of all the point-to-point covariances between a sample point and all the points located within the block. In practice, to obtain a representative arithmetic average between a point and a block, we select a certain minimum number of points within a block and calculate the point-to-point covariance between the selected points and the sample location; then, we take the arithmetic average. Schematically, this approach is shown in **Fig. 4.17.** In this figure, the covariance between a sampled location and 16 points selected within the block is calculated, and the arithmetic average of these 16 values represents the covariance between the point and a block. Mathematically,

$$C_v(\vec{u}_0, \vec{u}_i) \approx \frac{1}{n_b} \sum_{j=1}^{n_b} C\left(\vec{u}_{0_j}, \vec{u}_i\right), \quad \text{..............} \quad (4.15)$$

where n_b = the selected number of points within a block, and $C\left(\vec{u}_{0_j}, \vec{u}_i\right)$ = the covariance between the point, $\vec{u}_{0_j}$, within the block and the sampled point, $\vec{u}_i$.

The number n_b must be selected, so a representative covariance between a block and a point can be obtained. Numerical experiments suggest that for a 2D gridblock, 16 points within the gridblock will suffice, whereas for a 3D gridblock, 64 points will suffice.[4] The points should be evenly distributed within a gridblock.

Once the covariances between sampled points and the unsampled block are calculated, Eq. 4.15, in matrix form, is written as

$$\begin{bmatrix} C(\vec{u}_1, \vec{u}_1) & \cdots & C(\vec{u}_1, \vec{u}_n) \\ \vdots & & \vdots \\ C(\vec{u}_n, \vec{u}_1) & \cdots & C(\vec{u}_n, \vec{u}_n) \end{bmatrix} \begin{pmatrix} \lambda_1 \\ \vdots \\ \lambda_n \end{pmatrix} = \begin{bmatrix} C_v(\vec{u}_0, \vec{u}_1) \\ \vdots \\ C_v(\vec{u}_0, \vec{u}_n) \end{bmatrix}. \quad \text{..................} \quad (4.16)$$

Notice the similarity between Eqs. 4.16 and 4.10. The left sides are identical; the right sides are different. For point kriging, the right side represents the covariance between a sampled point and an unsampled location. For block kriging, the right side represents the covariance between a sampled point and an unsampled block. By solving Eq. 4.16, we obtain the values of λ_i which can be used to estimate the block value. In addition, error variance is calculated as

$$\hat{\sigma}_E^2 = C_{vv}(\vec{u}_0, \vec{u}_0) - \sum_{i=1}^{2} \lambda_i C_v(\vec{u}_i, \vec{u}_0), \quad \text{.........} \quad (4.17)$$

where $C_{vv}(\vec{u}_0, \vec{u}_0)$ = block variance. The value of block variance is calculated by obtaining point-to-point covariance between several representative pairs within the block and then taking the arithmetic average of the covariances. Similar to point kriging, we also obtain the error variance for the block estimation.

Eqs. 4.13 and 4.17 are more efficient to use when estimating block values, rather than a scaled-down point kriging procedure and a calculation for the arithmetic average of point values. This is true because to obtain point estimates (i.e., 16 locations within the block), point kriging must use Eq. 4.10 16 times, thus, inverting the matrix 16 times. In contrast, block kriging uses Eq. 4.16 one time. The block estimate is obtained by inverting the matrix only once. The additional computational costs, associated with estimating the covariance between a block and a sample point, is much smaller, compared to inverting the matrix numerous times. Therefore, as long as the small-scale values can be arithmetically averaged to estimate block values, block kriging is more efficient.

Numerical Example 4.2. Estimate the porosity value and the corresponding error variance for the block shown in **Fig. 4.18.** Assume that the variogram is defined as

$$\gamma(L) = 0.1 + 0.3\left[\frac{3}{2}\left(\frac{L}{800}\right) - \frac{1}{2}\left(\frac{L}{800}\right)^3\right] \text{ for } L \leq 800$$

$$= 0.4 \text{ for } L \geq 800.$$

The coordinates and the corresponding porosities at the sampled points are given in **Table 4.2.**

What is the porosity value of the block located within the larger block (shown by the dotted line)? Assume four points,

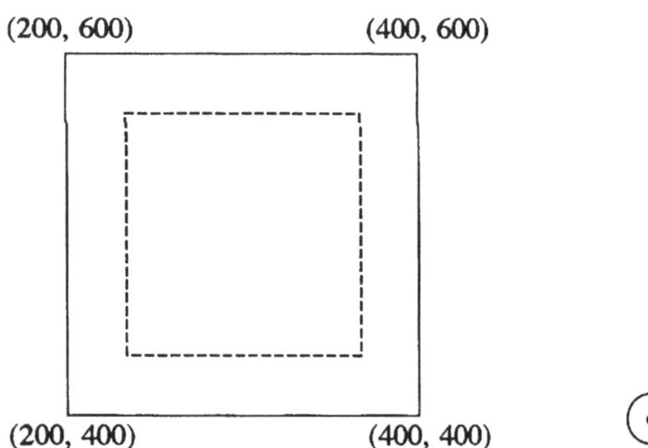

2

Fig. 4.18—Sample configuration for Numerical Example 4.2.

within the gridblock, are used to estimate the covariance between a point and a block.

Solution—Part 1. We estimate the block value with equation

$$X_v^*(\vec{u}_0) = \lambda_0 + \sum_{i=1}^{3} \lambda_i X(\vec{u}_i),$$

where $n = 3$ in this particular case. Our goal is to calculate the values of λ_i. The matrix equation is written as

$$\begin{bmatrix} C(\vec{u}_1, \vec{u}_1) & C(\vec{u}_1, \vec{u}_2) & C(\vec{u}_1, \vec{u}_3) \\ C(\vec{u}_2, \vec{u}_1) & C(\vec{u}_2, \vec{u}_2) & C(\vec{u}_2, \vec{u}_3) \\ C(\vec{u}_3, \vec{u}_1) & C(\vec{u}_3, \vec{u}_2) & C(\vec{u}_3, \vec{u}_3) \end{bmatrix} \begin{bmatrix} \lambda_1 \\ \lambda_2 \\ \lambda_3 \end{bmatrix} = \begin{bmatrix} C_v(\vec{u}_1, \vec{u}_0) \\ C_v(\vec{u}_2, \vec{u}_0) \\ C_v(\vec{u}_3, \vec{u}_0) \end{bmatrix}.$$

The left side of the matrix is similar to the point kriging matrix. Once the distance between the two locations is calculated, based on the variogram model, covariance values are calculated.

$\gamma(\vec{u}_1,\vec{u}_1) = \gamma(\vec{u}_2,\vec{u}_2) = \gamma(\vec{u}_3,\vec{u}_3) = 0$; therefore, $C(\vec{u}_1,\vec{u}_1) = C(\vec{u}_2,\vec{u}_2) = C(\vec{u}_3,\vec{u}_3) = 0.4$. The distance between any two points is calculated by knowing the coordinate locations. For example,

$C(\vec{u}_1,\vec{u}_2) = C(\vec{u}_2,\vec{u}_1) = C(0) - \gamma(\vec{u}_1,\vec{u}_2) = 0.4 - \gamma(403.1) = 0.0924.$

Similarly,

$L_{1,3}$ = 715.9 ft; therefore, $C(\vec{u}_1,\vec{u}_3) = C(\vec{u}_3,\vec{u}_1) = 0.4 - \gamma(715.9) = 0.0048.$

$L_{2,3}$ = 538.5 ft; therefore, $C(\vec{u}_2,\vec{u}_3) = C(\vec{u}_3,\vec{u}_2) = 0.4 - \gamma(538.5) = 0.0428.$

TABLE 4.2—COORDINATES OF DATA POINTS FOR NUMERICAL EXAMPLE 4.2

Data Point	u, ft	v, ft	ϕ
1	250	200	0.13
2	600	400	0.25
3	400	900	0.34

Mean Porosity = 28%.

These values allow us to define the left side of the matrix.

The right side of the matrix requires covariance between a point and a block. We assume the average of four point-to-point covariances will provide a representative covariance between a point and a block. If we select four points within the larger block, as shown in **Fig. 4.19,** we can calculate the covariance between the sample points and these four points, and calculate the arithmetic average. For example, for sample point 1,

$L_{1,01}$ = 350 ft; therefore, $C(\vec{u}_1,\vec{u}_{01}) = 0.1157.$

$L_{1,02}$ = 364 ft; therefore, $C(\vec{u}_1,\vec{u}_{02}) = 0.1094.$

$L_{1,03}$ = 250 ft; therefore, $C(\vec{u}_1,\vec{u}_{03}) = 0.1640.$

$L_{1,04}$ = 269.3 ft; therefore, $C(\vec{u}_1,\vec{u}_{04}) = 0.1542.$

We then calculate,

$$C_v(\vec{u}_1, \vec{u}_0) = \frac{1}{4}\,(0.1157 + 0.1094 + 0.1640 + 0.1542)$$
$$= 0.1358.$$

Similarly,

$$C_v(\vec{u}_2, \vec{u}_0) = \frac{1}{4}\,(0.1020 + 0.1433 + 0.1141 + 0.1614)$$
$$= 0.1302.$$

(200, 600) (400, 600)

01 (250, 550)

02 (350, 550)

03 (250, 450)

04 (350, 450)

(200, 400) (400, 400)

Fig. 4.19—Choosing points for point-block covariance (Numerical Example 4.2).

Fig. 4.20—Modified configuration of samples for Numerical Example 4.2.

$$C_v(\vec{u}_3, \vec{u}_0) = \frac{1}{4}\,(0.1020 + 0.1141 + 0.0645 + 0.0725)$$

$$= 0.0883.$$

Substituting, we write,

$$\begin{pmatrix} 0.4 & 0.0924 & 0.0048 \\ 0.0925 & 0.4 & 0.0428 \\ 0.0048 & 0.0428 & 0.4 \end{pmatrix} \begin{bmatrix} \lambda_1 \\ \lambda_2 \\ \lambda_3 \end{bmatrix} = \begin{pmatrix} 0.1358 \\ 0.1302 \\ 0.883 \end{pmatrix}.$$

Solving for λ_i, $\lambda_1 = 0.2818$, $\lambda_2 = 0.24$, and $\lambda_3 = 0.1917$.

Therefore,

$$\lambda_o = m\left(1 - \sum_{i=1}^{n} \lambda_i\right)$$

$$= 0.28(1 - 0.2818 - 0.24 - 0.1917)$$

$$= 0.0803.$$

$$x^*(u_0) = 0.0803 + (0.2818)(0.13) + (0.24)(0.25)$$

$$+ (0.1917)(0.34)$$

$$= 0.242.$$

To calculate the error variance, we need $C_v(\vec{u}_0, \vec{u}_0)$. This value is calculated with equation

$$C_{vv}(\vec{u}_0, \vec{u}_0) = \frac{1}{n_b^2} \sum_{i=1}^{4} \sum_{j=1}^{4} C\left(\vec{u}_{0_i}, \vec{u}_{0_j}\right),$$

where the covariance between every pair is calculated, and the arithmetic average is taken. Because, in our particular example, all four points within the block are symmetrical, we only calculate the covariance with respect to a point and the remaining three points, and then we take an average, which is

$$C_{vv}(\vec{u}_0, \vec{u}_0)$$

$$= \frac{1}{4}[C(\vec{u}_{01}, \vec{u}_{01}) + C(\vec{u}_{01}, \vec{u}_{02}) + C(\vec{u}_{01}, \vec{u}_{03}) + C(\vec{u}_{01}, \vec{u}_{04})].$$

Knowing the distances, the covariances are computed,

$$C_{vv}(\vec{u}_0, \vec{u}_0) = \frac{1}{4}(0.4 + 0.244 + 0.244 + 0.2214)$$

$$= 0.2774.$$

The error variance is calculated as

$$\hat{\sigma}_E^2 = C_{vv}(\vec{u}_0, \vec{u}_0) - \sum_{i=1}^{3} \lambda_i C(\vec{u}_i, \vec{u}_0)$$

$$= 0.2774 - 0.2818(0.1358) - 0.24(0.1302)$$

$$- 0.1917(0.0883)$$

$$= 0.191.$$

Solution—Part 2. If we estimate the block value, as shown by the dotted line, we must consider four points within that block, as shown in **Fig. 4.20.** Using the same procedure as before, we calculate

$$C_v(\vec{u}_0, \vec{u}_1) = \frac{1}{4}\,(0.1268 + 0.1233 + 0.1509 + 0.1465)$$

$$= 0.1369,$$

$$C_v(\vec{u}_0, \vec{u}_2) = \frac{1}{4}\,(0.1165 + 0.1381 + 0.1233 + 0.1465)$$

$$= 0.1311,$$

and

$$C_v(\vec{u}_0, \vec{u}_3) = \frac{1}{4}\,(0.0957 + 0.1013 + 0.0763 + 0.0808)$$

$$= 0.0885.$$

The matrix equation is

$$\begin{pmatrix} 0.4 & 0.0924 & 0.0048 \\ 0.0924 & 0.4 & 0.0428 \\ 0.0048 & 0.0428 & 0.4 \end{pmatrix} \begin{bmatrix} \lambda_1 \\ \lambda_2 \\ \lambda_3 \end{bmatrix} = \begin{pmatrix} 0.1369 \\ 0.1311 \\ 0.0885 \end{pmatrix}.$$

Solving for λ_i, $\lambda_1 = 0.2814$, $\lambda_2 = 0.2416$, $\lambda_3 = 0.192$, and $\lambda_3 = 0.079$.

Therefore,

$$x^*(u_0) = 0.079 + (0.2814)(0.13) + (0.2416)(0.25)$$

$$+ (0.192)(0.34)$$

$$= 0.242.$$

Although the weights are slightly different, the estimate has not changed.

The error variance is calculated as

$$\hat{\sigma}_E^2 = C_{vv}(\vec{u}_0, \vec{u}_0) - \sum_{i=1}^{3} \lambda_i C(\vec{u}_i, \vec{u}_0)$$

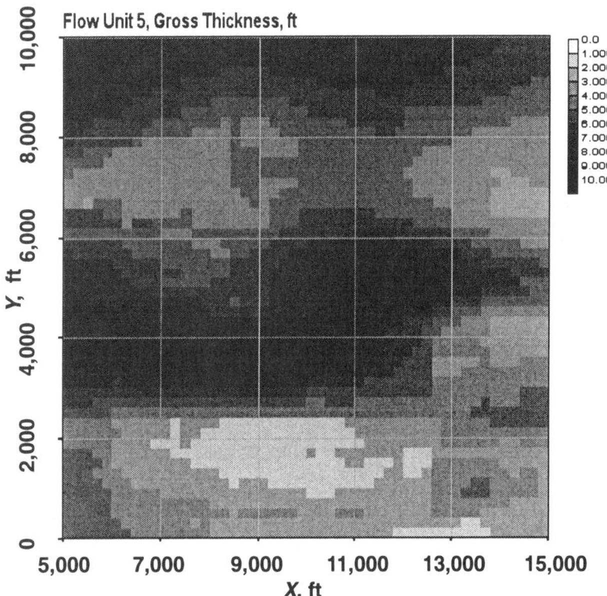

Fig. 4.21—Simple block kriging map for Flow Unit 5 gross thickness.

$$= 0.3010 - 0.2814(0.1369) - 0.2416(0.1311)$$
$$- 0.192(0.0885)$$
$$= 0.2138.$$

The value of $C_{vv}(\vec{u}_0, \vec{u}_0)$ is calculated using the same procedure as shown before for a larger block. The value is greater for a smaller block than a larger block. This is expected because the smaller the block, the larger is the variability of the values. The variance is also larger for a smaller block, indicating the impact of support effect on the estimated error variance.

Field Example 4.3—Simple Block Kriging To Generate Gross Thickness Maps. This field example illustrates the use of simple block kriging to generate a gross thickness map for Flow Unit 5. Block kriging accounts for changes in spatial correlation, resulting from differences between the data support scale and volume of the gridblock.

The following information is given:

• Conditioning data and variograms are given in Field Example 4.1.

• The optimum search neighborhood found in Field Example 4.1 consists of an anisotropic search radius based on the variogram parameters and 16 sample data points with a maximum of four samples per quadrant.

• The grid definition is given in Field Example 4.2.

Calculate the following:

• Generate a map of gross thickness for Flow Unit 5 using simple block kriging with 16 block samples within each gridblock.

• Compare statistics of conditioning data, simple, and block kriging maps.

Solution. Simple block kriging was used to generate a gross thickness map for Flow Unit 5 using 16 block samples to generate an average covariance for the gridblocks. The block covariance is calculated using the arithmetic average of covariances between data points and the 16 samples within each block. The 16 samples are regularly spaced in the x and y directions. The gross thickness map, generated with block kriging, is shown in **Fig. 4.21.** The major trends of thin and thick areas in this map are similar to the simple kriging results (Field Example 4.2). However, the block kriging map appears to have even less variability at small-scale than simple kriging.

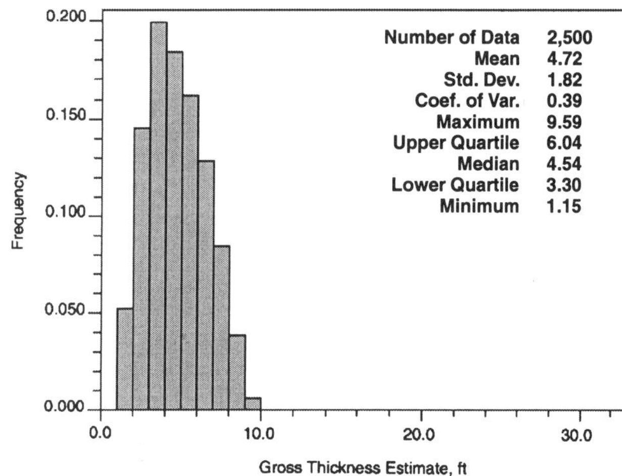

Fig. 4.22—Histogram of simple block kriging estimates for Flow Unit 5 gross thickness.

The histogram for simple block kriging estimates is shown in **Fig. 4.22.** The statistics of simple kriging (Fig. 4.16) and block kriging results are very similar. The standard deviation for simple block kriging estimates is 2.5 times smaller than the conditioning data. The histogram for simple block kriging is narrower and more symmetric than the histogram for the conditioning data (Fig. 4.7).

4.2.2 Ordinary Kriging. In a simple kriging procedure, we assume the mean value, $m(\vec{u})$, is known. By assuming first-order stationarity, $m(\vec{u})$ reduces to m. We must know the value of m before we use a simple kriging expression. In practice, however, the true global mean is rarely known, unless we assume the sample mean is the same as the global mean. Further, as seen before, the local mean within the search neighborhood may vary over the region of interest. As a result, the assumption of first-order stationarity may not be strictly valid. An ordinary kriging procedure overcomes this problem by redefining the equation for estimation.

Consider Eq. 4.4 used for simple kriging.

$$X^*(\vec{u}_o) = \lambda_o + \sum_{i=1}^{n} \lambda_i X(\vec{u}_i). \quad \text{(4.4)}$$

However, the unbiased condition requires

$$E[X^*(\vec{u}_o) - X(\vec{u}_o)] = 0. \quad \text{(4.5)}$$

If we assume $E[X^*(\vec{u}_o)] = E[X(\vec{u}_i)] = m(\vec{u}_o)$, where $m(\vec{u}_o)$ represents the mean within the search neighborhood of location $\vec{u}_o$, we write

$$\lambda_o = m(\vec{u}_o)\left(1 - \sum_{i=1}^{n} \lambda_i\right). \quad \text{(4.18)}$$

However, we force λ_o to be zero, if we assume

$$\sum_{i=1}^{n} \lambda_i = 1. \quad \text{(4.19)}$$

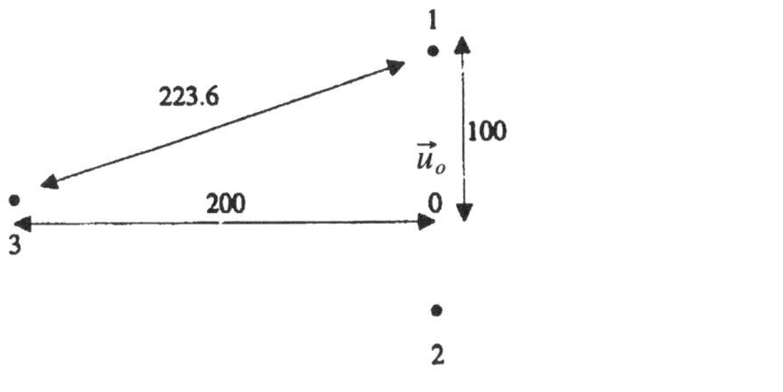

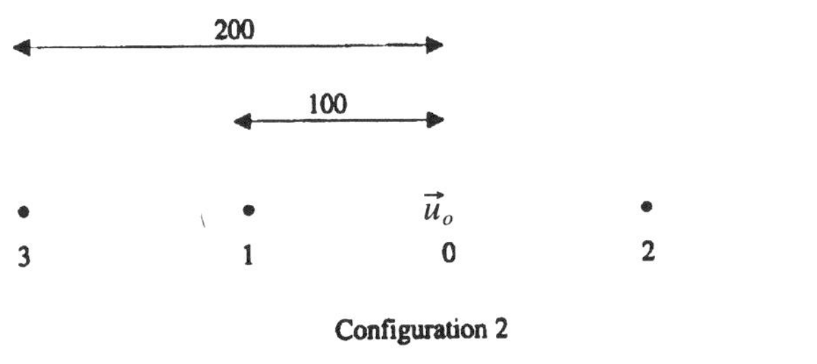

Fig. 4.23—Two configurations for Numerical Example 4.3.

The estimate is written as

$$X^*(\vec{u}_o) = \sum_{i=1}^{n} \lambda_i X(\vec{u}_i), \quad \ldots\ldots\ldots\ldots\ldots\ldots (4.20)$$

with a constraint defined by Eq. 4.19. By forcing λ_o to be zero, the necessity for requiring the mean value is eliminated. Further, we relax the assumption that first-order stationarity be strictly satisfied, by assuming the local mean is dependent on the location.[9]

In addition to the unbiased condition, we also are required to satisfy the minimum variance condition. Minimizing the variance with a constraint in Eq. 4.19 results in

$$\sum \lambda_j C(\vec{u}_i, \vec{u}_j) + \mu = C(\vec{u}_i, \vec{u}_o) \text{ for } i = 1,\ldots,n, \quad \ldots\ldots\ldots\ldots (4.21)$$

where μ is called a Lagrange parameter, and C represents the covariance. For details regarding the derivation of Eq. 4.21, see Appendix D.

In matrix form, Eq. 4.21 is written as

$$\begin{bmatrix} C(\vec{u}_1, \vec{u}_1) & \ldots & C(\vec{u}_1, \vec{u}_n) & 1 \\ \vdots & & \vdots & \vdots \\ C(\vec{u}_n, \vec{u}_1) & \ldots & C(\vec{u}_n, \vec{u}_n) & 1 \\ 1 & \ldots & 1 & 0 \end{bmatrix} \begin{bmatrix} \lambda_1 \\ \vdots \\ \lambda_n \\ \mu \end{bmatrix} = \begin{bmatrix} C(\vec{u}_1, \vec{u}_0) \\ \vdots \\ C(\vec{u}_n, \vec{u}_0) \\ 1 \end{bmatrix}. \quad \ldots\ldots\ldots\ldots (4.22)$$

To solve λ_i, we write, in matrix form,

$$|\Lambda| = |C|^{-1}|c|. \quad \ldots\ldots\ldots\ldots\ldots\ldots (4.23)$$

Once λ_i is calculated, the estimate, $X^*(\vec{u}_o)$, is obtained with Eq. 4.20. We also estimate the error variance as

$$\hat{\sigma}_E^2 = C(\vec{u}_o, \vec{u}_o) - \sum_{i=1}^{n} \lambda_i C(\vec{u}_i, \vec{u}_o) - \mu. \quad \ldots\ldots\ldots (4.24)$$

Numerical Example 4.3. Consider the configurations in **Fig. 4.23**:

$x(\vec{u}_1) = 20,\ x(\vec{u}_2) = 50,\ x(\vec{u}_3) = 30,$ and $x(\vec{u}_4) = 100.$

Assume the following variogram models:

(a) Isotropic Variogram

$$\gamma(\vec{L}) = 100 M_{s_{500}}(\vec{L}),$$

(b) Isotropic Variogram

$$\gamma(\vec{L}) = 50 + 50 M_{s_{500}}(\vec{L}),$$

and

(c) Anisotropic Variogram

$$\gamma(\vec{L}) = 100 M_{s_{500}}(\vec{L}) \text{ east/west direction,}$$

and

$$\gamma(\vec{L}) = 100 M_{s_{250}}(\vec{L}) \text{ north/south direction.}$$

Estimate the value and the error variance at location $\vec{u}_o$ with the previous three models for both configurations.

Solution—Model 1. We assume that the variogram model is given by

$$\gamma(\vec{L}) = 100\left[\frac{3}{2}\left(\frac{L}{500}\right) - \frac{1}{2}\left(\frac{L}{500}\right)^3\right] \text{ for } L \le 500$$

$$= 100 \quad L \ge 500.$$

Configuration 1. Because there are four sample points, the covariance matrix is 5×5. We know

$$C(1,1) = C(2,2) = C(3,3) = C(4,4) = C(0) = 100.$$

We calculate

$$C(1,2) = C(0) - \gamma(1,2) = C(2,1)$$

$= 100 - \gamma(200)$ because the distance between one and two is 200 ft and

$$= 100 - 100\left[\frac{3}{2}\left(\frac{200}{500}\right) - \frac{1}{2}\left(\frac{200}{500}\right)^3\right].$$

Similarly, we compute values of other terms in the covariance matrix. The right side of the equation requires the covariance between the unsampled location and the sample point. For example,

$$C(1,0) = C(0) - \gamma(1,0) = C(0) - \gamma(100) = 70.4.$$

In matrix form, we write

$$\begin{bmatrix} C(1,1) & C(1,2) & C(1,3) & C(1,4) & 1 \\ C(2,1) & C(2,2) & C(2,3) & C(2,4) & 1 \\ C(3,1) & C(3,2) & C(3,3) & C(3,4) & 1 \\ C(4,1) & C(4,2) & C(4,3) & C(4,4) & 1 \\ 1 & 1 & 1 & 1 & 0 \end{bmatrix} \begin{bmatrix} \lambda_1 \\ \lambda_2 \\ \lambda_3 \\ \lambda_4 \\ \mu \end{bmatrix} = \begin{bmatrix} C(1,0) \\ C(2,0) \\ C(3,0) \\ C(4,0) \\ 1 \end{bmatrix}.$$

By substituting all the values, we obtain

$$\begin{bmatrix} 100.0 & 43.2 & 37.4 & 37.4 & 1 \\ 43.2 & 100.0 & 37.4 & 37.4 & 1 \\ 37.4 & 37.4 & 100.0 & 5.6 & 1 \\ 37.4 & 37.4 & 5.6 & 100.0 & 1 \\ 1 & 1 & 1 & 1 & 0 \end{bmatrix} \begin{Bmatrix} \lambda_1 \\ \lambda_2 \\ \lambda_3 \\ \lambda_4 \\ \mu \end{Bmatrix} = \begin{bmatrix} 70.4 \\ 70.4 \\ 43.2 \\ 43.2 \\ 1 \end{bmatrix}.$$

Solving for λ_i and μ, by inverting the matrix, we obtain

$\lambda_1 = 0.429$, $\lambda_2 = 0.429$, $\lambda_3 = 0.071$, $\lambda_4 = 0.071$,

and $\mu = 3.632$.

Once the weights are obtained, notice that the sum of all weights equals one. We estimate the value at $x^*(\vec{u}_o)$ as

$$x^*(\vec{u}_o) = 0.429 \times 20 + 0.429 \times 50 + 0.071 \times 30 + 0.071 \times 100 = 39.24.$$

The error variance is estimated as

$$\hat{\sigma}_E^2 = C(\vec{u}_o, \vec{u}_o) - \sum_{i=1}^{4} \lambda_i C(\vec{u}_i, \vec{u}_o) - \mu$$

$$= 100 - (.429 \times 70.4 + .429 \times 70.4 + .071 \times 43.2 + .071 \times 43.2) - 3.632$$

$$= 29.81.$$

Notice that higher weights are assigned to the closest points, and symmetry of the configuration is maintained in the weights as well.

Configuration 2. In this configuration, we have the four data points situated differently. The relative distances from the unsampled location are the same as Configuration 1. The difference between this configuration and the first is the relative locations of the sample points with respect to each other.

Following similar computations as before, we write the equation in the matrix form as

$$\begin{bmatrix} 100.0 & 43.2 & 70.4 & 20.8 & 1 \\ 43.2 & 100.0 & 20.8 & 70.4 & 1 \\ 70.4 & 20.8 & 100.0 & 5.6 & 1 \\ 20.8 & 70.4 & 5.6 & 100.0 & 1 \\ 1 & 1 & 1 & 1 & 0 \end{bmatrix} \begin{Bmatrix} \lambda_1 \\ \lambda_2 \\ \lambda_3 \\ \lambda_4 \\ \mu \end{Bmatrix} = \begin{bmatrix} 70.4 \\ 70.4 \\ 43.2 \\ 43.2 \\ 1 \end{bmatrix}.$$

By solving these equations, we obtain

$\lambda_1 = 0.518$, $\lambda_2 = 0.518$, $\lambda_3 = -0.018$, $\lambda_4 = -0.018$,

and $\mu = -2.14$.

Note, the weights assigned to samples three and four are slightly negative. This results from the screening of sample points three and four from the unsampled locations. Because sample points three and four are directly behind points one and two, respectively, the unsampled location does not "see" these samples, reducing their influence on the estimate. This can result in negative weights assigned to those samples.

As discussed in Sec. 4.2.1, unlike conventional interpolation techniques (i.e., the inverse squared distance method), kriging accounts for the relative locations of the samples among themselves.

The error variance is estimated as

$$\hat{\sigma}_E^2 = 100 - (0.518 \times 70.4 + 0.518 \times 70.4 - 0.018 \times 43.2 - 0.018 \times 43.2) + 2.14$$

$$= 30.8.$$

The error variance for this configuration is higher, compared to the first configuration. This is expected, considering Configuration 2 does not provide as much information as the first configuration. As Fig. 4.23 shows, in Configuration 1, the sample data are well distributed surrounding the unsampled location. In contrast, in Configuration 2, all samples are in one line. One would expect more uncertainty when estimating the value for Configuration 2, compared to Configuration 1 because of less information. This is reflected in the estimated error variance.

This illustrates that error variance is a reflection of surrounding data configuration. More and well distributed samples will result in a smaller error variance. Fewer and clustered samples will result in a bigger error variance.

Solution—Model 2. We assume that the variogram model is provided by

$$\gamma(\vec{L}) \begin{cases} = 0 \text{ for } \vec{L} = 0 \\ = 50 + 50\left[\frac{3}{2}\left(\frac{\vec{L}}{500}\right) - \frac{1}{2}\left(\frac{\vec{L}}{500}\right)^3\right] \text{for } 0 < L \leq 500 \\ = 100 \text{ for } \vec{L} \geq 500. \end{cases}$$

The only difference between Models 1 and 2 is the nugget effect.

Configuration 1. The procedure for calculating the covariance model is the same as stated before. The final covariance matrix equation is written as

$$\begin{bmatrix} 100.0 & 21.6 & 18.7 & 18.7 & 1 \\ 21.6 & 100.0 & 18.7 & 18.7 & 1 \\ 18.7 & 18.7 & 100.0 & 2.8 & 1 \\ 18.7 & 18.7 & 2.8 & 100.0 & 1 \\ 1 & 1 & 1 & 1 & 0 \end{bmatrix} \begin{Bmatrix} \lambda_1 \\ \lambda_2 \\ \lambda_3 \\ \lambda_4 \\ \mu \end{Bmatrix} = \begin{bmatrix} 35.2 \\ 35.2 \\ 21.6 \\ 21.6 \\ 1 \end{bmatrix}.$$

Solving for λ_i, we obtain

$\lambda_1 = 0.309$, $\lambda_2 = 0.309$, $\lambda_3 = 0.191$, $\lambda_4 = 0.191$,

and $\mu = 9.56$.

The estimated value, $x^*(\vec{u}_o)$, = 46.43, and the error variance, $\hat{\sigma}_E^2$, = 79.53.

By comparing the weights, using this model, with Model 1 for the same configuration, one notices that the weights assigned to the closest two sample points are reduced, whereas the weights assigned to the two farthest points are increased. This is consistent with the model used. A large nugget in the model indicates uncertainty about the spatial relationship. Because we know less about the spatial relationship, the weights assigned to individual points are only partially influenced by the relative distances between sample points and the unsampled location. In the limiting case, if the variogram model is described by a pure nugget, weights assigned to all the points will be equal because we have no prior information about the spatial relationship.

The error variance is also higher than the corresponding error variance for Model 1. This also reflects a lack of spatial relationship. The more we know about the spatial relationship, the more we can use the information from surrounding sample points to estimate the value at an unsampled location. The more we are informed, the less we will be uncertain. This is reflected in a lower error variance. In contrast, in the presence of a high nugget value, the spatial information is partially lacking. This results in a higher error variance.

In the extreme case, if the variogram is represented by a pure nugget, the error variance is represented by

$$\hat{\sigma}_E^2 = C(\vec{u}_o, \vec{u}_o) - \sum_{i=1}^{4} \lambda_i C(\vec{u}_i, \vec{u}_o) - \mu$$

$$= C(\vec{u}_o, \vec{u}_o) - \mu$$

because the value of $C(\vec{u}_i, \vec{u}_o) = 0$ for all samples. The value of μ is shown as

$$\mu = -\frac{C(\vec{u}_o, \vec{u}_o)}{n},$$

where n is the number of sample points. Substituting,

$$\hat{\sigma}_E^2 = C(\vec{u}_o, \vec{u}_o)\left(1 + \frac{1}{n}\right) = C(o)\left(1 + \frac{1}{n}\right).$$

That is, error variance is an addition of variance of the data and error, associated with the estimation of the mean, $C(o)/n$.

For a large value of n, if the variogram is represented by a pure nugget, then the estimated value is the arithmetic mean of the samples, and the error variance is the variance of the samples. As discussed in Sec. 4.2.1, in the absence of spatial information, this observation is consistent with our intuitive understanding of first- and second-order stationarities.

Configuration 2. For this configuration, the covariance matrix is

$$\begin{bmatrix} 100.0 & 21.6 & 35.2 & 10.4 & 1 \\ 21.6 & 100.0 & 10.4 & 35.2 & 1 \\ 35.2 & 10.4 & 100.0 & 2.8 & 1 \\ 10.4 & 35.2 & 2.8 & 100.0 & 1 \\ 1 & 1 & 1 & 1 & 0 \end{bmatrix} \begin{bmatrix} \lambda_1 \\ \lambda_2 \\ \lambda_3 \\ \lambda_4 \\ \mu \end{bmatrix} = \begin{bmatrix} 35.2 \\ 35.2 \\ 21.6 \\ 21.6 \\ 1 \end{bmatrix}.$$

Solving for λ_i, we obtain

$$\lambda_1 = 0.317,\ \lambda_2 = 0.317,\ \lambda_3 = 0.183,\ \lambda_4 = 0.183,$$

and $\mu = -11.68$.

By using the appropriate equations, $x^*(\vec{u}_o) = 46.0$, and $\hat{\sigma}_E^2 = 81.5$. As explained in the previous model, for Configuration 2, the weights assigned to the two closest points increase, compared to the two farthest points. This is because of the shielding of the two farthest points. However, in contrast to Model 1, the differences in weight, compared to Configuration 1, are small. This is expected because we do not know as much about the spatial information. The error variance compared to Configuration 1 is higher. This is also consistent, considering the sample configuration is not as well distributed as Configuration 1.

Solution—Model 3. We assume the following anisotropic models:

(a) north/south

$$\gamma(\vec{L}) = 100\left[\frac{3}{2}\left(\frac{\vec{L}}{250}\right) - \frac{1}{2}\left(\frac{\vec{L}}{250}\right)^3\right] \text{for } \vec{L} \leq 250$$

$$= 100 \text{ for } \vec{L} \geq 250,$$

and (b) east/west

$$\gamma(\vec{L}) = 100\left[\frac{3}{2}\left(\frac{\vec{L}}{500}\right) - \frac{1}{2}\left(\frac{\vec{L}}{500}\right)^3\right] \text{for } \vec{L} \leq 500$$

$$= 100 \text{ for } \vec{L} \geq 500.$$

Knowing these two models, a variogram in any direction can be estimated with the procedure provided in Chap. 3.

Configuration 1. For this configuration, the covariance matrix is written as

$$\begin{bmatrix} 100.0 & 5.6 & 24.2 & 24.2 & 1 \\ 5.6 & 100.0 & 24.2 & 24.2 & 1 \\ 24.2 & 24.2 & 100.0 & 5.6 & 1 \\ 24.2 & 24.2 & 5.6 & 100.0 & 1 \\ 1 & 1 & 1 & 1 & 0 \end{bmatrix} \begin{bmatrix} \lambda_1 \\ \lambda_2 \\ \lambda_3 \\ \lambda_4 \\ \mu \end{bmatrix} = \begin{bmatrix} 43.2 \\ 43.2 \\ 43.2 \\ 43.2 \\ 1 \end{bmatrix}.$$

Notice that the covariances between the sample points and the unsampled location are identical. Because we have different ranges in different directions, the relative influences of sample points one and two, and three and four are identical. The range in the north/south direction is smaller; therefore, the influence of sample points in that direction diminishes over a shorter distance. Solving the matrix problem, we obtain

$$\lambda_1 = \lambda_2 = \lambda_3 = \lambda_4 = 0.25, \text{and } \mu = 4.702.$$

The estimated value, $x^*(\vec{u}_o)$, is 50, and the error variance, $\hat{\sigma}_E^2$, is 52.1. The weight assigned to each of the sample points is identical. This is because the variogram distance for all the four points (which is indicated by the variogram value between the sample point and the unsampled location) is identical.

The error variance, 52.1, is higher than the error variance for the same configuration for Model 1. Recall that Model 1 was an isotropic variogram model with a range of 500. We obtained the error variance of 29.81. Although the sample configuration is the same, the range in the north/south direction is half in Model 3, compared to Model 1. As a result, the neighboring values in that direction do not provide as much information. Recall that the longer the range, the stronger is the relationship between neighboring values at a given distance. The lack of spatial information in the north/south direction results in a higher error variance, compared to the error variance obtained with Model 1.

Configuration 2. The matrix equation is written as

$$\begin{bmatrix} 100.0 & 43.2 & 70.4 & 20.8 & 1 \\ 43.2 & 100.0 & 20.8 & 70.4 & 1 \\ 70.4 & 20.8 & 100.0 & 5.6 & 1 \\ 20.8 & 70.4 & 5.6 & 100.0 & 1 \\ 1 & 1 & 1 & 1 & 0 \end{bmatrix} \begin{bmatrix} \lambda_1 \\ \lambda_2 \\ \lambda_3 \\ \lambda_4 \\ \mu \end{bmatrix} = \begin{bmatrix} 70.4 \\ 70.4 \\ 43.2 \\ 43.2 \\ 1 \end{bmatrix}.$$

Solving the matrix problem, we obtain

$$\lambda_1 = 0.518,\ \lambda_2 = 0.518,\ \lambda_3 = -0.0181,\ \lambda_4 = -0.0181,$$

and $\mu = -2.14$.

$x^*(\vec{u}_o) = 33.92$, and $\hat{\sigma}_E^2 = 30.76$.

These answers are similar to Model 1 for this configuration. This is because the variogram in the east/west direction is the same in both models, and all the sample points are located in that direction.

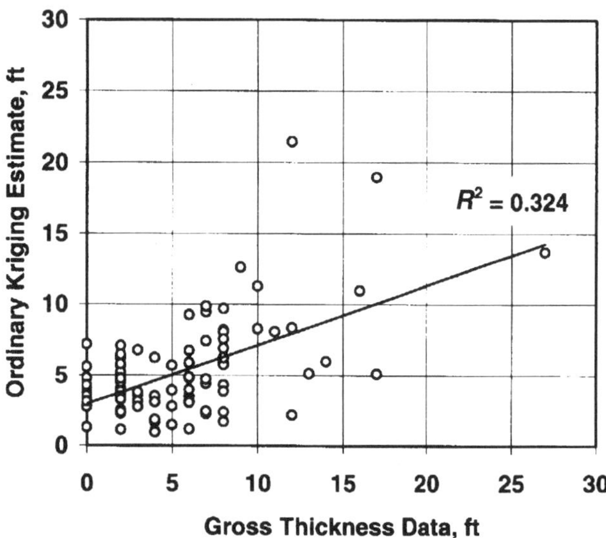

Fig. 4.24—Ordinary kriging cross validation estimates for Flow Unit 5 gross thickness.

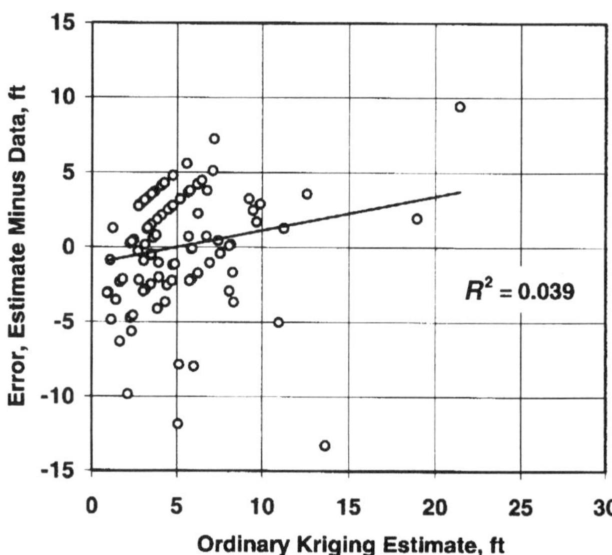

Fig. 4.25—Ordinary kriging cross validation errors for Flow Unit 5 gross thickness.

The only comparison worth pointing out is the difference in the error variance for this configuration vs. the error variance in the first configuration. Unlike Model 1, for this model, the error variance is higher for Configuration 1, compared to Configuration 2. Although the data seem to be better distributed for Configuration 1, compared to Configuration 2, the information obtained in Configuration 2 is better than that for Configuration 1. This is because the two closest points from the unsampled location in Configuration 1 are aligned in the direction of minimum continuity, thereby providing less information. In contrast, in Configuration 2, the two closest points are aligned in the direction of maximum continuity. As a result, they are much more informative. This results in a smaller error variance. Note, unlike Model 1, the right side of the matrix equation for the two configurations is different. The covariance indicates the extent of the relationship between the data points. Higher values of covariances in Configuration 2 indicate a stronger relationship between the unsampled location and the sample points.

To summarize, the error variance is a quantitative indicator of the surrounding sample configuration. However, in considering the sample configuration, we must examine the relative variogram distances among the sample points rather than the physical distances. If more points are located in the direction of maximum continuity, they may provide more information about the unsampled location than apparently well distributed samples, many of which may be in the direction of minimum continuity. In comparing the two configurations, an unsampled location surrounded by well distributed (and more informative) samples in terms of variogram distance has a smaller error variance than a corresponding configuration surrounded by less informative samples.

Field Example 4.4—Ordinary Kriging Cross Validation. In this field example, ordinary kriging is used to cross validate the gross thickness of Flow Unit 5. In ordinary kriging, the mean changes from location to location, while in simple kriging the mean is fixed and equal to the global mean of the data. In practice, it is not common for the mean to change across the field; therefore, the assumption of constant mean or first-order stationarity is not adequate.

The following information is given:

• Conditioning data and variograms are given in Field Example 4.1.

• The optimum search neighborhood, found in Field Example 4.1, consists of an anisotropic search radius based on the variogram parameters and 16 sample data points with a maximum of four samples per quadrant.

Make the following calculations:

• Calculate ordinary kriging cross-validation estimates for gross thickness data of Flow Unit 5.

• Compare ordinary and simple kriging cross-validation estimates.

Solution. The ordinary kriging cross-validation estimates are compared to the data for Flow Unit 5 gross thickness in **Fig. 4.24.** Ordinary kriging provides good estimates even for values that are much larger than the mean of the data (5.05 ft). The correlation coefficient between the data and ordinary kriging is slightly smaller than for simple kriging (0.57 vs. 0.61). **Fig. 4.25** shows the cross-validation errors. The correlation coefficient indicates that there is a small trend where ordinary kriging tends to overestimate gross thickness, as the magnitude of data increases.

The simple kriging estimates from Field Example 4.1 and the ordinary kriging estimates are compared in **Fig. 4.26.** Both techniques provide similar estimates of gross thickness. Simple kriging estimates are slightly better than ordinary kriging for values smaller than the mean of the data, while ordinary kriging works better for large values greater than the mean. Simple kriging does not estimate large values that are significantly greater than the mean very well.

Field Example 4.5—Ordinary Kriging To Generate Gross Thickness Maps. The following information is given:

• Conditioning data and variograms are given in Field Example 4.1.

• The grid definition is given in Field Example 4.2.

• The optimum search neighborhood found in Field Example 4.1 consists of an anisotropic search radius based on the variogram parameters and 16 sample data points with a maximum of four samples per quadrant.

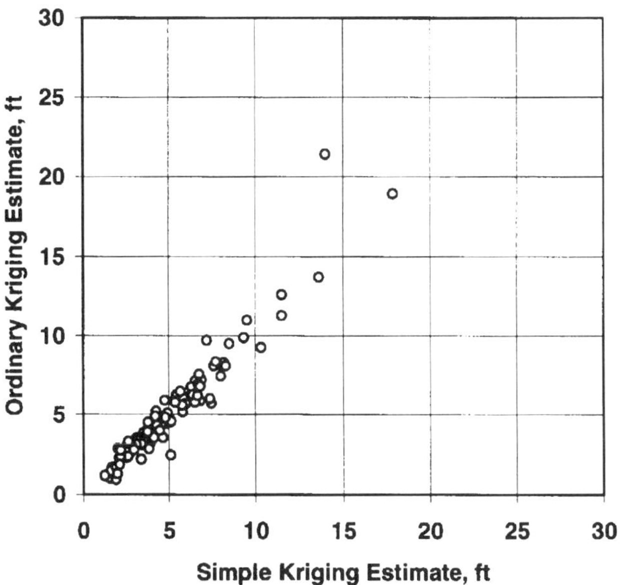

Fig. 4.26—Comparison of simple and ordinary kriging estimates for Flow Unit 5 gross thickness.

Make the following calculations:

• Generate a map of gross thickness for Flow Unit 5 with ordinary kriging.

• Compare statistics of conditioning data, simple and ordinary kriging maps.

Solution. Flow Unit 5 gross thickness map, generated with ordinary kriging, is shown in **Fig. 4.27.** In this field example, ordinary and simple kriging (Fig. 4.14) yield similar results. **Fig. 4.28** shows a closer comparison between simple and ordinary kriging and indicates that most differences occur for gross thickness between 4 and 7 ft.

The histogram for ordinary kriging of gross thickness estimates is shown in **Fig. 4.29.** The statistical parameters of ordinary kriging are very similar to those for simple kriging (Fig. 4.16).

Field Example 4.6—Ordinary Kriging To Estimate Original Oil in Place. One of the common applications of kriging is to estimate reserves. This field example shows how maps

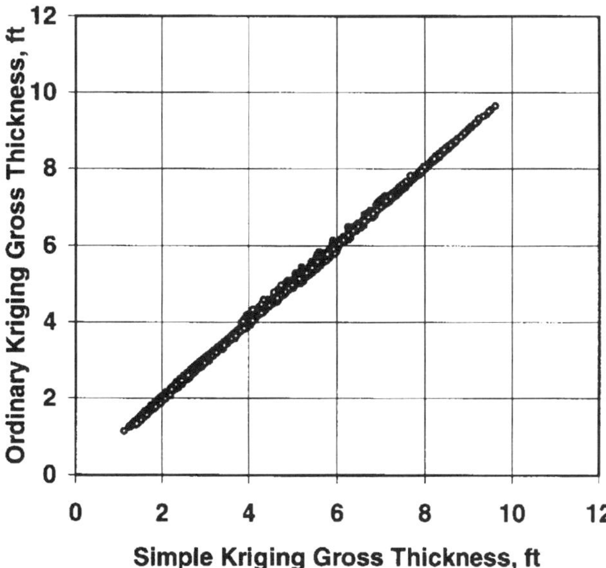

Fig. 4.28—Ordinary and simple kriging estimates for Flow Unit 5 gross thickness.

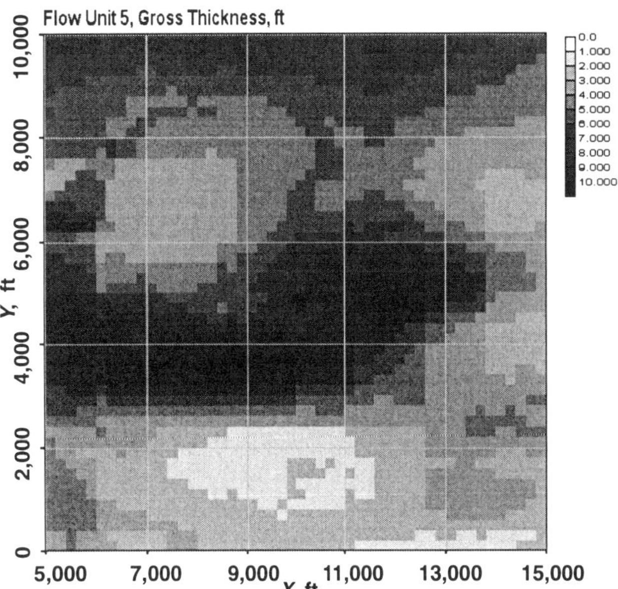

Fig. 4.27—Ordinary kriging map for Flow Unit 5 gross thickness.

of porosity and gross thickness are used to estimate original oil in place (OOIP).

The following information is given:

• Conditioning data for porosity of Flow Unit 5 (Appendix A). The location of the conditioning data is the same as the data in Field Example 4.1 (Fig. 4.6).

• Variogram models for Flow Unit 5 porosity are

$$\gamma(\vec{L}) = 10 + 30M_{s_{4,000}}(\vec{L}), \text{ principal direction east/west,}$$

and

$$\gamma(\vec{L}) = 10 + 30M_{s_{10,000}}(\vec{L}), \text{ minor direction north/south.}$$

These models were calculated with the methods described in Chap. 3. The principal and minor directions coincide with those found for gross thickness (Field Example 4.1).

• The grid definition is given in Field Example 4.2.

Make the following calculations:

• Generate a map for Flow Unit 5 porosity with ordinary kriging.

• Compare the statistics of porosity data and an ordinary kriging map.

• Generate a map of OOIP for the area covered by the grid.

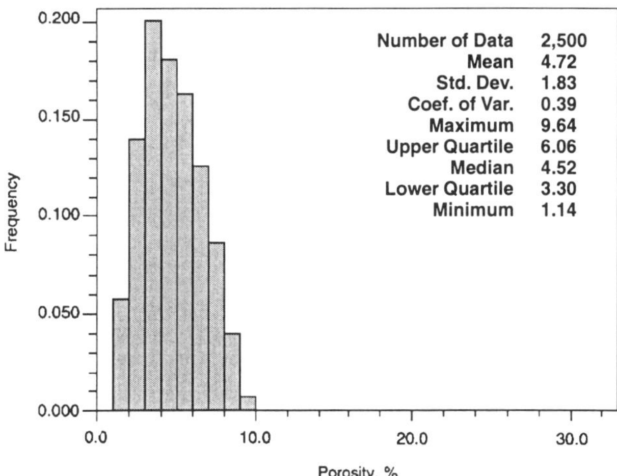

Fig. 4.29—Histogram of ordinary kriging estimates for Flow Unit 5 gross thickness.

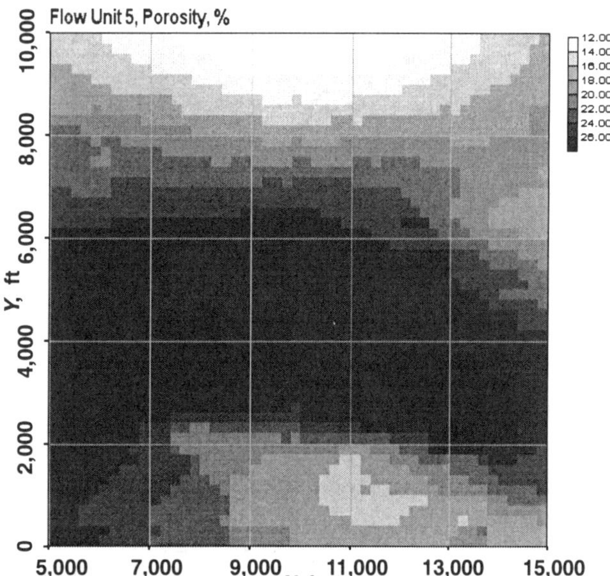

Fig. 4.30—Ordinary kriging porosity map for Flow Unit 5.

• The other properties needed to calculate OOIP are assumed to be constant in this area of the field. The net to gross ratio is 0.70, initial connate water saturation is 20%, and formation volume factor is 1.2 res bbl/STB. Alternatively, if these properties vary significantly across the field, and if enough data are available, kriging can also be used to generate maps of all the properties needed to calculate OOIP.

Solution. The porosity map generated with ordinary kriging is shown in **Fig. 4.30.** Porosity has spatial trends similar to the gross thickness map (Fig. 4.27), resulting from similarities in the data configuration and variogram directions. The highest porosities are in the middle of the map and extend east/west. Histograms of porosity data and kriging estimates are shown in **Fig. 4.31.** The mean, variance and median of the kriging estimates are similar to the conditioning data. However, extreme values of the data are not recognized by kriging.

OOIP is calculated with gross thickness (Fig. 4.27) and porosity maps (Fig. 4.30) generated with ordinary kriging. OOIP in stock tank barrel (STB) for each gridblock in the map is given by

$$\text{OOIP} = \frac{A \times h \times ng \times (\phi/100)(1 - Sw)}{5.615 \times Bo}.$$

A = the surface area of a block (200 × 200 = 40,000 ft^2), h = gross thickness (ft), ng = net to gross ratio, ϕ = porosity (%), Sw = the water saturation and Bo = the formation volume factor (res bbl/STB). The OOIP for Flow Unit 5 calculated with the previous equation is shown in **Fig. 4.32.** The map shows that the highest OOIP is in the middle section of the map, which coincides with the areas of greatest thickness and porosity. The total OOIP for the map area is calculated by adding the values of all gridblocks. The total OOIP is 8,502,974 STB.

Block Estimation. Block estimation involves the estimation of gridblock values using an ordinary kriging procedure. The basic estimation equation is written as

$$X_v^*(\vec{u}_o) = \sum_{i=1}^{n} \lambda_i X(\vec{u}_i), \quad \text{(4.25)}$$

Conditioning Data

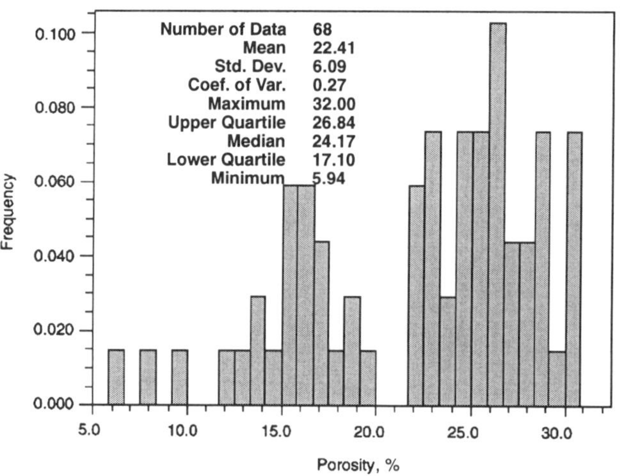

Ordinary Kriging

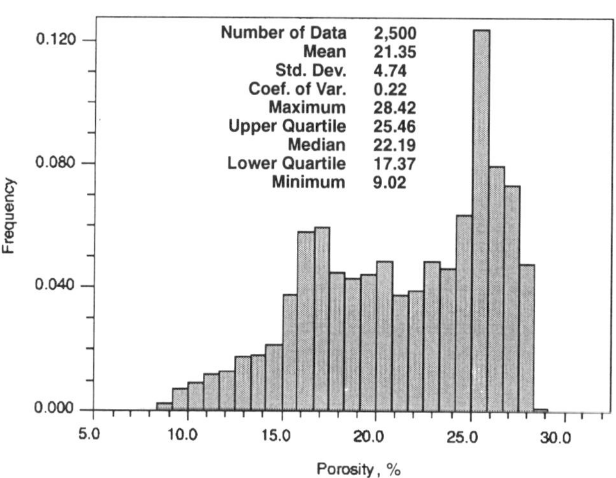

Fig. 4.31—Histograms of Flow Unit 5 porosity data and ordinary kriging estimates.

where the subscript v = the block estimate. If we apply the unbiased condition and minimum variance, we ultimately develop an equation in matrix form.

$$\begin{bmatrix} C(\vec{u}_1, \vec{u}_1) & \cdots & C(\vec{u}_1, \vec{u}_n) & 1 \\ \vdots & & \vdots & \vdots \\ C(\vec{u}_n, \vec{u}_1) & \cdots & C(\vec{u}_n, \vec{u}_n) & 1 \\ 1 & \cdots & 1 & 0 \end{bmatrix} \begin{bmatrix} \lambda_1 \\ \vdots \\ \lambda_n \\ \mu \end{bmatrix} = \begin{bmatrix} C_v(\vec{u}_1, \vec{u}_0) \\ \vdots \\ C_v(\vec{u}_n, \vec{u}_0) \\ 1 \end{bmatrix}. \quad \text{(4.26)}$$

The left side of this equation is identical to the ordinary point kriging equation. The right side of the equation represents the covariance between a point and a block. The calculation is similar to the one explained for simple block kriging (Eq. 4.15).

The error variance is estimated as

$$\hat{\sigma}_E^2 = C_{vv}(\vec{u}_o, \vec{u}_o) - \sum_{i=1}^{n} \lambda_i C_v(\vec{u}_i, \vec{u}_o) - \mu, \quad \text{(4.27)}$$

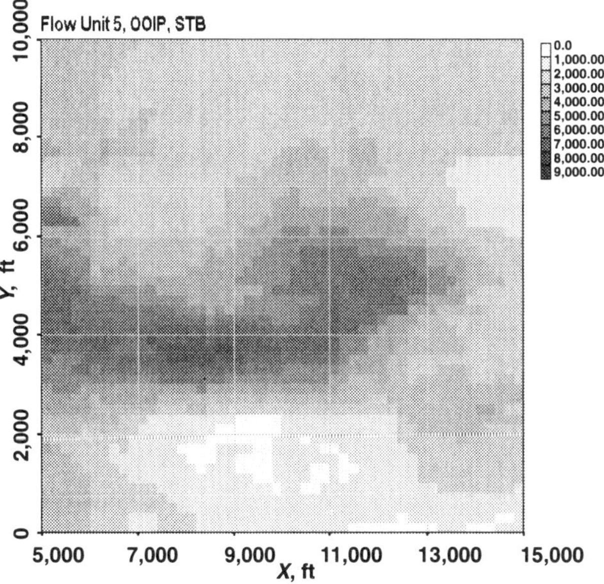

Fig. 4.32—Map of OOIP (STB) for Flow Unit 5.

where $C_{vv}(\vec{u}_o, \vec{u}_o)$ = the block variance and μ = the Lagrange parameter.

The overall block kriging procedure remains very similar to point kriging, with the exception of point-block covariance estimates. The limitation, described for block kriging, still applies. It can only be applied to variables that can be arithmetically averaged.

4.2.3 Cokriging. Cokriging is used to estimate one variable value with other sampled variables. Unlike ordinary kriging, where we estimate the value at the unsampled location with surrounding samples of the same variable, the cokriging procedure also uses the surrounding samples of other related variables. The goal in cokriging is to improve the estimate and reduce the uncertainty in the estimate with the help of spatial information available from other variables. There is an implicit assumption in the process. The variable of interest and the other variables are spatially related to each other.

Two common examples where cokriging is applied to improve estimates in reservoir description are the estimation of permeability using porosity data and the estimation of porosity using the seismic data. Typically, only a few wells are cored; however, almost all the wells are logged. If we establish a spatial relationship between the permeability and the porosity data, the permeability estimate, at the unsampled location, is improved by the surrounding porosity data. Similarly, during the early development stage, we may have an abundance of seismic data but limited well control. If the well porosity data is spatially related to seismic data, we can improve the estimates of interwell porosity with seismic information. Both applications make the cokriging technique relatively popular.

Before we discuss the necessary equations for cokriging estimation, we must understand its limitations.

First, the cokriging technique presumes that variables are linearly related to each other. If we make a crossplot of two variables, measured at the same location, we expect a linear relationship. Covariance only quantifies the spatial relationship, if the relationship is linear; otherwise, for a non-linear relationship, covariance might not provide meaningful information. It is critical to check the relationship between the variable of interest (principal variable) and the supporting variables (covariables). The relationship must be linear before we use covariables for estimation. As an example, it is more appropriate to estimate the log of permeability with the porosity data because we typically assume a linear relationship between the log of permeability and the porosity data.

Second, the relationship between the principal and covariable must be strong. Because the word "strong" is qualitative, we must determine the strength of the relationship. First, we measure the strength by the correlation coefficient between two variables, and the number of sample points used to estimate the correlation coefficient. The higher the correlation coefficient, and the more sample points that are used, the better is the relationship. Second, we check if the relationship is based on physical factors. For example, two-way travel time is related to depth of the formation. We might observe a similar physical basis between the principal variable and the covariable. Finally, we examine historical evidence, when the covariable is used to estimate the principal variable. We may answer some relevant questions. Have we used the covariable successfully in the past to estimate the principal variable for a similar geological environment? Has there been empirical evidence of a strong relationship? A good example of such an empirical relationship is between log permeability and porosity. Strong empirical evidence exists that such a relationship is observed; however, no strong physical basis exists for such a relationship. If all three factors are favorable, then we use the covariable to estimate the primary variable. On the other hand, if the three factors indicate a weak relationship, the covariable must be used with caution. It is important to understand the cokriging estimate is not indifferent to the covariable, if covariable sampling is extensive. In some instances, the covariable can misguide the direction of continuity for the principal variable. Therefore, before we decide to use the covariable to improve the estimate of the primary variable, we must ensure that a basis exists.

Third, the application of cokriging requires a substantial spatial modeling effort. The more covariables we add, the more extensive is the modeling effort. For example, if we have one principal (X) and one covariable (Y), we must model four covariance functions: C_X, C_Y, C_{XY} and C_{YX}. If we assume that $C_{XY} = C_{YX}$, then, we only model three covariance functions. To generalize, if we have a total of K variables, we model $K(K+1)/2$ variogram functions. This requires that we assume symmetry with respect to the cross-covariance function. For example, if we use four variables, we will model 10 variograms. Further, in modeling those variograms, we must satisfy the condition of positive definiteness. As explained in Chap. 3, such modeling requires careful analysis and significant effort.

Fourth, solving the system of cokriging requires additional computational effort compared to an ordinary kriging system. This is a result of a bigger matrix, which must be inverted to obtain the appropriate weights assigned to the principal and the covariable samples. As the number of variables increases, it produces a larger matrix to invert.

Knowing the amount of effort involved for modeling and estimation, we question if cokriging is worth the effort instead of conventional kriging. The overall consensus is that it depends. If the variable of interest is considerably undersampled, and there is a good spatial relationship between the principal and the covariables, the estimate can be improved by cokriging. On the other hand, if both variables are adequately sampled, the additional effort involved with cokriging may not significantly improve the estimate or reduce the uncertainty.

To qualitatively explain, if a variable is undersampled, we might not have sufficient spatial information about that variable. This is reflected in a large nugget value in the variogram model. If a covariable is adequately sampled, the variogram model for the covariable has a relatively small or no nugget. The nugget for the crossvariogram falls in between. If the two variables are spatially related to each other, the estimation of the principal variable is improved by the supplementary spatial information available from the covariable. Because of the smaller nugget for the covariable (indicating more spatial information), an appropriate large weight is assigned to the covariable samples, increasing their influence on the estimate, as well as the error variance. On the other hand, if the principal variable is well sampled, the spatial model will reflect that sampling. Further, in estimating the value of the principal variable, the nearby principal variable samples have the most influence on the estimate, screening (shielding) the sampled covariable values. This essentially reduces cokriging to a conventional kriging procedure with no improvement in the estimation. The numerical example illustrates this point following the development of the equations.

In developing the appropriate equations, we first must determine to follow either a simple kriging approach or an ordinary kriging approach. The final equations will slightly differ, depending on the kriging approach chosen.

This section follows the ordinary kriging approach. As explained in the previous section, ordinary kriging is more commonly practiced than the simple kriging procedure. To simplify our equations further, we assume to have only one covariable to estimate the value of the principal variable at the unsampled locations. Appendix D gives a more generalized expression for multiple covariables.[10,11]

Let us assume that we are interested in estimating a value, $x^*(\vec{u}_o)$, at the unsampled location, $\vec{u}_o$. Within our search neighborhood, we have n samples of the principal variable and m samples of the covariable Y. We can write the estimation equation as

$$X^*(\vec{u}_o) = \sum_{i=1}^{n} \lambda^*_{X_i} X(\vec{u}_{X_i}) + \sum_{k=1}^{n} \lambda^*_{Y_k} Y(\vec{u}_{Y_k}), \quad \text{....... (4.28)}$$

where λ_{X_i} is the weight assigned to the sample, $X(\vec{u}_{X_i})$, located at $\vec{u}_{X_i}$, and λ_{Y_k} is the weight assigned to the sample, $Y(\vec{u}_{Y_k})$, located at $\vec{u}_{Y_k}$. It is possible to observe both the principal and the covariable at the same location. However, to maintain the generality, we assumed different notations to indicate the locations of the two variables.

By applying the unbiased condition, we obtain

$$E[X^*(\vec{u}_o) - X(\vec{u}_o)] = 0. \quad \text{.................... (4.5)}$$

Substituting Eq. 4.28 in Eq. 4.5, we observe

$$m_X \sum_{i=1}^{n} \lambda_{X_i} + m_Y \sum_{k=1}^{n} \lambda_{Y_k} - m_X = 0, \quad \text{.......... (4.29)}$$

where m_X and m_Y = the expected values of the X and Y variables, respectively.

To satisfy Eq. 4.29, we write

$$\sum_{i=1}^{n} \lambda_{X_i} = 1 \text{ and } \sum_{k=1}^{n} \lambda_{Y_k} = 0. \quad \text{............... (4.30)}$$

Eq. 4.30 ensures that the unbiased condition is satisfied. Other alternate methods can also be used to satisfy the unbiased condition. See Appendix D.

In addition, we must also satisfy the condition of minimizing variance. Mathematically, minimize

$$\mathrm{Var}\left[X(\vec{u}_o) - \sum_{i=1}^{n} \lambda_{X_i} X(\vec{u}_{X_i}) - \sum_{k=1}^{n} \lambda_{Y_k} Y(\vec{u}_{Y_k})\right]. \quad \text{.................. (4.31)}$$

Minimizing Eq. 4.31, with the two constraints defined in Eq. 4.30, yields

$$\sum_{j=1}^{n} \lambda_{X_j} C_X(\vec{u}_{X_i}, \vec{u}_{X_j}) + \sum_{k=1}^{m} \lambda_{Y_k} C_C(\vec{u}_{X_i}, \vec{u}_{Y_k}) + \mu_X = C_X(\vec{u}_0, \vec{u}_{X_i}) \text{ for } i = 1, \ldots, n, \quad \text{.............. (4.32)}$$

and

$$\sum_{i=1}^{n} \lambda_{X_i} C_C(\vec{u}_{X_i}, \vec{u}_{X_i}) + \sum_{l=1}^{m} \lambda_{Y_l} C_Y(\vec{u}_{Y_k}, \vec{u}_{Y_l}) + \mu_Y = C_C(\vec{u}_0, \vec{u}_{Y_k}) \text{ for } k = 1, \ldots, m. \quad \text{............. (4.33)}$$

In Eqs. 4.32 and 4.33, C_X and C_Y = the covariances for the X and Y variables, respectively, and C_C = the cross-covariance between the two variables: μ_X and μ_Y = Lagrange parameters, respectively.

In matrix form, the equation is written as

$$\begin{bmatrix} C_X(\vec{u}_{X_1}, \vec{u}_{X_1}) & \ldots & C_X(\vec{u}_{X_1}, \vec{u}_{X_n}) & C_C(\vec{u}_{X_1}, \vec{u}_{Y_1}) & \ldots & C_C(\vec{u}_{X_1}, \vec{u}_{Y_m}) & 1 & 0 \\ \vdots & & \vdots & \vdots & & \vdots & \vdots & \vdots \\ C_X(\vec{u}_{X_1}, \vec{u}_{X_n}) & \ldots & C_X(\vec{u}_{X_n}, \vec{u}_{X_n}) & C_C(\vec{u}_{X_n}, \vec{u}_{Y_1}) & \ldots & C_C(\vec{u}_{X_n}, \vec{u}_{Y_m}) & 1 & 0 \\ C_C(\vec{u}_{X_1}, \vec{u}_{Y_1}) & \ldots & C_C(\vec{u}_{X_n}, \vec{u}_{Y_1}) & C_Y(\vec{u}_{Y_1}, \vec{u}_{Y_1}) & \ldots & C_Y(\vec{u}_{Y_1}, \vec{u}_{Y_m}) & 0 & 1 \\ \vdots & & \vdots & \vdots & & \vdots & \vdots & \vdots \\ C_C(\vec{u}_{X_1}, \vec{u}_{Y_m}) & \ldots & C_C(\vec{u}_{X_n}, \vec{u}_{Y_m}) & C_Y(\vec{u}_{Y_m}, \vec{u}_{Y_1}) & \ldots & C_Y(\vec{u}_{Y_m}, \vec{u}_{Y_m}) & 0 & 1 \\ 1 & \ldots & 1 & 0 & \ldots & 0 & 0 & 0 \\ 0 & \ldots & 0 & 1 & \ldots & 1 & 0 & 0 \end{bmatrix}$$

$$\begin{bmatrix} \lambda_{X_1} \\ \vdots \\ \lambda_{X_n} \\ \lambda_{Y_1} \\ \vdots \\ \lambda_{Y_m} \\ \mu_X \\ \mu_Y \end{bmatrix} = \begin{bmatrix} C_X(\vec{u}_0, \vec{u}_{X_1}) \\ \vdots \\ C_X(\vec{u}_0, \vec{u}_{X_n}) \\ C_C(\vec{u}_0, \vec{u}_{Y_1}) \\ \vdots \\ C_C(\vec{u}_0, \vec{u}_{Y_m}) \\ 1 \\ 0 \end{bmatrix}.$$

On the left side, we have an $(n+m+2)$ size matrix that must be inverted to obtain the values of λ_{X_i} and λ_{Y_i}. Compared to ordinary kriging, the matrix size is significantly bigger. This results in additional computational effort.

We can also obtain an expression for the error variance as

$$\hat{\sigma}_E^2 = C(\vec{u}_o, \vec{u}_o) - \sum_{i=1}^{n} \lambda_{X_i} C_X(\vec{u}_o, \vec{u}_{X_i})$$

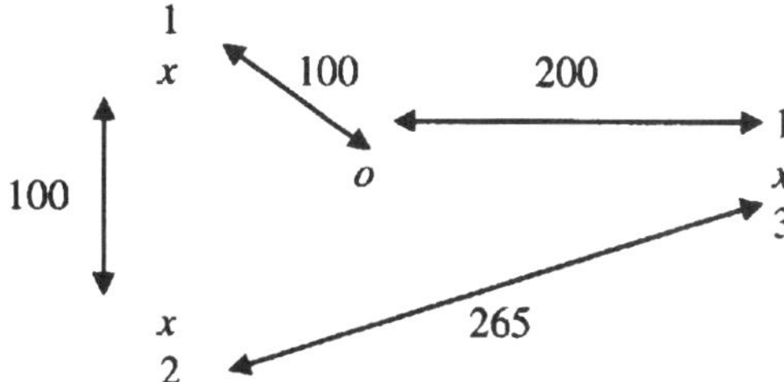

Fig. 4.33—Sample configuration for Numerical Example 4.4.

$$-\sum_{k=1}^{m}\lambda_{Y_k}C_C\left(\vec{u}_o,\vec{u}_{Y_k}\right)-\mu_X. \quad \ldots\ldots\ldots\ldots\ldots\ldots (4.34)$$

As in the case of ordinary kriging, the error variance is an indication of relative sample configuration surrounding the unsampled data.

Numerical Example 4.4. Consider the following configuration, as shown in **Fig. 4.33.**

We have three sample locations surrounding the unsampled location. Estimate the value at the unsampled location.

(a) The variogram models are given as

$$\gamma_X(L) = 15 + 10M_{s_{500}}(L),$$

$$\gamma_Y(L) = 0.2 + 1M_{s_{500}}(L),$$

and

$$\gamma_C(L) = 1 + 3M_{s_{500}}(L),$$

where X is the principal variable and Y is the covariable. The following sample values are provided:

$x(1) = \text{missing},\ y(1) = 0.8,$

$x(2) = \text{missing},\ y(2) = 0.6,$

and

$x(3) = 5,\ y(3) = 1.2.$

(b) Use the variogram model from (a) and the samples of the x variable, located at one and two. They are given as

$x(1) = 3,$

and

$x(2) = 10.$

(b) Use the sample data from (b) and the variogram model,

$$\gamma_X(L) = 6 + 19M_{s_{500}}(L).$$

All the other models remain the same.

The models are already checked for consistency of the positive definite condition.

Solution—Case 1. In this case, we have four sample points—$x_3, y_1, y_2,$ and y_3. Therefore, for cokriging, the matrix equation is written as

$$\begin{bmatrix} C_X(\vec{u}_3,\vec{u}_3) & C_C(\vec{u}_3,\vec{u}_1) & C_C(\vec{u}_3,\vec{u}_2) & C_C(\vec{u}_3,\vec{u}_3) & 1 & 0 \\ C_C(\vec{u}_3,\vec{u}_1) & C_Y(\vec{u}_1,\vec{u}_1) & C_Y(\vec{u}_1,\vec{u}_2) & C_Y(\vec{u}_1,\vec{u}_3) & 0 & 1 \\ C_C(\vec{u}_3,\vec{u}_2) & C_Y(\vec{u}_2,\vec{u}_1) & C_Y(\vec{u}_2,\vec{u}_2) & C_Y(\vec{u}_2,\vec{u}_3) & 0 & 1 \\ C_C(\vec{u}_3,\vec{u}_3) & C_Y(\vec{u}_3,\vec{u}_1) & C_Y(\vec{u}_3,\vec{u}_2) & C_Y(\vec{u}_3,\vec{u}_3) & 0 & 1 \\ 1 & 0 & 0 & 0 & 0 & 0 \\ 0 & 1 & 1 & 1 & 0 & 0 \end{bmatrix}$$

$$\begin{bmatrix} \lambda_{X_1} \\ \lambda_{Y_1} \\ \lambda_{Y_2} \\ \lambda_{Y_3} \\ \mu_X \\ \mu_Y \end{bmatrix} = \begin{bmatrix} C_X(\vec{u}_3,\vec{u}_0) \\ C_C(\vec{u}_1,\vec{u}_0) \\ C_C(\vec{u}_2,\vec{u}_0) \\ C_C(\vec{u}_3,\vec{u}_0) \\ 1 \\ 0 \end{bmatrix}.$$

Because the X and Y variables are located at the same point, the subscripts from the distance vectors are removed. Once we estimate the covariance values, we can solve for λ and μ. The computation of covariance is similar to the ordinary kriging procedure. For example,

$$C_X(\vec{u}_3,\vec{u}_3) = C_X(0) = 25,$$

$$C_C(\vec{u}_1,\vec{u}_3) = C_C(0) - \gamma_C(\vec{u}_1,\vec{u}_3) = C_C(0) - \gamma_C(265)$$

$$= 4 - \left[1 + 3M_{s_{500}}(265)\right] = 0.838,$$

and

$$C_Y(\vec{u}_1,\vec{u}_2) = C_Y(0) - \gamma_Y(\vec{u}_1,\vec{u}_2) = 1.2 - \gamma_C(100)$$

$$= 1.2 - \left[0.2 + 1M_{s_{500}}(100)\right] = 0.704.$$

Similarly, for the right side,

$$C_X(\vec{u}_3,\vec{u}_0) = C_X(0) - \gamma_X(\vec{u}_3,\vec{u}_0)$$

$$= 25 - \left[15 + 10M_{s_{500}}(200)\right] = 4.32.$$

After calculating the covariance, we obtain

$$\begin{bmatrix} 25.0 & 0.838 & 0.838 & 4.0 & 1 & 0 \\ 0.838 & 1.2 & 0.704 & 0.279 & 0 & 1 \\ 0.838 & 0.704 & 1.2 & 0.279 & 0 & 1 \\ 4.0 & 0.279 & 0.279 & 1.2 & 0 & 1 \\ 1 & 0 & 0 & 0 & 0 & 0 \\ 0 & 1 & 1 & 1 & 0 & 0 \end{bmatrix} \begin{bmatrix} \lambda_{X_1} \\ \lambda_{Y_1} \\ \lambda_{Y_2} \\ \lambda_{Y_3} \\ \mu_X \\ \mu_Y \end{bmatrix} = \begin{bmatrix} 4.32 \\ 2.11 \\ 2.11 \\ 1.30 \\ 1 \\ 0 \end{bmatrix}.$$

Solving, we obtain

$$\lambda_{X1} = 1.0, \lambda_{Y1} = 1.248, \lambda_{Y2} = 1.248, \lambda_{Y3} = -2.496,$$

and

$$\mu_X = -12.78, \mu_Y = -0.41.$$

Substituting in the estimation equation,

$$x^*(u_0) = 1 \times 5 + 1.248 \times 0.8 + 1.248 \times 0.6 - 2.496$$

$$\times 1.2 = 3.75,$$

and

$$\hat{\sigma}_E^2 = 25 - 1 \times 4.32 - 1.248 \times 2.11 - 1.248 \times 2.11$$

$$+ 2.496 \times 1.3 + 12.78 = 31.83.$$

Intuitively, the weight, assigned only to the X sample point, must be one because all weights must equal one. The weights assigned to Y_1 and Y_2 should be, and are, identical. These weights are positive and large. In contrast, the weight assigned to Y_3 is negative. This is partly because the weights assigned to the Y samples must equal zero. More important, Y_3 is screened by X_3, resulting in a negative weight.

The error variance = 31.83. This is larger than the sill or variance of the variable X, which is 25. This is because, in ordinary kriging, we assume that the mean is unknown. In the presence of only one sample, the error variance also reflects the uncertainty with respect to mean, in addition to the variance of the data.

Compared to cokriging, ordinary kriging uses only one sample, and the estimate is calculated by

$$x^*(\vec{u}_0) = 1 \times 5 = 5,$$

and the error variance = 41.36.

In the presence of one sample, both the estimate and the error variance are significantly affected by cokriging. The error variance is reduced significantly by using the information related to variable *Y*.

Solution—Case 2. In this case, we have the same variogram models but also have three samples each for both *X* and *Y* variables. The matrix equation is written as

$$\begin{bmatrix} C_X(\vec{u}_1,\vec{u}_1) & C_X(\vec{u}_1,\vec{u}_2) & C_X(\vec{u}_1,\vec{u}_3) & C_C(\vec{u}_1,\vec{u}_1) & C_C(\vec{u}_1,\vec{u}_2) & C_C(\vec{u}_1,\vec{u}_3) & 1 & 0 \\ C_X(\vec{u}_2,\vec{u}_1) & C_X(\vec{u}_2,\vec{u}_2) & C_X(\vec{u}_2,\vec{u}_3) & C_C(\vec{u}_2,\vec{u}_1) & C_C(\vec{u}_2,\vec{u}_2) & C_C(\vec{u}_2,\vec{u}_3) & 1 & 0 \\ C_X(\vec{u}_3,\vec{u}_1) & C_X(\vec{u}_3,\vec{u}_2) & C_X(\vec{u}_3,\vec{u}_3) & C_C(\vec{u}_3,\vec{u}_1) & C_C(\vec{u}_3,\vec{u}_2) & C_C(\vec{u}_3,\vec{u}_3) & 1 & 0 \\ C_C(\vec{u}_1,\vec{u}_1) & C_C(\vec{u}_1,\vec{u}_2) & C_C(\vec{u}_1,\vec{u}_3) & C_Y(\vec{u}_1,\vec{u}_1) & C_Y(\vec{u}_1,\vec{u}_2) & C_Y(\vec{u}_1,\vec{u}_3) & 0 & 1 \\ C_C(\vec{u}_2,\vec{u}_1) & C_C(\vec{u}_2,\vec{u}_2) & C_C(\vec{u}_2,\vec{u}_3) & C_Y(\vec{u}_2,\vec{u}_1) & C_Y(\vec{u}_2,\vec{u}_2) & C_Y(\vec{u}_2,\vec{u}_3) & 0 & 1 \\ C_C(\vec{u}_3,\vec{u}_1) & C_C(\vec{u}_3,\vec{u}_2) & C_C(\vec{u}_3,\vec{u}_3) & C_Y(\vec{u}_3,\vec{u}_1) & C_Y(\vec{u}_3,\vec{u}_2) & C_Y(\vec{u}_3,\vec{u}_3) & 0 & 1 \\ 1 & 1 & 1 & 0 & 0 & 0 & 0 & 0 \\ 0 & 0 & 0 & 1 & 1 & 1 & 0 & 0 \end{bmatrix}$$

$$\begin{bmatrix} \lambda_{X_1} \\ \lambda_{X_2} \\ \lambda_{X_3} \\ \lambda_{Y_1} \\ \lambda_{Y_2} \\ \lambda_{Y_3} \\ \mu_X \\ \mu_Y \end{bmatrix} = \begin{bmatrix} C_X(\vec{u}_1,\vec{u}_0) \\ C_X(\vec{u}_2,\vec{u}_0) \\ C_X(\vec{u}_3,\vec{u}_0) \\ C_C(\vec{u}_1,\vec{u}_0) \\ C_C(\vec{u}_2,\vec{u}_0) \\ C_C(\vec{u}_3,\vec{u}_0) \\ 1 \\ 0 \end{bmatrix}.$$

We estimate the values of covariances, as explained in Case 1. Solving for λ and μ, we obtain

$$\lambda_{X1} = 0.333, \lambda_{X2} = 0.333, \lambda_{X3} = 0.333, \lambda_{Y1} = 0.124,$$

$$\lambda_{Y2} = 0.124, \lambda_{Y3} = -0.249, \mu_X = -5.11,$$

and

$$\mu_Y = -0.37.$$

The estimate, at the unsampled location, is

$$x^*(\vec{u}_0) = 0.333 \times 3 + 0.333 \times 10 + 0.333 \times 5 + 0.124 \times 0.8 + 0.124 \times 0.6 - 0.249 \times 1.2 = 5.87.$$

The error variance = 24.14.

The weights, assigned to the three *X* variable samples, are equal. In the presence of a large nugget for the *X* variable, the weights must be similar to each other. The weights, assigned to *Y* samples, are reduced significantly, reflecting the importance of *X* samples at the same locations. Compared to error variance in Case 1, it is reduced significantly in the presence of additional *X* samples.

Solution—Case 3. In this case, there are three samples each for the *X* and *Y* variables. In addition, the variogram model for *X* is different. It has a much smaller nugget, compared to the prior model, reflecting the available additional spatial information. In practice, as more *X* samples are collected, we are able to model the variogram better. Relatively smaller nugget values reflect sampling of additional data.

In the presence of six samples, the matrix structure remains the same, as in Case 2. The covariance values are calculated with variogram functions. Some of the covariance values, compared to Case 2, change because of differences in the variogram model for the *X* variable. After solving the matrix, we obtain

$$\lambda_{X1} = 0.38, \lambda_{X2} = 0.38, \lambda_{X3} = 0.24, \lambda_{Y1} = -0.0338,$$

$$\lambda_{Y2} = -0.0338, \lambda_{Y3} = 0.067,$$

and

$$\mu_X = 1.62, \mu_Y = -0.365.$$

By substituting the values of λ and μ in the estimation equation, we obtain that $x^*(\vec{u}_0) = 6.17$, and the error variance = 14.26.

The higher weights, assigned to X_1 and X_2, reflect a better variogram model. Therefore, the closer points are assigned a higher weight. Compared to Case 2, the weights, assigned to Y_1, Y_2, and Y_3, are much smaller, reflecting the diminishing importance of the covariables in the presence of the better variogram model for the *X* variable and the additional samples. The lower value of error variance, compared to Case 2, results solely from the better variogram model for the *X* variable. Significant reduction in the error variance is obtained by better spatial information.

As a comparison, the value at the unsampled location is estimated with the ordinary kriging procedure. Only the variogram model for *X* and the three samples points for the *X* variable are used. The weights assigned are

$$\lambda_{X1} = 0.37, \lambda_{X2} = 0.37, \lambda_{X3} = 0.26, \text{and } \mu = -2.25.$$

The estimated value = 6.12, and the error variance = 15.2. Comparing these values with the cokriging estimate, the weights assigned to the *X* variable samples are very similar for both the kriging and cokriging procedures. Further, the estimate and the error variance at the unsampled location are very close to each other for cokriging and ordinary kriging procedures. A small reduction in error variance, in the presence of cokriging, reflects little improvement in the reduction of uncertainty.

In Case 3, cokriging did not significantly improve the results compared to ordinary kriging. Neither the estimate nor the error variance are affected by the presence of the covariable. As the number of principal variable samples increases, the spatial model improves, and the weights assigned to covariable samples decreases dramatically (almost by two orders of magnitude). This reaffirms that cokriging is useful, only if there is sparse sampling of the primary variable, and relatively limited information about the spatial relationship, with respect to that variable. Then, better spatial information about the covariable and the presence of additional covariable samples can help to better estimate the principal variable and reducte error variance at the unsampled location.

Field Example 4.7—Cokriging To Estimate Permeability Maps Using Initial Potential. In this field example, cokriging is used to generate maps of the permeability and thickness product (kh) for all flow units in the reservoir, using initial potential (IP) as the secondary variable. IP represents the initial productivity of a well, and it is correlated to kh of the interval open to flow. Usually, kh data are available only for a few

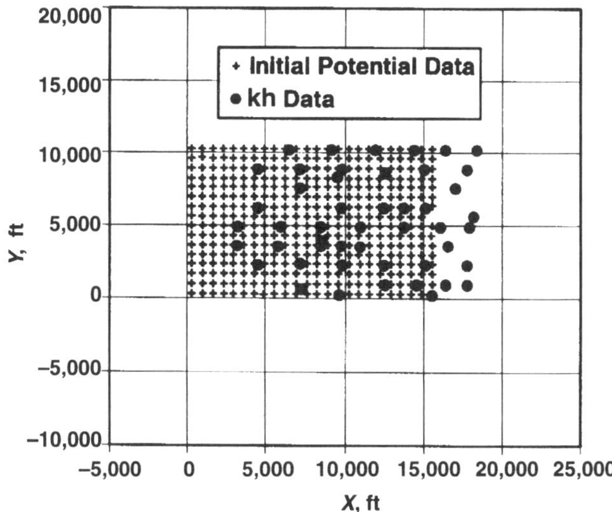

Fig. 4.34—Location of conditioning data for 48 wells with kh data and 382 wells with initial potential data.

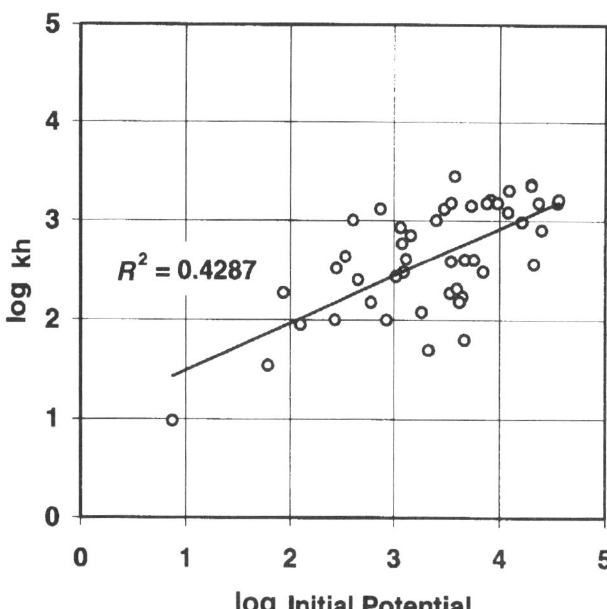

Fig. 4.35—log kh vs. log IP data.

cored wells, and IP is available in all production wells. Cokriging uses IP and the cross-correlation between kh and IP to bring additional information into the kh description. The logarithm of kh is used as the primary variable. As described in Chap. 3, the logarithm operator transforms data with highly skewed distributions and a large range of values into distributions that are closer to Gaussian.

The following information is given:

• Conditioning data for log kh and log IP (**Fig. 4.34**).

• Variogram and cross-variogram models for log kh and log IP are given as

$\gamma(L) = 0.15 + 0.58 M_{G_{7,700}}(L)$, for log kh variogram,

$\gamma(L) = 0.06 + 0.30 M_{G_{7,700}}(L)$, for log IP variogram,

and

$\gamma(L) = 0.06 + 0.30 M_{G_{7,700}}(L)$, for log kh and log IP cross-variograms.

These variograms are isotropic. Chap. 3 describes the calculation of these variograms.

• The grid definition is given in Field Example 4.2.

Make the following calculations:

• Generate a map of log kh with cokriging.

• Generate a map of log kh with ordinary kriging.

• Compare the cokriging and ordinary kriging estimates.

Solution. The field locations of 48 wells with kh and IP data and 382 additional wells with IP data are shown in Fig. 4.31. The wells fall in the same area used in Field Example 4.1. **Fig. 4.35** shows that the logarithms of kh and IP have a strong positive correlation based on the 48 wells. The correlation coefficient = 0.66. The abundance of IP data and positive correlation with log kh makes IP a good candidate for a secondary variable in cokriging.

Fig. 4.36 shows the map of log kh estimated with cokriging, and **Fig. 4.37** shows ordinary kriging. The spatial distributions of log kh are significantly different for the two kriging methods. The map, generated with ordinary kriging, varies smoothly throughout the field because the Gaussian variogram model heavily influences the spatial correlation. The map generated with cokriging is not as smooth as ordinary kriging. The influence of the IP data adds some unique fea-

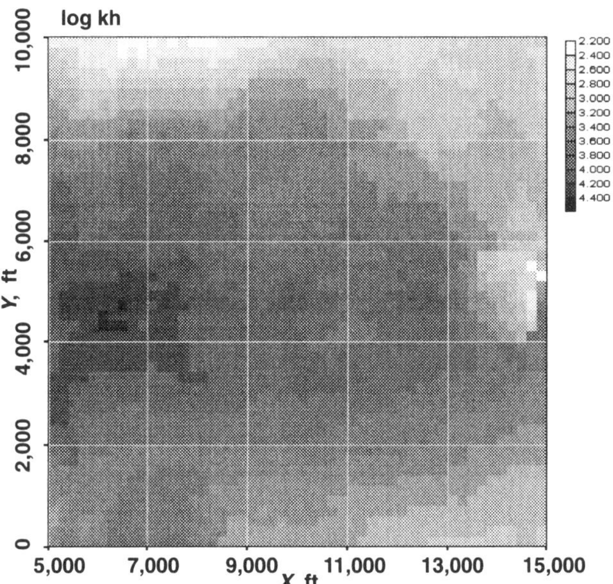

Fig. 4.36—Map of cokriging estimates for log kh.

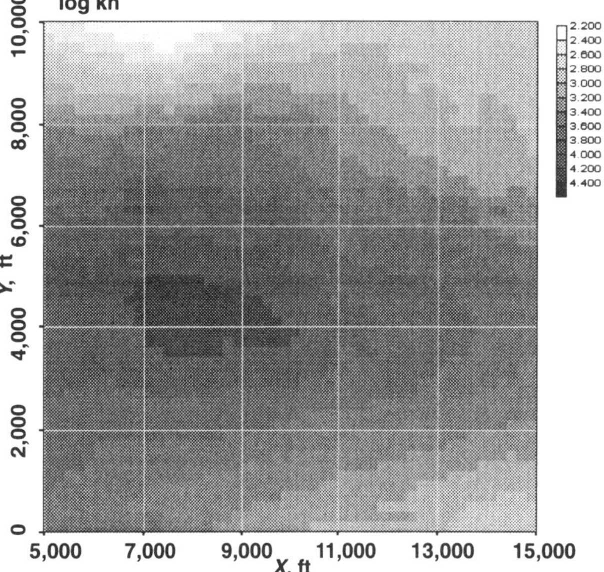

Fig. 4.37—Map of ordinary kriging for log kh.

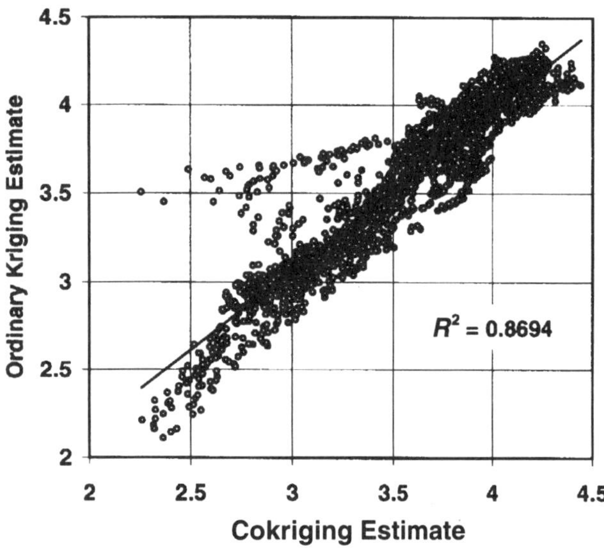

Fig. 4.38—Cokriging and ordinary kriging estimates of log kh.

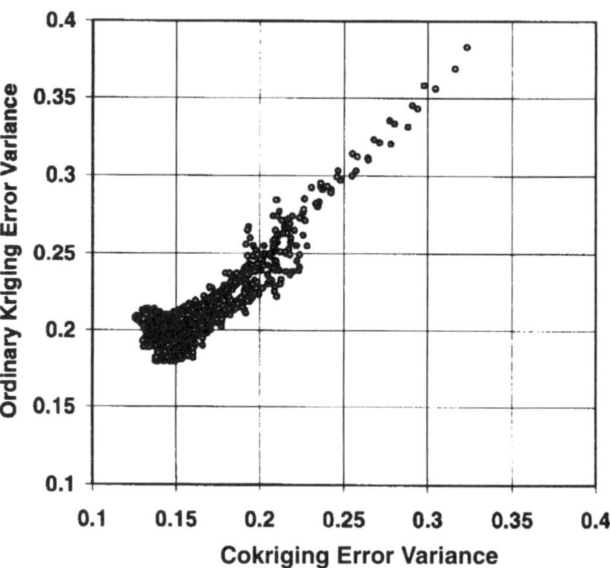

Fig. 4.39—Cokriging and ordinary kriging error variance of log kh.

tures to the kh map that ordinary kriging cannot capture with a single variogram model. The cokriging and ordinary kriging estimates differ significantly, as shown in **Fig. 4.38.** Ordinary kriging tends to overestimate log kh. **Fig. 4.39** shows that the cokriging error variance is smaller than the error variance estimated with ordinary kriging. This shows that additional information used in cokriging reduces the error variance in the estimates.

One of the difficulties typically encountered in the cokriging approach is that the estimate at the unsampled location can be overwhelmed by the covariable samples within the search neighborhood. If a large number of covariable samples is available within the search neighborhood, only covariable samples may be selected for estimation purposes. Therefore, we may not select any principal variable samples. To avoid such conditions, we define different search neighborhoods for the principal and the covariables. For sparsely sampled primary variable data, we define a large search neighborhood. For extensively sampled covariables, we define a smaller search neighborhood. Further, we restrict the maximum number of samples we select within each of the search neighborhoods. By defining a larger maximum number of samples for the principal variable than the covariable, we ensure that the influence of the surrounding principal variable samples is accounted for properly.

One limiting case, which restricts the covariable samples, is the collocated cokriging method.[12] In this method, only the covariable sample measured at the unsampled location $[Y(\vec{u}_o)]$ is used to estimate the principal variable value at the unsampled location $[X^*(\vec{u}_o)]$. The other covariable samples, away from the unsampled locations, are assumed to be screened by the covariable sample at the unsampled location $\vec{u}_o$.

Collocated cokriging requires the covariable sample to be available at every location where the principal variable must be estimated. In practice, this is possible when estimating porosity values with seismic data. In effect, collocated cokriging uses only one covariable sample for the estimation of the principal variable. By using only one covariable sample, the matrix size of the cokriging system is reduced. Further, spatial modeling of the covariable is not necessary. By making additional assumptions, the spatial model for the cross-covariance can be inferred from the covariance model for the principal variable. This considerably reduces the modeling effort. (See Appendix D for details.) This method is effective when integrating seismic data with well information.[13, 14]

4.2.4 Universal Kriging. The universal kriging procedure estimates the value of a variable in the presence of a trend. If we have a trend in sampled information, the assumption of first-order stationarity is not satisfied. The local mean varies in the direction of the trend and is not reasonably close to the global mean. This type of data must not be subjected to kriging techniques, unless the data are preprocessed to remove the trend, so the assumption of first-order stationarity is re-established.

One option, which incorporates the trend of the data, redefines the variable as

$$X(\vec{u}) = m(\vec{u}) + R(\vec{u}), \quad \ldots\ldots\ldots\ldots\ldots\ldots (4.35)$$

where $X(\vec{u})$ = the variable of interest at location, $\vec{u}$, $m(\vec{u})$ = the drift or mean, and $R(\vec{u})$ = the residual. If we divide the variable into drift and residual, by removing the drift (or mean) from the data, we obtain the values of residuals that satisfy the first-order stationarity requirements.

Consider a simple example. If we have a trend in our data, for each data point, we draw a search neighborhood within which the trend is minimal. We calculate the average of all the samples within that neighborhood, and we assume that value is the drift or the local mean. Mathematically,

$$m(\vec{u}) = \sum_{\text{neighborhood}} X(\vec{u}_i). \quad \ldots\ldots\ldots\ldots\ldots\ldots (4.36)$$

If we define similar search neighborhoods, we observe the mean to be higher for sample points, where all the surrounding samples are high. The mean is lower if surrounding samples are low. If we subtract the local mean from individual samples, we obtain

$$R(\vec{u}) = X(\vec{u}) - m(\vec{u}). \quad \ldots\ldots\ldots\ldots\ldots\ldots (4.37)$$

The residual values, with the local trend removed, do not exhibit any trend. By definition, the mean of residuals = zero, and the first-order stationarity assumption is satisfied. In principle, by applying the kriging technique to the residuals and adding back the trend, we can estimate the values at the unsampled locations.

It is worth discussing the estimation of variograms for residuals at this point. Because we apply the kriging technique to residuals, we also must estimate and model the variograms for the residuals. Several options are available to estimate and model the variogram for the residuals.

• By using moving neighborhoods for each sample, the local mean (drift) is estimated. By subtracting the local mean from the sample value, the residuals are obtained at each location, and the variograms of the residuals can be estimated.

• Find the direction where the trend is not significant. Assuming that first-order stationarity is reasonably satisfied in that direction, the variogram for the original variable is estimated in that direction and applied in the direction of the trend.

Both options require a subjective understanding of the trend and a decision that determines the local stationarity data. As a result, the variogram model requires careful analysis.

Next, once the variogram is estimated and modeled, we must assume that a specific type of trend is present in the data. This assumption is required for universal kriging. In general, we define drift with the equation

$$m(\vec{u}) = \sum_{i=0}^{L} a_l f_l(\vec{u}), \quad \ldots\ldots (4.38)$$

where a_l = the coefficient of function $f_l(\vec{u})$. By definition, we always assume $f_0(\vec{u})$ is equal to one. For example, if we have a linear trend, we write

$$m(\vec{u}) = a_0 + a_1\vec{u}, \quad \ldots\ldots (4.39)$$

or for a quadratic trend, we write

$$m(\vec{u}) = a_0 + a_1\vec{u} + a_2\vec{u}^2. \quad \ldots\ldots (4.40)$$

As a practical matter, rarely is a trend higher than a second order considered. During the estimation process, we assume that the type of trend is known—linear, quadratic, etc.

The estimation process for universal kriging begins with the same equation used in ordinary kriging.

$$X^*(\vec{u}_0) = \sum_{i=1}^{n} \lambda_i X(\vec{u}_i). \quad \ldots\ldots (4.41)$$

Application of the unbiased condition requires that

$$\sum_{i=1}^{n} \lambda_i f_l(\vec{u}_i) = f_l(\vec{u}_0) \text{ for } l = 0,\ldots,L. \quad \ldots\ldots (4.42)$$

Equation 4.42 imposes $L+1$ constraints. Notice for $l=0$, we have a constraint, which is used for ordinary kriging—the sum of all weights must equal one. This is true because we always define $f_0(\vec{u})$ as equal to one.

Applying the condition of minimum variance, we obtain (see Appendix D for details)

$$\begin{bmatrix} C(\vec{u}_1,\vec{u}_1) & \ldots & C(\vec{u}_1,\vec{u}_n) & f_0(\vec{u}_1) & \ldots & f_L(\vec{u}_1) \\ \vdots & & & \vdots & & \\ C(\vec{u}_n,\vec{u}_1) & \ldots & C(\vec{u}_n,\vec{u}_n) & f_0(\vec{u}_n) & \ldots & f_L(\vec{u}_n) \\ f_0(\vec{u}_1) & \ldots & f_0(\vec{u}_n) & 0 & \ldots & 0 \\ \vdots & & & \vdots & & \vdots \\ f_L(\vec{u}_1) & \ldots & f_L(\vec{u}_n) & 0 & \ldots & 0 \end{bmatrix} \begin{bmatrix} \lambda_1 \\ \vdots \\ \lambda_n \\ \mu_0 \\ \vdots \\ \mu_L \end{bmatrix} = \begin{bmatrix} C(\vec{u}_1,\vec{u}_0) \\ \vdots \\ C(\vec{u}_n,\vec{u}_0) \\ f_0(\vec{u}_0) \\ \vdots \\ f_L(\vec{u}_0) \end{bmatrix}. \quad \ldots\ldots (4.43)$$

The left side of the matrix has a size of $n+L+1$. We have $(L+1)$ Lagrange parameters corresponding to $(L+1)$ constraints.

By solving Eq. 4.43, we obtain the values for λ_i and μ_i. By substituting λ_i and μ_i in Eq. 4.40, we obtain the error variance estimate at the unsampled location. The error variance is estimated as

$$\hat{\sigma}_E^2 = C(\vec{u}_0,\vec{u}_0) - \sum_{i=1}^{n} \lambda_i C(\vec{u}_i,\vec{u}_0) - \sum_{l=0}^{L} \mu_l f_l(\vec{u}_0). \quad \ldots\ldots (4.44)$$

Because universal kriging accounts for a trend in the data, if we extrapolate information, the estimated values can be either greater than or smaller than the observed data. We cannot capture this feature with any other kriging technique.

If we are only interested in the interpolation of data, it is possible to obtain comparable results with ordinary kriging. Although ordinary kriging does not explicitly account for the trend in data, by localizing the neighborhood, and assuming that the local mean is unknown, the variations for the local mean are implicitly accounted for.[15] This results in very similar estimates using either universal kriging or ordinary kriging—the extra assumptions and size of the matrix do not lead to estimate improvement.

Another difficulty in applying the universal kriging technique is a lack of understanding about the trend. Unless we have *a priori* knowledge about the type of trend, we may not be able to apply universal kriging. Two options commonly used to circumvent this problem are either the use of intrinsic random functions of order k[16] or the method of external drift.[17]

The use of the intrinsic random function of order k method assumes a more generalized covariance model of order k. The order k is determined by the order of the trend. If the order is zero (no trend), the generalized covariance model reduces to a variogram model. In practice, the order k is determined with an automated procedure; a procedure similar to kriging is applied with a generalized covariance model.

Another alternative uses the method of external drift.[17] In this method, we assume that the trend in the data is provided by a secondary variable. For example, seismic data could provide information about the trend in the top of the interval; however, the only hard data is well data. If $\vec{u}_0$ is the unsampled location, we assume

$$m(\vec{u}_o) = a_0 + a_1 Y(\vec{u}_0), \quad \ldots\ldots (4.45)$$

where $Y(\vec{u}_0)$ is the secondary variable value at the unsampled location. The subsequent development of equations is relatively easy (see Appendix D). The advantage of using the external drift method is that we do not have to assume a specific trend model; it is solely governed by the secondary data. The disadvantage of the method is that secondary data is required to be available at every unsampled location, and the relationship between the drift and the secondary variable must be linear.

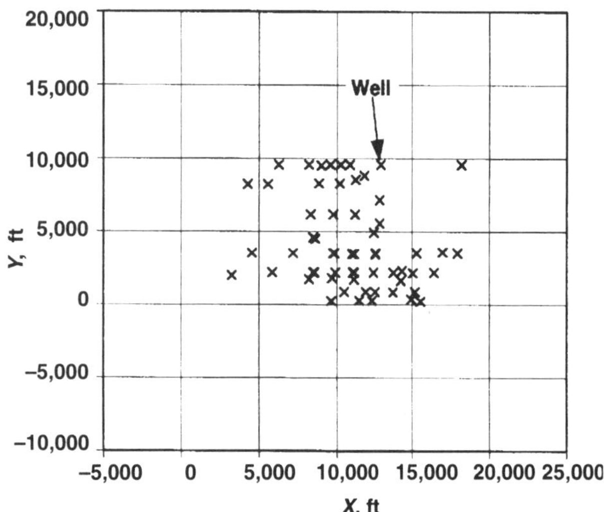

Fig. 4.40—Location of reservoir top well data.

Field Example 4.8—Universal Kriging To Estimate Top of Reservoir. In this field example, universal kriging is used to map the top of the reservoir. Universal kriging accounts for trends observed in the data. In this field example, the top of the reservoir exhibits a linear trend that runs east/west.

The following information is given:

• The locations for the top of reservoir depth conditioning data are shown in **Fig. 4.40.**

• Variogram models for residuals between the top of reservoir depth and a linear trend are

$$\gamma(\vec{L}) = 200M_{s_{2,500}}(\vec{L}),\ \text{principal direction east/west,}$$

and

$$\gamma(\vec{L}) = 200M_{s_{2,500}}(\vec{L}),\ \text{minor direction north/south.}$$

Universal kriging requires variogram models for the residuals of the data. Residuals are equal to the data minus the trends. The models for reservoir top residuals are isotropic with a correlation range = 2,500 ft. The original data indicate that in the direction with no trend (north/south), variograms of the data and residuals are similar, while in the direction with a trend (east/west), variograms differ significantly. The variogram calculations are described in Chap. 3.

• The grid definition is given in Field Example 4.2.

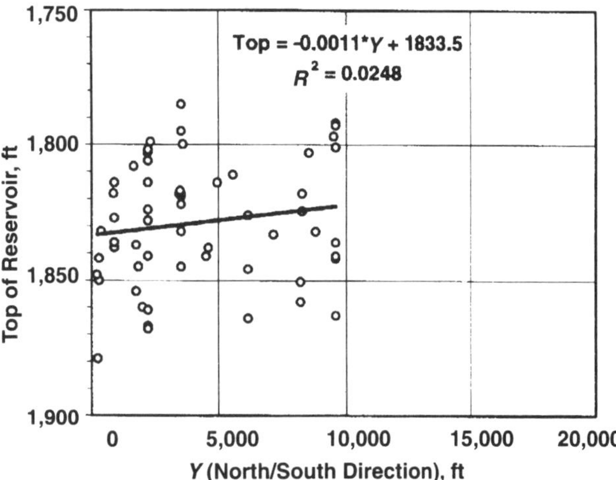

Fig. 4.42—Reservoir top running north/south.

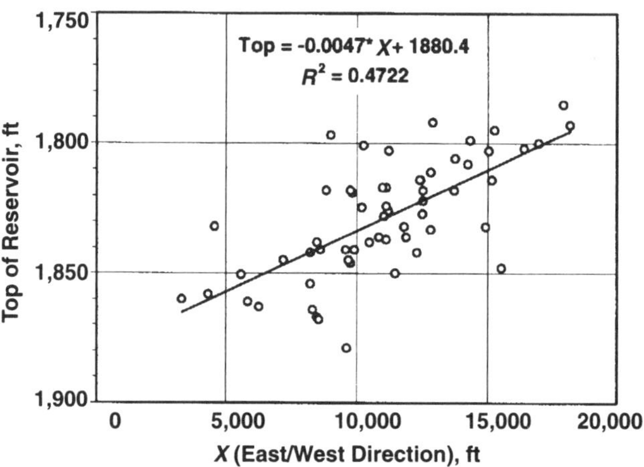

Fig. 4.41—Reservoir top running east/west.

Make the following calculations:

• Compare cross validation results for universal kriging (with a linear trend) to cross validation results of simple and ordinary kriging.

• Generate a map of top of reservoir depth using universal kriging with a linear trend.

Solution. The locations of the 59 wells with reservoir top data (Fig. 4.40) are in the same area as the ones from the prior field examples. Trends in the data are identified by plotting transects in different directions, such as the ones shown in **Figs. 4.41** and **4.42** for the east/west and north/south directions, respectively. These transects indicate that the reservoir top dips east/west with a strong linear trend. This trend is modeled with universal kriging. In the north/south direction, the reservoir top does not show any trend.

Cross validations for universal, simple and ordinary kriging are shown in **Fig. 4.43.** Universal and ordinary kriging estimates are similar, and both are close to the data. The similarities can be explained because ordinary kriging accounts for the trend by estimating the mean at each location. In this field, ordinary kriging captures changes in the mean very well because of the good coverage of the conditioning data. However, in cases where there is significant extrapolation away from data control, ordinary and universal kriging differ significantly. Simple kriging does not account for trends; therefore, the estimates are not as good as ordinary and universal kriging.

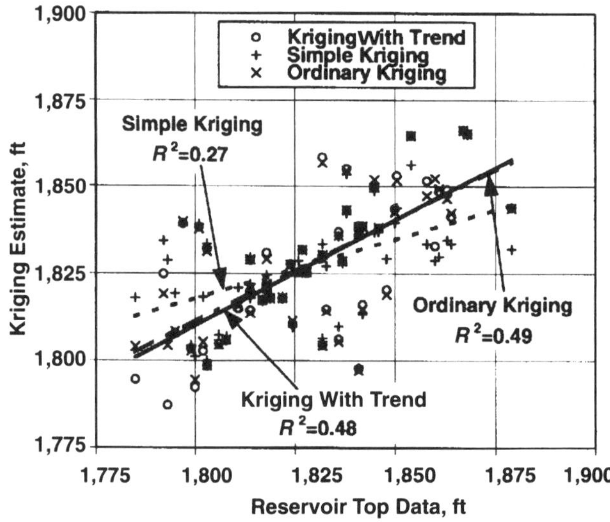

Fig. 4.43—Cross-validation estimates for reservoir top using universal, simple, and ordinary kriging.

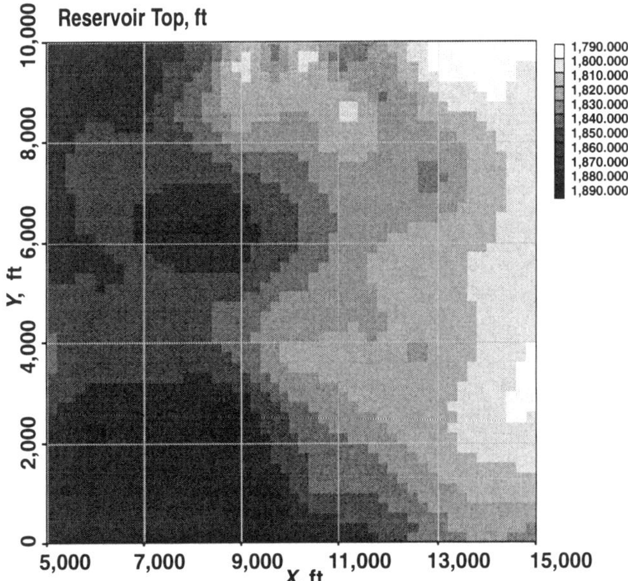

Fig. 4.44—Map of reservoir top generated with universal kriging.

Simple kriging estimation errors are greater at the extremes farther away from the mean of the reservoir top (1,828.7 ft).

The map of the top of the reservoir, generated with universal kriging, is shown in **Fig. 4.44.** This map clearly shows that the linear trend of the reservoir top runs east/west.

Field Example 4.9—Ordinary Kriging With an External Trend To Estimate Permeability Using a Porosity Trend. This field example uses kriging with an external trend to map a reservoir property based on additional information from a secondary variable correlated to the primary variable. Often, in reservoir description applications, trend maps are available from a reservoir attribute derived from seismic data. Maps of the top of the reservoir, gross thickness, and porosity are generated from seismic data. Seismic-derived properties can be correlated to other reservoir properties, depending on the depositional environment and indications from well data. In this field example, porosity is used as a trend to guide the permeability maps. Kriging with an external trend assumes that there is a linear relationship between the trend and the primary variable. Because the primary variable is usually only available at widely spaced wells, it is very valuable to use the trend to define the correlation more precisely in areas with poor well control.

The following information is given:

• The conditioning data for Flow Unit 5 permeability is available at the same well locations, as in Field Example 4.1 (Fig. 4.6). The histogram of permeability is shown in **Fig. 4.45.**

• Variogram models for the log permeability of Flow Unit 5 are

$$\gamma(\vec{L}) = 0.9M_{s_{7,000}}(\vec{L}), \text{ principal direction } 112.5°\text{N},$$

and

$$\gamma(\vec{L}) = 0.9M_{s_{4,000}}(\vec{L}), \text{ minor direction } 22.5°\text{N}.$$

• The grid definition is given in Field Example 4.2.

• The porosity map used for external trend is generated in Field Example 4.6, and it is shown in Fig. 4.30.

Make the following calculations:

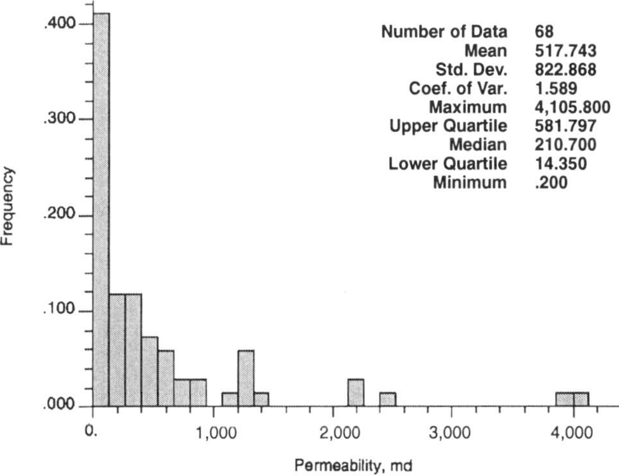

Fig. 4.45—Histogram of Flow Unit 5 permeability data.

• Generate a permeability map for Flow Unit 5 using ordinary kriging with a porosity trend.

• Compare permeability maps generated in the previous step with ordinary kriging.

• Compare statistics of sample permeability data with estimated permeability values in the first step.

Solution. Porosity and permeability, based on core measurements for Flow Unit 5, exhibit a very strong correlation. As shown in **Fig. 4.46,** the correlation coefficient between porosity and logarithm of permeability exceeds 0.90. This strong correlation makes porosity a good secondary variable candidate to guide permeability mapping. The logarithm transformation is needed to honor the linear relation between porosity and permeability. The logarithm transformation is applied to the permeability conditioning data, and the logarithms of permeability are used as the primary variable for kriging.

Fig. 4.47 shows the log permeability map generated with ordinary kriging with a porosity trend. The spatial correlation of log permeability is similar to the porosity map (Fig. 4.28). The external trend appears to have greater influence in the spatial correlation than the variogram models of permeability. The spatial trend in the porosity map is controlled by the principal direction of the porosity variograms (Field Example 4.6) that are slightly different than for the permeability variograms. The log permeability map, generated with ordinary kriging without a trend, is shown in **Fig. 4.48.** The ordinary

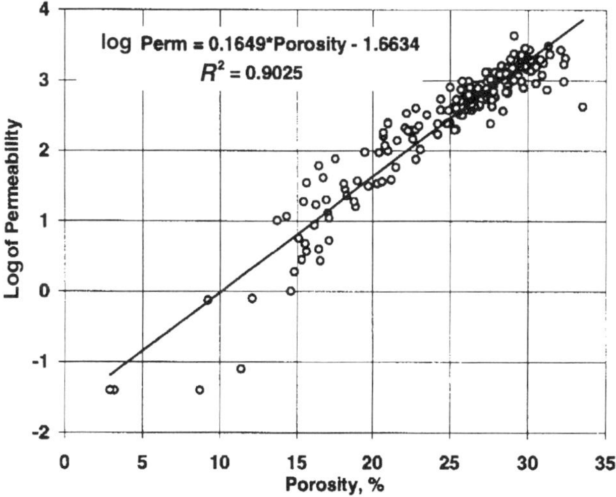

Fig. 4.46—Porosity vs. logarithm of permeability for Flow Unit 5.

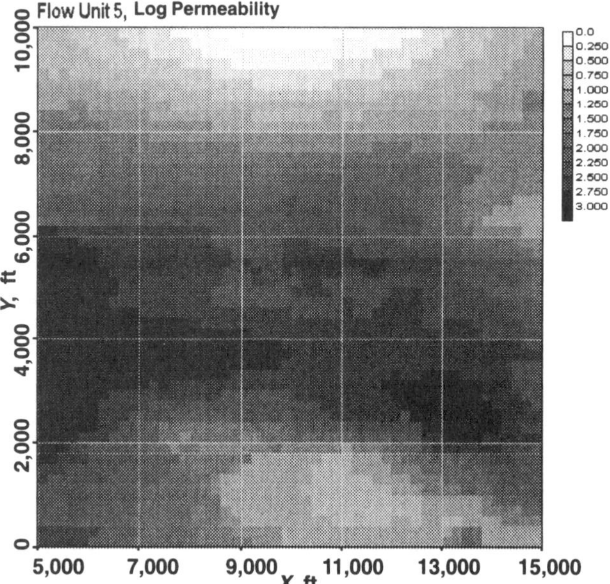

Fig. 4.47—Ordinary kriging with a trend map for logarithm of permeability of Flow Unit 5.

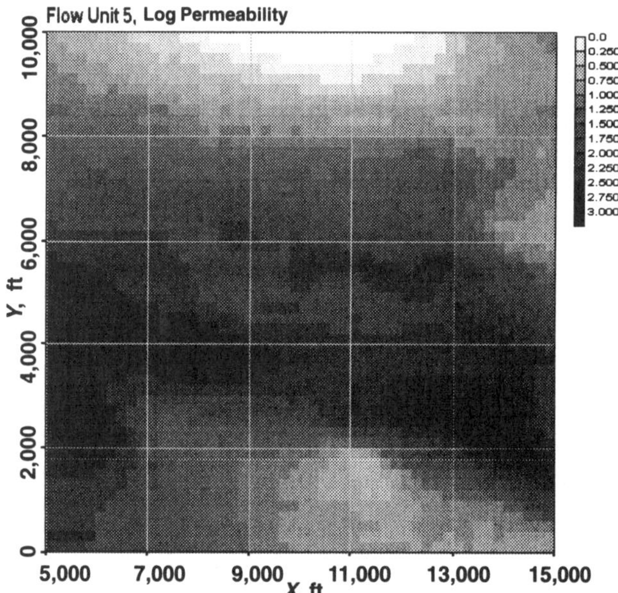

Fig. 4.48—Ordinary kriging map for logarithm of permeability of Flow Unit 5.

kriging map has a smoother spatial variation than the map generated with the trend, and the direction of most continuity is clearly aligned with the principal direction of the variogram. The main contribution of the external trend is that it provides additional information about the small and large scale spatial correlation of the property, which is not included in the variogram models.

Histograms for ordinary kriging with trend estimates and conditioning data for log permeability are shown in **Fig. 4.49.** The histogram for the kriging estimates is narrower than for the conditioning data. Similar to the other kriging types, kriging with a trend does not reproduce extreme values of the conditioning data very well. The standard deviation for kriging estimates is smaller than the standard deviation for the data.

4.3 Nonlinear Kriging Techniques

The term "nonlinear" is a misnomer in the sense that the kriging procedure itself is still based on linear regression principles. The difference, however, is that the variable values are transformed before the actual estimation procedure is implemented. Unlike conventional linear kriging, in which the spatial relationship (based on the variable samples) is used to estimate the value of the variable (at the unsampled locations), nonlinear kriging first transforms the variable into another domain before applying the spatial modeling, as well as the estimation process. The transformation is usually nonlinear. Once the variable is transformed, the spatial model is estimated and modeled, and by using the linear regression procedure, a value at the unsampled location is estimated. The estimated value is then transformed back to the original domain. Because the transformation, as well as back transformation, is a nonlinear process, the estimated value of the original sample is no longer linearly related to the nearby samples. The linear relationship only exists in the transformed domain. Because the relationship, between the estimated value and the surrounding samples, is nonlinear in the original domain, the name nonlinear kriging is used to describe these kriging techniques.

Several reasons exist to apply the nonlinear kriging technique. Principally among them are (a) generation of a more stable spatial relationship; (b) easier estimation of uncertainty at unsampled locations, based on a few parameters; and (c) flexibility in using various types of data. As we go through some of these nonlinear kriging procedures, we discuss the specific advantages that attribute to a specific method.

4.3.1 Log-Normal Kriging. Log-normal kriging is one of the simplest nonlinear kriging procedures. In log-normal kriging, the original variable is transformed by taking the log of the sample; the spatial relationship for the transformed variable is established; the value is estimated at the unsampled location, using the spatial relationship of the transformed variable; and the estimated value is back-transformed, using an appropriate equation. The two reasons log-normal kriging is used are (a) log transform reduces the variability of the sample points, thus better estimating the spatial relationship; and (b) by taking a log-transform, many of the variables, which are log-normally distributed, are now distributed normally. Normal distribution makes it easier to describe the uncertainty of estimation by only knowing the error variance. These two advantages, however, do not come without a cost. The major disadvantage of log-normal kriging is that the estimation is sensitive to the sill of the variogram for the transformed variable.

If we assume that variable $X(\vec{u})$ is subjected to log-normal estimation, we first must define

$$Y(\vec{u}) = \log[X(\vec{u})], \qquad (4.46)$$

where $Y(\vec{u})$ = the transformed variable. Once all the samples are transformed to the log values, an appropriate spatial relationship is established with the transformed values. Depending on whether simple or ordinary kriging is used for estimation, we can estimate the value of $Y(\vec{u}_0)$ with equations

$$Y^*(\vec{u}_0) = \lambda_0 + \sum_{i=1}^{n} \lambda_i Y(\vec{u}_0), \qquad (4.47)$$

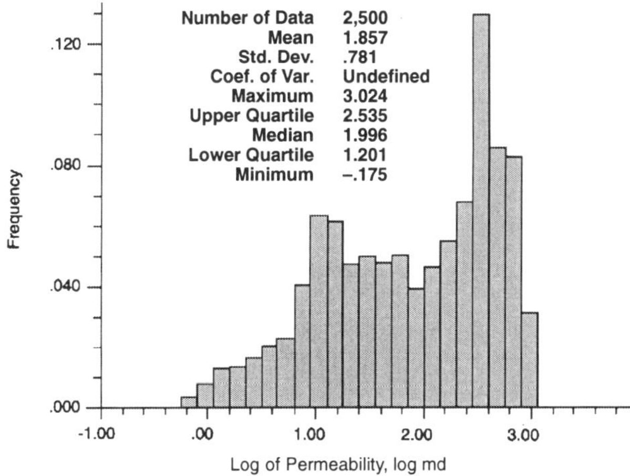

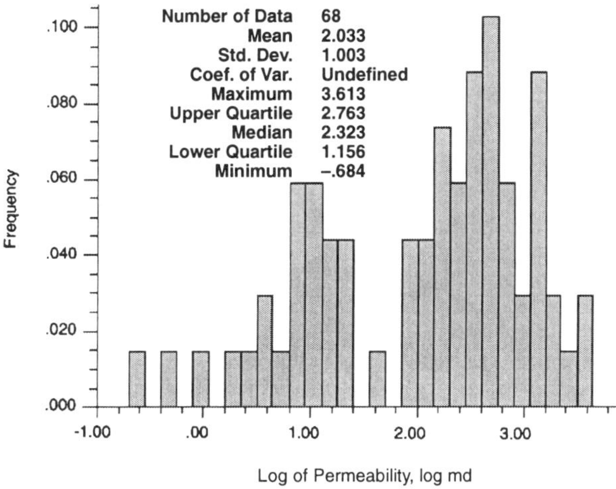

Fig. 4.49—Histograms of ordinary kriging with trend estimates and conditioning data for logarithm of permeability of Flow Unit 5.

for simple kriging, and

$$Y^*(\vec{u}_0) = \sum_{i=1}^{n} \lambda_i Y(\vec{u}_0) \quad \ldots\ldots (4.48)$$

for ordinary kriging. Note, for simple kriging, the value of λ_0 is estimated with equation

$$\lambda_0 = m\left(1 - \sum_{i=1}^{n} \lambda_i\right), \quad \ldots\ldots (4.7)$$

where m is the mean of the variable Y.

Along with the estimate, we can also estimate the error variance, $\hat{\sigma}_E^2$, which is an indication of surrounding sample configuration with respect to variable Y. Once we estimate the value of Y and the associated error variance at the unsampled location, we back-transform the estimate to the original domain. It is shown for simple kriging,[12] where the value of $X(\vec{u}_0)$ is calculated as[18]

$$X^*(\vec{u}_0) = \exp\left[Y^*(\vec{u}_0) + \hat{\sigma}_E^2/2\right], \quad \ldots\ldots (4.49)$$

and, for ordinary kriging, the estimate is obtained as

$$X^*(\vec{u}_0) = \exp\left[Y^*(\vec{u}_0) + \hat{\sigma}_E^2/2 + \mu\right], \quad \ldots\ldots (4.50)$$

where μ = the LaGrange parameter.

The estimated value at the unsampled location not only depends on the estimate in the transformed domain, but also on the associated error variance, which means the estimated value depends on the surrounding samples, as well as the sill value of the variogram in the transformed domain. Further, by taking an exponent, any error in the estimation of $Y^*(\vec{u}_0)$, as well as $\hat{\sigma}_E^2/2$, is magnified as we estimate the value of $X^*(\vec{u}_0)$. As discussed in the previous sections on kriging, the error variance is able to capture the effect of surrounding sample configuration; however, it might not be able to capture the local uncertainty (Sec. 4.4). Because the error variance might not be a true representation of the local uncertainty, the use of the error variance in estimating the value of $X^*(\vec{u}_0)$ could render the estimation meaningless. Because of this difficulty, log-normal kriging is not a very popular technique. Instead, it is being replaced by the two techniques we describe in the next two sections.

4.3.2 Multi-Gaussian Kriging. Multi-Gaussian kriging transforms the original data into Gaussian space, using the normal score transform; models the variogram of the transformed variable; estimates the value at the unsampled location, using the modeled variogram; and back-transforms the estimated value into original space. As in the case of log-normal transform, multi-Gaussian kriging reduces the variability of the original data and allows us to better define the estimation uncertainty in the transformed space by knowing the estimation error variance.

In Chap. 3, we discussed the transformation of continuous variable data and estimation of the variogram in the Gaussian space. Similar to continuous variable data, discrete variable data can also be transformed into a Gaussian domain. For example, if we collect facies data at the well location, by numerically identifying each facies, we can generate a cumulative distribution of facies, as shown in **Fig. 4.50.** This is a stepwise distribution with each step representing the fraction of each facies. To transform the data into Gaussian space, we truncate the Gaussian distribution, as shown in Fig. 4.50.

When transforming the data from a discrete to a continuous variable, we must assign a range of values of continuous variables for a given discrete variable. In Fig. 4.50, for the given facies, K, a corresponding range of possible values in the continuous domain is shown. To ensure proper back transform, we truncate the Gaussian distribution to certain minimum and maximum values (e.g., for facies A, the transformation in the Gaussian space is restricted to between -3 and A_t. Similar truncation is needed for facies D, as well. When sample data at a given location are identified as a certain facies, we know the corresponding range of values in the Gaussian space. To assure a continuous distribution, we randomly pick any value with that range and assign it to that location. A more sophisticated technique could assign a value that reflects surrounding samples as well.

Once the data are transformed and the spatial relationship is modeled, the value at the unsampled location is estimated by

$$Y^*(\vec{u}_0) = \lambda_0 + \sum_{i=1}^{n} \lambda_i Y(\vec{u}_i), \quad \ldots\ldots (4.51)$$

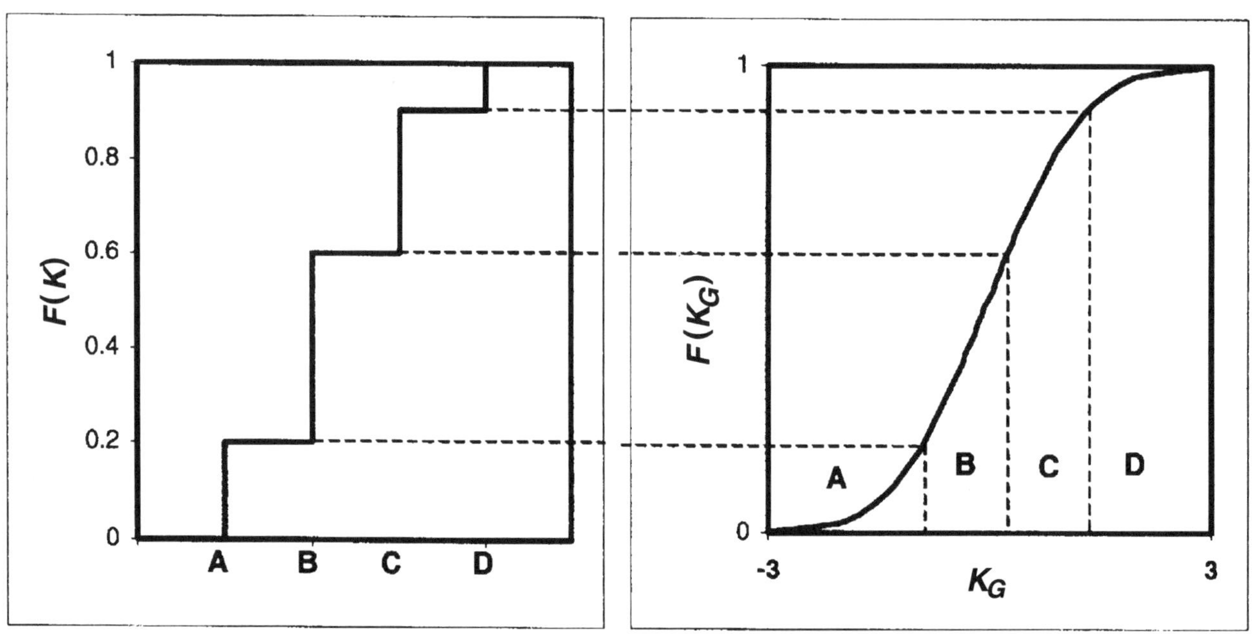

Fig. 4.50—Gaussian transform of discrete data.

where $Y(\vec{u}_i)$ = the transformed value of the original sample, $X(\vec{u}_i)$. We also assume that a simple kriging procedure is used to estimate the value. Along with the estimate, we also calculate the error variance with equation

$$\hat{\sigma}_E^2 = C_t(\vec{u}_0, \vec{u}_0) - \sum_{i=1}^{n} \lambda_i C_t(\vec{u}_i, \vec{u}_0), \quad \text{.......... (4.52)}$$

where $C_t(\vec{u}_0, \vec{u}_0)$ represents the variance of the transformed data, and $C_t(\vec{u}_i, \vec{u}_0)$ represents the covariance between the sample located at $\vec{u}_i$ and the unsampled location at $\vec{u}_0$ in the transformed domain.

Once the estimated value and the error variance are known, by assuming Gaussian distribution, we can completely describe the uncertainty at the unsampled location. Note, transforming the data into a Gaussian domain does not automatically guarantee that the assumption of a multivariate Gaussian distribution is valid. There are methods in the literature to check this assumption independently.[2] However, in most instances, the assumption of a multivariate Gaussian distribution is reasonable and can be used to describe uncertainty.

Using the estimated value and the error variance, we bracket the uncertainty in the transformed domain. Because there is a one-to-one correspondence between the transformed domain and the original data, for a continuous variable, we can also define the uncertainty data into the original domain. **Fig. 4.51** represents a schematic representation for estimating the uncertain region in the original domain if the estimated value and the error variance are known. The figure illustrates the calculation of the range with 95% probability that a true value will fall in that region. Knowing that this range is defined by $Y^*(\vec{u}_0) \pm 2\hat{\sigma}_E$, we can estimate the corresponding region in the original sample domain.

A similar method is used to calculate the uncertain region for any other confidence interval. Alternately, the method can also be used to estimate the probability that the estimated value is greater than a certain threshold or cutoff,[19] or as we discuss in the chapter on conditional simulation, the method can also be used for sampling a realization at an unsampled location.

For a result of a discrete variable back transformation, once a value in the Gaussian space is estimated, a corresponding discrete variable can be assigned at the unsampled location by knowing the range in which it falls.

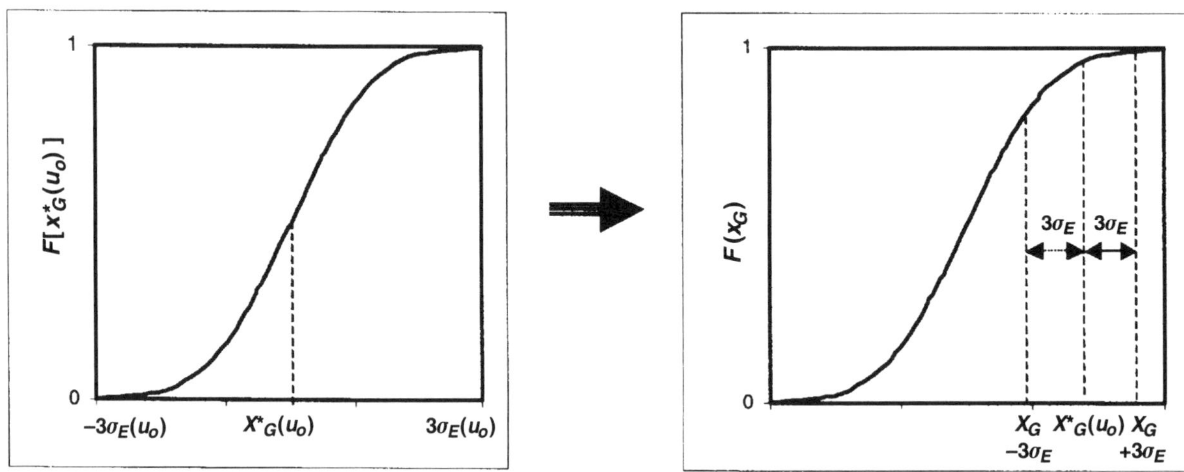

Fig. 4.51—Uncertainty in original domain.

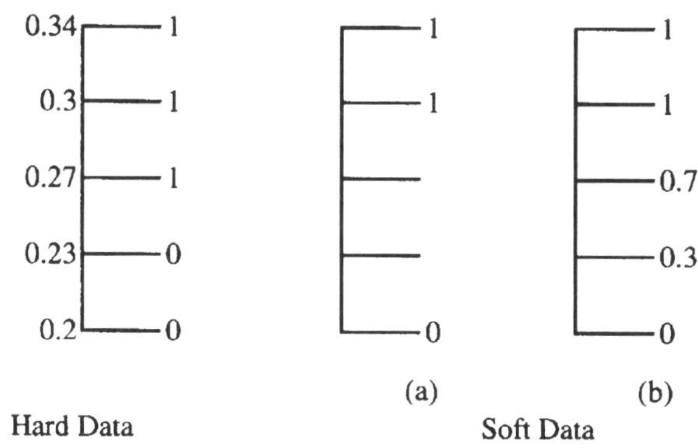

Fig. 4.52—Indicator transformation for continuous variables.

4.3.3 Indicator Kriging. We briefly discussed the indicator transform in Chap. 3. In this section, we discuss how kriging techniques are implemented when using indicator transform. One key difference between indicator kriging and other nonlinear kriging techniques is that the final result of indicator kriging represents the uncertain distribution at the unsampled location. However, there is no provision for a simple back transform into the original domain based on the estimated values. Instead, we must apply a conditional simulation technique to sample a value at the unsampled location. As we proceed with the discussion in this section, the reasons for this difference will become more clear.

Similar to other nonlinear transforms, indicator transform minimizes the variability of the sample data; thus, it results in a more stable variogram estimation. In addition, indicator transform also allows the use of incomplete or soft information; it provides more flexibility in spatial modeling of the original variable, and finally, it provides an estimate of uncertainty without assuming any particular distribution function.

For a continuous variable, the equation for an indicator transform is written as[20,21]

$$I(\vec{u}_j, x_t) = 1, \text{if } X(\vec{u}_j) \leq x_t$$
$$= 0, \text{if } X(\vec{u}_j) > x_t, \quad \text{.................} \quad (4.53)$$

where $I(\vec{u}_j, x_t)$ = the indicator value; $X(\vec{u}_j)$ = the value of the variable at $\vec{u}_j$; and x_t = the threshold value. Depending on the value of $X(\vec{u}_j)$, the indicator variable can take either a value of one or zero.

Similar to the continuous variable, an equation for a discrete variable is written as

$$I(\vec{u}_j, K_t) = 1, \text{if } K(\vec{u}_j) = K_t$$
$$= 0, \text{if } K(\vec{u}_j) \neq K_t. \quad \text{................} \quad (4.54)$$

K_t represents a threshold value. Depending on whether the sample value is equal or not equal to the threshold value, the indicator variable can take either a value of zero or one.

It is helpful to understand the indicator variable in terms of the confidence in the sample value. If we are 100% confident about a sample, indicator values are defined in terms of zero or one. On the other hand, a value between zero and one represents the uncertainty in the sampled value. This understanding also provides us with flexibility when assigning indicator values if the information about a particular sample point is incomplete.

Briefly, we refer to the data with 100% certainty as hard data and with less than 100% uncertainty as soft data. Through indicator variable transformation, we have the ability to assign indicator values to both the hard and the soft data.

Consider the illustration in **Fig. 4.52,** where both hard and soft data are represented for a continuous variable.

In this figure, the indicator transformation for both the hard and soft data are shown. For illustration purposes, assume that porosity data in a field must be represented by five threshold values. These threshold values are 0.2, 0.23, 0.27, 0.3, and 0.34, respectively. If at a particular sample location, a porosity value is observed to be 0.24, with the indicator transform equation, we can assign the indicator values at each threshold, as shown in Fig. 4.52, which represents the hard data. The zero values in Fig. 4.52 represent 0% probability that a value is less than a given threshold. Alternately, there is 100% probability that a value is greater than a given threshold. Because 0.24 is greater than both 0.2 and 0.23, the indicator value at those two thresholds should be zero. An indicator value of one represents 100% probability that a value is less than a given threshold. 0.24 is less than thresholds 0.27, 0.3, and 0.34; therefore, the corresponding indicator values should be one. By examining the indicator values, we know that there is 0% probability that a sample value is less than 0.23, and there is 100% probability that a sample value is less than 0.27. Therefore, the sample value is within an interval of 0.23 to 0.27. By simply knowing the indicator values, we will not be able to know the exact value, except to state that it falls within a given interval. When we transform a continuous variable into a discrete indicator variable, there is a loss of information within class values. Whether porosity is 0.24, 0.25, or 0.26, the indicator transforms for all three values would be identical, as shown in Fig. 4.52. The subtle within class differences can be reestablished if we divide the data in additional classes. For example, if we define an additional threshold of 0.25, we can define the difference between 0.24 and 0.26. However, more spatial modeling and estimation effort is required. Further, if we do not have enough samples within each class, through variogram analysis, as well as during the estimation step (kriging), it will be hard to distinguish between the classes. A reasonable compromise defines the number of thresholds, so those thresholds can make a difference in characterizing the reservoir. For example, if reservoir permeabilities vary between .001 md and 50 md, it is much more critical to capture an order of magnitude permeability difference than the small changes in permeability, such as the distinction between 0.3 md and 0.8 md.

The figure also illustrates the representation of the soft data. To illustrate, for the same threshold value as the hard data, if we know that at a particular location the porosity value falls between 0.21 and 0.28, we can represent this interval information through an indicator variable. This information could be obtained through seismic interpretation. Because of a weak relationship between the porosity and seismic interpretation, we can only define an interval estimate. However, because we know that porosity is greater than 0.2, we can assign an indicator value of zero at the threshold. We also know that porosity is less than 0.3 and 0.34. Therefore, we can assign a value of one to those thresholds. We do not have any information about the thresholds 0.23 and 0.27, so we can not assign any indicator values at those thresholds. On the other hand, if we also know that the porosity value falls within the interval between 0.21 and 0.28, and the distribution of porosity is known, we can assign the probability that a value of po-

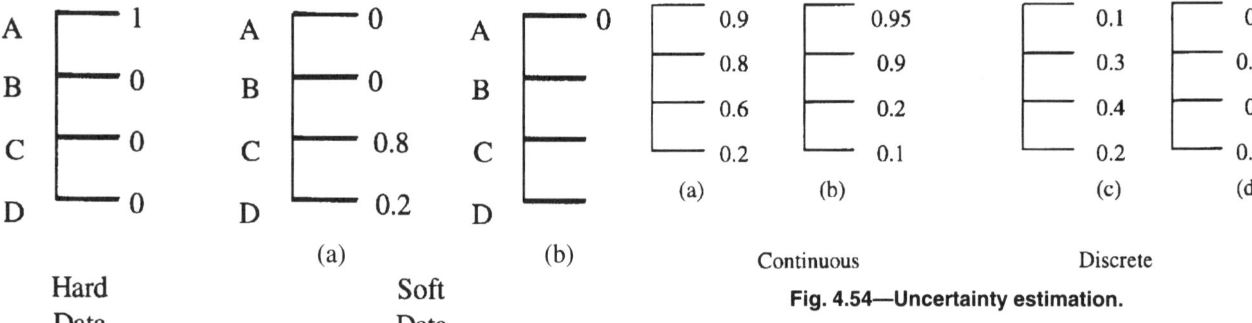

Fig. 4.53—Discrete variable transform.

Fig. 4.54—Uncertainty estimation.

rosity will be less than a given threshold. If we know that there is 30% probability that porosity is less than 0.23, and there is 70% probability that porosity is less than 0.27, we can assign those probability values as indicator values at those thresholds. In this sense, indicator values represent the degree of confidence we have with respect to a particular sample value.

A similar example for a discrete variable is provided in **Fig. 4.53.** For four geological facies, if we know that, at a particular sample location, we have facies A, we can assign indicator values of 1, 0, 0, 0, indicating 100% probability that facies A is present and 0% probability that facies B, C or D are present. This is hard data. If a geologist is not certain about other locations, then we can assign values between zero and one. For example, if the geologist is 80% certain that, at a particular location, facies C exists, and 20% chance that facies D exists, we can assign indicator values of 0.8 and 0.2, respectively, at those thresholds. Naturally, we assign a value of zero for facies A and B because we know with certainty that neither A nor B exists at that location. This is an example of soft data, where we do not know exact facies with 100% certainty. Another example of soft data is where the geologist only knows that facies A does not exist at a sampled location. We can then assign an indicator value of zero for threshold A and leave blank indicator values at other thresholds. Because indicator values represent the probability that certain facies exist, the sum of the indicator value at a given location should equal one. For the second example of soft data, the sum of the indicator values for thresholds B, C and D should equal one.

Once the indicator values at each threshold are defined, the next step is to estimate the spatial relationships or variograms. For indicator kriging, the number of variograms that must be modeled, depends on the number of thresholds. For continuous variables, if high permeability values exhibit different continuity than low permeability values, indicator approach provides the flexibility to model different levels of permeability with different variograms. For discrete variables, indicator variables allow the flexibility to model each of the thresholds separately. For example, different geological facies can be modeled differently, reflecting their architecture.

The next step in indicator kriging is the estimation of indicator values at the unsampled locations. The approach is the same as the one for conventional kriging, except the kriging procedure must repeat at each threshold. For continuous variables, we write

$$I^*(\vec{u}_0, x_t) = \lambda_0 + \sum_{j=1}^{n} \lambda_j I(\vec{u}_j, x_t), \quad \text{(4.55)}$$

if we use simple kriging, and

$$I^*(\vec{u}_0, x_t) = \sum_{j=1}^{n} \lambda_j I(\vec{u}_j, x_t), \quad \text{(4.56)}$$

if we use ordinary kriging.

For discrete variables, we write

$$I^*(\vec{u}_0, K_t) = \lambda_0 + \sum_{j=1}^{n} \lambda_j I(\vec{u}_j, K_t), \quad \text{(4.57)}$$

if we use simple kriging, and

$$I^*(\vec{u}_0, K_t) = \sum_{j=1}^{n} \lambda_j I(\vec{u}_j, K_t), \quad \text{(4.58)}$$

if we use ordinary kriging.

Because the weights assigned to sample points fall between zero and one, and the indicator values are between zero and one, the estimate is logically going to fall between zero and one.

As seen from the previous two equations, the procedure must be repeated at each threshold. Each threshold could have a different spatial relationship, different search neighborhood, and different sample points, if soft data are used.

After all the unsampled points are visited, we have indicator values at each location. The estimate depends on whether we use indicator kriging for continuous or discrete variables. **Fig. 4.54** shows the possible estimates we can obtain for continuous and discrete variables. For a continuous variable, we obtain the cumulative distribution function values at each threshold. For example, location (a) for the continuous variable indicates that there is a 20% probability that the value at that location is less than the first threshold. A similar explanation is given for other thresholds. By examining the probability values for location (b), we can state that the probability that a sample value will be between the second and the third threshold is $0.9 - 0.2 = 0.7$. This represents a high probability that the value is going to fall within that interval.

For discrete variables, the indicator values represent a discrete variable, which is assigned a particular threshold value. For example, location (a) represents the probability of 0.2 that threshold A (or 1) will be assigned at that location, whereas at location (b), the probability is 0.7 that threshold A will be assigned at that location. By examining these probability values, we can understand our degree of confidence with respect to assigning a particular threshold value at the unsampled location.

Obviously, because these indicator values represent the cumulative distribution function, or probability density function, they must satisfy certain requirements. For a continuous variable, Eq. 4.59 must be satisfied.

$$I(\vec{u}_j, x_{t1}) \leq I(\vec{u}_j, x_{t2}), \text{ if } x_{t1} < x_{t2}. \quad \text{(4.59)}$$

This ensures that a cumulative distribution function is nondecreasing for the discrete variable,

$$\sum_{l=1}^{L} I\left(\vec{u}_j, K_{t_l}\right) = 1, \quad \ldots\ldots\ldots\ldots\ldots\ldots (4.60)$$

where L = the total number of thresholds. This ensures that the sum of the probabilities for the density function equals one. In practice, if these conditions are not satisfied, adjustments must be made to correct the values. Discussion on these corrections is deferred until Chap. 6.

Once the indicator values at each threshold are estimated, they provide us with uncertain estimates, with respect to each threshold at each unsampled location. With this information, we generate the probability contour maps to evaluate the quality of the reservoir. For example, if porosity less than 4% represents a nonreservoir rock, we can contour, say, 90% of the probability intervals, and the porosity will be less than 4% within that interval. We might decide to stay away from these intervals, in terms of future in-fill drilling.

Other than probability intervals, the actual back transformation to the original domain is not straight-forward. The common method used for such back transformation is the conditional simulation technique. We discuss that technique in Chap. 6.

Field Example 4.10—Indicator Kriging To Generate Permeability-Thickness Maps. In this field example, indicator kriging is used to generate map permeability-thickness (kh) indicators. Indicator kriging is a useful technique to describe variables with highly skewed distributions or with distributions that depart significantly from the Gaussian method. Usually, permeability or kh has highly skewed histograms, with a shape similar to log-normal distribution. Indicators are used to describe the shape of any arbitrary distribution. Data are transformed to indicators using a specified number of threshold values. The number of thresholds should be just enough to capture the shape of the distribution. It is important not to use more than the necessary thresholds to avoid excessive variogram modeling and computational requirements. Simple or ordinary kriging is used to generate maps of indicator variables for each threshold. Unlike simple or ordinary kriging, which provides a single map, indicator kriging generates one map for each threshold. These maps represent the probability of exceeding threshold values, and at each location, these are used to estimate the local conditional probability distribution.

The following information is given:

• The kh conditioning data is the same as the data used in Field Example 4.7.

• Three thresholds are selected to describe kh. The thresholds are equal to the lower quartile (952.9 md-ft), median (3,482.2 md-ft) and third quartile (7,904.5 md-ft).

• Variogram models for the first threshold (lower quartile) are

$$\gamma\left(\vec{L}\right) = 0.25M_{s_{13,000}}\left(\vec{L}\right), \text{ principal direction East, 22.5°S;}$$

and

$$\gamma\left(\vec{L}\right) = 0.25M_{s_{13,000}}\left(\vec{L}\right), \text{ minor direction East, 22.5°N.}$$

Variogram models for the second threshold (median) are

$$\gamma\left(\vec{L}\right) = 0.3M_{s_{10,000}}\left(\vec{L}\right), \text{ principal direction east/west;}$$

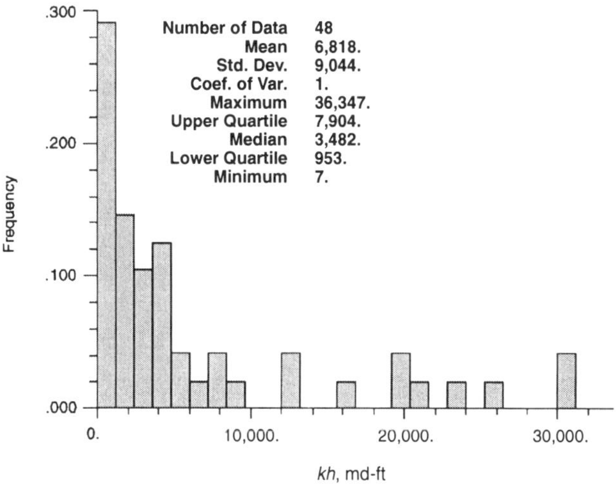

Fig. 4.55—Histogram of *kh* conditioning data.

and

$$\gamma\left(\vec{L}\right) = 0.3M_{s_{5,000}}\left(\vec{L}\right), \text{ minor direction north/south.}$$

Variogram models for the third threshold (upper quartile) are

$$\gamma\left(\vec{L}\right) = 0.25M_{s_{14,000}}\left(\vec{L}\right), \text{ principal direction east/west;}$$

$$\gamma\left(\vec{L}\right) = 0.25M_{s_{8,000}}\left(\vec{L}\right), \text{ minor direction north/south.}$$

• The grid definition is given in Field Example 4.2.

Make the following calculations:

• Generate indicator maps for the lower, median, and third quartile thresholds of kh with indicator kriging.

• Compare the statistics of indicator kriging to the conditioning data.

• Compare the indicator kriging and cokriging maps generated in Field Example 4.7.

Solution. Indicator maps are generated for three thresholds selected from the histogram of the conditioning data shown in **Fig. 4.55.** The three thresholds correspond to the lower, median, and upper quartile, and these are 952.9, 3,482.2, and 7,904.5 md-ft, respectively. Even though the maximum data is 36,347.4 md-ft, the largest threshold is smaller because there is only a small proportion of data in the long tail of the histogram. The data is converted to indicators using the transformation described in Chap. 3.

The indicator kriging maps for the three thresholds shown in **Figs. 4.56 through 4.58** represent the probability that kh will be less than the threshold. The map for the lower quartile threshold shows that most areas have low probabilities of having kh less than 952.9 md-ft. Areas in the northeast and southeast corners of the map have the highest probability of low kh. The median threshold map shows that areas with a high probability of having kh less than 3,482.2 md-ft are larger than for the lower quartile. The upper quartile map shows that areas in the center of the map have a high probability that kh will exceed 7,904.5 md-ft.

The cokriging map (Field Example 4.7) shows similar trends of low and high kh values observed in the indicator maps. One advantage of indicator kriging over other kriging methods is that variogram models are different for each threshold. As seen in this field example, different thresholds or range of kh values have different correlation ranges and principal directions. For permeability, preserving the connectivity of low and high values is an important property because

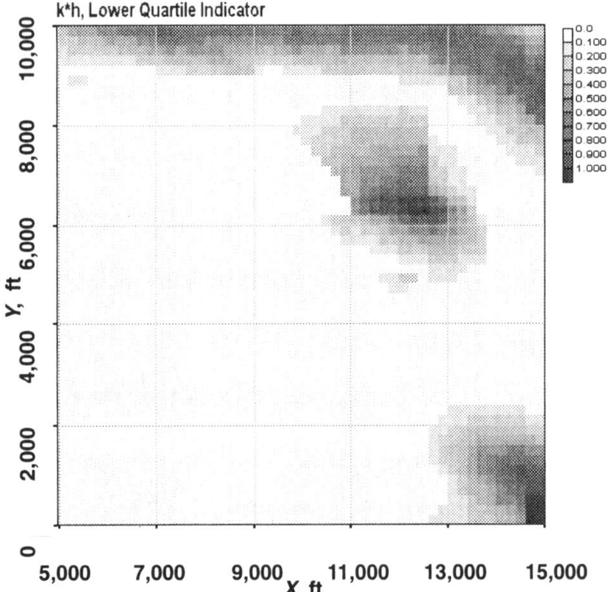

Fig. 4.56—Map of indicator kriging estimate for lower quartile.

it affects fluid movement in the reservoir. Also, unlike the other kriging methods that provide a single estimate, indicator kriging provides estimates for each threshold, and together these describe the conditional probability distribution in each gridblock.

Histograms of the indicator maps are shown in **Fig. 4.59.** Because indicators represent cumulative probability, the mean of the estimates increases as thresholds increase. The shape of the histograms shifts from values close to zero for the lower threshold to values close to one for the upper threshold.

4.3.4 Probability Kriging. Probability kriging involves the transformation of data from an original domain into rank index values. If you recall, in Chap. 3, we defined the rank index as

$$R_i = \frac{i}{n+1}, \quad \ldots\ldots\ldots\ldots\ldots\ldots\ldots\ldots\ldots\ldots (4.61)$$

where i is the ranking of sample i after the samples are arranged in ascending order, and n is the number of sample points. The rank index falls between zero and one, thus reducing the variability of original samples.

Once the rank index is estimated and corresponds to each sample point, a variogram for the rank index is estimated and modeled. After that, a kriging procedure is applied, and the rank index is estimated at the unsampled location,

$$R^*(\vec{u}_0) = \lambda_0 + \sum_{i=1}^{n} \lambda_i R(\vec{u}_i), \quad \ldots\ldots\ldots\ldots\ldots\ldots (4.62)$$

if simple kriging is used. A similar equation is written for ordinary kriging. The estimated rank index typically falls between zero and one. Therefore, by knowing the rank index value, we can back transform it to the original domain as shown in **Fig. 4.60.**

In addition to minimizing the variability of data, probability kriging has another useful application. Probability kriging provides values between zero and one at every unsampled location. Using these values, we sample from local uncertainty estimations to assign a value at the unsampled location. These local, uncertain estimations are constructed by other

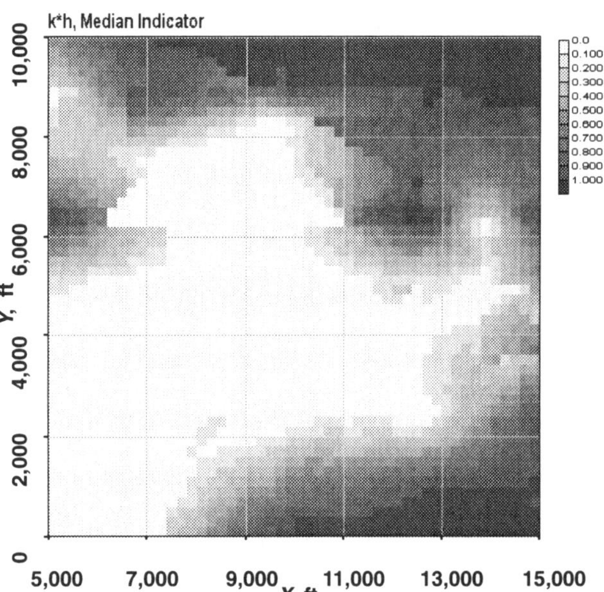

Fig. 4.57—Map of indicator kriging estimate for median.

means.[22] This is a very useful method for constructing alternate images of a desired variable very quickly. More of its usefulness is discussed in Chap. 6.

4.4 Estimation of Uncertainty

In the last three sections, we discussed various estimation methods. Depending on the type of method, we obtain a different estimation. In general, we write the equation for estimation as

$$x^*(\vec{u}_0) = \sum_{i=1}^{n} \lambda_i x(\vec{u}_i), \quad \ldots\ldots\ldots\ldots\ldots\ldots\ldots (4.63)$$

where the weights to individual samples are determined by solving a matrix equation, and the values at the unsampled location are estimated. Slightly modified forms of Eq. 4.63 are used for most of the kriging techniques This equation indicates that the estimated value is dependent on the samples within the neighborhood.

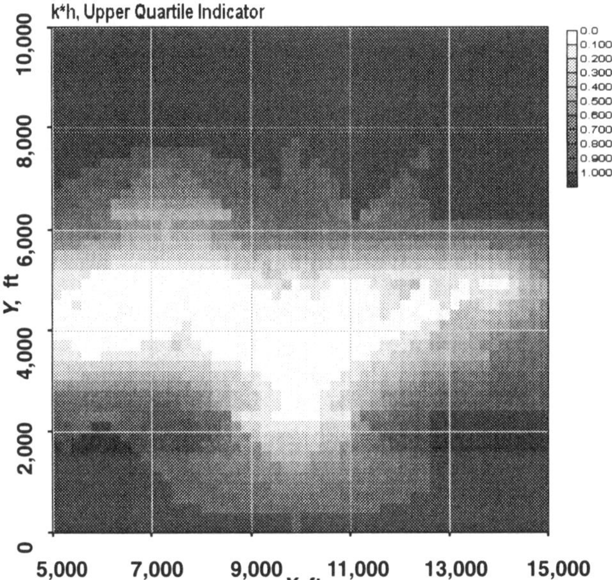

Fig. 4.58—Map of indicator kriging estimate for upper quartile.

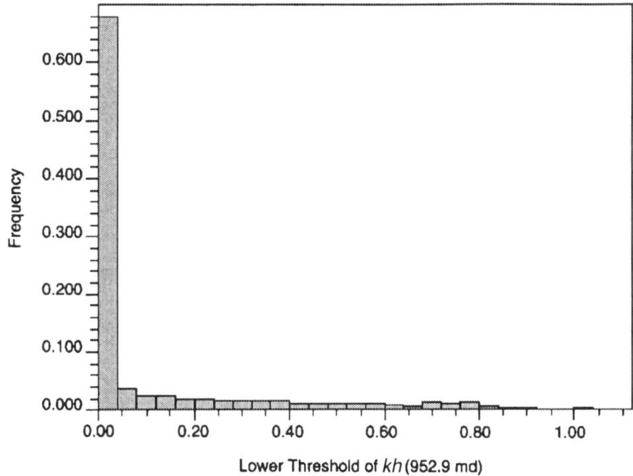

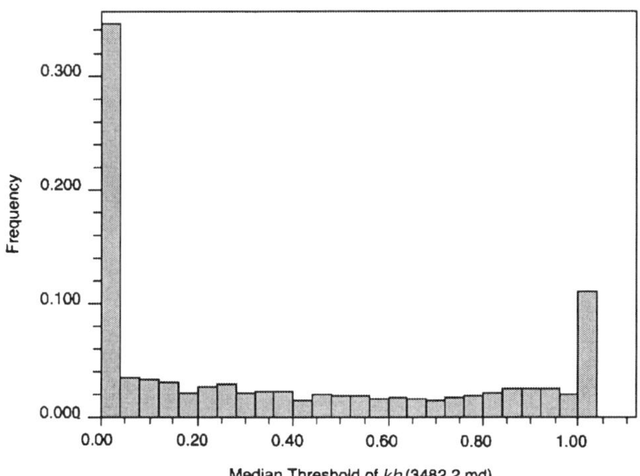

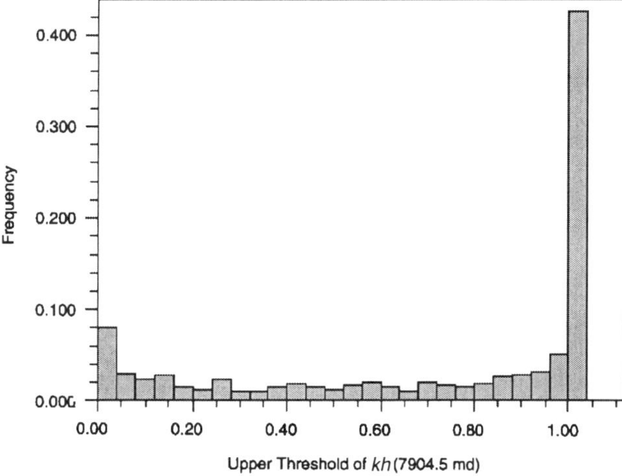

Fig. 4.59—Histograms of indicator kriging estimates for lower, median, and upper quartile thresholds.

In contrast, the equation for error variance is written as

$$\hat{\sigma}_E^2 = C(\vec{u}_0, \vec{u}_0) - \sum_{i=1}^{n} \lambda_i C(\vec{u}_i, \vec{u}_0). \quad \text{(4.64)}$$

Equation 4.64 is for simple kriging; however, a form similar to Eq. 4.64 is written for other types of kriging techniques. An important distinction between Eqs. 4.64 and 4.63 is that the estimation at the unsampled location is dependent on the samples, whereas the error variance is independent.

As explained in Numerical Example 4.3, error variance provides information related to the surrounding sample configuration. In the absence of spatial relationships, Eq. 4.64 reduces to

$$\hat{\sigma}_E^2 = C(\vec{u}_0, \vec{u}_0). \quad \text{(4.65)}$$

The error variance is the sample variance. The uncertainty with respect to estimate is represented by the variance of the sample. As samples with spatial relationships are added, the error variance is reduced. Notice that the reduction in error variance is related to both the weight assigned to the sample and the covariance between the sample and the unsampled location. As discussed, the weight assigned to the sample depends on its proximity to the unsampled location, as well as its relation to other samples. If it is a part of clustered samples, the weight assigned will be small; if it is isolated, the weight assigned will be large. Obviously, if a sample point has a strong spatial relationship (large covariance) and is isolated from other samples (large weight), it provides the most information. Therefore, it reduces the most uncertainty. Eq. 4.64 proves this to be true. If the samples are well distributed (unclustered) and strongly related to the unsampled location, the error variance will be smaller. On the other hand, if the samples are clustered and weakly spatially related, error variance will approach the variance of the sample data. Intuitively, this explanation is logical and easy to understand. Further, if we assume Gaussian distribution, the estimation and the associated error variance will quantify the entire distribution of uncertainty. For example, we write

$$p\left[X(\vec{u}_0) \text{ E } X^*(\vec{u}_0) \pm 2\hat{\sigma}_E\right] = 0.95. \quad \text{(4.66)}$$

The probability that a true value falls within a certain interval is defined, once we make an assumption for Gaussian distribution. Thus, in principle, with the kriging technique, we can quantify the uncertain distribution at any unsampled location.

Unfortunately, in practice, error variance, by itself, is unable to quantify local uncertainty distribution at the unsampled location. The error variance is a good measure of surrounding sample configuration; however, it does not capture local uncertainty for the following reasons:

• Kriging techniques assume first- and second-order stationarities. In simple terms, this translates into local means, and the variance do not change substantially from global means and variance. In practice, however, the local means and local variance can change. This happens because we are

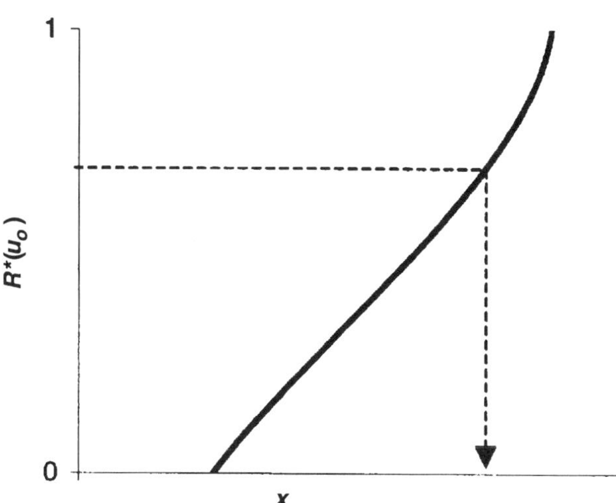

Fig. 4.60—Back transform using probability kriging.

forced to define a much larger region as the region of stationarity because of limited data points. The fewer the number of wells, the larger is the region of stationarity to obtain statistically meaningful information. Local variability within reservoir is observed; however, instead of defining smaller regions of stationarity, we define a larger region and try to address local variability problems by modifying the procedure. For example, we modify the variogram equation to account for variations in local means and local variances. We also use search neighborhoods in kriging to use the most relevant samples in estimating the value at unsampled locations. By restricting the samples within a search neighborhood, we ensure that only the samples that reflect the local variability are included in the estimation process. In contrast, the error variance is independent of local samples. It only depends on the covariance or spatial relationship. In other words, error variance, according to kriging equations, is independent of surrounding samples.

The assumption that error variance is independent of surrounding samples and, in turn, of the estimated value, is called the assumption of homoscedasticity. This assumption is rarely satisfied in field data and is rarely required by the user. Consider an example where an estimate of permeability at a particular location is 52 md, whereas the true value is 50 md. At another location, the estimate is 0.3 md, whereas the true value is 1.0 md. The better estimate, agreed by most, is 52 md, a much better estimate than 0.3 md. This is because relative to the true value, the first estimate is much better. Although the absolute error is bigger for the first estimate, the relative error is smaller. We accept, intuitively, a bigger absolute error, as our estimate increases, as long as the relative error remains small. Absolute error is dependent on the estimate. The traditional error variance, however, requires absolute error to be independent of the estimate—an unrealistic requirement which is not satisfied in practice.

Therefore, if we have, for example, two regions with different local variabilities, as long as a surrounding sample configuration is identical, we will estimate the same error variance. For example, in a low permeability region, we estimate a permeability value of 2.0 md, with an error variance of 4.0 md^2, whereas in a high permeability region, we estimate a permeability value of 200.0 md with an error variance of 4.0 md^2. The estimates differ because the estimate depends on the surrounding samples. The higher the values of samples within a search neighborhood, the higher are the estimates. The error variance is the same because the surrounding samples have the same configuration with respect to the unsampled location. For a given covariance model, the error variance is the same. The error variance might not represent the local uncertainty for either estimate. 4.0 md^2 is an error variance probably too large for a 2.0 md estimate and too small for a 200.0 md estimate. We must use a better method for quantifying local uncertainty.

- When defining uncertainty, with respect to the estimate, we assume that the uncertainty can be captured with Gaussian distribution. For Gaussian distribution, once the estimate (mean) and the error variance is known, the entire probability density function is known. The assumption of Gaussian distribution makes it very convenient to define the uncertainty by using only two parameters. However, we do not independently verify the validity of our assumption. If distribution is non-Gaussian, the error variance may not be enough to represent the uncertainty distribution.

- Quantification of uncertainty is dependent on the type of problem at hand. Some uncertainty can be tolerated—even accepted—whereas some uncertainty is undesirable. For example, if the minimum production required from a well is 50 RB/D, any production less than 50 RB/D is considered detrimental to the decision, much more so than a production greater than 50 RB/D. That is, the uncertainty is weighed differently depending on the problem. The error variance could provide us with a range of uncertainty, but might not provide a true quantitative impact of uncertainties. More sophisticated techniques could be used to quantify such an impact.

In the following sections, we provide alternate techniques of quantifying local uncertainties. We also discuss the advantages and disadvantages of the techniques.

4.4.1 Parametric Estimations. A simple technique for quantifying the uncertainty, at the same time maintaining the local variability, uses the parametric technique. This technique defines the uncertainty with Gaussian distribution but accounts for local changes.

As discussed in Numerical Example 4.1, the estimation at the unsampled location is not affected by the sill of the variogram model, as long as the other parameters are not changed. For example, if a variogram model is defined as

$$\gamma(\vec{L}) = 1.0 + 4.0M_{s_{2,000}}(\vec{L}), \qquad (4.67)$$

where the total sill value is 5.0, we can normalize the variogram by dividing the variogram equation by five.

$$\gamma_N(\vec{L}) = 0.2 + 0.8M_{s_{2,000}}(\vec{L}). \qquad (4.68)$$

The normalized variogram has a sill of one, but otherwise, it is the same model. If we use a normalized variogram, our estimates at unsampled locations are not affected; however, the error variance is different. With Eq. 4.64, we calculate

$$\left(\hat{\sigma}_E^2\right)_N = 1.0 - \sum_{i=1}^{n} \lambda_i C(\vec{u}_i, \vec{u}_0). \qquad (4.69)$$

The maximum value of normalized covariance for simple kriging is one, with appropriate reduction in the error variance, depending on the surrounding sample configuration. We calculate the error variance at every unsampled location, with Eq. 4.68 or something similar for other kriging techniques, and we scale the value appropriately, based on local variance. For example, if the variance of samples within a search neighborhood is s_l^2, we calculate the localized error variance as

$$\hat{\sigma}_E^2 = s_l^2 \cdot \left(\hat{\sigma}_E^2\right)_N. \qquad (4.70)$$

Eq. 4.70 reduces to local variance, if normalized error variance is one. This is consistent with the fact that maximum local error variance is equal to the variance of the sample. If the surrounding samples are well distributed, the error variance will be proportionately reduced.

Once the local error variance is calculated, if we assume Gaussian distribution, the local uncertainty is defined.[23]

Another possibility when using Gaussian distribution to describe uncertainty is to transfer the data into a Gaussian domain with normal score transform. Once the data are transferred into a Gaussian domain, we can further assume that uncertainty can also be described by Gaussian distribution.

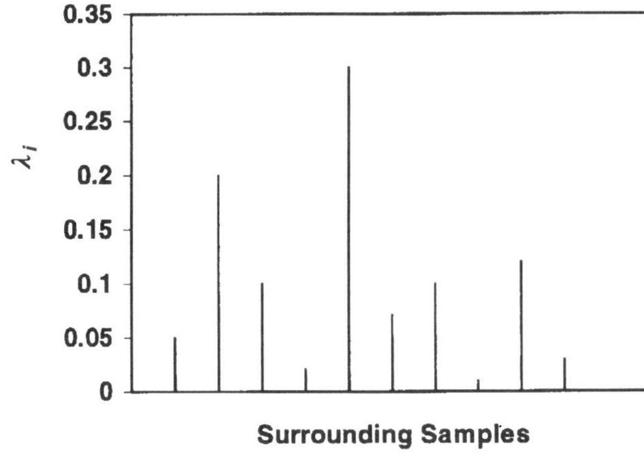

(a)

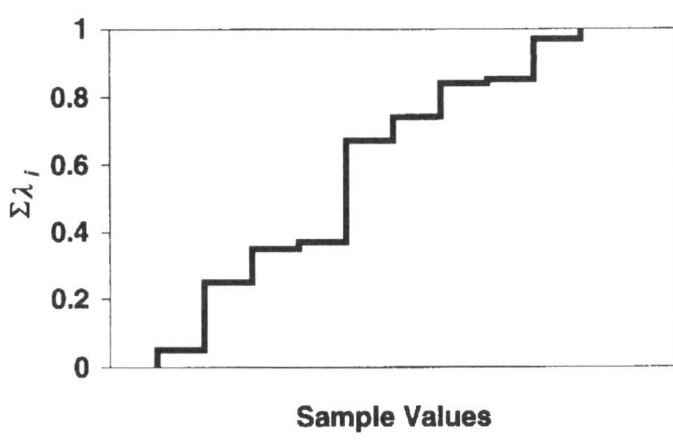

(b)

Fig. 4.61—Nonparametric estimates using kriging weights.

This assumption must be tested, and procedures are available in the literature.[24-25]

In practice, however, once the data are transformed into a Gaussian domain, it is assumed that the uncertainty can also be described by Gaussian distribution.

In the transformed domain, once the value and the error variance is estimated, we assume that

$$p\left\{X_G(\vec{u}_0)\varepsilon\left[X_G^*(\vec{u}_0) \pm 2\hat{\sigma}_E\right]\right\} = 0.95, \qquad (4.71)$$

where $X_G(\vec{u}_0)$ is the true value at the unsampled location and $X_G^*(\vec{u}_0)$ is the estimated value. We back-transform this uncertain region into an original domain to capture the local uncertainty at that location (Fig. 4.48). By knowing the uncertainty in the Gaussian domain, we can completely describe the uncertainty in the original domain.

Although easy to implement, the proposed method still requires an assumption of stationarity in the Gaussian domain. If we only transform data into a Gaussian domain, there is no guarantee that local variability will be eliminated, but the assumption of stationarity is satisfied. Despite this limitation, this technique remains popular because of its simplicity and speed.

4.4.2 Nonparametric Estimations. If the assumption of Gaussian distribution is too restrictive, an alternate approach is to use methods that do not require assumptions about the type of distribution that describes the uncertainty.

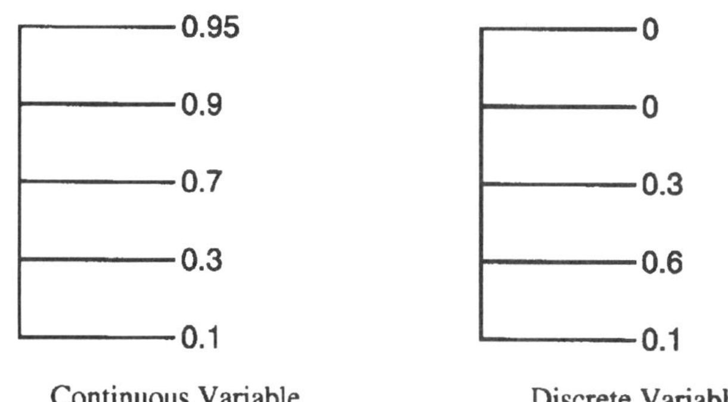

Fig. 4.62—Uncertainty representation using indicator values.

One approach, which describes the uncertainty, uses surrounding samples within a search neighborhood to capture the estimation of uncertainty. Recall that the value at the unsampled location is estimated by equation

$$x^*(\vec{u}_0) = \sum_{i=1}^{n} \lambda_i x(\vec{u}_i), \qquad (4.72)$$

where n samples are selected, and weights are assigned to each. These weights represent the probability that an unsampled location will assume that value. That is, because we require that the sum of all weights must equal one. We plot the probability histogram as shown in **Fig. 4.61.** The left side represents the weight associated with each sample, and the right side represents cumulative weights as a function of a sample value, starting with the smallest value. The right side is the cumulative distribution function of uncertainty associated with the unsampled location. The error variance is calculated as

$$\hat{\sigma}_E^2 = \frac{\sum \lambda_i [X_i - X^*(\vec{u}_0)]^2}{\sum \lambda_i}. \qquad (4.73)$$

Note, in the absence of any spatial relationship, the estimate is the local mean (all weights are equal), and the error variance is the local variance. This is consistent with our understanding of the limiting values. Instead of discrete distribution, if we require continuous distribution, we can use piece-wise linear interpolation between the discrete values.

This technique definitely has the advantage. It captures local variability much more precisely; however, it is much more cumbersome to use. Further, the uncertain distribution could depend on the search neighborhood, and an ad-hoc procedure might be needed to adjust the negative weights assigned to the samples.

Another procedure of quantifying uncertainty uses indicators. We discussed indicator kriging in the previous section. As described, indicator values represent the probability that the value at the unsampled location will be within a certain interval. For continuous variables, the indicator values represent the cumulative probability distribution, whereas for discrete variables, the indicator values represent the probability mass function, as shown in **Fig. 4.62.** For continuous variables, indicator values represent the probability that a value at that location will be less than a certain threshold. For discrete variables, indicator values represent the probability that a value will be equal to a particular category.

By separately modeling and estimating each indicator, continuity at each threshold is preserved. Because the indicator estimates depend on the samples within a search neighborhood, the local variability is captured in the estimates.

The disadvantage of indicator estimations is that the order of indicator values might not satisfy the probability requirements. For example, for continuous variables, indicator values might not be in nondecreasing order; the sum of the probabilities for discrete variables might not equal one; or the indicator values could be less than zero or greater than one. An arbitrary correction is required to satisfy the probability requirements. Further, for continuous variables, indicator transforms could create discontinuities and might not reproduce local uncertainties as well as Gaussian transform.[26]

4.4.3 Loss Functions. One of the questions associated with uncertainty is the best estimate at the unsampled location. The best estimate at the unsampled location is a function of uncertainty, as well as the procedure at hand. In the kriging procedure, we assume that the best estimate is the one that minimizes error variance. However, depending on a particular problem, we might come up with a different estimate. For example, if we are interested in drilling an exploration well, we would like to drill a well in a region with a high probability of success. If we would like to have a 90% chance of success, then the best estimate, at a particular location, would be the one that provides 90% probability of success. For Gaussian distribution of error, the best estimate corresponds to a 50% probability of success. Therefore, once the uncertain range is understood, we can calculate a value that corresponds to a 90% success rate. On the other hand, if we are interested in cleaning up a contaminated environmental site, to be conservative, we might prefer the best estimate to be the one with 5% probability of concentration to exceed a threshold value. We would like to clean up any site with even a small possibility of threshold concentration because the consequences of not cleaning are much more severe than cleaning an area that is already clean.

This type of best estimate is calculated with the concept of loss function. Loss function assigns a weight or penalty to the uncertain distribution. The more severe the consequences, the higher is the penalty assigned to that probability. The less severe the consequences, the less severe is the penalty. The objective is to minimize the penalty function. The value that minimizes the loss function is the best estimate.

For example, if we assume that x^* is the best estimate, we can define the loss function, $\psi(x^* - X)$, where X is the random variable. We calculate the expected value of the loss function as

$$E[\psi(x^* - X)] = \int_{-\infty}^{\infty} \psi(x^* - X) f(x) dx, \quad \ldots\ldots\ldots (4.74)$$

where $f(x)$ is the probability density function that represents the uncertainty at the unsampled location. For Gaussian distribution, the function is well known. If we have nonparametric, discrete, uncertain distribution, we can calculate the expected value as

$$E[\psi(x^* - X)] = \sum_{K=1}^{K} \psi(x^* - \bar{x}_k)[F(x_{k+1}) - F(x_k)], \quad \ldots\ldots\ldots (4.75)$$

where $\bar{x}_k$ is the average value within a class k, $F(x_k)$ is the cumulative distribution function for class k, and x_k is the threshold value of class k. The best estimate, x^*, is the one that minimizes the expected loss. Depending on the definition of loss function, the best estimate changes. For example, if $\psi(x^* - x) = |x^* - x|$ (i. e., absolute error), the best estimate is the median value.[27] If the loss function is

$$\begin{aligned} \psi(x^* - x) &= a(x^* - X) \text{ for overestimation} \\ &= b(x^* - X) \text{ for underestimation,} \end{aligned}$$

the best estimate is the value that corresponds to the cumulative distribution function of $b/a + b$. A good, practical example where such a loss function is used is provided in Ref. 28. If analytical solutions are not available, a numerical solution for the best value can be obtained through an iterative process.

Summary

This chapter covers the basic estimation techniques used in geostatistics. Depending on the type of data available, a particular method is used. If the mean value is known, simple kriging is used; if local variability is a problem, ordinary kriging is most applicable; if secondary variable data are available and related to primary data,[29] cokriging improves the estimation; if a trend in the data is visible, universal kriging accounts for the trend in the estimation process.

In addition to these four techniques, we discuss various nonlinear kriging techniques. These techniques provide improved solutions in the transformed domain. The advantages range from reducing the variability of the sample data set to better describing the local uncertainty.

In the last section, we provide different methods to improve the local uncertain estimates. We use either parametric or nonparametric techniques to estimate the uncertainty. If uncertainty is properly quantified, we can also obtain the best estimate at the unsampled location, depending on the objective of the estimation.

A common characteristic in all kriging techniques is the apparently smooth pictures the kriging process creates. After all, the kriging procedure is a filtering technique where small variabilities are eliminated and large-scale features are preserved. A potential difficulty in the kriging technique is its inability to properly capture the local uncertainties. Although some alternate methods are presented, they might not be adequate enough. In the next chapter, we introduce the technique of conditional simulation, which addresses some of these problems.

Nomenclature

A = surface area of a block
a_l = coefficient of $f_l(\vec{u})$
B_o = formation volume factor
$|C|$ = covariance matrix, as defined in Eq. 4.10
$|c|$ = covariance vector, as defined in Eq. 4.10
$C(0)$ = variance of sample data
$C(\vec{u}_i, \vec{u}_j)$ = covariance between two points located at $\vec{u}_i$ and $\vec{u}_j$, respectively
$C(\vec{u}_i, \vec{u}_o)$ = covariance between sample point i and unsampled location o
$C_c(\vec{L})$ = cross-covariance at distance $\vec{L}$
$C_t(\vec{L})$ = covariance of transformed variable at distance $\vec{L}$
C_{vv} = covariance of two block values
$E[\,]$ = expected value of bracketed term
$\underline{E}$ = expected value
$e(\vec{L})$ = error between estimate and observed value at $\vec{u}$
$e(\vec{u})$ = estimation error at location $\vec{u}_i$
$f_l(\vec{u})$ = trend function used in universal kriging procedure, Eq. 4.38
$f(x)$ = probability density function of variable x

$h =$ gross thickness
$I_R =$ index, rank
$I(\vec{u},x) =$ indicates transform of continuous variable
$I(\vec{u},K) =$ indicator transform of discrete variable
$i =$ dummy index
$K =$ discrete variable
$\underline{L} =$ lag distance, L, ft
$\vec{L} =$ lag distance vector
$m =$ mean
$m(\vec{u}) =$ local mean within a search neighborhood
$M_{s_a}(\vec{L}) =$ spherical variogram with range a
$n =$ number of sample points
$n_g =$ net to gross ratio
$n_b =$ number of points selected within a gridblock for calculating point-block covariance
$p =$ probability
$R(\vec{u}) =$ residual at $\vec{u}$
$s_l =$ variance of sample within a search neighborhood
$S_w =$ water saturation
Var[ ] = variance of bracketed term
$X(\vec{u}) =$ random variable X at $\vec{u}$
$x(\vec{u}) =$ realization of random variable x at $\vec{u}$
$Y(\vec{u}) =$ random variable Y at $\vec{u}$
$\gamma(L) =$ variogram at lag distance, L
$\phi =$ porosity
$\lambda =$ weight
$\mu =$ Lagrange multiplier
$\sigma^2[x] =$ variance of random variable x
$\hat{\sigma}_E^2 =$ error variance
$\psi =$ loss function
$|\Lambda| =$ vector of weights, λ_i

Subscripts

$G =$ Gaussian transform
$i =$ sampled location i
$N =$ normalized
$o =$ unsampled location
$t =$ threshold
$v =$ block estimate
$x =$ variable x
$y =$ variable y

Superscript

$* =$ estimate

References

1. Matheron, G.: "The Theory of Regionalized Variables and its Applications," Paris School of Mines, Cah. Cent. Morphologic Math., Fontainbleau, France (1965) **5.**
2. Deutsch, C.V. and Journel, A.G.: *GSLIB: Geostatistical Software Library and User's Guide,* Oxford University Press, New York City (1992) 30–34.
3. Journel, A.G.: *Geostatistics for the Environmental Sciences,* Project Report (No. CR 811893), Environmental Monitoring Systems Laboratory, Las Vegas, Nevada (1987) Chap. 6.
4. Isaaks, E.H. and Srivastava, R.M.: *An Introduction to Applied Geostatistics,* Oxford University Press, New York City (1989) Chaps. 14–15, 338–68.
5. Davis, B.M.: "Uses and Abuses of Cross-Validation in Geostatistics," *Math Geology* (1987) **19,** No. 3, 241–248.
6. Knudsen, H. and Kim, Y.C.: "Application of Geostatistics to Roll Front Type Uranium Deposits," 107th AIME Annual Meeting, Preprint No. 78-AR-94 (1978).
7. Journel, A.G.: "Geostatistics: Models and Tools for the Earth Sciences," *Math Geology* (1986) **18,** No. 1, 119–140.
8. Olea, R.A.: *Optimum Mapping Techniques Using Regionalized Variable Theory,* Kansas Geological Survey, Lawrence, Kansas (1975), 10–61.
9. Journel, A.G. and Rossi, M.E.: "When Do We Need a Trend Model in Kriging?" *Math Geology* (1989) **21,** No. 7, 715–738.
10. Myers, D.E.: "Matrix Formation of Co-Kriging," *Math Geology* (1982) **14,** 249–257.
11. Carr, J., Myers, D.E., and Glass, C.: "Cokriging: A Computer Program," *Computers and Geosciences* (1985) **11,** 111–127.
12. Xu, W. *et al.*: "Integrating Seismic Data in Reservoir Modeling: The Collocated Cokriging Alternative," paper SPE 24742 presented at the 1992 SPE Annual Technical Conference and Exhibition, Washington, DC, 4–7 October.
13. Araktingi, U.G. *et al.*: "Integration of Seismic and Well Log Data in Reservoir Modeling," in *Reservoir Characterization III,* B. Linville (ed.), PennWell Publishing, Tulsa (1993) 515–554.
14. Chambers, R.L., Zinger, M.A., and Kelly, M.C.: "Constraining Geostatistical Reservoir Descriptions with 3-D Seismic Data to Reduce Uncertainty," *Stochastic Modeling and Geostatistics,* J. Yarus and R.L. Chambers (eds.), AAPG Computer Applications in Geology, Tulsa (1994) No. 3, 143–157.
15. Journel, A.G. and Rossi, M.: "When Do We Need A Trend Model in Kriging?" *Math Geology* (1989) **21,** No. 7, 715–739.
16. Wackernagel, H.: *Multivariate Geostatistics,* Springer Verlag, Berlin (1995) Chap. 29.
17. Marechal, A.: "Kriging Seismic Data in Presence of Faults," *Geostatistics for Natural Resources Characterization,* G. Verly *et al.* (eds.), Reidel, Dordrecht, The Netherlands (1984) 271–294.
18. Journel, A.G.: "The Log Normal Approach to Predicting Local Distributions of Selective Mining Unit Grades," *Math Geology* (1980) **12,** No. 4, 285–303.
19. Verly, G.: "The Multi-Gaussian Approach and Its Applications to the Estimation of Local Reserves," *Math Geology* (1983) **15,** No. 2, 259–286.
20. Alabert, F.: "Stochastic Imaging of Spatial Distributions Using Hard and Soft Data," MS Thesis, Stanford University, Stanford, California (1987).
21. Journel, A. and Alabert, F.: "Non-Gaussian Data Expansion in the Earth Sciences," *Terra Nova* (1989) **1,** 123–134.
22. Srivastava, R.M.: "Reservoir Characterization With Probability Field Simulation," paper SPE 24753 presented at the 1992 SPE Annual Technical Conference and Exhibition, Washington, DC, 4–7 October.
23. Isaaks, E.H. and Srivastava, R.M.: *An Introduction to Applied Geostatistics,* Oxford Press, New York City (1989) Chap. 20, 489–524.
24. Borgman, L.: "New Advances in Methodology for Statistical Tests Useful in Geostatistical Studies," *Math Geology* (1988) **20,** No. 4, 383–403.
25. Journel, A. and Posa, D.: "Characteristic Behavior and Order Relations for Indicator Variograms," *Math Geology* (1990) **22,** No. 8, 1011–1025.
26. Hansen, K.M.: "The Use of Sequential Indicator Simulation to Characterize Geostatistical Uncertainty," Sandia Report SAND91-0758, Sandia National Laboratories, Albuquerque, New Mexico (1992).
27. Journel, A.G.: *Fundamentals of Geostatistics in Five Lessons,* American Geophysical Union (1989) **8,** Chap. 4.
28. Srivastava, R.M.: "An Application of Geostatistical Methods for Risk Analysis in Reservoir Management," paper SPE 20608 presented at the 1990 Annual Technical Conference and Exhibition, New Orleans, 23–26 September.
29. Kalkomey, C.T.: "Potential Risks When Using Seismic Attributes as Predictors of Reservoir Properties," *The Leading Edge* (March 1997) 247–51.

SI Metric Conversion Factor

ft × 3.048* E − 01 = m

*Conversion factor is exact.

Chapter 5
Conditional Simulation Techniques

Chap. 5 provides a bridge between conventional estimation techniques and conditional simulation techniques that are discussed in the next two chapters. This chapter also highlights the similarities and differences between conventional estimation techniques and simulation techniques and offers a preview of the various types of simulation techniques that have been used to describe reservoir properties.

5.1 Definition

The conditional simulation technique, conditioned by prior information, is a procedure that simulates various attributes at unsampled locations. Note the difference between conventional estimation techniques when attributes are estimated rather than simulated. In simulation techniques, the overall goal is to simulate reality rather than to obtain a picture of the reservoir, which minimizes error variance.

These techniques constitute a part of a broader class of simulation techniques and are called Monte Carlo simulations. Sometimes, a distinction is made between unconditional and conditional simulations. However, the word, "unconditional," is a misnomer. No simulation is strictly unconditional because some prior information is always available or assumed. In the absence of prior information, the solution domain is too large to obtain meaningful results. Instead, we can assume that the prior univariate distribution or prior spatial information is known, and based on this information, we attempt to construct the reservoir description. When a particular method in the literature is referred to as an unconditional simulation, it simply means that hard, prior data, at sampled locations, are not honored. In that sense, it is unconditional to prior sample data. Even unconditional simulations, in most cases, assume the prior univariate distribution of the sample, and the spatial relationships are known.

In general, the power of any conditional simulation technique is derived from its ability to condition, with as much information as possible, and, at the same time, minimize the computation that is required to simulate attribute values. These two requirements are often in conflict. As a result, depending on the quality and type of information desired to be honored, an appropriate conditional simulation procedure is selected. That method will emphasize a particular type of information, while ignoring or minimizing other types of information. To date, we do not have a procedure that properly integrates all the available information from a reservoir. As a compromise, we must select a conditional simulation procedure that prioritizes the type of information we consider most valuable.

5.2 Distinguishing Features

Three important features distinguish conditional simulation techniques from conventional estimation techniques. The next three sections explain these features.

5.2.1 Sample Variability. Traditional estimation techniques typically assume a linear relationship between the value to be estimated and the nearby samples. A relationship between the estimated value and nearby samples is given by

$$X^*(\vec{u}_o) = \sum_{i=1}^{n} \lambda_i X(\vec{u}_i), \quad \text{(5.1)}$$

where $X^*(\vec{u}_o)$ = the estimated value; $X(\vec{u}_i)$ = the sample located at $\vec{u}_i$; and λ_i = the weight assigned to the sample located at $\vec{u}_i$. In Chap. 4, we observed that the sum of the weights, assigned to individual samples, might be required to equal one, or

$$\sum_{i=1}^{n} \lambda_i = 1. \quad \text{(5.2)}$$

This requirement makes the weights, assigned to individual samples, less than one.

As a result of Eqs. 5.1 and 5.2, when a value is estimated at an unsampled location, it never reproduces the extreme values observed in the sample data. Even though one of the samples is extreme, the weight assigned to that sample is less than one, thus reducing its effect. When added to the weighted values of the other samples, the result is both less than the extreme high value and greater than the extreme low value. The estimated values always fall within the range of the extreme values. The only exception is when only one sample is within

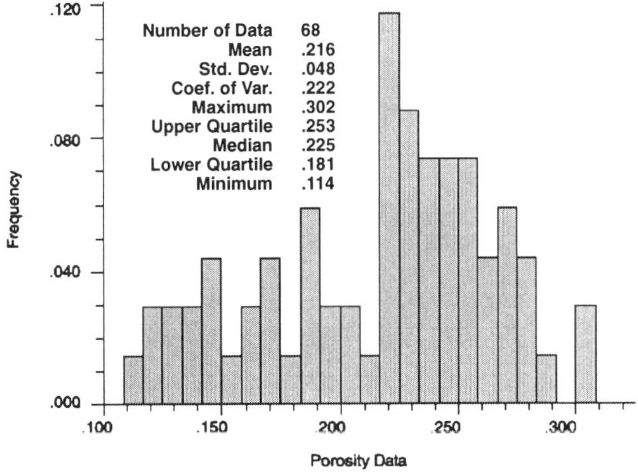

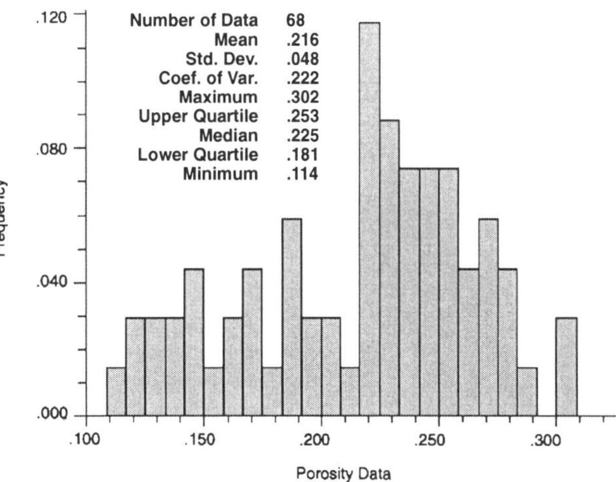

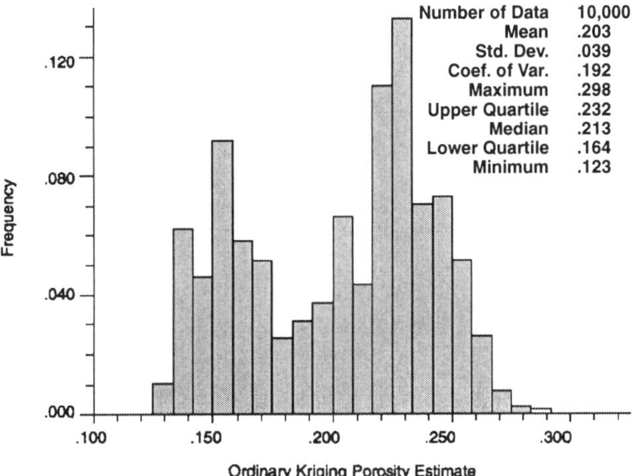

Fig. 5.1—Comparison between sample data distribution and kriging estimates for porosity in Flow Unit 3.

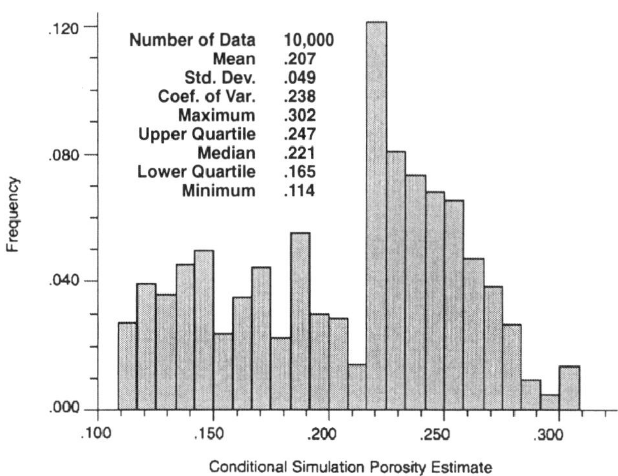

Fig. 5.2—Comparison between sample data distribution and conditional simulation realization for porosity in Flow Unit 3.

a search neighborhood, and it is an extreme value. In this case, the weight assigned to the sample point equals one, and the estimated value is equal to the extreme value.

Fig. 5.1 shows a comparison of the univariate distribution obtained from actual samples vs. the estimated values obtained with ordinary kriging. Note, although the overall distribution of the sample data might be well preserved, the maximum and minimum values observed in the sample data are not properly captured. These extreme values might constitute only a very small part of the overall distribution. Preservation of such a small part could become important, if these extreme values have a disproportionate impact on reservoir performance.

Consider an example of permeability data. If extreme permeability values are not preserved, we miss out on some very important information. High permeability values could form a thief zone between an injector and a producer. Although only a very small part of the reservoir, it influences the breakthrough of the injected fluid and the recycling effort that is required to operate the field efficiently. On the other hand, if we do not preserve the extremely low values of permeability, we will not be able to reproduce a vertical permeability barrier that isolates two producing zones from each other. Obviously, it is critical to preserve as much variability as possible.

The problem is compounded even further if the true univariate distribution of reservoir properties is unknown. We typically assume that the property distribution is obtained from the well data. This might not be true if the well locations are biased or partially sampled. Further, additional bias is created if core samples are taken only from selected intervals, rather than a random selection. It is important to consider the uncertainty associated with the unknown initial sample distribution.

Conditional simulation techniques offer this type of flexibility. The techniques honor the initial sample distribution, if desired. See **Fig. 5.2** for the comparison between the sample and the estimated values, using the conditional simulation technique. In addition, if the user desires a distribution of the estimated values, which is wider than the sample distribution, conditional simulation procedures will create a distribution of the estimated values, which mimics the input distribution. If preservation of the desired sample variability is important, conditional simulation techniques offer this choice.

5.2.2 Sample Spatial Relationship. In conventional geostatistical estimations, we input a spatial relationship based on the sample variogram analysis. This variogram is estimated based on the sample data and is modeled using appropriate constraints. After estimation at unsampled locations, if we use the estimated values to construct the variogram, we ob-

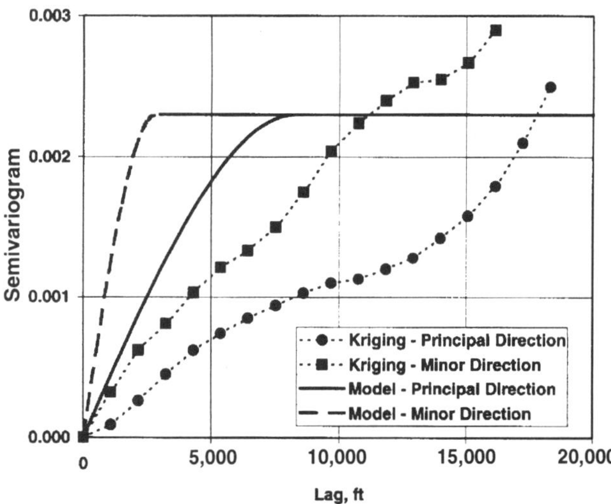

Fig. 5.3—Comparison between input variogram model and variogram of kriging estimates for porosity in Flow Unit 3.

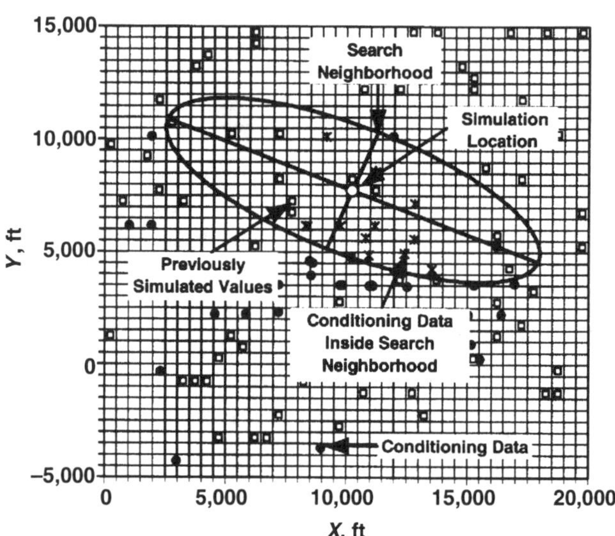

Fig. 5.4—Search neighborhood for the sequential simulation technique.

serve that the variogram, based on the estimated values, does not match the sample variogram. See **Fig. 5.3**, which illustrates the comparison between the variogram, based on the estimated values, vs. the input variogram. Notice that the variogram, based on the estimated values, has a smaller sill than the input variogram. Further, the structure of the input variogram is not precisely observed in the variogram, based on the estimated values.

Two reasons provide an explanation for this apparent discrepancy. One, the sill of a variogram is dictated by the variance of the attribute. The higher the variance, the higher is the sill. As we stated previously, the estimated values do not preserve the extremes. In other words, the estimated values exhibit less variability than the sample values. Because the estimated values have a smaller variance than the sample values, the sill value for the variogram, based on estimated values, is also smaller. Fig. 5.3 reflects this characteristic.

The second reason for this discrepancy is the way we estimate the value at the unsampled location in traditional techniques. Examine **Fig. 5.4,** where we are interested in estimating a value at the unsampled location denoted by "0." The figure also shows a search neighborhood and selected samples denoted by "*." In addition to these samples, the search neighborhood also contains prior estimated values within the search neighborhood. These estimated values are represented by "❑." These unsampled locations were already visited, and the values of the attributes were determined. When estimating the value at the unsampled location, we always honor the spatial relationship between the unsampled location and the sampled points. In geostatistical analysis, this is accomplished through the right side of the matrix equation. (See Chap. 4 for details.) Therefore, when we estimate the value at the unsampled location, we ensure that the spatial relationship between the unsampled location and the samples within the search neighborhood are accounted for when the weights to the sampled locations are assigned. In contrast, prior estimated values, within the search neighborhood, are ignored when the value at the unsampled location is estimated. Therefore, the spatial relationship between the estimated value and the prior estimated values within the search neighborhood is not explicitly accounted for in the estimation process. However, because the estimates, at prior estimated locations, are based on nearby samples, some aspect of the spatial relationship between the prior estimated values and the current one (through the use of overlapped samples within search neighborhoods) is implicitly included. This might not, however, be sufficient when we generate a variogram based only on the estimated values. As a result, when variograms based on the estimated values and the one input in the estimation process are compared, they show differences.

Conditional simulation techniques account for the spatial relationships among the estimated values by modifying the conventional estimation process or by employing different types of algorithms. See **Fig. 5.5** for an example. We discuss some of these techniques in the next two chapters.

5.2.3 Quantification of Uncertainty. One of the desired goals, when applying geostatistical techniques, is to ascertain the uncertainty associated with the estimation. We discussed the concept of uncertainty using conventional geostatistical techniques, where the uncertainty is quantified through the error variance. If the assumption of a Gaussian distribution is valid, then the uncertainty at any unsampled location is com-

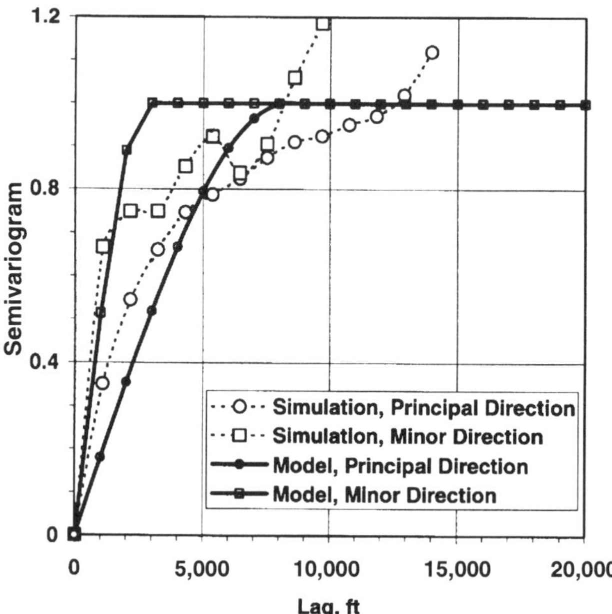

Fig. 5.5—Comparison between input variogram model and variogram of conditional simulation realization for porosity in Flow Unit 3.

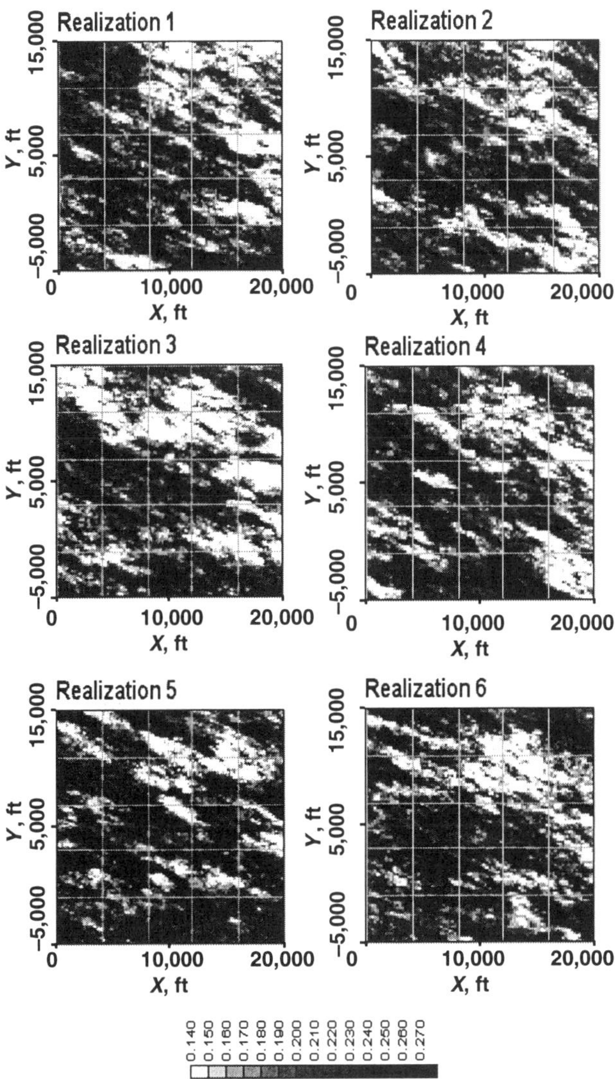

Fig. 5.6—Six equiprobable realizations generated with conditional simulation for porosity in Flow Unit 3.

pletely described through the estimate and the error variance. Unfortunately, as we discussed in Chap. 4, the error variance is a good indication of surrounding sample configurations, but it might not be a good indication of local uncertainty. In other words, it might not account for the local uncertainty, which resulted from local variabilities within the region of interest.

Conditional simulation provides for the local variability by creating alternate, equiprobable images. Instead of defining an estimate and the associated uncertainties, uncertainty is characterized by multiple possibilities that exhibit local variation.

A good analogy, which describes the application of conditional simulation, is a police-artist's rendering of a robber, based on a witness' description. If a bank robbery takes place during broad daylight and several witnesses see the robber leaving the scene of the crime, the police can collect a significant amount of information from them. When this information (i.e., brown eyes, dark hair, etc.) is provided to the various resident artists, they can draw a picture of the robber. The information, which is certain, must be honored, no matter who draws the picture. Information that is missing is filled in with the artists' imaginations. If five artists' pictures of the robber are compared, many similarities will be present, as well as some differences. The more information provided, the more similar are the pictures. If the robbery takes place at night, and only one witness sees the robber in the dark, the same five artists' pictures will look quite a bit different because there is more artistic freedom in drawing an image with limited information.

For reservoir description purposes, the conditional simulation procedure acts as an artist. It allows for the alternate drawings of the reservoir. The more information we have about the reservoir, the more alike the images will look. The less information we have, the fewer are the constraints we can impose on the reservoir description and, thus, the more different the various images are from one another. Just like the artists' renditions, all the images created by the conditional simulation are equally likely to occur. Just like each artist's work, each of the images is created based on the same amount of information. Because all images honor the same constraints, we cannot predict one image as more likely to occur than any other.

By constructing multiple images, we can visualize the differences among images. See **Fig. 5.6,** which exhibits six images of the same distribution. If the local uncertainty is small, a number of images will exhibit similar values at a particular location (narrow distribution), whereas if the local uncertainty is large, the differences, in simulated values at a particular location, will vary over a wide range. By examining the images, as well as the local distribution, we can ascertain the uncertainty at each location. If we create a sufficient number of images, one of the images could match the true case. This is another difference between the conventional estimation process and the simulation process. The conventional estimation process never reproduces reality; it is a filtering process, which removes high frequency information. See **Fig. 5.7** for a one-dimensional (1D) example and **Fig. 5.8** for a two-dimensional (2D) example. On the other hand, in the conditional simulation process, the high frequency information, as well as the low, is preserved (Fig. 5.7). If we create sufficient images similar to ones shown in Fig. 5.7, we might produce a true image—the conditional simulation technique could reproduce a true image. A less ambitious, but nevertheless important, goal is to create a sufficient number of images of the reservoir property, so the possible, required outcomes, from various images, can bind a true image. For example, if we construct alternate images of the property so that all images honor prior, historical production data, future predictions, based on the various images, will exhibit some variability. If all future predictions are plotted on the same graph, they will form bounds of the possible solutions. If our quantification of uncertainty is representative, the true image will fall within the bounds.

This ideal of capturing a true image within the bounds of multiple images is not easily achieved. The difficulty results from the same problems we discussed in Chap. 4, regarding the quantification of local uncertainties. Effectively, the conditional simulation method is a sampling from a multivariate distribution, which represents the distribution of individual, local, attribute values, with reference to other attribute values. If, for example, we have 10,000 gridblocks, then various properties, at each gridblock, are related to each other. Therefore, we have a 10,000-variate distribution for each property. Conditional simulation samples from this distribution for each of the realizations or images. The knowledge of such a 10,000-variate distribution might not be precisely

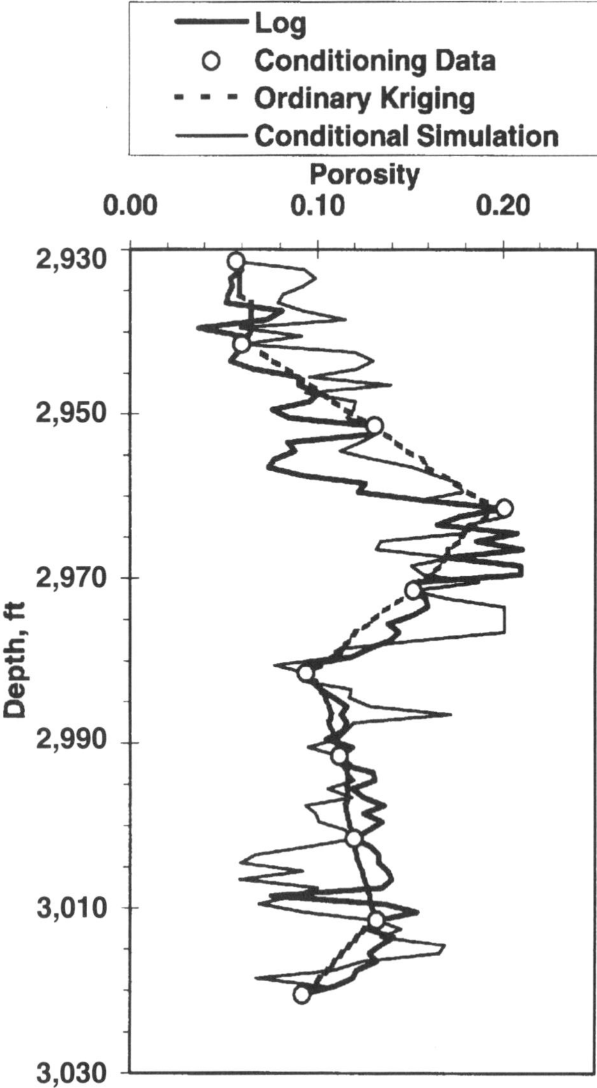

Fig. 5.7—Comparison of true distribution for the porosity log of well 50-30, kriging estimates, and conditional simulation realization.

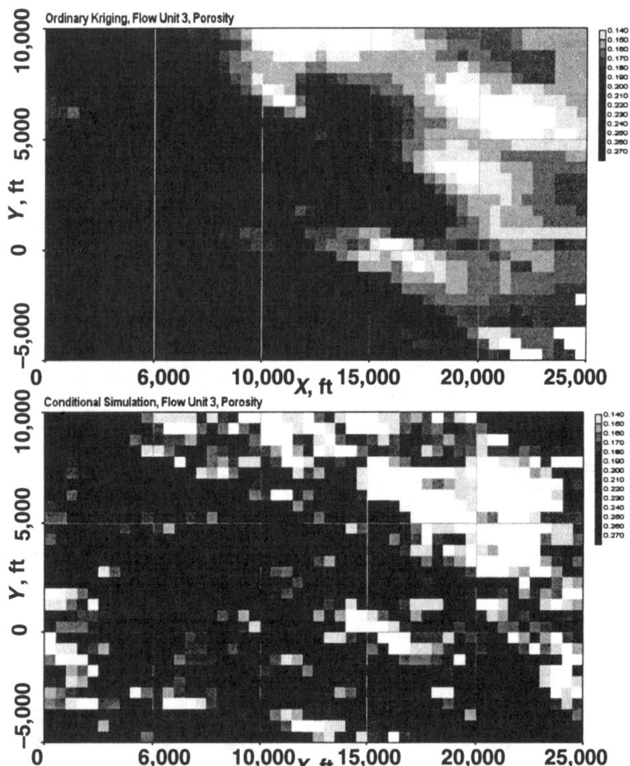

Fig. 5.8—Comparison of kriging estimates and conditional simulation realization for porosity in Flow Unit 3.

known. As a result, even if we create multiple images and the multiple images correctly represent the multivariate distribution, they might not bind a true image.

Because we may never be able to assess the multivariate distribution precisely, our goal is to incorporate as much information as possible in the overall reservoir description process. By forcing competing constraints to be satisfied, our images are forced closer to reality. Not necessarily reality, in the sense of truth, but a less modest goal, which ensures the range of our desired outcome is reasonably captured. For example, if our goal is to capture the future production performance of the reservoir, then if we generate the images that reasonably capture the range of future performances, we are satisfied. Our distribution of properties in the various images could be very different from the truth, but the truth may never be known. However, in terms of production performance, these images provide the range of possible uncertainties.

5.3 Simulation Methods

Several types of conditional simulation methods are reported in this text. Many of these methods are detailed in the next two chapters. Sec. 5.3.1 briefly discusses two broad classifications of conditional simulation techniques that describe reservoir properties.

5.3.1 Grid-Based Simulation Methods. As the name indicates, grid-based simulation techniques assume the reservoir can be discretely divided into several gridblocks. Within individual gridblocks, we assume the properties are homogeneous. Therefore, the gridblocks must be adequately fine, if we desire a more detailed description.

Although unnecessary, the majority of grid-based simulation techniques honor at least two types of prior constraints. The first is the univariate distribution of reservoir properties, and the second is the bivariate distribution of the reservoir properties through spatial relationships. In addition, information such as relationships among multiple attributes, multivariate spatial relationships, and relationships between the static and dynamic data, can also be honored. Grid-based simulation techniques are flexible enough to simulate either discrete variables (i.e., geological facies), or continuous variables (i.e., porosity or permeability). Grid-based simulation techniques can also honor the various local relationships among various attributes through a procedure commonly known as the cosimulation technique. For example, **Fig. 5.9** shows 2D images of the reservoir for three different attributes. The first image represents the geological facies; the second image represents the porosity data, and the third image represents the permeability. The dependence of porosity on the geological facies and the dependence of permeability on the porosity are already captured in these images.

The major disadvantage of grid-based simulation techniques is the assumption that all reservoir properties can be conveniently defined with gridblock structures. This assumption might not hold true for geological objects with complex shapes and sizes. These objects also have certain spatial distribution arrangements, with respect to other objects. For ex-

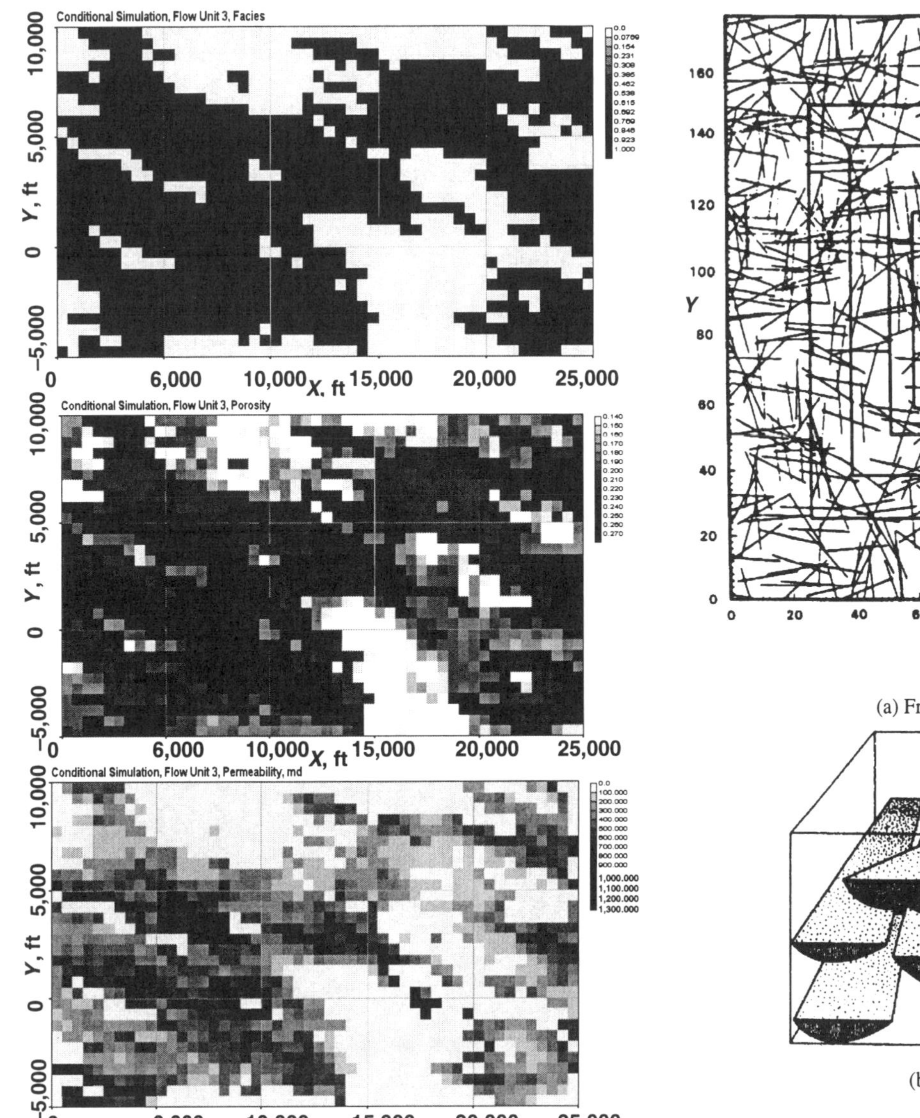

Fig. 5.9—Conditional simulation of facies, porosity, and permeability for Flow Unit 3.

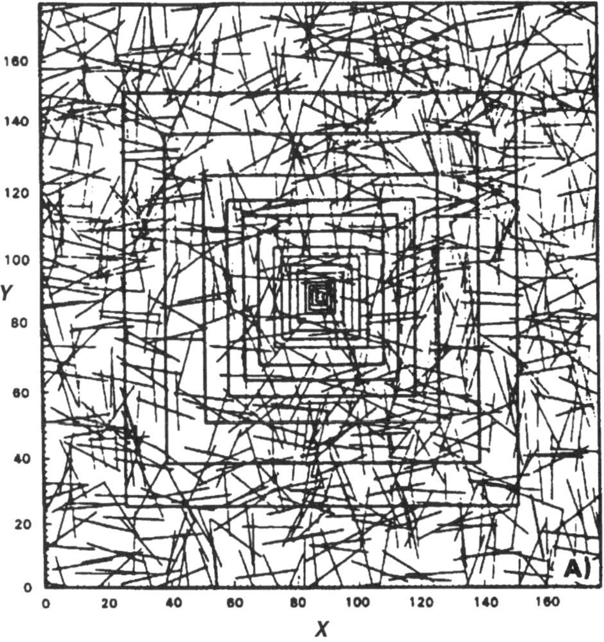

(a) Fracture Network

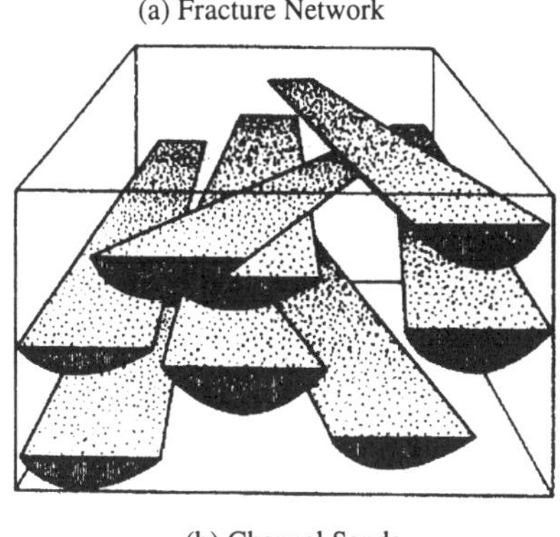

(b) Channel Sands

Fig. 5.10—Examples of stochastic modeling (from Ref. 1).

ample, a crevasse splay is always attached to a channel sand, or fractures, perpendicular to each other, either merge into one another or intersect each other, depending on the stress fields. Although, in principle, any object can be simulated with fine grids; however, the question still remains whether this is the most efficient way to capture the geological architecture of the reservoir. For example, consider a complex stack of channel sands. These channel sands overlap each other, and they could have similar shapes. To capture the shapes of these channels, conventional spatial relationships, such as variograms, could be inadequate. Another example is a naturally fractured reservoir. A reservoir, comprising natural intersecting fractures through a relatively low permeable matrix, is difficult to capture with grid-based simulation techniques. To capture very fine fractures, very fine grids must be used; however, a fine grid could inefficiently describe the matrix. We cannot define the grid beforehand either because we do not know where the fractures will be simulated before we start the conditional simulation process. By forcing the fit of nonorthogonal objects into grids, the shapes could be improperly reproduced and, at the same time, make the simulation process inefficient.

5.3.2 Object-Based Simulation Techniques. Unlike grid-based simulation techniques, where attributes are defined on a gridblock scale, object-based simulation techniques allow the simulations of various objects with different shapes and sizes. Depending on the type of simulation, objects are either two-dimensional (2D) or three-dimensional (3D).

The information, typically honored by object-based simulation, includes size, shape distribution of various geological objects, proportions of various objects in a 3D space, and various spatial arrangements of objects with respect to each other. For example, if one is interested in generating a deltaic sand environment, the simulation honors the channel sands width, height, and sinuosity distributions; the proportions of sand vs. shale; and the requirement of crevasse splay attachment to channel sands. See **Fig. 5.10** for an illustration.

The major advantage of object-based simulations is their ability to honor the shape and size distributions of various objects. Once these are known, we can populate 3D space very quickly, until the desired proportions of individual objects are reached. By being unconfined to gridblocks, the appropriate scales and shapes (i.e., fractures vs. matrix) are preserved and generated very efficiently. Further, by properly accounting for the spatial arrangements among the various objects, the geological architecture is preserved in a consistent manner.

Here, we describe the major disadvantages of object-based simulation.

• It is difficult to condition data at various well locations. Because the shape and size of an object are selected, based on the prior distribution function, it is difficult to ensure that the object will pass through the well at the appropriate location, so the information at the well is honored. Various procedures have been established to accommodate this shortcoming.

• We assume that the shape and size distributions of objects are known. This information is collected from well data, outcrop studies, geological knowledge, and other relevant information. The object size might be well-preserved in the subsurface environment. However, it is possible that diagenetic overprinting is significant, so the petrophysical properties within the objects are mostly influenced by diagenetic effects and not by the type of object. If diagenetic effects have a significant impact on altering the shape or size of the original objects, then the alterations of those shapes and sizes are difficult to reproduce.

• Ultimately, if our goal is to use flow simulation to predict reservoir performance, we must define petrophysical properties at a gridblock scale. Transformation techniques from object-based simulation to gridblock-based petrophysical properties within each object are known, and the objects are eventually fit into individual gridblocks through approximation. Obviously, this problem is eliminated, if we also conduct a flow simulation in an object-based mode. If the availability of such a simulator is absent, we must go through an additional step, which transforms object-based simulations into grid-based properties.

Summary

Chap. 5 introduces the concept of conditional simulation techniques. This procedure constructs alternative images of a reservoir. All images are constrained to prior information. Conditional simulation techniques have three major advantages over conventional interpolation techniques. They are the ability to preserve sample variability, preserve spatial relationships, and quantify uncertainty through multiple images.

The two conditional techniques that we commonly use with the reservoir description process are grid-based simulation and object-based simulation. Grid-based methods emphasize the generation of properties consistent with discretized gridblocks; object-based methods emphasize the generation of properties consistent with the shapes and sizes of geological objects. Both methods offer unique advantages and disadvantages. We provide additional details regarding these techniques in the next two chapters. Ultimately, choosing the most appropriate method depends on the overall goal of the reservoir description.

Nomenclature

$i =$ sample i
$n =$ number of samples within search neighborhood
$x =$ variable value
$\lambda_i =$ weight assigned to sample i
$\vec{u}_i =$ location of sample i
$\vec{u}_o =$ unsampled location

Subscript

$i =$ sample i

Superscript

$* =$ estimate

Reference

1. Haldorsen, H.H., Brand, P.J., and MacDonald, C.J.: "Review of Stochastic Nature of Reservoirs," *Mathematics in Oil Production,* S.S. Edwards and P.R. King (eds.), Clarendon Press, Oxford, U.K. (1988).

SI Metric Conversion Factor

ft × 3.048* E−01 = m

*Conversion factor is exact.

Chapter 6
Grid-Based Simulation Methods

In this chapter we discuss grid-based simulation methods. These methods allow the simulation of attributes at gridblock locations. Several methods are reported in the literature. We will, however, only discuss the techniques that receive significant attention and are used in practice. The two techniques we cover are sequential simulation methods and simulated annealing. Both techniques are widely used in the literature and have demonstrated their utility for field studies. After reviewing the basic principles of these techniques, we discuss the relative advantages and disadvantages.

6.1 Sequential Conditional Simulation Methods

Sequential conditional simulation methods are kriging-based methods, in which unsampled locations are sequentially visited in random order until all unsampled points are visited. At each unsampled location, a value of the desired variable is simulated, based on the estimate, as well as the local uncertainty. Multiple realizations are possible through the order in which unsampled locations are visited, as well as the way in which a value is sampled (simulated) at the unsampled location.

The two methods most commonly used for sequential simulation are sequential indicator simulation (SIS)[1,2] and sequential Gaussian simulation (SGS).[3-5] Both methods are extensively used for simulating discrete and continuous variables. The overall steps involved with both procedures are similar. Therefore, we discuss both methods simultaneously, stating the relative advantages and disadvantages of each technique. After discussing these techniques for a single variable, in the subsequent section, we discuss the application of sequential simulation methods for multiple variables. The application to multiple variables is called cosimulation, which is closely akin to cokriging.

In addition to indicator and Gaussian transforms, another method used to transform the original data is the probability transform.[6] Because of its simplicity, this method is also becoming popular. We briefly review probability simulation as well.

6.1.1 Single Variable Simulation. In this section, we discuss the sequential simulation of a single variable. The application of this method requires five steps. These steps involve transforming the original data into a new domain, modeling variograms in the transformed domain, determining a randomly-selected path to visit all the unsampled locations, sequentially estimating values at the unsampled locations with kriging-based techniques, and back-transforming the values to the original domain. A detailed discussion of these steps follows.

Transformation Into a New Domain. This step involves transforming the original data into a new domain. The primary purpose of this step transforms the data into a domain in which the uncertainty at the unsampled location is properly quantified. The two most popular transformations are indicator transform and Gaussian transform. We discussed both of these transforms in Chap. 4, in the section on nonlinear kriging. Indicator transforms are defined as

$$\begin{aligned} I(\vec{u}_j, x_t) &= 1, \text{if } X(\vec{u}_j) \le x_t \\ &= 0, \text{if } X(\vec{u}_j) > x_t \end{aligned} \quad \text{.................. (6.1)}$$

for a continuous variable, and

$$\begin{aligned} I(\vec{u}_j, K_t) &= 1, \text{if } K(\vec{u}_j) = K_t \\ &= 0, \text{if } K(\vec{u}_j) \ne K_t \end{aligned} \quad \text{.................. (6.2)}$$

for a discrete variable. Both of these transformations are used for the hard data (known values with certainty) and the soft data (incomplete information). Indicator values represent a degree of certainty with respect to measured values. For a continuous variable, the indicator values, at different thresholds, capture the cumulative distribution function, whereas for a discrete variable, the indicator values, at different thresholds, represent the probability density function. **Fig. 6.1** represents the examples of indicator transforms for both continuous and discrete variables. As shown, for continuous variables, the assignment of either zero or one indicates a complete certainty in the value. A value of one indicates 100% probability that the sample value is less than a given threshold, whereas a value of zero indicates 100% probability that it is greater than a given threshold. The overall sequence of values represents a cumulative distribution function, always in a nondecreasing order.

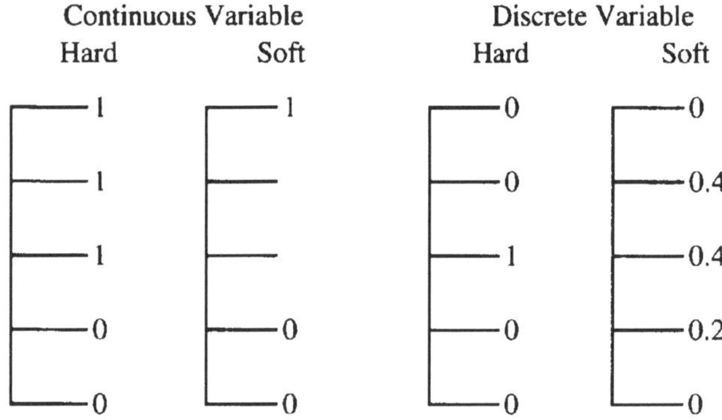

Fig. 6.1—Indicator transform of hard and soft data.

For a discrete variable, an indicator value of one indicates 100% certainty that a particular category of the discrete variable is present, whereas a value of zero indicates 100% certainty that a particular category is absent. A value between zero and one indicates the degree of certainty with respect to a particular category. The overall sequence of indicator values represents a probability density function, with the sum of all values equaling one.

Similar to the indicator transform, the Gaussian transform is also possible for both continuous and discrete variables. For continuous variables, all sample values are arranged in ascending order, and a cumulative distribution function (CDF) of the samples is constructed. The CDF varies between zero and one. For example, a sample that is ranked i among the n samples has a CDF value of

$$F(x) = \frac{i}{n+1}. \qquad (6.3)$$

Once the CDF of the sample value is constructed, the transform into a Gaussian domain is straightforward, and is illustrated in **Fig. 6.2.**

For a discrete variable, the transformation into Gaussian space first requires a transformation of the probability density function into a cumulative distribution function. Once a CDF is constructed, the transformation into Gaussian space is similar to the procedure for a continuous variable. Schematically, this procedure is shown in **Fig. 6.3.**

To explain Fig. 6.3, assume that the discrete variable is represented by A, B, C, and D (e.g., geological facies). Each category has an associated probability density (mass) function, $p(K)$. Once the order of the categories is determined—this is arbitrary—we calculate the cumulative distribution function, $F(K)$. In our example, the order is A, B, C, and D. It can be something else as well, as long as we know the order during the back transformation.

The transformation from cumulative distribution function to Gaussian variable follows the procedure similar to a continuous variable, except, in this case, we use a truncated Gaussian distribution. We define a minimum and maximum value in the Gaussian space. For example, it could be -3 and $+3$. We do not allow the value in the transformed domain to be less than -3 or greater than $+3$. Once the limits are defined, we can easily define the boundaries of each category in the Gaussian space. For example, as shown in Fig. 6.3, category A corresponds to values in the Gaussian space between -3 and -1.2, category B corresponds to values in the Gaussian space between -1.2 and 0, etc. With these limits, we randomly assign any value in the Gaussian space to represent a particular category. This type of random assignment could create random neighborhood values. This randomness can be avoided by using correlated numbers for sampling.

In addition to indicator and Gaussian transforms, another transform is also possible. This transform is called probability transform. This transform is simpler than either indicator or Gaussian transforms. For a continuous variable, once the cumulative distribution function is created (see Eq. 6.3), we use the CDF as a transform variable with values between zero and one. For discrete variables, once a cumulative distribution function is constructed (see **Fig. 6.4**), a value is randomly sampled within the limits and assigned. For example, for category A, once its CDF is constructed, we can randomly assign a value between 0 and 0.2; for category B, we randomly assign a value between 0.2 and 0.5, etc.

Comparing the three transforms discussed, we can make some general comments about these transforms. In general, the indicator transform is more flexible than the other two transforms. It has the capability to incorporate hard as well as soft data when assigning values at the unsampled locations. The indicator transform, however, has the disadvantage of losing some information when continuous data are transformed into indicator values. Once the data are transformed, we might know the class interval where those data are located; however, the precise value within a given class is not known. In contrast, for both Gaussian and probability transforms, there is no loss of information when either discrete or continuous data are transformed. The back transform from these two transforms provides the same original value of the variable. However, both Gaussian and probability transforms do not allow incorporation of soft information. Further, in going from a discrete to a Gaussian transform, or to a probability transform, artificial randomness might be introduced. We must consider these relative advantages and disadvantages when applying a particular transform.

Spatial Modeling in a Transformed Domain. Once the data are transformed into a new domain, the second step involves modeling the spatial relationships in that domain. This step is similar to any conventional variogram analysis in which the variogram, in the transformed domain, is estimated and modeled with an appropriate relationship.

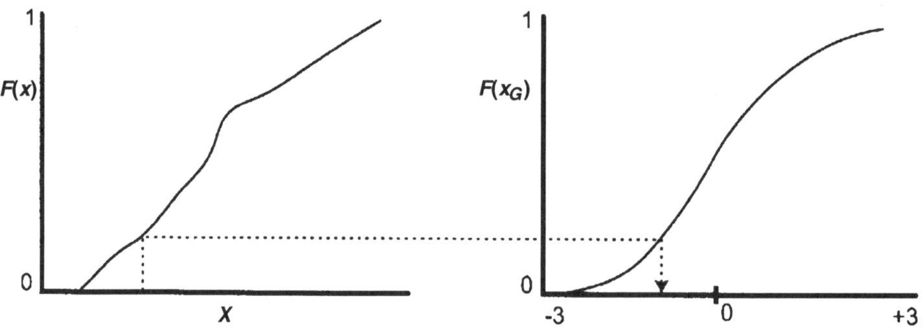

Fig. 6.2—Gaussian transform of a continuous variable.

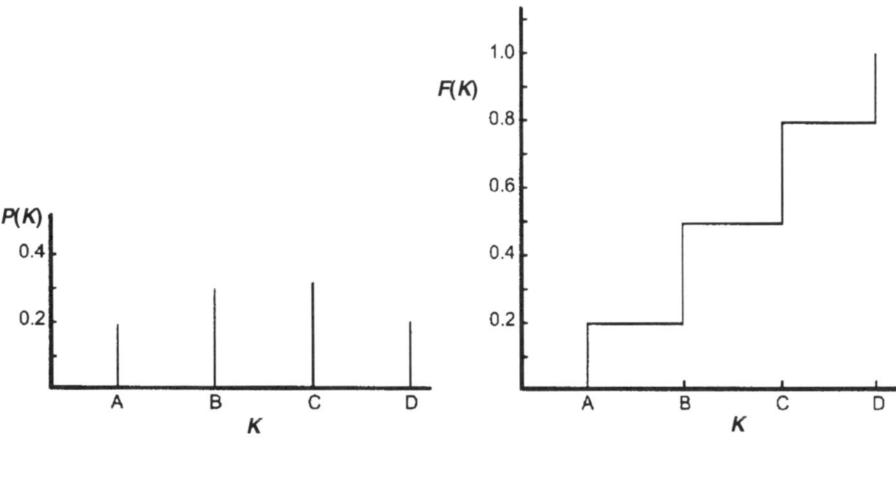

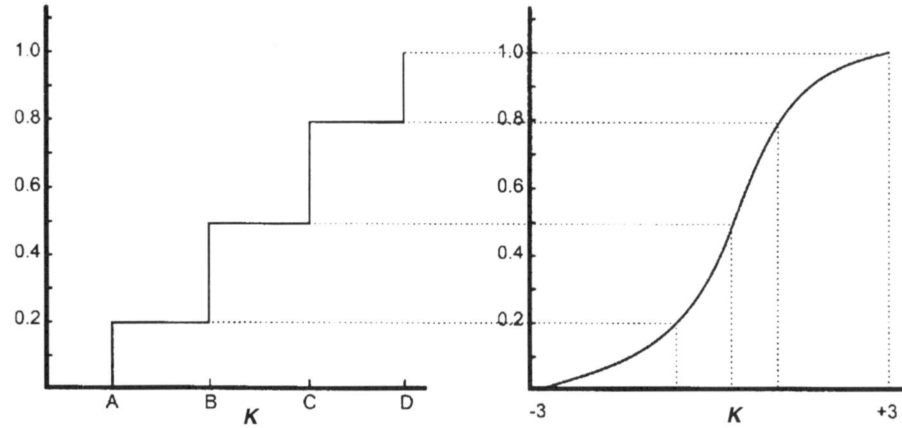

Fig. 6.3—Gaussian transform of a discrete variable.

If an indicator transform is used, the number of variograms, which must be estimated and modeled, is equal to the number of threshold values. For discrete variables, the number of threshold values is equal to the number of categories. For continuous variables, the number could vary, depending on the number of sample points, as well as on any external physical reason, which requires a particular division. In contrast to the indicator transform, for both Gaussian and probability transforms, only one variogram must be modeled.

The simplicity in modeling only one variogram for both Gaussian and probability transforms is contrasted with the flexibility of modeling different variograms for different thresholds with the indicator transform. If different thresholds exhibit different spatial relationships, the indicator transform allows us to capture these relationships through the use of different variograms at each threshold. On the other hand, if spatial relationships are not significantly different at different thresholds, Gaussian and probability transforms offer the simplicity of estimating and modeling a single variogram.

Random Path Selection. Once the variograms are estimated and modeled, the third step in sequential simulation involves a selection of a path in which every unsampled location is visited. Typically, with the help of a random number generator, a sequence of random numbers corresponding to

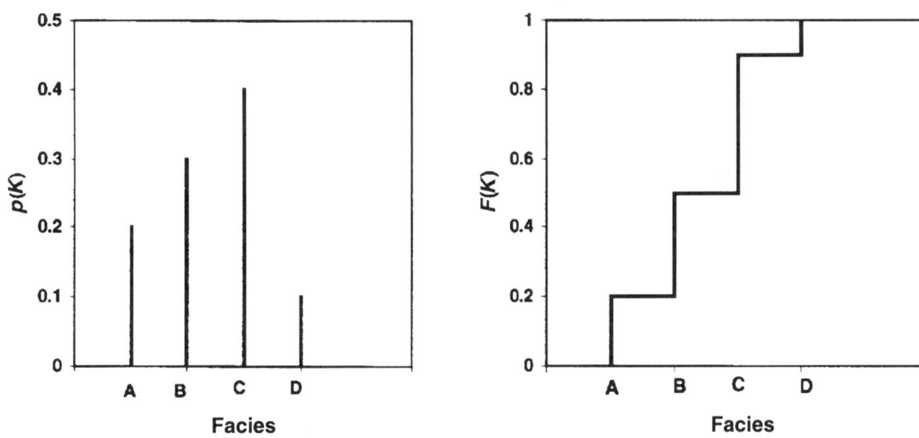

Fig. 6.4—Probability transform of a discrete variable.

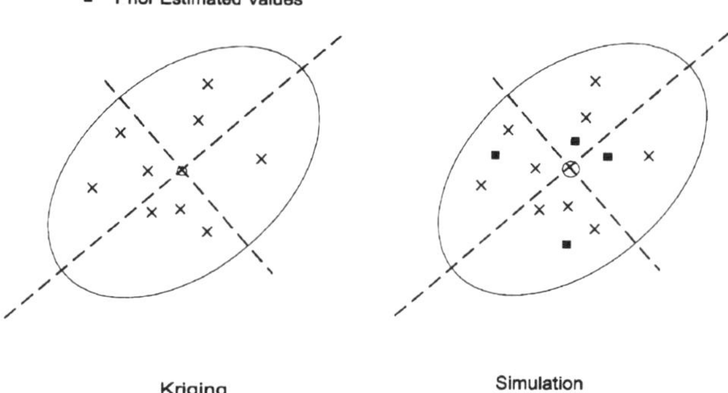

Fig. 6.5—Selection of samples.

the total number of gridblocks is generated, and based on the order, a path is selected in which all unsampled locations are visited. The reason for selecting a random path, when visiting the unsampled locations, rather than visiting all the unsampled locations in a sequential order, becomes clear in the next step.

Estimation at the Unsampled Location. When a randomly selected unsampled location is visited, a search neighborhood is defined, as in conventional kriging, and the sampled points within the search neighborhood are gathered. An important distinction between conventional kriging and the sequential simulation technique is the selection of samples within the search neighborhood. In addition to selecting the sampled points within the search neighborhood, sequential simulation also selects previously simulated values. See **Fig. 6.5** for the difference between the conventional kriging techniques and the conditional simulation technique. As illustrated, the conventional kriging technique only selects the original sample points within the search neighborhood. In contrast, conditional simulation techniques select the original samples as well as prior simulated values within the search neighborhood.

The difference in the selection of samples between conventional kriging vs. the conditional simulation technique has two consequences. First, by selecting the previous simulated values within the search neighborhood, the spatial relationship among the simulated values is explicitly honored in the conditional simulation technique. This is not possible in a conventional kriging technique. Second, the selection of prior simulated values makes the estimation at unsampled locations dependent on the order in which unsampled locations are visited. For example, if we start in an area where our original sample data exhibit high values, the simulated values will also be high. Progressing from that area into other areas results in estimates that are high, reflecting the prior simulated high values. This type of bias, to some extent, is avoided by randomly selecting the sequence in which the unsampled points are visited. In addition, we further control this bias by selecting a certain maximum number of prior simulated values within a search neighborhood. By maintaining a good balance between the original samples and prior simulated values within a search neighborhood, we ensure that our estimation is not overwhelmed by prior simulated values as the simulation progresses.

Once the samples within the search neighborhood are selected (original samples plus the prior simulated values), the estimation at the unsampled location requires an application of conventional kriging techniques. If we use ordinary kriging, for the indicator transform, we estimate the value at the unsampled location by

$$I^*(\vec{u}_0, x_t) = \sum_{j=1}^{n} \lambda_j I(\vec{u}_j, x_t) \quad \text{.................... (6.4)}$$

for a continuous variable, and

$$I^*(\vec{u}_0, K_t) = \sum_{j=1}^{n} \lambda_j I(\vec{u}_j, K_t) \quad \text{.................. (6.5)}$$

for a discrete variable. In Eqs. 6.4 and 6.5, $I(\vec{u}_j, x_t)$ and $I(\vec{u}_j, K_t)$, respectively, represent the original samples, as well as prior simulated values. The procedure must be repeated at each threshold with an appropriate spatial relationship. For example, for three thresholds, we must estimate three indicator values with three distinct indicator variograms.

We estimate, at an unsampled location, a direct representation of the local uncertainty, if we use the indicator transform. **Fig. 6.6** represents some examples of indicator estimates at the unsampled locations.

This figure illustrates that, for a continuous variable, we obtain a cumulative distribution function representing a local uncertainty, and for a discrete variable, we obtain a local probability density (mass) function. By comparing the three estimates (a, b, and c), we can conclude that "a" and "c" represent less uncertainty than "b." Estimate "b" represents significant local uncertainty because all the possibilities, at the unsampled location, have significant probabilities associated with them. In contrast, for estimates "a" and "c," a 90% probability is associated with only one outcome. For a continuous variable, in "a," there is a 10% probability that the value is less than the first threshold, whereas in "c," there is a 90% probability that the value is greater than the last threshold. Similar observation is made for the discrete variable estimates, in which estimates "a" and "c" represent a 90% probability that the outcome is the category with an indicator value of 0.9

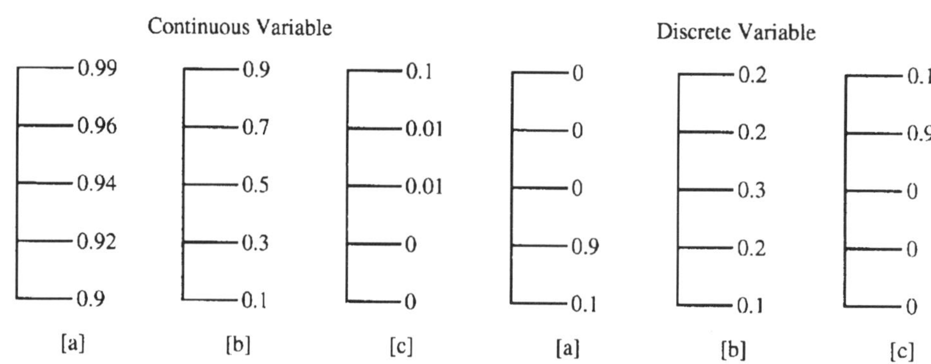

Fig. 6.6—Estimated indicator values at the unsampled locations.

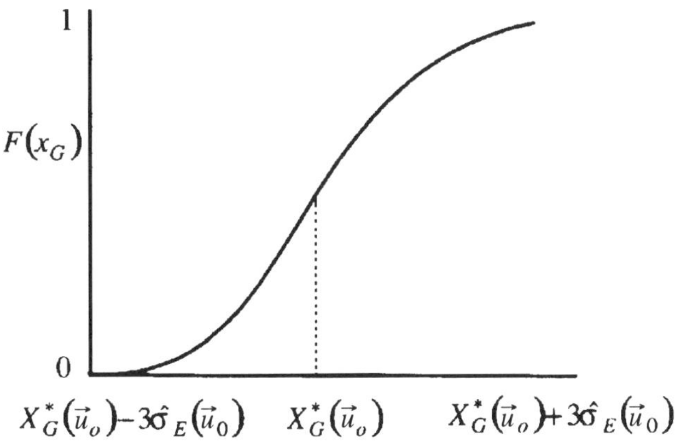

Fig. 6.7—Local uncertainty at $\vec{u}_0$.

associated with it. In contrast to estimates "a" and "c," estimate "b" represents a case where all the thresholds have a reasonable probability associated with them. This represents more uncertainty.

In contrast to the indicator transform, the Gaussian transform data requires only one estimate at the unsampled location. With ordinary kriging, we use

$$X_G^*(\vec{u}_0) = \sum_{i=1}^{n} \lambda_i X_G(\vec{u}_i) \quad \text{(6.6)}$$

for a continuous variable, and

$$K_G^*(\vec{u}_0) = \sum_{i=1}^{n} \lambda_i K_G(\vec{u}_i) \quad \text{(6.7)}$$

for a discrete variable. With the modeled variogram and selected samples (original plus prior simulated values), we estimate the value of the transformed variable in the Gaussian space. For the Gaussian transform, in addition to estimating the value, we also estimate the error variance with conventional kriging equations. If we assume that local uncertainty is represented by the Gaussian distribution, by knowing the estimate and the associated error variance, we capture the local uncertainty. For example, if we know the estimate $X_G^*(\vec{u}_0)$ and the error variance is $\hat{\sigma}_E^2(\vec{u}_0)$, we can define the cumulative distribution function at the location at $\vec{u}_0$, as indicated in **Fig. 6.7.** A similar figure is drawn for a discrete variable as well. This allows us to capture the local uncertainty.

For the probability transform, the value is estimated with a similar procedure to the Gaussian transform. The procedure that estimates the local uncertainty is not well established. One possibility is to use kriging weights to assign the local uncertainty. Because in an ordinary kriging process all weight values are required to add to one, we use the weight assigned to each sample within the search neighborhood as the probability mass function. As an example, if we have six samples in a search neighborhood, the probability mass function is represented, as shown in **Fig. 6.8.** In this figure, the y axis represents the kriging weights and the x axis represents the sample values. The weights fall between zero and one. This type of local uncertainty definition matches the constraint that the mean represents the estimated value.

By comparing the three transforms, it is clear that we must estimate a value at the unsampled location, as well as a local uncertainty. The indicator transform, through different thresholds, provides local uncertainty measurements nonparametrically. However, it is computationally more demanding because it requires the estimates of indicator values at each threshold. On the other hand, the Gaussian transform requires only one estimate; however, the quantification of uncertainty requires that we assume a Gaussian distribution, with respect to local uncertainty. Similar to the Gaussian transform, the probability transform also requires one estimate; however, the uncertainty is captured through nonparametric means with kriging weights.

Back Transform. Once all unsampled locations are visited and values, in the transformed domain, are estimated along with their local uncertainties, the next step is to simulate a value from the estimate and the local uncertainty and back transform the simulated value.

For indicator transformed data, the back transform is relatively easy. The estimated indicator values represent the uncertainty and, hence, the posterior distribution. By knowing the distribution, we can simulate a value with a uniform random number generator.

Before we sample a value, however, we must ensure that the posterior distribution satisfies the basic statistical requirements. For a continuous variable, because the posterior distribution represents a cumulative distribution function, one critical requirement is that it must be a nondecreasing function, starting from the lowest threshold. Because each indicator value is estimated independently, based on the variogram for that threshold, it is possible that this condition might not be satisfied. We must correct this problem before the sampling process starts. One common method is shown in **Fig. 6.9.** If we assume that the estimated values are represented by "a," they violate the rule of a nondecreasing distribution as the threshold increases. To correct this, we start at the top of the threshold and work downward. If the indicator value is smaller or equal to the higher threshold, we retain it; otherwise, we assign the same value as the higher threshold. We repeat the process from the smallest threshold. Working upward, if the indicator value is equal to or greater than the indicator corresponding to a lower threshold, we accept it; otherwise, we

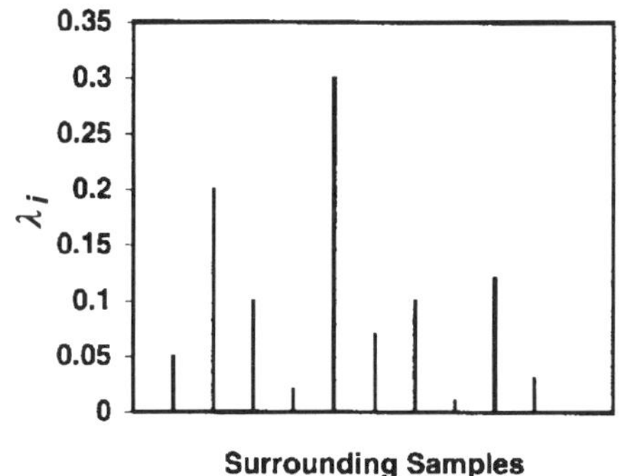

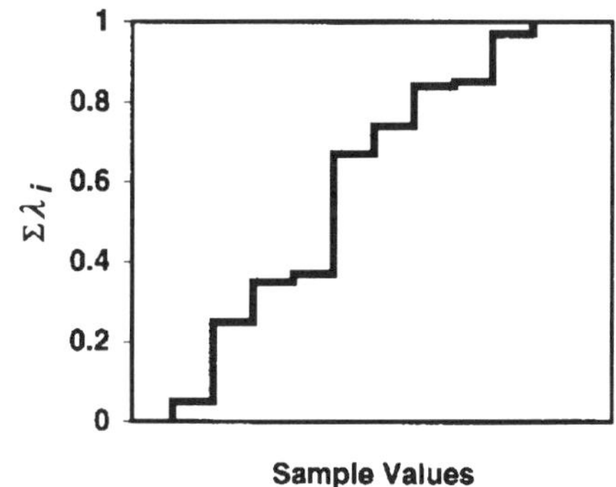

Fig. 6.8—Local uncertainty for probability transform.

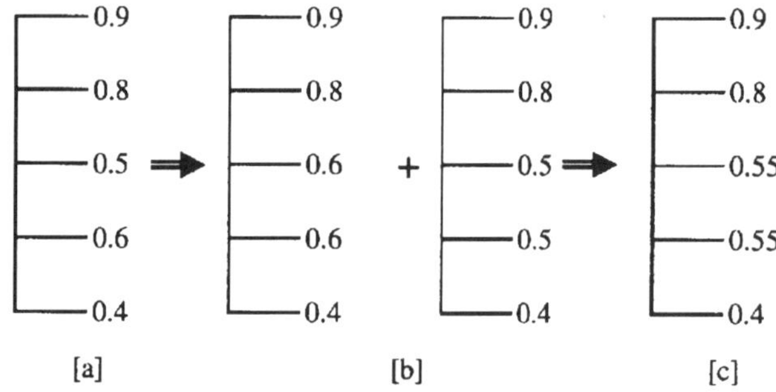

Fig. 6.9—Order relation correction for a continuous variable.

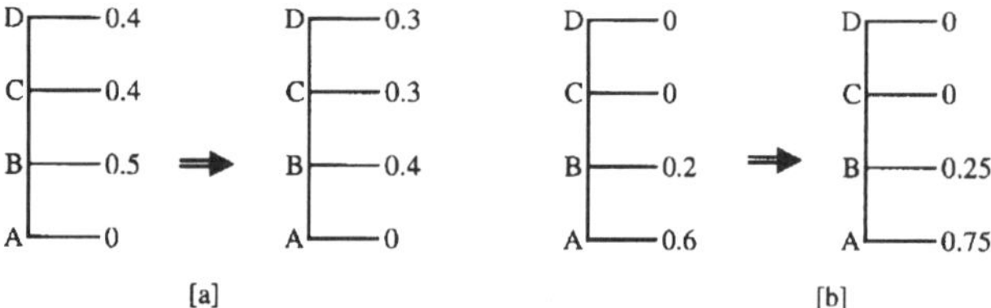

Fig. 6.10—Order relation correction for a discrete variable.

assign the same value as the lower threshold. This step is shown in the estimation of the value represented by "b." Once the corrected indicator values are calculated, we take an arithmetic average of two indicator values at each threshold (see the estimate represented by "c"). This average is used to sample a back-transformed value.

For a discrete variable, the correction is required if the sum of the probability density function does not equal one. That is, the sum of the indicator values at each threshold should equal one. If they do not, a simple correction is to normalize indicator values by dividing each indicator value by a cumulative sum of all the indicator values. This is shown in **Fig. 6.10.**

In both cases, in Fig. 6.10, the individual indicator values are divided by the sum of the indicator values. In the first case, the sum = 1.3, whereas in the second case, the sum = 0.8.

In addition to these corrections, the indicator values should always be greater than or equal to zero and less than or equal to one. The corrections are completely random for both continuous and discrete variables. Fortunately, in most cases, the required order relation correction is very small and, therefore, acceptable.

Once the order relation correction is made, a realization can be sampled from the posterior distribution. As shown in **Fig. 6.11,** for a continuous variable, a random number is selected from a uniform random number generator, and, depending on a selected value, an appropriate class is selected. In Fig. 6.11, a selected random value = 0.7; therefore, we must assign a value between thresholds three and four. Because we know the thresholds, we can sample from within the class distribution. As the simplest case, we linearly interpolate between two thresholds to assign a value at the unsampled location.

For a discrete variable, we first generate a cumulative distribution function from the probability density function. Once this distribution function is generated, we can sample from this distribution similar to the way shown in **Fig. 6.12.** Because we transform from one discrete distribution to another, unlike for a continuous variable, no interpolation is necessary. For example, if the random number 0.7 is selected, we can assign facies D at that location.

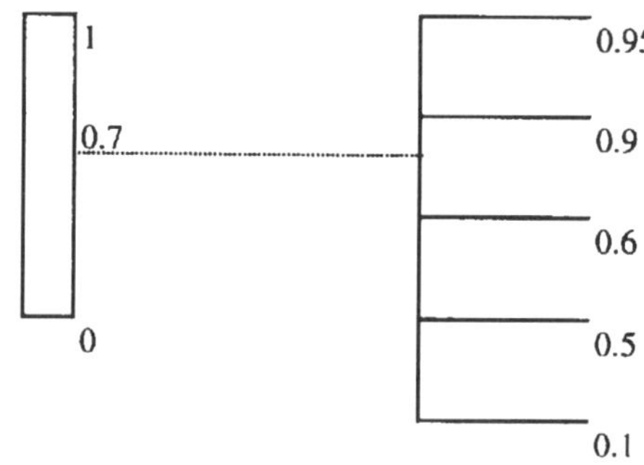

Fig. 6.11—Selection of a sample for a continuous variable.

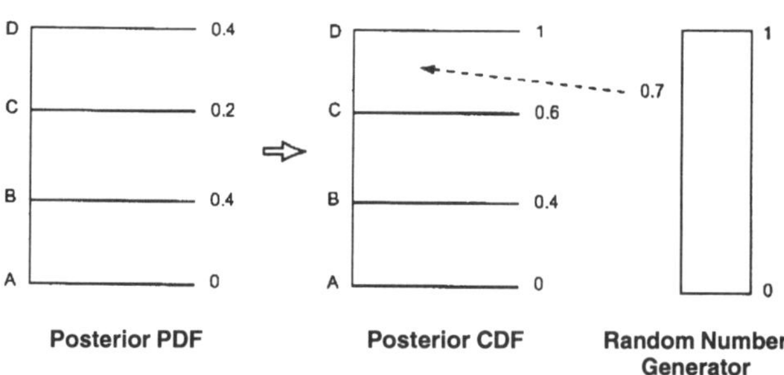

Fig. 6.12—Sampling from a discrete distribution.

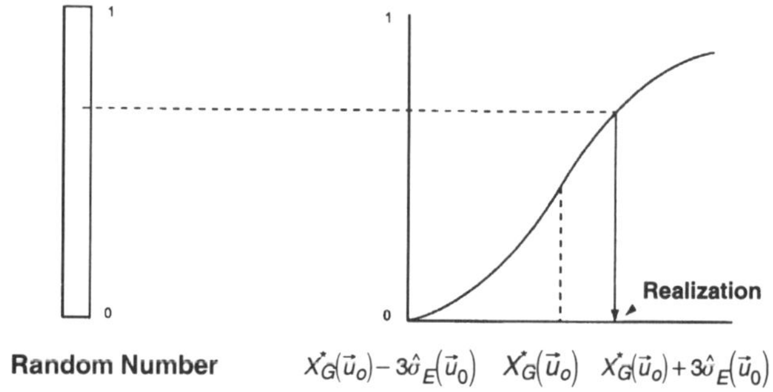

Fig. 6.13—Selection of a sample in a Gaussian transform.

The process of assigning the values is repeated until all unsampled locations are visited, and a realization is sampled at each location.

For a Gaussian transformed variable, we estimate a value and the associated error variance. This defines the uncertainty in the estimated value. The process of sampling a value from this distribution in a Gaussian domain is the same for both continuous and discrete variables. As shown in **Fig. 6.13,** with a uniform random number generator, a value from the Gaussian distribution is picked. In Fig. 6.13, based on a random number, the realization $x_G(\vec{u}_0)$ is picked. Several realizations can be generated in a similar manner. A value of the realization, with 99% probability, falls between $X^*_G(\vec{u}_0) \pm 3\hat{\sigma}_E(\vec{u}_0)$.

Once the value $x_G(\vec{u}_0)$ is sampled, it can be back transformed to the original domain. For a continuous variable, as shown in **Fig. 6.14,** it is straightforward. It is the reverse of Fig. 6.2. With the original sample distribution, a value is plotted in the Gaussian space, and for the same value of cumulative distribution function, a back transformed value is obtained. This is one realization at location $(\vec{u}_0)$.

For a discrete variable, the transform is done in two ways. A simple approach uses the original distribution (see Fig. 6.3)

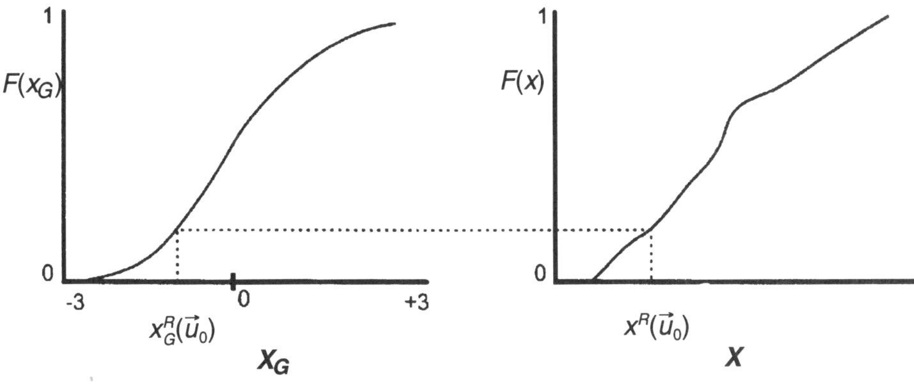

Fig. 6.14—Back transform from Gaussian space for a continuous variable.

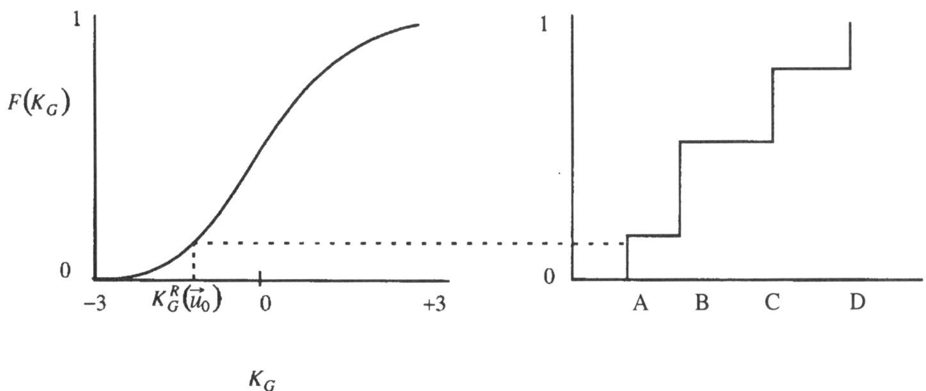

Fig. 6.15—Back transform from Gaussian space for a discrete variable.

and the reverse process to obtain the value for a particular category. This is shown in **Fig. 6.15.** In this case, based on the value in the transformed domain, category A is assigned at the unsampled location.

Another approach controls the back transform, depending on the vertical location. This is important from a geological perspective. Consider **Fig. 6.16;** a cross section between the two wells is shown. The geologist identified marker beds to define isochronal intervals. In one simplified example, the top zone contains facies A and B, and the bottom zone contains C and D. The geologist might be uncertain about the exact location of the isochronal interval between the two wells but is certain about the fact that facies A and B are deposited after C and D. If, while back transforming the data from the Gaussian domain, we use the scheme in Fig. 6.15, then it is always a possibility—albeit small—that facies C or D could be assigned above A or B, or facies A or B could be assigned below C or D. This clearly violates the principle of superposition, which states that a facies deposited in a prior time cannot be on top of the facies deposited at a later time.

To overcome this problem, define proportions of individual facies that can be used to back transform the data. For example, in the upper interval, based on the proportions of the individual facies, we define the proportions of A and B as 60 and 40%, respectively, with no presence of C or D. This will result in a different back transform, as shown in **Fig. 6.17.** The relative proportions of the different facies vary, depending on the depth. In this way, we are forced to sample a facies that is consistent with a particular isochronal interval. With Fig. 6.16 as a reference, as we go deeper, the proportions of the individual facies change. At the lower interval, it changes from A and B to C and D. The method of forcing the back transform to certain prior defined proportions is used extensively to create geologically consistent realizations. The changing vertical proportions as a function of depth are called proportion curves. This type of post-processing definitely offers more flexibility for the Gaussian transform, which is not possible for the indicator transform.

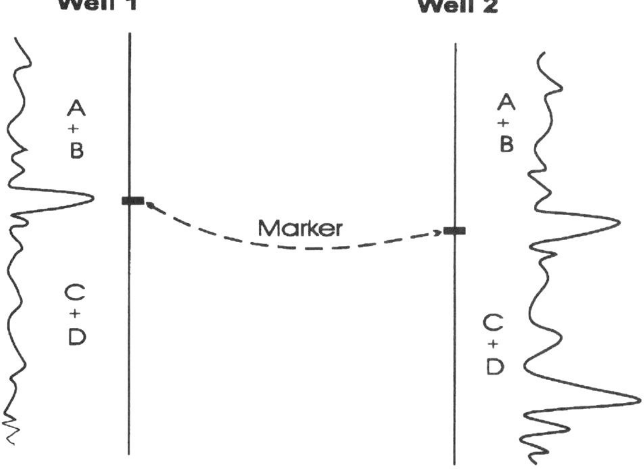

Fig. 6.16—Geological isochronal intervals.

For the probability transform, sampling a value from a posterior probability is similar for both discrete and continuous variables. The probability (density) mass function is provided by the kriging weights and the nearby samples; therefore, we can sample a value from that distribution with a uniform random number generator. This is shown in **Fig. 6.18.**

In this figure, the cumulative distribution is constructed by starting with the smallest transformed sampled value to the largest value and adding the kriging weights cumulatively. Once $x_p(\vec{u}_0)$ is determined, the back transform to the original

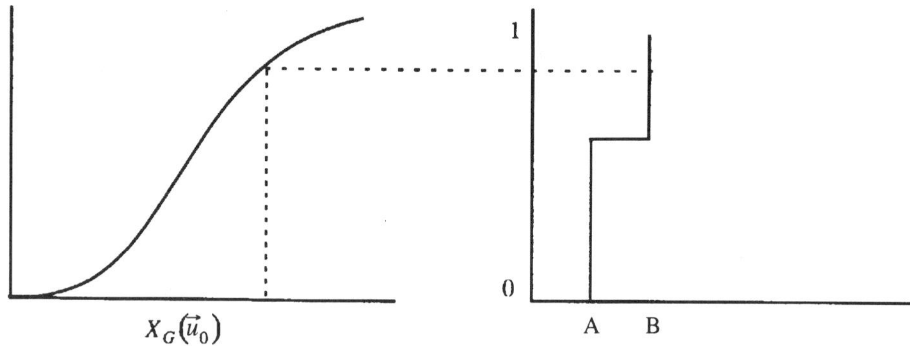

Fig. 6.17—Back transform using proportion curves.

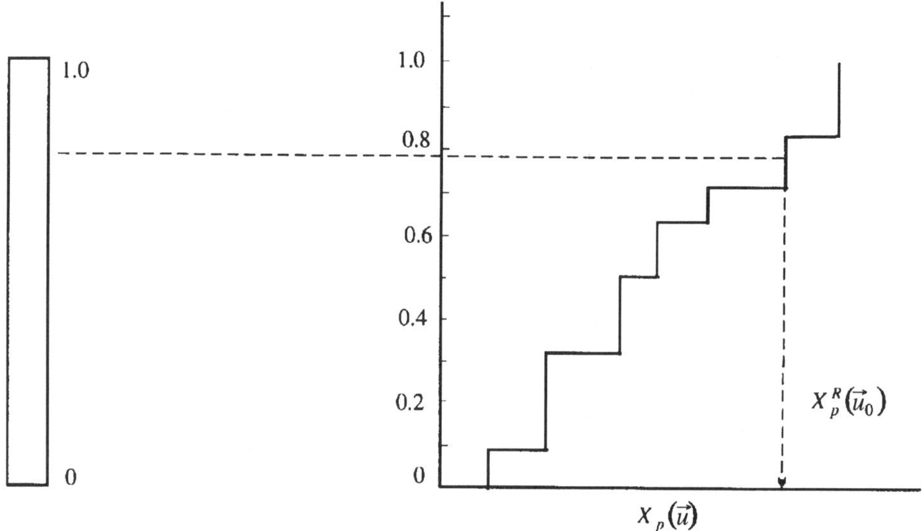

Fig. 6.18—Sampling from posterior probability distribution.

domain is straightforward for both the continuous and discrete variables, as shown in **Fig. 6.19.**

Irrespective of the type of transform used, we are able to generate multiple realizations with the help of a random number generator. The transform that will work best is difficult to determine; the choice strongly depends on the type of information available. As an approximate rule, it is safe to say that for geological facies, an indicator transform might be more appropriate, whereas for continuous variables, such as porosity and permeability, the Gaussian or probability transform might be best. This is true because, for geological facies, transforming the data to an indicator function does not result in a loss of information. Further, the architecture of the different facies is best captured with a different variogram for each facies.[7] In contrast, for a continuous variable, defining a single variogram, as in the case of the Gaussian transform, is adequate enough to capture the spatial relationship. Further, because we are transforming from one continuous variable to another, information is not lost within class distributions, as in the case of an indicator transform. This general rule can change for some special cases. For example, if the vertical proportion of geological facies is very critical, a Gaussian transform for the facies might be more appropriate. On the other hand, if different classes of the continuous variable exhibit different spatial relationships, an indicator transform for this continuous variable is appropriate.

In the literature, to take advantage of both methods, they are sometimes combined.[8] For example, one criticism against indicator simulation is that it is too discontinuous and does not adequately capture the smoothness of estimated values. One way to overcome this problem is with indicator kriging, which defines the proportion curves and sequential Gaussian simulation for the sampling of those curves to assign geological facies. This way, the spatial relations of the geological facies are honored, and the smoothness, which comes from a continuous variable transform (i.e., Gaussian or probability), is also preserved. Instead of a Gaussian transform, a probability transform can also be used to sample the facies from the indicator kriging-derived proportion curves.[6]

Another possible modification, in the sequential simulation process, samples a value and back transforms it into the original domain after every new unsampled location is visited. This means that steps four and five are carried out simultaneously. Every time an unsampled location is visited, an estimate in the transformed domain is made, and based on a random number generator, a value is sampled and back transformed before visiting the next unsampled location. This method appears to create a smoother distribution, compared to the first method, although clear evidence of the superiority of one alternative over another is lacking.

To summarize, the sequential simulation process provides a computationally efficient procedure to generate multiple realizations. Different nonlinear transforms are used to transform the original data into a new domain. These transformed domains allow us to better quantify local uncertainties, which are used to sample a realization at the unsampled location.

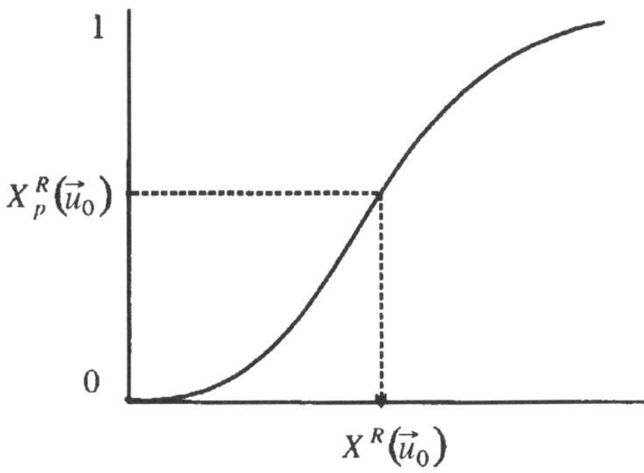

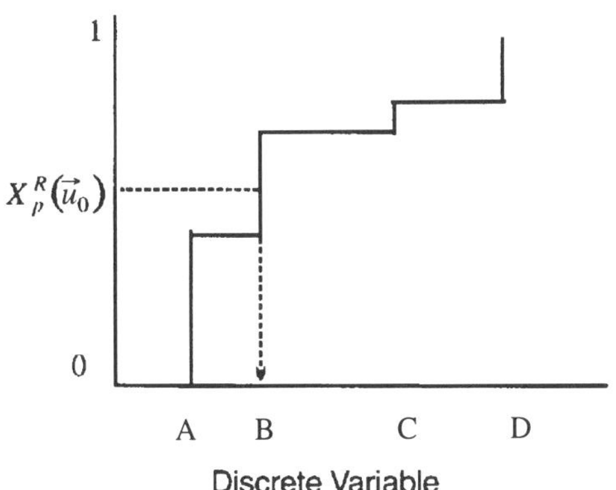

Fig. 6.19—Back transform for probability transform.

Field Example 6.1—Sequential Gaussian Simulation.

• Generate descriptions of porosity for Flow Unit 3 with sequential Gaussian simulation (SGS).

• Compare the statistics of the conditioning data and SGS description.

Solution. The following information is given:

• The porosity conditioning data for Flow Unit 3 is shown in **Fig. 6.20.** It consists of the *x* and *y* coordinates and average porosity of Flow Unit 3 at the well locations.

• The variogram model for the normal scores of porosity is shown in **Table 6.1.**

The calculation and modeling of variograms are described in Chap. 3.

TABLE 6.1—MODELING PARAMETERS FOR FIELD EXAMPLE 6.1

Model Type	Spherical
Principal Direction, deg.	112.5° (from North direction)
Nugget	0
Sill	1.0
Principal Range, ft	8,300
Minor Range, ft	2,800
Anisotropy Ratio	2.96
X and *Y* Origin	0 and –5,000 ft
X and *Y* Number of Blocks	100 and 100
X and *Y* Block Size	200 and 200 ft

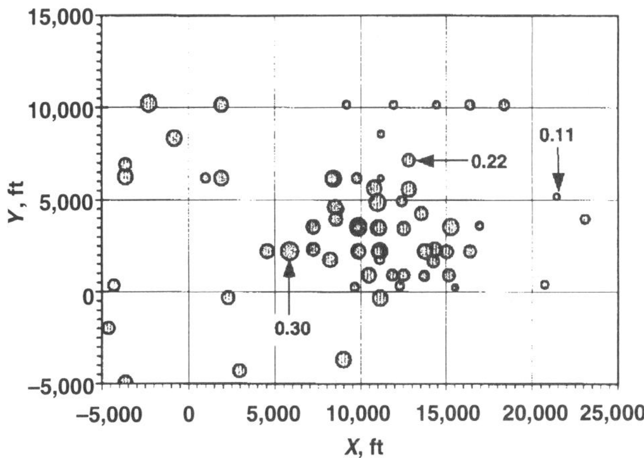

Fig. 6.20—Porosity conditioning data for Flow Unit 3.

• The location of the simulation grid is shown in Fig. 6.20. The grid covers an area of 20,000 × 20,000 ft. It is described in Table 6.1.

The porosity realization generated with SGS is shown in **Fig. 6.21.** The most significant feature of this realization is the southeast/northwest trend of low and high porosities. This results from the large anisotropy and its alignment with the principal variogram direction (112.5°). The relative arrangement of high and low porosity values is influenced by the conditioning data, which are seen by comparing Figs. 6.20 and 6.21. **Fig. 6.22** shows a second SGS realization generated with another random number seed. The overall appearance of both realizations is similar. Most differences occur at locations with no conditioning data. For example, the northeast corner of the first realization has high porosities, while the second realization has low porosities. The main application of conditional simulation quantifies uncertainty, at locations away from the conditioning data, with multiple realizations generated from different random number seeds.

The experimental variograms, shown in **Fig. 6.23,** indicate that the SGS realization is slightly more continuous than specified by the models. The small arrows, in Fig. 6.23, show the estimated correlation ranges for the realization. For the principal direction, the range is close to that of the model, but the sill is smaller. For the minor direction, the range is greater

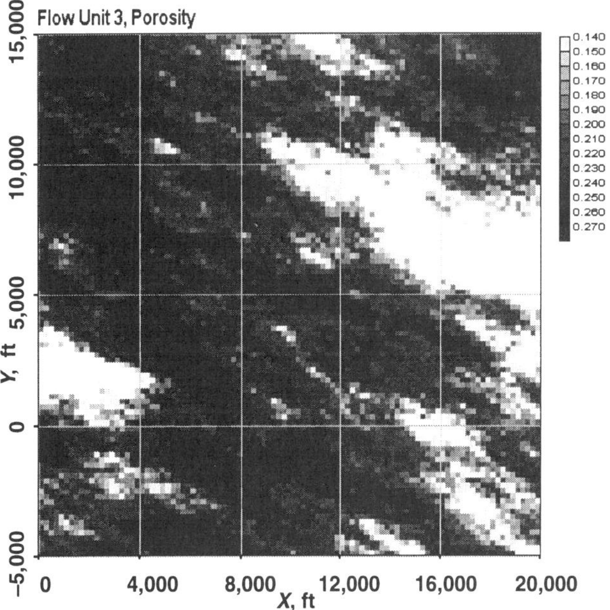

Fig. 6.21—SGS first realization of porosity for Flow Unit 3.

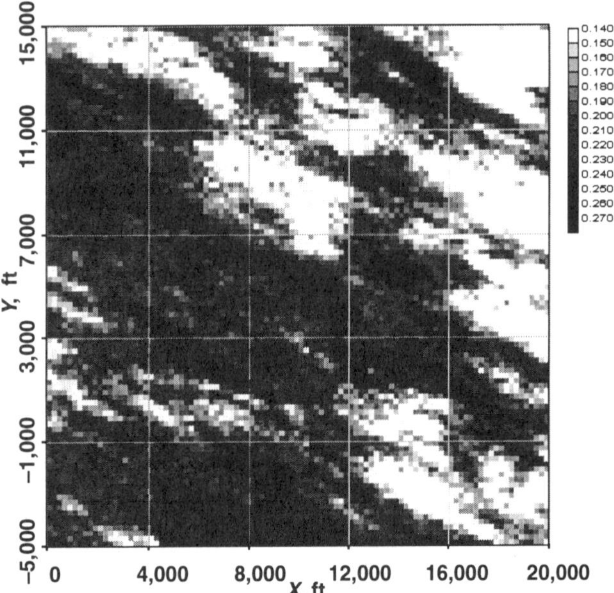

Fig. 6.22—SGS second realization of porosity for Flow Unit 3.

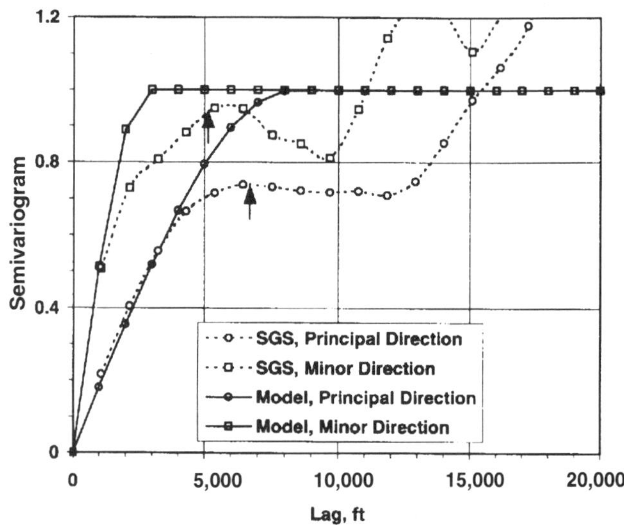

Fig. 6.23—Variograms of SGS first realization of porosity and models.

than the model's, but the sill is close to the model's sill. The additional continuity, created by SGS, is a result of the strong nonstationary trend of high porosities, aligned southeast/northwest, present in the conditioning data.

Figs. 6.24 and 6.25 show the statistics and histogram of the first SGS realization (Fig. 6.21) and conditioning data, respectively. Despite of the strong bimodal character of the conditioning data, SGS is capable of reproducing the distribution very well. The ability to closely reproduce the basic univariate statistics of the conditioning data is one of the best properties of SGS.

Field Example 6.2—Sequential Indicator Simulation for Continuous Variables.

• Generate descriptions of porosity for Flow Unit 3 with sequential indicator simulation (SIS) for continuous variables.

• Compare the statistics of the conditioning data and SIS description.

Solution. The following information is given:

• The porosity conditioning data for Flow Unit 3 is the same as in Field Example 6.1 (Fig. 6.20).

• Three indicator variables are used to describe porosity. The thresholds are 0.181 (lower quartile), 0.225 (median), and 0.253 (upper quartile).

• The variogram models for the porosity indicators are shown in **Table 6.2.**

• The simulation grid is the same as in Field Example 6.1.

The porosity realization, generated with SIS, is shown in **Fig. 6.26.** The major correlation trend, which runs southeast/northwest, is similar to that observed in the SGS realization (Fig. 6.21). The trend is controlled by the orientation of the indicator variogram models, which are similar to the normal score variograms (Field Example 6.1). One advantage of SIS is that it describes the spatial correlation of different thresholds with different variograms. This property of SIS is important, when it is necessary to preserve the connectivity of low and high values. In this field example, the first indicator variable, which represents the smallest porosity values, has the biggest range and anisotropy. Indeed, the realization in Fig. 6.26 shows that the smallest porosity values (for threshold 0.181) are correlated for longer distances, rather than the biggest values (for threshold 0.253), which have a shorter variogram range. SIS is designed to reproduce the spatial correlation for the class of values represented by each indicator. Within the range of each threshold, the simulated values exhibit greater variability because values are drawn at random

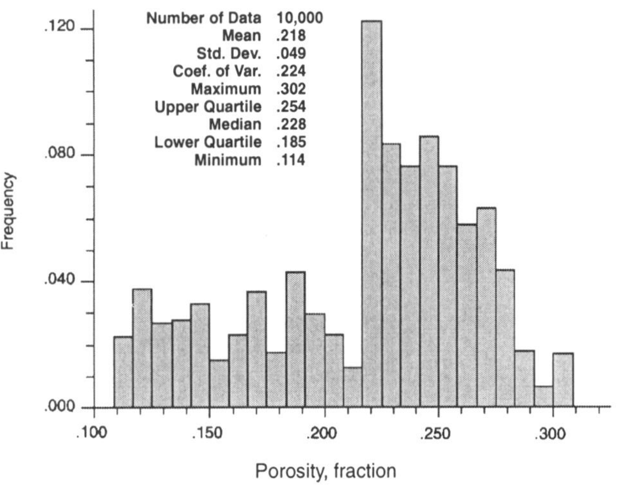

Fig. 6.24—Histogram of SGS first realization of porosity.

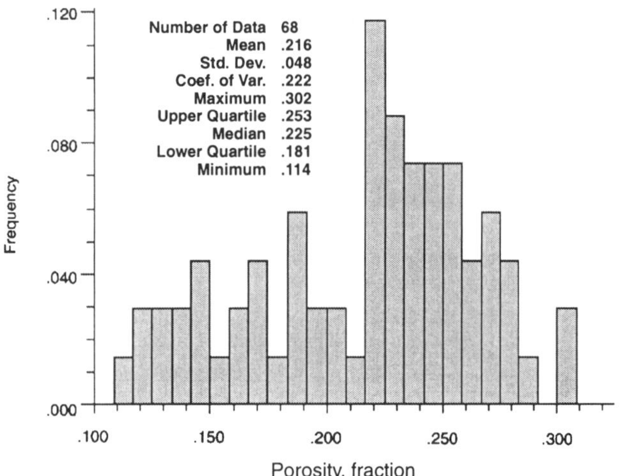

Fig. 6.25—Histogram of porosity data for Flow Unit 3.

TABLE 6.2—VARIOGRAM MODELS FOR FIELD EXAMPLE 6.2			
	Indicator 1	Indicator 2	Indicator 3
Threshold	0.181	0.225	0.253
Model Type	Spherical	Spherical	Spherical
Principal Direction, Deg.	112.5	112.5	112.5
Nugget	0	0.16	0.14
Sill	0.19	0.10	0.04
Principal Range, ft	11,000	8,000	5,000
Minor Range, ft	3,000	2,500	2,000
Anisotropy Ratio	3.67	3.2	2.5

TABLE 6.3—INDICATOR THRESHOLDS FOR FIELD EXAMPLE 6.3			
Discrete Class	Lower Threshold	Upper Threshold	Proportion of Data
1	0.114 (Minimum)	0.181 (Lower Quartile)	0.25
2	0.181	0.225 (Median)	0.25
3	0.225	0.253 (Upper Quartile)	0.25
4	0.253	0.302 (Maximum)	0.25

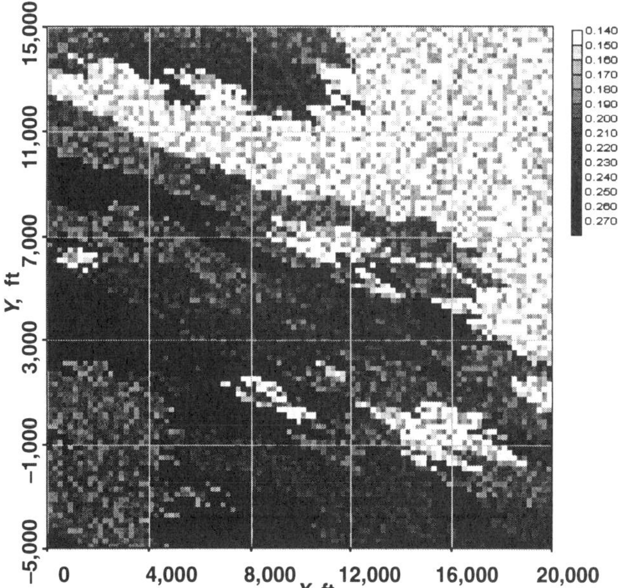

Fig. 6.26—SIS realization of porosity for Flow Unit 3.

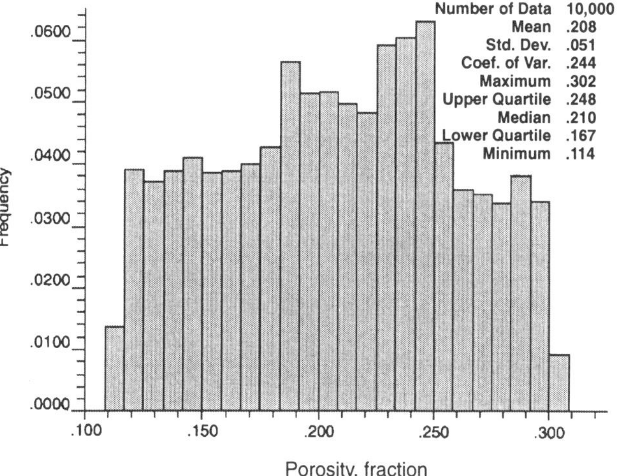

Fig. 6.27—Histogram of SIS realization of porosity.

from the local conditional distribution. Additional indicator thresholds are used to reduce the variability within thresholds. However, as the number of indicators increases, the required number of variogram models and the amount of computation also increase.

Fig. 6.27 shows that the univariate statistics of the SIS realization, including the quartiles, are in close agreement with the conditioning data (Fig. 6.25). SIS does not reproduce the shape of the conditioning data histogram very well. More indicator thresholds are needed to honor the complex shape of the histogram in this field example.

Field Example 6.3—Sequential Indicator Simulation for Discrete Variables.

• Generate descriptions of porosity classes for Flow Unit 3 with SIS for discrete variables.

• Compare the statistics of the conditioning data and SIS realization.

Solution. The following information is given:

• The porosity conditioning data for Flow Unit 3 is the same as in Field Example 6.1 (Fig. 6.20).

• Porosity is described with four discrete variables. The thresholds used to define the discrete variable are shown in **Table 6.3.**

These variables can be interpreted as representing reservoir sands with quality ratings of poor (1); intermediate (2) and (3); and good (4). In more rigorous reservoir studies, detailed geologic analyses of cores and logs are used to describe the discrete indicator variables of geologic facies. Then, several geologic concepts from analogs and outcrops are used as guides to model variogams and secondary information, for the conditional simulations of facies.

• The variogram models for the indicators of the discrete variable are shown in **Table 6.4.**

• The simulation grid is the same used in Field Example 6.1.

The realization of the discrete indicators generated with SIS is shown in **Fig. 6.28.** The large-scale trend in this description along the principal variogram direction is similar to the one in the SGS and SIS descriptions (Field Examples 6.1 and 6.2). SIS honors different correlation ranges for each dis-

TABLE 6.4—INDICATOR VARIOGRAMS FOR FIELD EXAMPLE 6.3				
	Indicator 1	Indicator 2	Indicator 3	Indicator 4
Discrete Class Value	1	2	3	4
Model Type	Spherical	Spherical	Spherical	Spherical
Principal Direction, Deg.	112.5	112.5	112.5	112.5
Nugget	0	0.12	0.10	0.14
Sill	0.19	0.10	0.12	0.06
Principal Range, ft	11,000	8,000	5,000	5,000
Minor Range, ft	3,000	5,000	5,000	5,000
Anisotropy Ratio	3.67	1.6	1.0	1.0

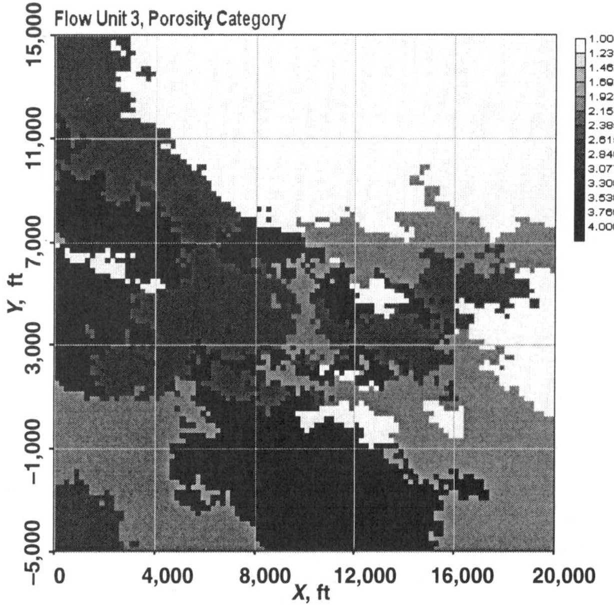

Fig. 6.28—SIS realization of porosity categories for Flow Unit 3.

crete variable. It is inferred from Fig. 6.28 that the range of class one is longer than that of class four, as specified by the variograms shown.

The histogram of the SIS realization is shown in **Fig. 6.29.** The small deviations from the specified global histogram, which equal 0.25 for all classes, are primarily caused by order-relation violations.

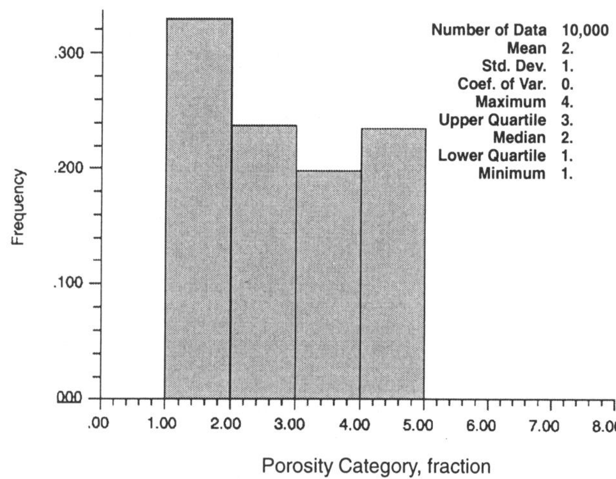

Fig. 6.29—Histogram of SIS realization of porosity discrete classes.

6.1.2 Sequential Cosimulation. Sequential cosimulation is a logical extension of the sequential simulation process. The process of cosimulation results in the simulation of multiple attributes simultaneously. The main advantage of cosimulation is its ability to honor the local relationships among the various attributes. Consider the example in which we use indicator simulation for a geological facies description. If the geological facies are shale and sand, respectively, then either sand or shale will be assigned to the individual gridblocks. For the same reservoir, if we use Gaussian simulation to estimate the porosity values, after the simulation, porosity values will be assigned to all individual gridblocks. When we compare the geological facies description with the porosity description, we might observe some inconsistencies. For example, if we honor both the facies distribution and the porosity distribution, we could observe a nonzero porosity assignment for shale and a zero porosity assignment for sand. This is possible because no local relationships between the geological facies and porosity are honored when simulating facies and porosity independently. Although there are some indirect local constraints (certain facies vs. porosity relationships are observed at the well locations, and if we honor the well data, it should also influence the local relationships away from the wellbore), when we independently assign facies and porosity at the gridblocks, it is possible to locally assign inconsistent values, unless a constraint imposing a strict relationship is incorporated.[9] Such a local constraint is only possible if we can simultaneously simulate both facies and porosity at the gridblock locations. This is the main purpose of cosimulation.

Cosimulation achieves this goal by simultaneously honoring the individual spatial relationships of the various attributes, as well as honoring the local relationship. Once we determine the order of the attributes to simulate, that order is followed at every unsampled location, to ensure the interrelationship among the various attributes is also honored. An example of such an order is illustrated in **Fig. 6.30.**

In this figure, we assume that we are interested in simulating the geological facies, porosity, and permeability values at a given gridblock. We assume that seismic attribute measurements are available at individual grid locations or that they are interpolated very easily because of extensive sampling. In this scenario, one seismic attribute could have a direct bearing on the geological facies, whereas the same or some other attribute could have bearing on the porosity values. Because seismic data are extensively sampled, we treat both the facies and porosity as dependent variables on the seismic data. In terms of facies vs. porosity, we assume geological facies influence the porosity values. That is, for each discrete facies, a certain porosity distribution exists. An extreme example is zero porosity for shale and a nonzero porosity distribution for sand. The permeability value at each gridblock, in turn, depends on the porosity, as well as the facies. That is, for each facies, a unique porosity vs. permeability relationship exists. This type of dependence is illustrated in **Fig. 6.31.** This figure shows that for each facies, a unique porosity distribution and a unique permeability vs. porosity relationship exists. Facies N represents a nonreservoir rock, where porosity and permeability are assigned values of zero.

By examining the interrelationships among different attributes, we determine the order in which various attributes are estimated. In our particular example, we first estimate the geological facies, followed by porosity and then the permeability values. Once the order is established, the procedure for sequential cosimulation can be explained.

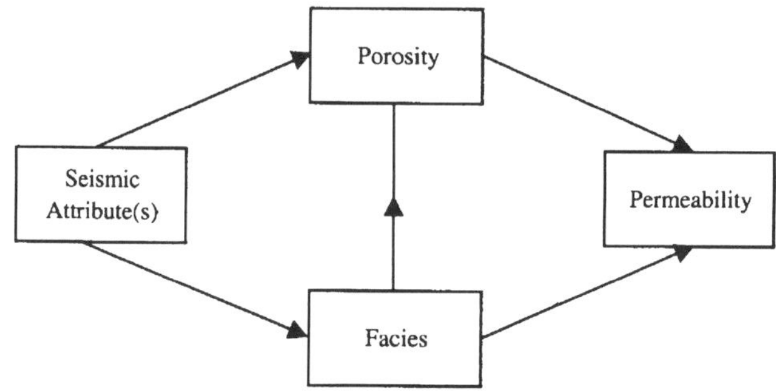

Fig. 6.30—Interrelated attributes at individual gridblocks.

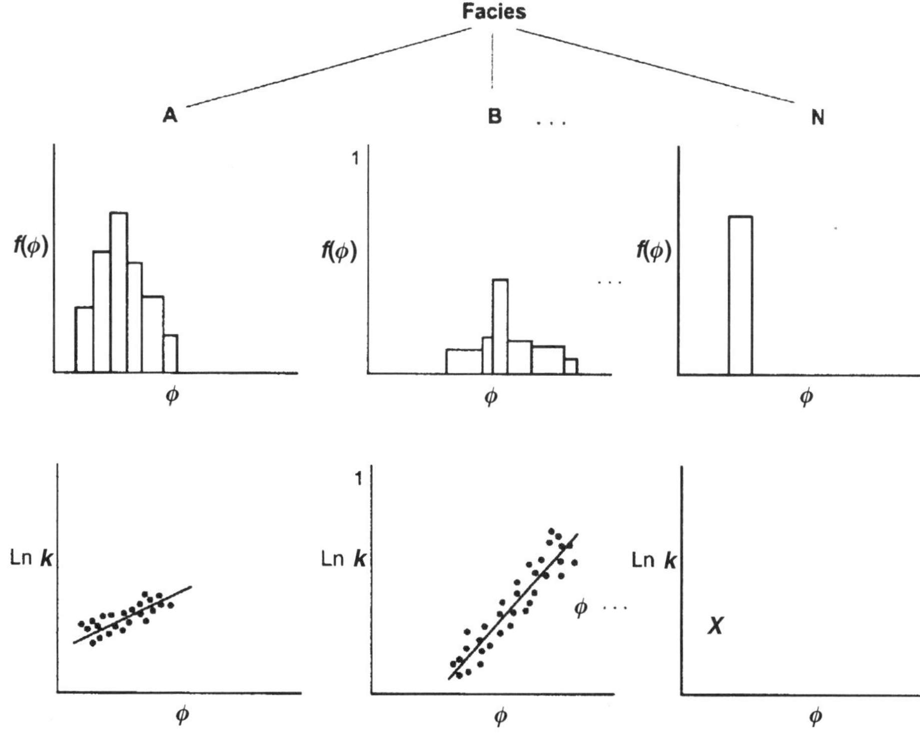

Fig. 6.31—Dependent relationship of various attributes.

Transformation Into a New Domain. Similar to sequential simulation, we transform the data into an appropriate domain. For example, we transform the facies in terms of indicator values and porosity and permeability in terms of Gaussian transforms. Any transform is appropriate, as long as the local relationships are approximately captured in the transformed domain. If a procedure similar to cokriging is used, the local relationship among the attributes must be linear in the transformed domain. This is true because cross-covariance can only quantify the extent of the relationship among the attributes if the relationship is linear.

Estimation and Modeling of Variograms. This step is also similar to sequential simulation. In the transformed domain, we estimate and model the variograms for all attributes. For example, if we assume that we have three geological facies transformed into an indicator domain and porosity and permeability in a Gaussian domain, then we must model three indicator variograms for facies and two Gaussian transformed variograms, one for porosity and one for permeability.

In addition to modeling the variograms, similar to cokriging, we are required to model the cross-variograms among various attributes. The number of cross-variograms to model depends on the number of possible dependencies, which must be honored at local gridblocks. For example, in the transformed domain, if we assume that the facies are related to the seismic data, then we must estimate the cross-variogram between the facies and the seismic attribute. If porosity is assumed to be related to the seismic attribute as well as the facies, then we need cross-variograms between the porosity and facies and between the porosity and seismic attribute. Similarly, we might need cross-variograms between permeability and porosity, as well as permeability and facies, etc. Please note that the modeling of cross-variograms is only necessary if the cosimulation procedure is based on a cokriging procedure. If the relationships, even in the transformed domain, are not linear, alternate procedures for honoring the local relationships are available and are discussed in later sections. Under these conditions, the modeling of cross-variograms might not be necessary.

Selection of a Random Path. Once all the necessary variograms and cross-variograms are modeled, a random path is selected for all unsampled gridblocks to be visited in a random sequence. Such a selection is necessary because bias could be created in the simulation process if we visit the unsampled locations in sequential order. This is similar to sequential simulation, which simulates the next gridblock value based on the original samples, as well as prior simulated values. A sequential order could add bias based on prior simulated values in a particular order. This is avoided by selecting a random path.

Simulation of Attributes. Unlike the detailed procedure described for sequential simulation, in cosimulation, the estimation and sampling of a realization is one step. In general, the value of a variable (attribute) is estimated along with its uncertainties. Based on the estimate and the related uncertainties, a value is sampled in the transformed domain and eventually back transformed into the original domain. Similar to sequential simulation, in addition to the original samples, prior estimated values are used to estimate a value at the unsampled location. The estimated values include the estimated values within a search neighborhood, as well as prior estimated values of different attributes at the same location. For example, to estimate the permeability value at an unsampled location, we use prior permeability estimates within a search neighborhood, as well as prior estimates of porosity and facies, if permeability is related to those two variables. Once an unsampled location is visited, all the unknown attributes are sampled in sequential order to preserve their relationship. Typically, the order is from the least dependent attribute to the most dependent attribute. For example, we simulate geological facies first, followed by porosity and then the permeability value.

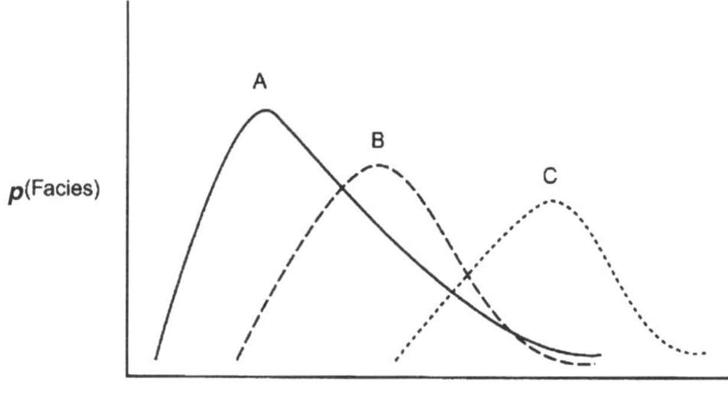

Fig. 6.32—Relationship between seismic attributes and facies.

The procedure, in which subsequent variables are sampled based on prior values depends on the type of local relationship honored. We describe several possibilities.

Sequential Gaussian Cosimulation.[10] If we assume the transformed domain is a Gaussian space and the relationships among various attributes are linear, by estimating a value and an error variance at each sampled location, we can quantify the uncertainties.

To estimate a value, we use the cokriging procedure. In general terms, to estimate the attribute $X_k(\vec{u}_0)$, we write the equation as

$$X_k^*(\vec{u}_0) = \sum_{i=1}^{n_k} \lambda_{i_k} X_k(\vec{u}_i) + \sum_{l=1}^{L} \sum_{j=1}^{n_l} \lambda_{j_l} X_l(\vec{u}_j), \quad \ldots\ldots (6.8)$$

where λ_{i_k} = the weight assigned to sample $X_k(\vec{u}_i)$ (original or a prior estimated value) within a search neighborhood. These are the samples of the same variable. λ_{j_l} = the weights assigned to the covariable samples $X_l(\vec{u}_j)$ in the search neighborhood, and L changes with the variable simulated at the unsampled location. With Fig. 6.30 as an example, for the estimations of facies, we could have $L=1$, which represents the measured seismic attribute. For the estimation of porosity, $L=2$, which includes measured seismic attributes and a prior simulated facies value. With the same logic, for the permeability estimation, $L=3$.

Equation 6.8 is similar to the one used for collocated cokriging, if $(\vec{u}_j)$ is restricted to $(\vec{u}_0)$ only. The cross-variograms between the variables are inferred from the variograms. This minimizes the spatial modeling effort.

This procedure is computationally very efficient. The disadvantage of the procedure is that it assumes the local uncertainty is captured by a Gaussian distribution, and the relationships among the various attributes are linear.

Bayes' Rule Cosimulation.[11-12] This procedure is based on Bayes' rule, which defines a relationship between the posterior probability and the prior probability. Mathematically, Bayes' rule is stated as

$$p(A|B) = \frac{p(A \cap B)}{p(B)}, \quad \ldots\ldots (6.9)$$

where $p(A|B)$ represents the probability of A, given B has occurred; this is the posterior probability. $p(A \cap B)$ represents the probability that both events, A and B, will occur. $p(B)$ represents the probability of event B. Similar to Eq. 6.9, we can also write $p(A \cap B)$ as

$$p(A \cap B) = p(B|A) \cdot p(A). \quad \ldots\ldots (6.10)$$

By substituting Eq. 6.10 in Eq. 6.9, we obtain

$$p(A|B) = \frac{p(B|A) \cdot p(A)}{p(B)}. \quad \ldots\ldots (6.11)$$

If we do not know the value of $p(B)$, we write

$$p(A|B) \ \alpha \ p(B|A) \cdot p(A). \quad \ldots\ldots (6.12)$$

That is, the posterior probability of an event A is related to the prior probability of event A. The proportional constant need not be known, if the sample is comprised of n and mutually-exclusive events A_i; therefore, we can write

$$\sum_{i=1}^{n} p(A_i|B) = 1. \quad \ldots\ldots (6.13)$$

By knowing Eqs. 6.12 and 6.13, we can calculate the posterior probabilities, if we know $p(B|A_i)$ for A_i.

This equation is used to update the probability of an event, given the prior probability and value of $p(B|A_i)$. To illustrate this procedure further, consider a specific example of the relationship between a seismic attribute and a geological facies. Seismic attributes tend to be continuous variables, whereas geological facies are discrete variables. The relationship between the seismic attribute and geological facies, based on well observations, is shown in **Fig. 6.32.**

In this figure, we measure the seismic attribute and the corresponding facies; we could observe that the presence of individual facies is related to the seismic attribute. For each facies, a particular seismic attribute distribution is defined. Obviously, there is some overlap of values—we do not have an exclusive relationship. However, we define, based on this observation, the probability that the seismic attribute will take a certain value, given a particular facies. That is, $p(S_i|K)$, where S_i = the value of the seismic attribute (it could be a class of seismic attribute values), and K_j = the facies. It is possible that $p(S_i|K_j)$ = zero, where a certain attribute value is not possible for a given facies. As a practical matter, if we divide the seismic attribute into several classes, for each class we estimate the value of $p(S_i|K_j)$, where S_i represents class i.

Once this information is known, we can use it with the Bayes' rule method to estimate the posterior probability of facies given a certain seismic attribute, $p(K_j|S_i)$. We write

$$p(K_j|S_i) \ \alpha \ p(S_i|K_j) p(K_j). \quad \ldots\ldots (6.14)$$

We estimate $p(K_j)$ from prior information. For example, if we use an indicator transform for geological facies, then the $p(K_j)$ values are obtained through sequential indicator simulation. These values are then updated with Eq. 6.14.

Numerical Example 6.1. A reservoir contains three geological facies. At an unsampled location, based on sequential indicator simulation, the indicator values for the three facies are observed as 0.2, 0.6, and 0.2, respectively. A seismic attribute is related to the geological facies. A measurement of that attribute(s) is available at that location. Based on the comparison of the well data and seismic attribute, we write

$$p(S|K_1) = 0.2; \ p(S|K_2) = 0.8, \text{and} \ p(S|K_3) = 0.0.$$

With this information, determine the posterior distribution of geological facies.

Solution. We are given

$$p(K_1) = 0.2, \ p(K_2) = 0.6, \ p(K_3) = 0.2;$$

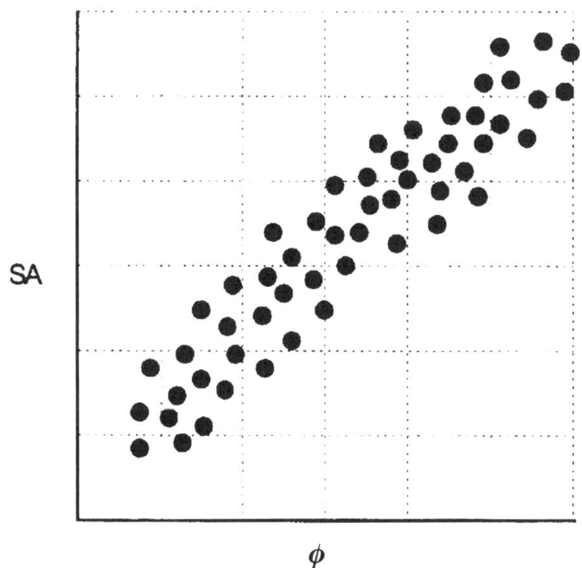

Fig. 6.33—Relationship between porosity and seismic attributes.

$$p(S|K_1) = 0.2,\ p(S|K_2) = 0.8; \text{and}\, p(S|K_3) = 0.0.$$

With this information, we write, based on Bayes' rule,

$$p(K_1|S)\ \alpha\ p(K_1)\cdot p(S|K_1) = 0.2 \times 0.2 = 0.04,$$

$$p(K_2|S)\ \alpha\ p(K_2)\cdot p(S|K_2) = 0.6 \times 0.8 = 0.48,$$

$$\text{and}\, p(K_3|S)\ \alpha\ p(K_3)\cdot p(S|K_3) = 0.2 \times 0.0 = 0.0.$$

We do not know the proportionality constant; however, we know

$$\sum_{j=1}^{3} p\left(K_j|S\right) = 1.$$

To satisfy this requirement, we must divide the product on the right by (0.04 + 0.48 = 0.52). That is,

$$p(K_1|S) = \frac{0.04}{0.52} = 0.077,$$

$$p\left(K_2|S\right) = \frac{0.48}{0.52} = 0.923,$$

and

$$p\left(K_3|S\right) = \frac{0.0}{0.52} = 0.$$

These values provide the posterior distribution of geological facies for a given seismic attribute. This is definitely an improvement over prior knowledge because it eliminates the probability that facies three will occur, and there are the probabilities that facies one will approach zero and facies two will approach one. That is, we improve our confidence in the estimation with seismic attribute information. It is important to emphasize that additional information, such as seismic data, will not always reduce uncertainty. If prior information is consistent with subsequent information, adding the subsequent information reduces uncertainty. This is illustrated in Numerical Example 6.1. However, if subsequent information is inconsistent or contradictory, with respect to prior information, adding more information increases uncertainty.

Although Numerical Example 6.1 illustrates the application for geological facies, it extends to any set of attributes, including continuous attributes. If there is a relationship between the porosity and seismic attribute data, we can use the same approach. **Fig. 6.33** shows a generalized relationship between seismic data and porosity in the Gaussian domain.

The porosity transforms in the Gaussian domain and plots against the seismic attribute. The relationship is clearly present, but it is not linear. However, if we divide the seismic attribute and the porosity data into several classes, we also develop values of $p\left(S_i|\phi_{N_j}\right)$, which represent the probability of the given seismic attribute class, given a porosity class in the Gaussian domain.

This relationship is used with porosity cosimulation. If the porosity is estimated with sequential Gaussian simulation, by estimating the error variance and the estimate, itself, we can easily estimate the probability that a value will fall within the certain class interval, $\left[p\left(\phi_{N_j}\right)\right]$. With this class interval and $p\left(S_i|\phi_{N_j}\right)$, we obtain the posterior distribution with the equation

$$p\left(\phi_{N_j}|S_i\right)\ \alpha\ p\left(\phi_{N_j}\right)\cdot p\left(S_i|\phi_{N_j}\right). \quad \text{.............. (6.15)}$$

Again, by requiring the sum of the posterior probabilities for all porosities to equal one, we can estimate the posterior probabilities of individual classes. This posterior distribution can sample a porosity value.

This procedure is not restricted to prior measurements at the unsampled locations; it is also used for prior estimated values. For example, in sequential cosimulation, we estimated a porosity value and the geological facies, so we can use sequential Gaussian simulation for permeability data and estimate a value and the associated uncertainties with the assumption of a Gaussian distribution. Once the prior distribution is known for a particular class, we can use a similar equation for permeability, given a particular porosity value,

$$p\left(k_{N_l}|\phi_{N_j}\right)\ \alpha\ p\left(k_{N_l}\right)\cdot p\left(\phi_{N_j}|k_{N_l}\right), \quad \text{.............. (6.16)}$$

where $p\left(k_{N_l}\right)$ represents the probability of class l of permeability, based on Gaussian simulation. $p\left(\phi_{N_j}|k_{N_l}\right)$ represents the probability of porosity class ϕ_{N_j} for a given class of permeability, k_{N_l}. This information is obtained from the permeability—the porosity relationship for a given geological facies. The subscript N represents the Gaussian (or normal) distribution. The important reason to use the normal distribution is for its ability to describe the uncertainties through error variance. If other methods are available to capture the uncertainty, the transformation to normal distribution is not necessary.

To summarize the Bayesian rule approach, it allows the updating of the prior probability to a posterior probability with new information. The method is efficient and does not require spatial relationship knowledge derived from cross-variograms. As an added advantage, it is also applicable to nonlinear relationships.

Proportion Curves Cosimulation. Another alternative approach to conducting cosimulation uses a procedure similar to proportion curves discussed in the section on sequential Gaussian simulation.[13-14] Recall, in sequential Gaussian sim-

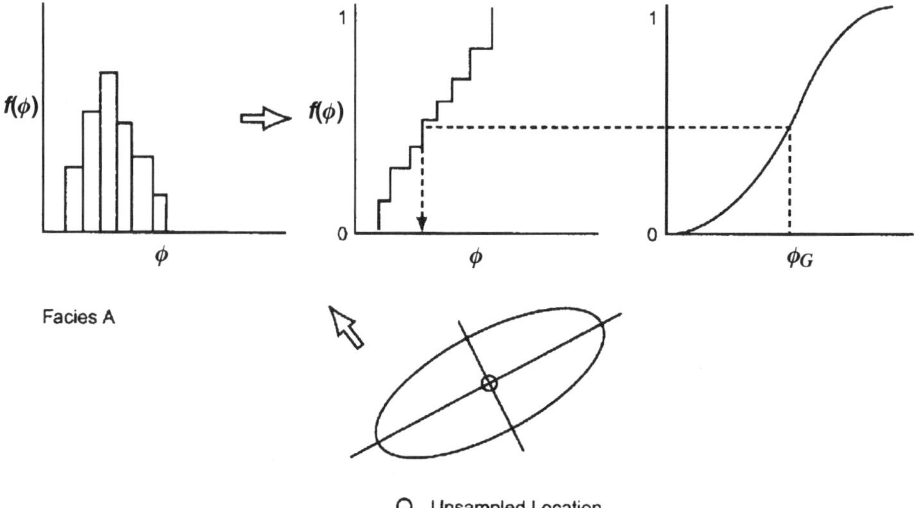

Fig. 6.34—Back transform using proportion curves.

ulation of geological facies, with the use of proportion curves, only certain geological facies are allowed to be sampled. The proportion curves represent the percentage of geological facies at a particular depth.

The same procedure is used to selectively sample a particular attribute. Let us illustrate this procedure with a simple example. If we simultaneously assign geologic facies and porosity values at an unsampled grid location, we will first assign a geologic facies at that location. In the next step, we must assign a porosity value at the same location. If we use sequential Gaussian simulation for estimating porosity, we can estimate the porosity and the associated uncertainty in the Gaussian domain. With a random number generator, we can sample a value from the posterior distribution. Up to this point, the procedure is similar to standard sequential Gaussian simulation. When we want to back transform, we must sample a porosity value based on the value in the Gaussian domain. We restrict this back transform to porosity values that only correspond to a particular facies. For example, if facies A is assigned to an unsampled location, we only need to sample a porosity value from that facies. This is illustrated in **Fig. 6.34.** As shown, the porosity distribution for facies A is grouped to generate a cumulative distribution function.

The sampled value in the Gaussian domain is back transformed with the cumulative distribution function for that facies. This ensures that the porosity value is consistent with the underlying geology.

This procedure is not unique to porosity. We can extend it to any attribute. As long as the conditional distribution corresponds to particular constraints, we can sample from that distribution with the proportion curves method. For example, permeability is sampled for a given porosity with the same procedure. Permeability is estimated with the Gaussian transform, and a value is sampled. The back transform is restricted to the conditional distribution of permeability for a given porosity class. This is schematically shown in **Fig. 6.35.**

The method of proportion curves has the advantage of not requiring a linear relationship. By restricting the back transform to only a certain distribution, the proportion curves method will always ensure that the conditional distributions of various attributes are honored. The method also does not require us to model the cross-variogram functions.

An alternative cosimulation method depends on the type of problem to be solved. Sequential Gaussian cosimulation is based on a mathematically rigorous basis but might not work for discrete variables. It also requires the inference of cross-variogram functions, as well as an assumption of linear relationships. The Bayesian rule method is simple to implement and is based on a sound probabilistic principle. Implementation is difficult if the two types of data provide contrary information. For example, if indicator simulation provides the probabilities of the three facies as 0.2, 0.8, and 0, and the seismic attribute provides the conditional probabilities of those three facies as 0, 0, and 1, we might not obtain an answer. This

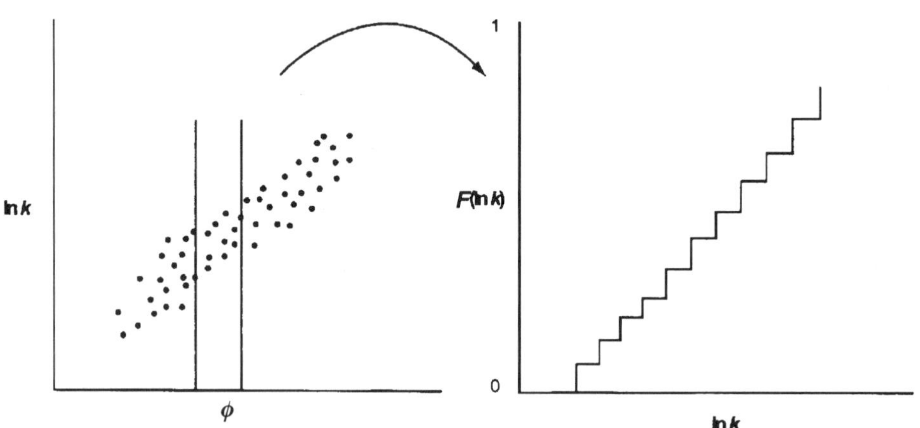

Fig. 6.35—Back transform of permeability.

TABLE 6.5—VARIOGRAM PARAMETERS FOR PERMEABILITY IN FIELD EXAMPLE 6.4	
Model Type	Spherical
Principal Direction, Deg.	112.5° (from North direction)
Nugget	0.3
Sill	0.7
Principal Range, ft	6,000
Minor Range, ft	2,500
Anisotropy Ratio	2.4

type of information is difficult to incorporate correctly. The method of proportion curves is the most robust of all three—it strictly honors the distributions of observed values. However, by restricting it only to the observed data, it could restrict the uncertainty domain much more than the other two methods. Further, by back transforming with a conditional distribution, the realization could exhibit too much variability in the simulated values, which might not be acceptable. Ideally, a combination of the three methods, depending on the type of problem being solved, might be the best solution.

Field Example 6.4—Sequential Gaussian Cosimulation.

• Generate descriptions of permeability for Flow Unit 3 with sequential Gaussian cosimulation (SGCOSIM).

• Compare the statistics of the conditioning data and SGCOSIM realization.

Solution. The following information is given:

• The permeability conditioning data for Flow Unit 3 are available at the same well locations shown in Fig. 6.20. At each well, permeability is estimated with the correlation between porosity and permeability derived from core measurements.

• The variogram model for the normal scores of the logarithm of permeability consists of the following parameters shown in **Table 6.5.**

• The secondary data is the porosity description generated in Field Example 6.1 (Fig. 6.21).

• The correlation coefficient between the normal scores of permeability and porosity from core measurements is 0.77.

• The simulation grid is the same as Field Example 6.1.

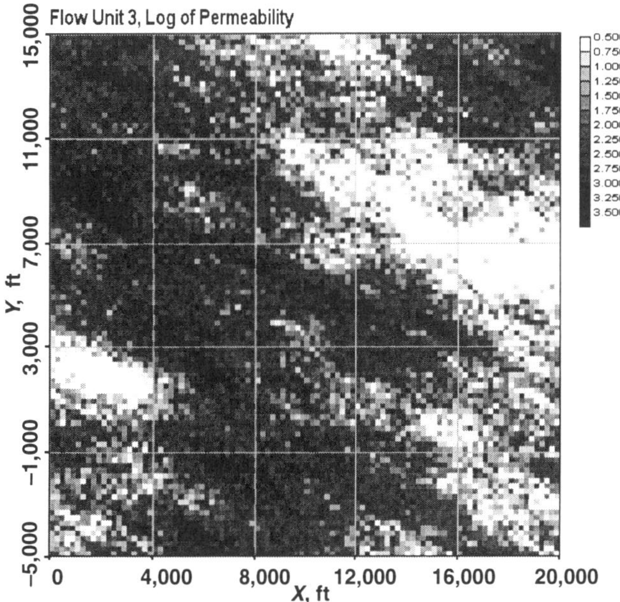

Fig. 6.36—SGCOSIM realization of logarithm of permeability for Flow Unit 3.

In this field example, SGCOSIM simulates permeability with porosity as the secondary variable. This means that SGCOSIM honors the correlation between permeability and porosity, in addition to the univariate and spatial properties honored by SGS. **Fig. 6.36** shows an SGCOSIM realization of the log of permeability. The major trends of low and high values and the overall appearance of this realization are close to the porosity description used for the secondary variable (Fig. 6.21). However, the log of permeability realization exhibits more spatial variability because it has smaller correlation ranges than porosity. **Fig. 6.37** is the same realization shown in Fig. 6.36 but with a linear color scale. Examining the realization with a linear scale gives another perspective of the connectivity of actual low and high permeabilities.

The statistics and histograms of the SGCOSIM realization and the permeability data are shown in **Figs. 6.38** and **6.39,** respectively. SGCOSIM closely reproduces the statistics and histogram of the conditioning data.

The plot between porosity from SGS and permeability from SGCOSIM is shown in **Fig. 6.40.** SGCOSIM closely reproduces the specified correlation coefficient. The large variation around the linear trend is caused by the large degree of freedom needed by SGCOSIM to simultaneously honor the spatial correlation models. The SGCOSIM results also closely reproduce the original porosity and permeability correlation, based on core measurements (**Fig. 6.41**). The simulated values have a slightly narrower range than the core measurements because of the difference in the scales between core plugs and flow units. The simulation conditioning data were generated by scaling-up core plug measurements to the flow unit scale at the wells.

6.2 Probability Field Simulation

One disadvantage of the sequential simulation process is the fact that once the posterior distributions are constructed, we must use a uniform random number generator to sample a value from that distribution. For two locations, where the estimated values as well as the posterior distributions are very similar, it is possible that the simulated values will be different. **Fig. 6.42** shows an example of two nearby sample locations that have similar indicator values at each threshold,

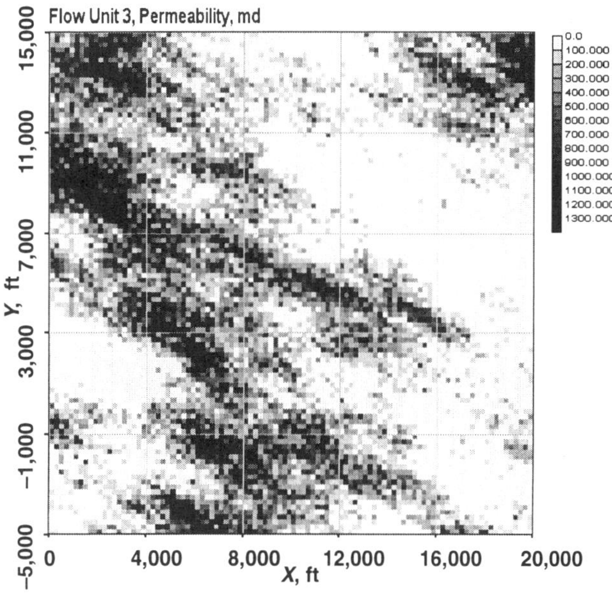

Fig. 6.37—SGCOSIM realization of permeability for Flow Unit 3.

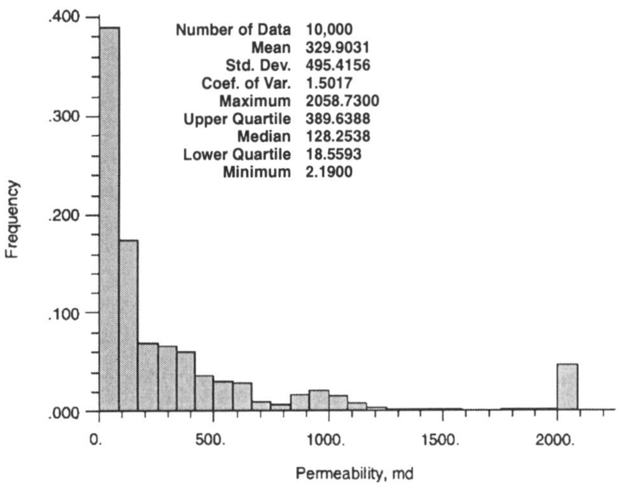

Fig. 6.38—Histogram of SGCOSIM realization of permeability.

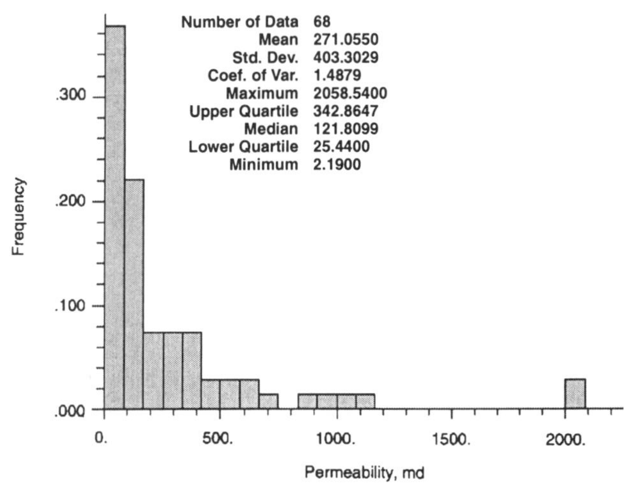

Fig. 6.39—Histogram of permeability data for Flow Unit 3.

which indicates similarities at both locations. Locations one and two are very close to each other. If we decide to simulate a value at each location with a random number generator, it is possible that a value of B will be simulated at location one and a value of D at location two. Because generated random numbers are not correlated with each other, it is possible to create images that exhibit local, short-scale variability, which is not realistic. By sampling two very different values at two locations, we lose the underlying similarities, which are indicated by the posterior distribution.

One option to overcome this problem uses correlated random numbers to simulate a value at each location. For example, if we use the random number 0.5 at location one to sample the value, we will use, for example, a number of 0.52 to sample the value at location two. The simulated values at the nearby locations exhibit similarities as implied by the posterior distribution. At the same time, by using different sequences of correlated values that fall between zero and one, we can generate multiple realizations that do not exhibit unnecessary short-scale variability.

Probability field simulation minimizes the short-scale variability by independently generating values that fall between zero and one and exhibiting spatial correlations of the desired length.

The process that creates these values is not important, as long as they cover the entire range of values from zero to one and exhibit a spatial relationship that is based on the underlying sample data. One possibility is to use probability kriging based on the original sample data. Kriging is done over a larger grid than the desired domain, and by shifting the grid over the domain, we can progressively sample multiple values at each location. An alternative is to use multi-Gaussian fields that have the desired correlation scale (spatial continuity) and transform them into values between zero and one by calculating a cumulative distribution function corresponding to each value. Irrespective of the way we create the correlated fields, once the posterior distributions are created, it is very easy to quickly generate multiple realizations.

Field Example 6.5—Probability Field Simulation.

- Generate descriptions of porosity for Flow Unit 3 with probability field simulation (PFSIM).
- Compare the statistics of the conditioning data and the PFSIM realization.

Solution. The following information is given:

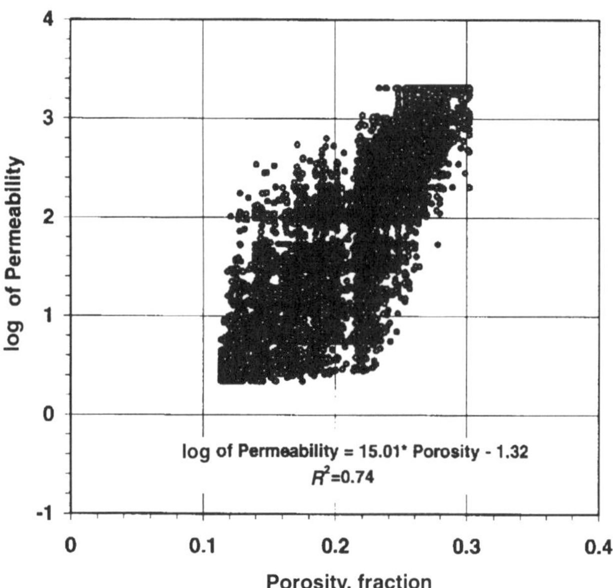

Fig. 6.40—Plot of SGSCOSIM permeability realization and porosity secondary data.

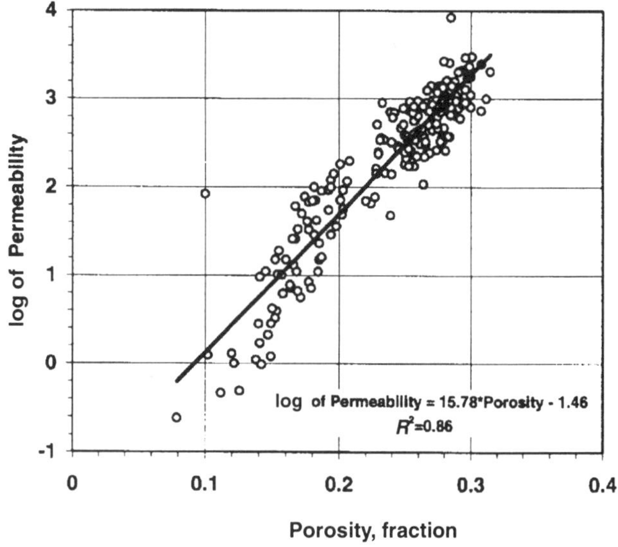

Fig. 6.41—Plot of permeability and porosity core data for Flow Unit 3.

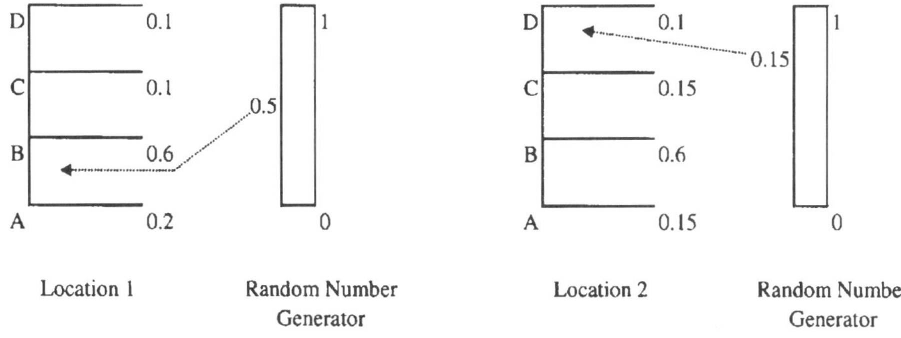

Fig. 6.42—Location 1 and 2—random number generator.

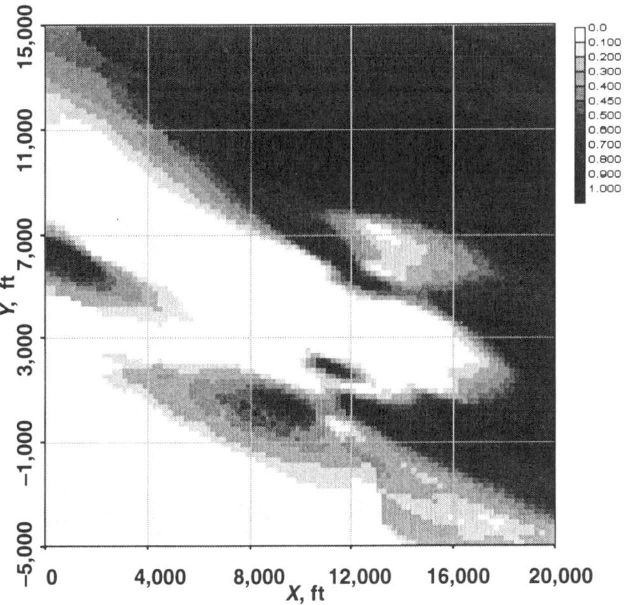

Fig. 6.43—Indicator kriging map of porosity for Flow Unit 3 for threshold = 0.181.

• The porosity is described with three continuous indicator variables. The thresholds are 0.181 (lower quartile), 0.225 (median), and 0.253 (upper quartile).

• The maps of cumulative probability, shown in **Figs. 6.43–6.45** for each threshold, were generated with ordinary indicator kriging. The data for these maps includes the indicator conditioning data (Fig. 6.20) and variogram models used in Field Example 6.2. Chap. 4 provides a description of the indicator kriging technique.

• The probability field is obtained from a realization of normal deviates, generated with SGS. The realization of normal deviates, shown in **Fig. 6.46,** is unconditional with a mean and variance of zero and one, respectively, and it honors the variogram of the normal scores of porosity (Field Example 6.1). The probability field used in PFSIM consists of the Gaussian cumulative probabilities, which corresponds to the normal deviates in Fig. 6.46. A similar field is generated with a probability transform as well.

• The simulation grid is the same as the one used in Field Example 6.1.

The porosity realization, generated with PFSIM, is shown in **Fig. 6.47.** The major correlation trend, which runs southeast/northwest in this realization along the principal direction of the variogram, is similar to SGS (Field Example 6.1) and SIS (Field Example 6.2) realizations. The overall appearance of the PFSIM realization is somewhere between the SGS and SIS realizations (Figs. 6.21 and 6.26). PFSIM does not exhibit the large variability observed in SIS within each indicator threshold because of the additional correlation imposed by the probability field (Fig. 6.46).

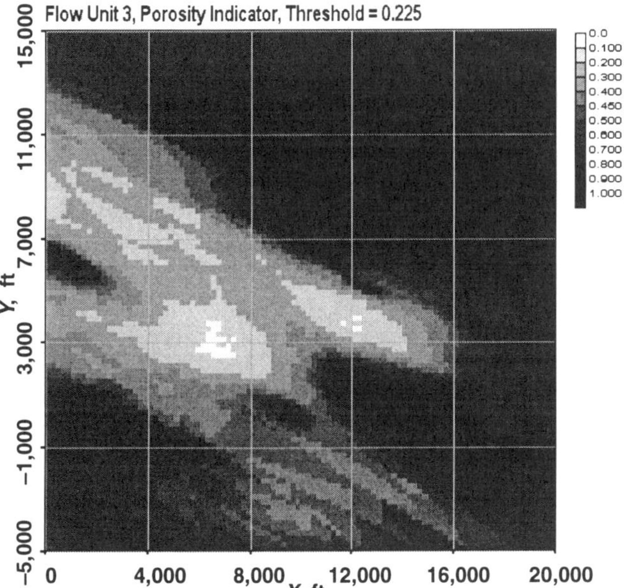

Fig. 6.44—Indicator kriging map of porosity for Flow Unit 3 for threshold = 0.225.

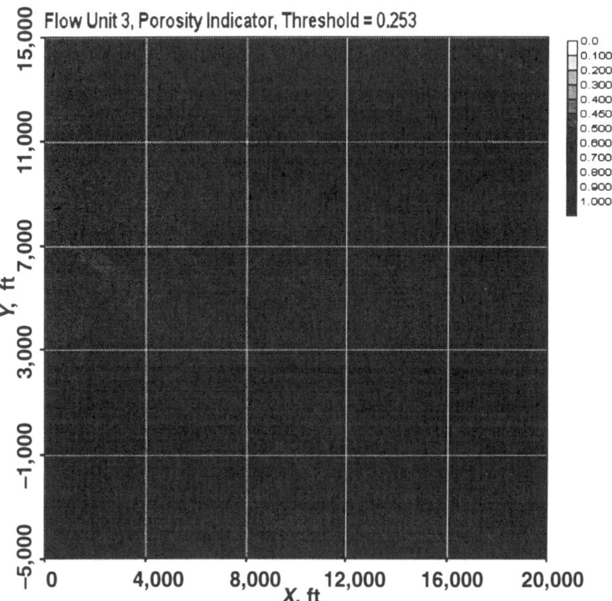

Fig. 6.45—Indicator kriging map of porosity for Flow Unit 3 for threshold = 0.253.

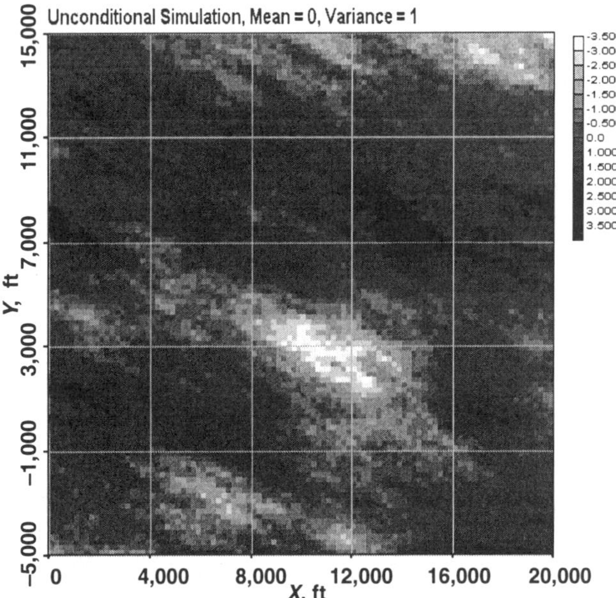

Fig. 6.46—SGS unconditional simulation used for probability field.

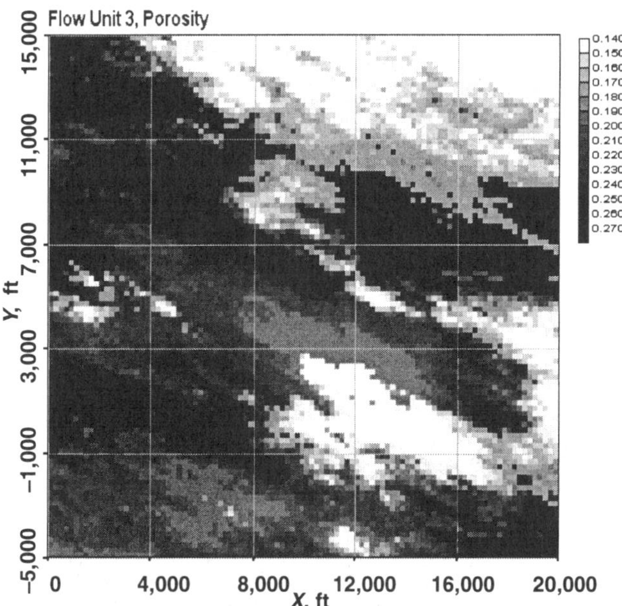

Fig. 6.47—PFSIM realization of porosity for Flow Unit 3.

The statistics and histogram of the PFSIM realization are shown in **Fig. 6.48.** PFSIM honors the univariate statistics very closely. However, it fails to reproduce the bimodal shape of the histogram of the conditioning data (Fig. 6.25) because only three thresholds are used to describe the probability distribution. Similar to the case of SIS (Field Example 6.2), more indicator thresholds are needed to better define the entire shape of the probability distribution.

The major advantage of PFSIM is the speed it generates for the realizations. It is effectively a post-processing algorithm. This makes PFSIM the fastest of the conditional simulation techniques described in this chapter. The disadvantage is that it requires significant work prior to the actual simulation to generate the probability distribution maps and probability fields.

6.3 Simulated Annealing

In this section, we describe another popular technique that is extensively used to integrate various reservoir properties. The method of simulated annealing is based on the principles that are useful for describing the cooling of molten metal. Similar to the sequential simulation method, it also honors the univariate statistics, spatial relationships, and relationships among various attributes. Unlike the sequential simulation method, the simulated annealing method has the flexibility to incorporate other constraints, which might be difficult to incorporate within a geostatistical framework. Therefore, the advantage of the simulated annealing process is its flexibility and robustness. The disadvantage of simulated annealing is its lack of speed in converging to a desired outcome. In general, simulated annealing tends to be much slower than conditional simulation techniques for the same type of constraints.

We divided this section into several subsections. In the first subsection, we explain the background of simulated annealing. In the second subsection, we discuss the simulated annealing algorithm and the underlying principles. In the last subsection, we explain the annealing process and its implementation. That section also includes an illustration of the process with a field example.

6.3.1 Background. As explained previously, the simulated annealing procedure was established, based on the physical process of cooling molten metal. In cooling molten metal, it is noted that the rate of cooling affects the quality of the frozen metal. Slower cooling results in better quality, whereas faster cooling results in a more defective product. The reasons for this difference are traced back to thermodynamic principles. Thermodynamic principles tell us that molten metal tries to achieve the most stable state (corresponding to the lowest energy), if possible. This requires the molecules to move randomly within the system, sometimes acquiring a state that temporarily results in a higher energy state. However, by allowing the molecules to search through many states—sometimes favorable and sometimes unfavorable—the solution domain for the most stable state can be properly explored, and local minima in the system are avoided. The motion of the molecules depends on the temperature; therefore, the higher the temperature, the more likely the molecules will move randomly within the system, thus reaching places of favorable and unfavorable states with great frequency. As the temperature is slowly reduced, the molecules, after exploring the solution domain, settle down in their most stable state, which eventually results in a less defective, more stable product. To

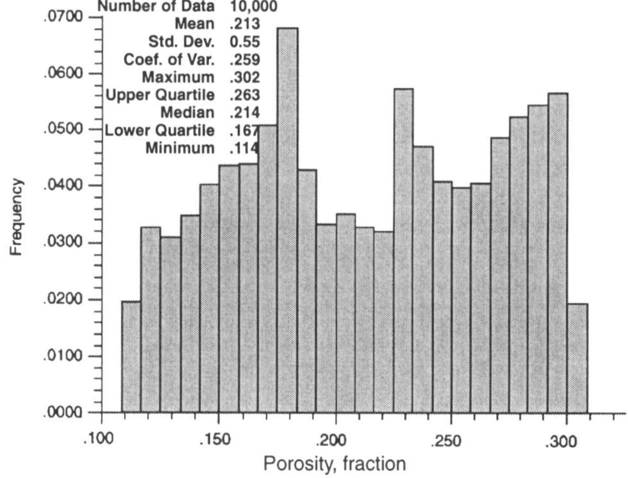

Fig. 6.48—Histogram of PFSIM realization of porosity.

properly achieve the balance between the economics of cooling more quickly and achieving a good product, the annealing schedule was developed, which was later adopted for other mathematical problems with similar characteristics.

According to Aarts and Korst,[15] the mathematical algorithm of simulated annealing was independently introduced by Kirkpatrick *et al.*[16] and Cerny.[17] Per Collins *et al.*,[18] some applications of simulated annealing include the placement of computer components, wiring of electronic components, and image processing. Several investigators used the traveling salesman problem (calculation of the shortest distance required to visit several geographical sites) as a benchmark for different implementations of simulated annealing. The first application of simulated annealing to describe reservoir properties was introduced by Farmer.[19] Subsequent to Farmer, several investigators used the concept of annealing to describe various reservoir properties.[20]

Because simulated annealing is founded on concepts from statistical mechanics and the annealing process, the implementation of simulated annealing requires knowing the three system component definitions that are analogous to the energy, temperature, and interactions of a molecular system. The objective function represents the energy of the system, and it is defined as the function to minimize (or maximize). The control parameter represents the temperature of the system, which is an independent parameter and not necessarily related to any other parameter of the problem. The interchange mechanism corresponds to the molecular interactions, and it usually consists of a finite set of perturbations to the independent variables, which produces a change of the objective function.

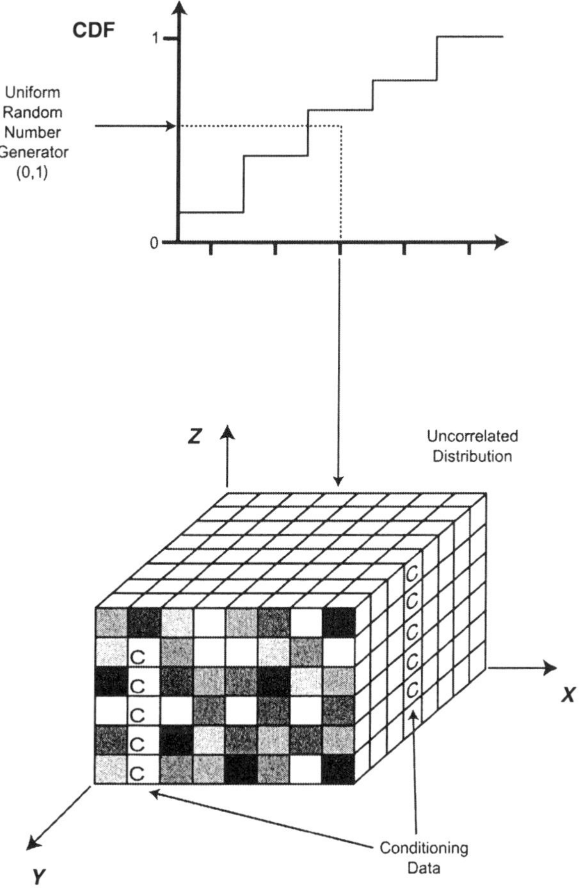

Fig. 6.49—Schematic of transformation method to generate initial distribution.

The goal of the simulated annealing method is to determine the configuration of the independent variables that yields a global minimum or maximum of the objective function. The simulation process consists of performing a specific number of changes to the independent variables with the interchange mechanism, and for each change, the objective function is evaluated. A change of variables is accepted and retained if it reduces the magnitude of the objective function. If a change increases the magnitude of the objective function, it is accepted or rejected and discarded according to the probability function proposed by Metropolis *et al.*[21] This probability function, introduced by Metropolis *et al.* to study the statistical mechanics of molecular systems, is a Boltzmann distribution, which depends on the energy and temperature of the system. At high temperatures, when the system's energy and degree of disorder are large, there is a high probability that a change will be accepted, which results in an increase of energy. As the temperature decreases, the energy decreases, and the system (molecules) becomes more orderly, and at this stage, there is a small probability that a change will be accepted that results in an increase of the systems energy. This probabilistic feature of simulated annealing allows the system to escape from the solutions that correspond to the local minima of the objective function.

The procedure that reduces the temperature or control parameter and specifies the number of iterations or changes required at each temperature is known as the annealing schedule. If the control parameter is reduced too fast, the system will reach a local minimum. At each level of the control parameter, the number of iterations or changes must be large enough for the energy function to achieve an equilibrium state. The control parameter is reduced until the stopping criteria are satisfied. Examples of stopping criteria are attained small-valued energy functions (in the case of a minimization process) and further changes at subsequent control parameter levels that have negligible effects on the changes in the energy function.

6.3.2 Simulation Algorithm. Similar to sequential simulation methods, the goal of a simulated annealing method is to generate a distribution of random variables with specific statistical properties in a three-dimensional (3D) grid. The constraints, which the distribution must satisfy, are normally captured in the objective function. The objective function described has enough flexibility, so, in principle, any constraint can be incorporated as a part of the objective function. Some examples of objective functions are described later.

Initial Distribution. In the first step of the simulation algorithm, we generate an initial distribution of random variables. Any initial distribution, deemed suitable, can be used to describe the initial distribution. In practice, an initial distribution, based on well data, is used to populate a 3D grid. **Fig. 6.49** shows the procedure to generate the initial distribution.

A random number, in the range of 0–1, is drawn from a random number generator, and a variable value is assigned to a gridblock, by sampling a value from the CDF of the variable. The CDF is derived from the well data, or it is defined based on outside information. The procedure is repeated until all gridblocks are assigned a variable value. The gridblocks with conditioning data (i.e., well locations) are not assigned values because the conditioning data represent that well location. If we are interested in generating the distribution of more than one variable, then the gridblock cells are populated with all the desired variables. For example, as in the case of cosimulation, if we are interested in assigned geological facies, porosity, and permeability to a gridblock, all three variables must be assigned to an individual gridblock.

Objective Function. The objective function or the energy function represents the function that must be minimized, as part of the annealing algorithm. The objective function can contain many terms representing various constraints. In general, we define the objective function as

$$O^k = \frac{1}{O_o}\sum_{i=1}^{n} w_i O_i^k, \quad (6.17)$$

where O_o is defined as the initial objective function,

$$\text{and } O_o = \sum_{i=1}^{n} w_i O_{oi}. \quad (6.18)$$

O_o = an individual component of the overall objective function, and w_i = the weight assigned to it. The superscript k represents the value of the objective function after the kth iteration.

The weight assigned to an individual component of the objective function is arbitrary and depends on the importance of a particular component. If more emphasis must be provided to a particular component, a larger weight is assigned to that component; if the component is less important, a smaller weight is assigned to it. In many instances, the exchange mechanism does not alter the individual components by the same amount. To compensate for such differences, the weights to individual components might need adjustment, so one or a few components do not dominate the minimization process.

To provide an idea about the flexibility of the annealing process, some examples of the objective function components are illustrated.

Spatial Relationship. Variogram models are matched by defining the objective function as

$$O_i^k = \sqrt{\sum_{j=1}^{N_d}\sum_{l=1}^{N_{h,j}}\left[\frac{\gamma_e(h_{l,j}) - \gamma_o(h_{l,j})}{\gamma_o(h_{l,j})}\right]^2}, \quad (6.19)$$

where

$$O_{io} = \sqrt{\sum_{j=1}^{N_d}\sum_{l=1}^{N_{h,l}}\left[\frac{\gamma_e^o(h_{l,j}) - \gamma_o(h_{l,j})}{\gamma_o(h_{l,j})}\right]^2}, \quad (6.20)$$

where O_i^k = the objective function after the iteration step k; N_d = the number of directions in which variogram models are defined; $\gamma_e(h_{l,j})$ = the estimated variogram after k iterations; $\gamma_o(h_{l,j})$ = the model variogram at lag distance h in direction j; and O_{io} = the objective function at iteration zero. $N_{h,j}$ represents the number of lags used in direction j.

The term in the denominator amplifies the effect of the variogram for smaller lags in the objective function. Although not necessary, this modification might be useful when magnifying the effect of conditional data.

Cumulative Distribution Function. It might be desirable for the cumulative distribution function, in different parts of the reservoir, to be different, or that each layer, in the reservoir, has a different distribution. This type of constraint is captured by

$$O_i^k = \sum_{l=1}^{n_c}\left[\frac{F_l^k(x_{t_l}) - F_{l_o}(x_{t_l})}{F_{l_o}(x_{t_l})}\right]^2, \quad (6.21)$$

where

$$O_{io} = \sum_{l=1}^{n_c}\left[\frac{F_l^o(x_{t_l}) - F_{l_o}(x_{t_l})}{F_{l_o}(x_{t_l})}\right]^2, \quad (6.22)$$

where $F_l^k(x_{t_l})$ = the cumulative distribution function at threshold x_{t_l}, after k iterations; $F_l^o(x_{t_l})$ = the desired cumulative distribution function at threshold value x_{t_l}; and O_{io} = the initial objective function at iteration zero. n_c is the number of classes in which the overall distribution is divided.

Correlation Between Two Variables. Sometimes it is desirable for the correlation coefficient between the two variables to reach a desired value. An example is a correlation coefficient between permeability and porosity at interwell locations, which might need to be the same as the one observed at the well locations. Alternatively, we might desire that the correlation coefficient, between seismic data and porosity at interwell locations, attain a desired value. This type of constraint is incorporated as

$$O_i^k = \sqrt{\left[\frac{r^k(x,y) - r_o(x,y)}{r_o(x,y)}\right]^2}, \quad (6.23)$$

where

$$O_{io} = \sqrt{\left[\frac{r^o(x,y) - r_o(x,y)}{r_o(x,y)}\right]^2}, \quad (6.24)$$

where $r^k(x,y)$ = the correlation coefficient between variables x and y after k iterations; $r_o(x,y)$ = the desired correlation coefficient between variables x and y, and $r^o(x,y)$ is the initial correlation coefficient value at iteration zero. If, instead of the correlation coefficient, the conditional distribution of variable y must be satisfied for a particular class of x, then a condition similar to the one described in the cumulative distribution function case can be used. See Eqs. 6.21 and 6.22. The only difference is that Eqs. 6.21 and 6.22 must be replaced for variable y for every class of x for which the conditional distribution must be honored.

Well Test Permeability. If we want to incorporate well test permeability as part of the constraint, we can incorporate it either by matching the pressure data directly or by using an effective average of near wellbore gridblock permeability values. If we assume that we know the effective averaging technique to estimate the well test permeability, we can define the objective function as[22]

$$O_i^k = \sqrt{\left(\frac{k_{wt}^k - k_{wt_o}}{k_{wt_o}}\right)^2}, \quad (6.25)$$

where

$$O_{io} = \sqrt{\left(\frac{k_{wt}^o - k_{wt_o}}{k_{wt_o}}\right)^2}, \quad (6.26)$$

where k_{wt}^k = the effective average of near wellbore permeability values after k iterations; k_{wt_o} = the well test permeability, based on evaluations of pressure data; and k_{wt}^o = the average of the initial distribution of near wellbore permeability values. The estimated value k_{wt}^k is calculated many different ways. For example, one possibility is

$$k_{wt} = \left(\prod_{l=1}^{n_{wt}} k_l^w \right)^{1/w}, \quad \ldots\ldots (6.27)$$

where k_l = the permeability of gridblock l, and w = a value between -1 and $+1$, except zero.

-1 represents the harmonic average, and $+1$ represents the arithmetic average. n_{wt} represents the number of gridblocks used for averaging. Eq. 6.25 is written for a single well test. If multiple well tests are available, this equation easily extends.

Production Data. Production data are also incorporated as part of the constraint. For example, if pressure data are available from many wells, and through a history matching process, we want to honor the pressure data, we can write the objective function as

$$O_i^k = \sqrt{\sum_{j=1}^{n_w} \sum_{l=1}^{n_t} \left[\frac{p_{l,j}^k - \left(p_{l,j}\right)_o}{\left(p_{l,j}\right)_o} \right]^2}, \quad \ldots\ldots (6.28)$$

where

$$O_{io} = \sqrt{\sum_{j=1}^{n_w} \sum_{l=1}^{n_t} \left[\frac{p_{l,j}^o - \left(p_{l,j}\right)_o}{\left(p_{l,j}\right)_o} \right]^2}, \quad \ldots\ldots (6.29)$$

where $p_{l,j}^k$ represents the simulated pressure value after iteration k for the lth time step and j^{th} well; $\left(p_{l,j}\right)_o$ = the observed pressure value; $p_{l,j}^o$ = the initial simulated pressure value at iterations zero; n_w = the number of wells for which data are available; and n_t is the number of time steps.

Many other constraints are considered as well, including multipoint statistics, as explained in Chap. 3. It is, however, important to understand that the more constraints imposed as part of the energy function, the more difficult it becomes to converge to a solution. Computationally, it also becomes more demanding as more constraints are added. One remark about the constraints is very relevant. The constraints that work best in simulated annealing are the ones that can be locally updated, when we perturb the system through the exchange mechanism. If every perturbation in the system requires recomputing the objective function, the procedure becomes extremely inefficient. On the other hand, if the objective function is updated, instead of requiring recomputation, the procedure becomes very efficient. Many of the constraints discussed in this section can be locally updated and, therefore, are the types of constraints that will be effective for the simulated annealing procedure. An exception to this local updating requirement is the incorporation of dynamic production data. If we use a flow simulator to generate the pressure data used in the objective function, the simulator has to run at each iteration to calculate the objective function. No local updating is possible. Therefore, incorporation of production data, as part of the constraint in simulated annealing, is very difficult and appears impractical for large field studies under present conditions. Production data can only be incorporated in an annealing procedure when we find a clever way to locally update a flow simulation after every perturbation. No such procedure is available at present. Fortunately, most of the objective functions that incorporate static constraints can be locally updated. For examples of objective functions that can be locally updated, see Appendix E.

Interchange Mechanism. Once the initial distribution and objective function are defined, the system can be perturbed or altered through various interchange mechanisms. One of the most common techniques swaps the values of the variables at two locations, selected at random in the simulation region. Any two locations can be selected, as long as those locations do not coincide with the locations of the conditioning data. After the locations are selected, but before performing the actual exchange, we calculate the objective function that would result if the exchange were completed. We call this energy function O'. After calculating O', we calculate the difference in the objective functions O' and O^k—the objective function value at the previous step is

$$\Delta O^k = O' - O^k. \quad \ldots\ldots (6.30)$$

If ΔO^k is negative, the objective function O' is smaller than O^k; therefore, the exchange is accepted, and the distribution is updated by executing the interchange. If ΔO^k is positive, the interchange could still be accepted, if the Metropolis condition is satisfied.

The Metropolis condition is defined as

$$p(\Delta O^k, T) = \exp\left(-\frac{\Delta O^k}{T} \right), \quad \ldots\ldots (6.31)$$

where p is the probability function; ΔO^k = the change in energy (or objective function); and T = a control parameter (temperature parameter). **Fig. 6.50** shows the effect of the control parameter on the probability function. To determine if we should accept or reject an exchange for a positive ΔO^k, we select a random number between zero and one from a uniform random number generator and compare it with the probability value determined with Eq. 6.31. If the random number is less than the probability value, we accept the swap; otherwise, we reject it.

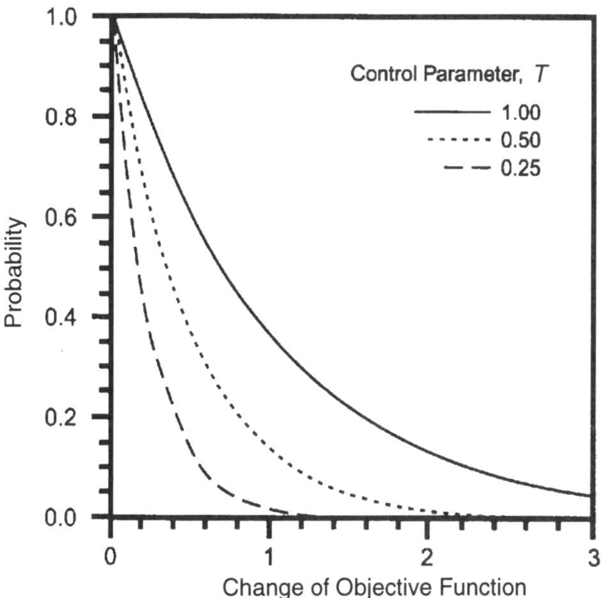

Fig. 6.50—Probability function for Metropolis condition.

The reason for applying the Metropolis condition is to avoid reaching a local minimum with respect to the objective function. By temporarily increasing the objective function, we eventually reach a global minimum with respect to the energy function. This procedure is repeated until a desired value of the objective function is reached. At that point, we stop the simulation. The distribution created satisfies the conditioning data, initial distribution, and any other constraints that are imposed.

Other interchange mechanisms are possible, as well. For example, instead of using a swap of two randomly selected points, a single point is perturbed at a time, and the objective function is calculated. The point is perturbed by randomly selecting a value from the cumulative distribution function to replace the original value or by selecting a value from the cumulative distribution function with a prior probability value associated with each value, depending on how favorable a change it creates. This type of exchange mechanism is called a heat-bath algorithm.[23]

Another modification of the exchange mechanism is to only accept perturbations that result in a favorable change in the objective function. Any perturbation that causes an unfavorable change is always rejected. This is called a greedy mechanism.[20] For generating reservoir descriptions with various constraints, this mechanism proves to be adequate to reach desired minimum values of objective functions.

6.4 Simulation Process

The simulation process in a simulated annealing method is quite involved because of the variety of input parameters and the types of stopping criteria one can use to terminate the simulation. In this section, we briefly describe the important parameters that control the process, and then we describe the process in detail.

6.4.1 Input Parameters. The five important parameters that control the simulation process are:

• Control Parameter, T.

The control parameter is analogous to temperature in the annealing process. This parameter determines the acceptance rate of unfavorable perturbations. The higher the value of the control parameter, the higher is the acceptance rate. Although the control parameter should be continually reduced as the system gets closer to the final desired objective function, in practice, the control parameter is changed discretely from one step to another. Different mechanisms for changing the control parameter from one step to another are available;[24] the simplest of these is

$$T^{r+1} = \alpha T^r, \quad \text{(6.32)}$$

where r represents the step number, and α represents a fraction between zero and one, which defines the ratio of the control parameter for the new step, $r+1$, to that of the previous step, r. The smaller the value of α, the faster the control parameter is reduced; the higher the value of α, the slower the control parameter is reduced. A good value of α is 0.5.[15]

The initial value of the control parameter must also be calculated at the first step. Although a value of 1.0 is commonly assumed, a better estimate of the initial control parameter is obtained by requiring a large number of unfavorable perturbations to be accepted at the initial stage. One possible method is shown in Appendix E.

• Maximum Accepted Swaps Per Step, M_a.

We must accept a certain number of swaps at each step before the control parameter is reduced. If we accept too few swaps, the control parameter is reduced too quickly, with the possibility of reaching a local minimum. If we accept too many swaps per step, the procedure is slow and inefficient.

If we define one cycle as the total number of gridblocks, then a good value of M_a is five cycles. For example, if we have 10,000 gridblocks, then we must accept 50,000 swaps at each step before moving to the next step.

• Maximum Number of Swaps at Each Step, M_t.

At the beginning of the simulation process, the total number of accepted swaps and the total number of swaps are approximately equal. As the control parameter is reduced, the percentage of accepted swaps becomes smaller as more swaps are rejected. To ensure that we can change the control parameter after a certain number of swaps, we provide an alternate mechanism to change the control parameter. This mechanism involves reaching a maximum number of total swaps, M_t. Once that number is reached, the control parameter is changed to the next step. The value of M_t changes, depending on the control parameter. The smaller the value of the control parameter, the larger is the value of M_t. One possible method for calculating M_t as a function of the control parameter is illustrated in Appendix E. In the alternate, a fixed value of M_t is assumed. A good number is 40 cycles.

• Objective Function Tolerance, E.

The simulation process is terminated when a desired tolerance is reached for the objective function. Because the objective function is already normalized with respect to the initial value of the objective function, the tolerance is defined with respect to the initial value of the objective function. In simulating the gridblock values, a starting value of E equal to 0.01, or even 0.1, is reasonable. If the process converges without any hitch to that value, higher tolerance values (e.g., 0.001 or 0.0001) are used to improve the quality of the results.

• Stopping Number, S.

An alternate mechanism for terminating the program is necessary when it becomes obvious that most of the swaps are rejected and better convergence is not possible. If we define the acceptance ratio as

$$A_r = \frac{m_r}{t_r}, \quad \text{(6.33)}$$

where m_r = the accepted number of swaps in step r, and t_r = the total number of swaps in step r, the value of the acceptance ratio should be respectable, to make sure that the process of simulated annealing is reasonable. For example, if we reach a value of A_r less than 0.01, less than one in 100 swaps is accepted. This is not very efficient. If the stopping number is three, then when the acceptance ratio reaches a value less than 0.01 in three successive steps, the program is terminated. A good value for the minimum acceptance ratio is 0.01, and a good value for the stopping number is three.

6.4.2 Process. The steps of the simulation process are:

• Calculate the initial distribution of the variable, and distribute it randomly within the region of stationarity.

• Calculate the initial energy function, O_o, with Eq. 6.18.

• Calculate the initial value of the control parameter, T^o.

• Calculate the maximum number of iterations at each step. This number, M_t, typically has units of cycles, where one cycle represents the total number of gridblocks within the region. Also, set the number of iterations to be accepted in each

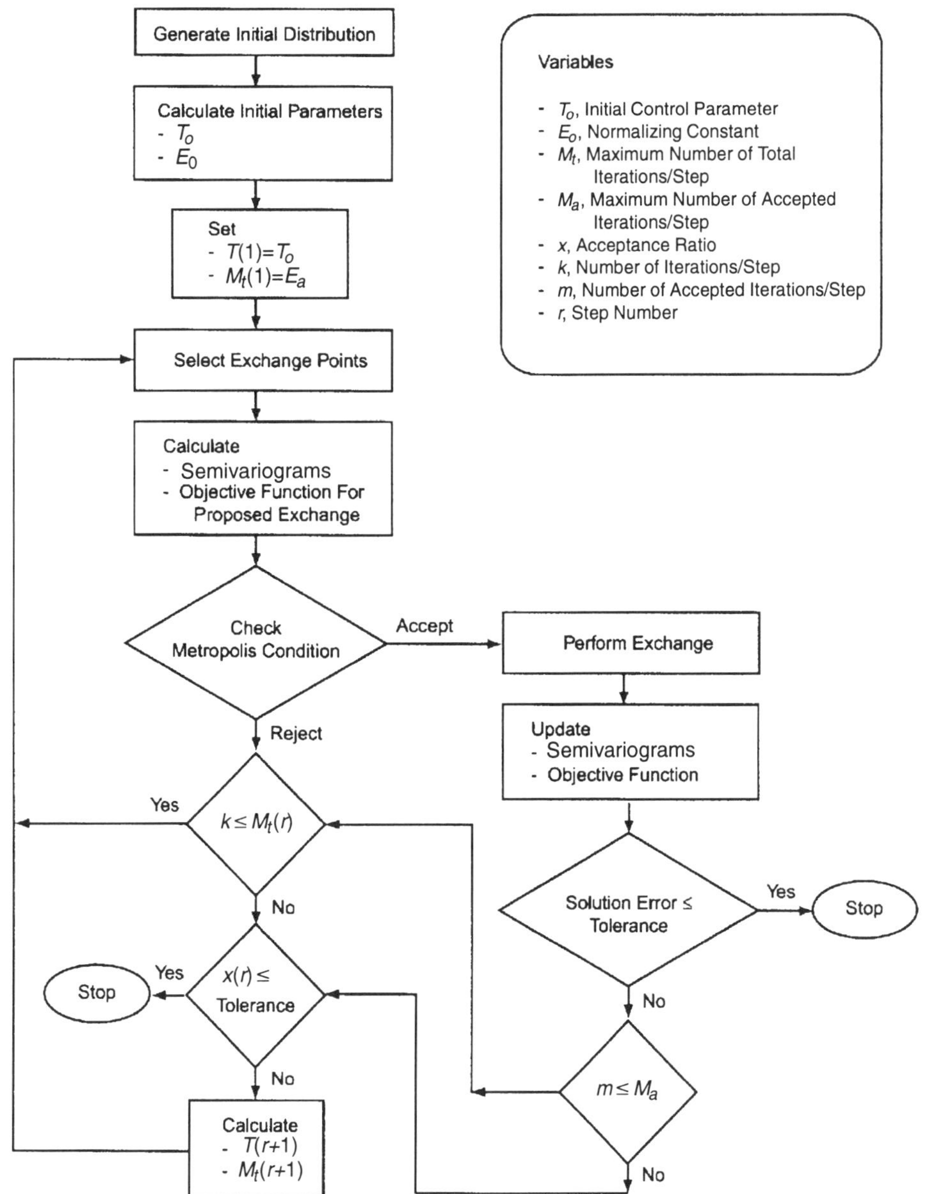

Fig. 6.51—Simulated annealing algorithm.

step. This number, M_a, is also defined in terms of cycles. A good assumption for this number is 5.0.

• Generate a perturbation in the system through the exchange mechanism. Calculate the energy function, O', and, hence, calculate ΔO^k. Apply the Metropolis condition. If the exchange is accepted, update the initial distribution, as well as the energy function, by defining

$$O^{k+1} = O', \quad \text{(6.34)}$$

where k is the previous iteration.

• Check for the tolerance of the objective function. If

$$O^{k+1} \leq E, \quad \text{(6.35)}$$

where E is a tolerance, with respect to the objective function, you can stop simulation.

• Repeat the procedure at a given control parameter, T, until the maximum number of swaps, M_t, or the maximum number of accepted swaps, M_a, is reached.

• Check for the acceptance ratio for the present step. The acceptance ratio for step r is given by

$$A_r = \frac{m_r}{t_r}. \quad \text{(6.36)}$$

If $A_r <$ the minimum fraction of accepted swaps (typically 0.01) for successive steps equal to the swapping number, the simulation is stopped. This indicates the futility of carrying out the simulation any further because most of the swaps are being rejected.

• Calculate the control parameter for the next step with Eq. 6.32.

• Start with the fourth step (perturbation) in the procedure.

The simulation procedure terminates when either Eq. 6.35 is satisfied or the stopping number is reached. The flow diagram of the simulation process is shown in **Fig. 6.51.**

Field Example 6.6—Simulated Annealing Simulation.

• Generate descriptions of porosity for Flow Unit 3 with simulated annealing simulation (SAS).

• Compare the statistics of the conditioning data and SAS description.

Solution. The following information is given:

TABLE 6.6—ANNEALING PARAMETERS FOR FIELD EXAMPLE 6.6	
Porosity	Variogram Model
Model Type	Spherical
Principal Direction, Deg.	112.5° (from North direction)
Nugget	0
Sill	0.0023
Principal Range, ft	8300
Minor Range, ft	2800
Anisotropy Ratio	2.96
Annealing Schedule	Parameters
Initial Temperature	1.0
Temperature Reduction Factor	0.2
Maximum Number of Swaps	500 × Number of Blocks
Maximum Number of Accepted Swaps	30 × Number of Blocks
Number of Times that Maximum Number of Swaps is Reached (Stopping Criterion)	10
Minimum Objective Function	0.001

- The porosity conditioning data for Flow Unit 3 is the same as in Field Example 6.1 (Fig. 6.20).
- The variogram model for porosity is shown in **Table 6.6.**
- The annealing schedule is shown in Table 6.6.
- The simulation grid is the same as the one used in Field Example 6.1.

The SAS porosity realization is shown in **Fig. 6.52.** SAS reproduces the major porosity trend along the variogram principal direction. However, the small-scale variability appears to be greater than that observed in the SGS (Field Example 6.1) and SIS (Field Example 6.2) realizations. **Fig. 6.53** shows the variograms of the SAS realization. Similar to the SGS case (Field Example 6.1), the SAS experimental variograms exhibit more continuity than those of the models. In theory, it is possible to design a better annealing schedule that yields a closer match between the experimental and model variograms.

Fig. 6.54 shows the histogram of the porosity realization. SAS honors the statistics and the histogram of the conditioning data (Fig. 6.25) better than the other conditional simulation techniques in this chapter.

SAS is the most general simulation technique considered in this chapter. It can simultaneously accommodate several constraints like conditioning data, histograms, variograms, indicator variograms, correlations with secondary variables, and secondary information, including seismic, well-test and production data. However, the widespread use of SAS is limited because it suffers from two major weaknesses. First, the annealing schedule varies for different conditions. It is difficult to design an optimum schedule that works for all cases. The general guidelines proposed in the literature for annealing schedules serve only as a starting point. Even experienced users must generate several preliminary trial simulations to compose a reasonable annealing schedule. Second, SAS is the slowest of the techniques considered in this chapter.

Summary

Chap. 6 discusses popular conditional simulation methods. Although other conditional simulation methods are available in the literature, in the last several years, the methods described in this chapter are the ones most often used. Sequential simulation methods are fast and efficient. However, these methods handle a limited number of constraints. Simulated annealing, on the other hand, is very flexible and robust. It incorporates

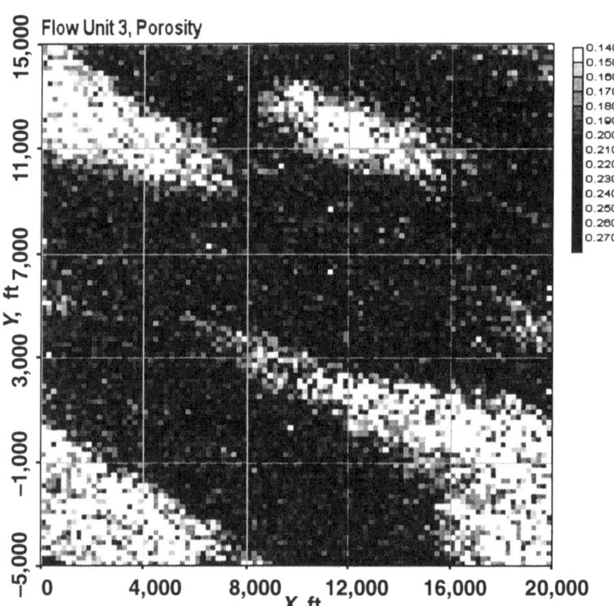

Fig. 6.52—SAS realization of porosity for Flow Unit 3.

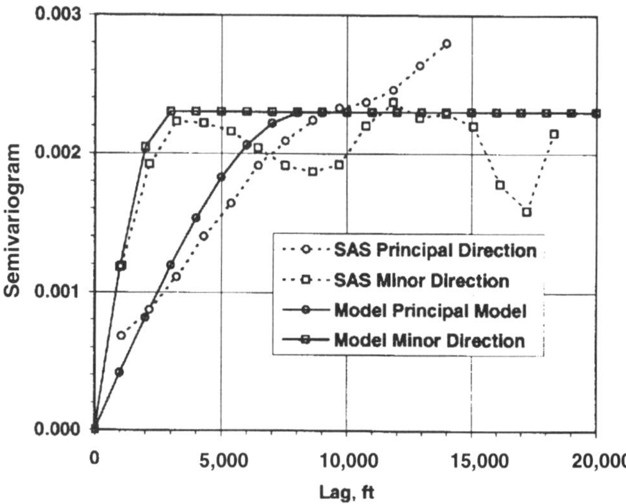

Fig. 6.53—Variograms of SAS realization of porosity and models.

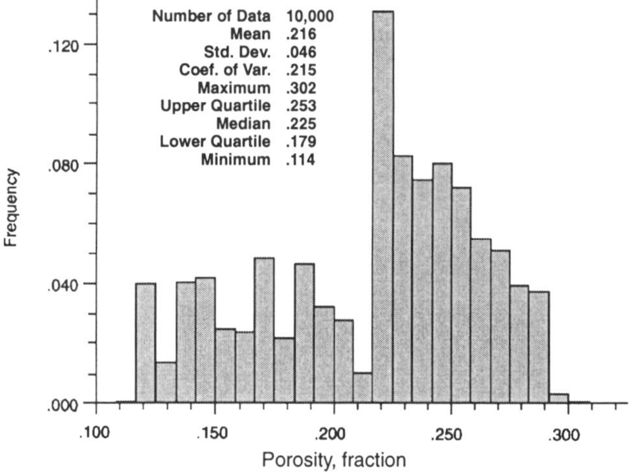

Fig. 6.54—Histogram of SAS realization of porosity.

varied constraints. However, it is computationally demanding. Ultimately, there is a possibility that no single method will be sufficient to generate a reservoir description for a particular case. A combination of methods might be needed before a satisfactory reservoir description is constructed.

Nomenclature

A,B = events in a random experiment
A_i = mutually exclusive event i
A_r = acceptance ratio in step r
E = tolerance in objective function
$F(x)$ = cumulative distribution function of variable x
h = lag distance
$I(\vec{u},x)$ = indicator transform of a continuous variable
$I(\vec{u},K)$ = indicator transform of a discrete variable
i = ranking of a sample
j = subscript; dummy variable
K = categorical variable
$K(\vec{u})$ = categorical variable at location $\vec{u}$
k = permeability
m = number of accepted swaps per step
M_a = maximum number of accepted swaps in each step
M_t = maximum number of total swaps in each step
m_r = accepted swaps in step r
N_d = number of directions in which variogram models are defined
$N_{h,j}$ = number of logs used in direction j
n = total number of samples
O = objective function
p = probability
r = correlation coefficient
S_i = an event that seismic value in class i exists
T = control parameter
t_r = total number of swaps in step r
$\vec{u}$ = location
$X(\vec{u})$ = random variable at location $\vec{u}$
x = realization
w_i = weight assigned to individual objective functions
y = coordinate axis
Y = coordinate axis
Z = coordinate axis
α = multiplier for control parameter in simulated annealing
ϕ = porosity
γ = variogram
λ = kriging weight
ω = permeability exponent in upscaling
$\hat{\sigma}_E^2$ = error variance
ΔO = incremental objective function

Subscripts

c = classes
e = estimated
G = Gaussian transform
j = indicates location j
k = dummy variable, attribute k
l = dummy variable
N = normal (Gaussian) transform
o = unsampled location; also model value in simulated annealing
p = probability transform
r = step r
t = threshold
wt = well test

Superscripts

$*$ = estimated value
k = kth iteration
o = initial value of objective function
r = step r

References

1. Journel, A.G. and Alabert, F.: "New Method for Reservoir Mapping," *JPT* (February 1990) 212.
2. Gomez-Hernandez, J. and Srivastava, R.: "ISIM3D: An ANSI-C Three Dimensional Multiple Indicator Conditional Simulation Program," *Computer and Geosciences* (1990) **16,** No. 4, 395.
3. Alabert, F.: "The Practice of Fast Conditional Simulation Through the LU Decomposition of the Covariance Matrix," *Math Geology* (1987) **19,** No. 5, 369.
4. Davis, M.: "Production of Conditional Simulations via the LU Decompositions of the Co-Variance Matrix," *Math Geology* (1987) **19,** No. 2, 91.
5. Galli, A. *et al.*: "The Pros and Cons of the Truncated Gaussian Method," *Geostatistical Simulations,* M. Armstrong and P.A. Dowd (eds.) Kluwer Academic Publishers (1994) 217.
6. Srivastava, R.M.: "Reservoir Characterization with Probability Field Simulation," paper SPE 24573 presented at the 1992 SPE Annual Technical Conference and Exhibition, Washington, DC, 4–7 October.
7. Jordan, D.L.: "An Application of Categorical Indicator Geostatistics for Facies Modeling in Sand-Rich Turbidite Reservoirs," paper SPE 30603 presented at the 1993 SPE Annual Technical Conference and Exhibition, Dallas, 22–25 October.
8. Xu, W. and Journel, A.G.: "GTSIM: Gaussian Truncated Simulations of Reservoir Units in a W. Texas Carbonate Field," paper SPE 27412 available from SPE, Richardson, Texas (1993).
9. Alabert, F.G. and Massonnat, G.J.: "Heterogeneity in a Complex Turbiditic Reservoir: Stochastic Modeling of Facies and Petrophysical Variability," paper SPE 20604 presented at the 1990 SPE Annual Technical Conference and Exhibition, New Orleans, 23–26 September.
10. Almeida, A.S. and Journel, A.: "Joint Simulation of Multiple Variables with a Markov-type Coregionalization Model," *Math Geology* (1994) **26,** No. 5, 565.
11. Doyen, D.M., Psaila, D.E., and Strandenes, S.: "Bayesian Sequential Indicator Simulation of Channel Sands from 3-D Seismic Data in the Oseberg Field," paper SPE 28382 presented at the 1994 SPE Annual Technical Conference and Exhibition, New Orleans, 25–28 September.
12. Doyen, P.M., Psaila, D.E., and Den Boer, L.D.: "Reconciling Data at Seismic and Well Log Scales in 3D Earth Modeling," paper SPE 38698 presented at the 1997 SPE Annual Technical Conference and Exhibition, San Antonio, Texas, 5–8 October.

13. Bahar, A.: "Co-Simulation of Lithofacies and Petrophysical Properties," PhD dissertation, U. of Tulsa, Tulsa (1997).
14. Bahar, A. and Kelkar, M.: "Integrated Lithofacies and Petrophysical Properties Simulation," paper SPE 38261 presented at the 1997 SPE Western Regional Meeting, Long Beach, California, 25–27 June.
15. Aarts, E. and Korst, J.: *Simulated Annealing and Boltzmann Machines: A Stochastic Approach to Combinatorial Optimization and Neural Computing,* John Wiley and Sons, Ltd., Chichester, U.K. (1989) 13.
16. Kirkpatrick, S., Gelatt, C.D. Jr., and Vecchi, M.P.: "Optimization by Simulated Annealing," *Science* (13 May 1983) 671.
17. Cerny, V.: "Thermodynamical Approach to the Traveling Salesman Problem: An Efficient Simulation Algorithm," *J. of Optimization Theory and Applications* (1985) **45,** No. 1, 41.
18. Collins, N.E., Eglese, R.W., and Golden, B.L.: "Simulated Annealing—An Annotated Bibliography," *Simulated Annealing (SA) and Optimization: Modern Algorithms With VLSI, Optimal Design, and Missile Defense Applications,* M.E. Johnson (ed.), American Sciences Press, New York City (1988) 209.
19. Farmer, C.L.: "Numerical Rocks, The Mathematical Generation of Reservoir Geology," paper presented at the 1989 Joint IMA/SPE European Conference, Cambridge U., 25–27 July.
20. Perez, G.: "Stochastic Conditional Simulation for Description of Reservoir Properties," PhD dissertation, U. of Tulsa, Tulsa (1991).
21. Metropolis, N. *et al.*: "Equation of State Calculations by Fast Computing Machines," *J. of Chemical Physics* (June 1953) 1087.
22. Deutsch, C.: "Annealing Technologies Applied to Reservoir Modeling and the Integration of Geological and Engineering (Well Test) Data," PhD dissertation, Stanford U., Stanford, California (1992).
23. Sen, M. *et al.*: "Stochastic Reservoir Modeling Using Simulated Annealing and Genetic Algorithms," *SPEFE* (October 1995) 49.
24. Geman, S. and Geman, D.: "Stochastic Relaxation, Gibbs Distributions, and the Bayesian Restoration of Images," *IEEE Transactions on Pattern Analysis and Machine Intelligence,* Vol. PAMI, No. 6 (November 1984) 721.

SI Metric Conversion Factor

ft × 3.048*   E − 01 = m

*Conversion factor is exact.

Chapter 7
Object-Based Simulation Techniques

Object modeling is another set of techniques in the arsenal of geostatistics tools. They are used to describe geological bodies, facies and lithofacies with objects of discrete geometry. Geoscientists use object modeling to build three-dimensional (3D) models of reservoirs filled with geologic bodies of preconceived shapes. It is a good tool to synthesize geologic information such as observations and measurements of formations in outcrops, modern analogues and analogues from other fields, geologic bodies inferred from seismic interpretation and core descriptions. Visually, object models are very appealing because they can mimic the look of sand channels and shales in outcrops. However, this technique is not universally applicable to all reservoirs. Most applications of object modeling are designed for the description of clastic reservoirs deposited in fluvial, deltaic and deep marine turbidite depositional environments. Several other criteria, to effectively use object modeling, are discussed in this chapter.

Chap. 7 describes the marked point process, which is the most widely used object modeling technique. The marked point process is a stochastic technique that generates multiple, equi-probable realizations of geologic objects such as sand channels, different types of shale, crevasse splays, levees and mouth bars. One or more object types can be included in a reservoir model, and these are constrained by probability functions that describe their dimensions and depositional directions, volume fractions, and well data. The hybrid simulation techniques, which combine object modeling and the pixel-based techniques described in previous chapters, have been extensively used for reservoir characterization. These techniques are illustrated in this chapter through several field case studies from the literature.

7.1 Marked Point Process Technique

The marked point process is the most widely used object modeling technique to describe geologic bodies that can be approximated by discrete shapes in reservoir models. Common objects used include, among many others, parallelepipeds to describe shale barriers, ellipsoids to describe calcite cements and ribbon-shaped half-cylinders to describe sand channels. The marked point process algorithm is simple to implement. It only requires a random number generator to select the locations of the objects within the reservoir and to sample the dimensions and directions of the objects. For each type of object, the target volume fraction is set to a constant value in the reservoir, and the dimensions and directions are specified with probability distribution functions. The volume fraction is defined as the total bulk volume for each type of object divided by the bulk reservoir volume.

7.1.1 Simulation Process. The simulation process starts with an empty volume, which covers the reservoir area of interest and zones. It is always assumed that background facies fill the space, and other objects replace the background one at a time. Objects of different types are randomly inserted into the volume with the marked point process. First, objects are inserted at conditioning data locations to honor the presence of observed geologic facies at wells. Then, objects are inserted at random locations within the reservoir volume until the target volume fractions are reached. The spatial relationships among the objects may also be honored. At the end of the simulation, the remaining volume that is not filled by objects is assigned to the background facies. **Fig. 7.1** illustrates the simulation process. The major steps in the marked point simulation process are:

1. Select a location from the conditioning data set to insert an object. The object locations are drawn one at a time from the conditioning data.
2. Insert an object at the location of the conditioning data selected in step one. Draw the dimensions and orientation of the object, and place its centroid at the conditioning data location.
3. Check if objects have been inserted at all conditioning data locations. If there are conditioning data locations without an object, return to step one. If all conditioning data locations have been populated with objects, proceed to step four.
4. Select the location of an object away from the conditioning data locations. Draw, at random, the coordinates for the centroid of the object.
5. Insert an object at the location selected in step four. Draw the dimensions and orientation of the object, and place its centroid at the location from step four.

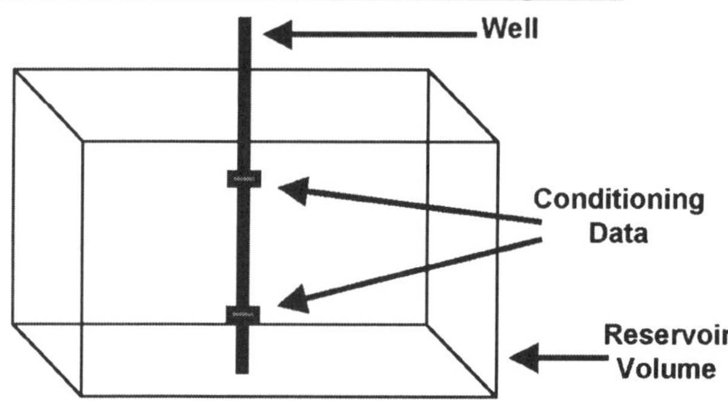

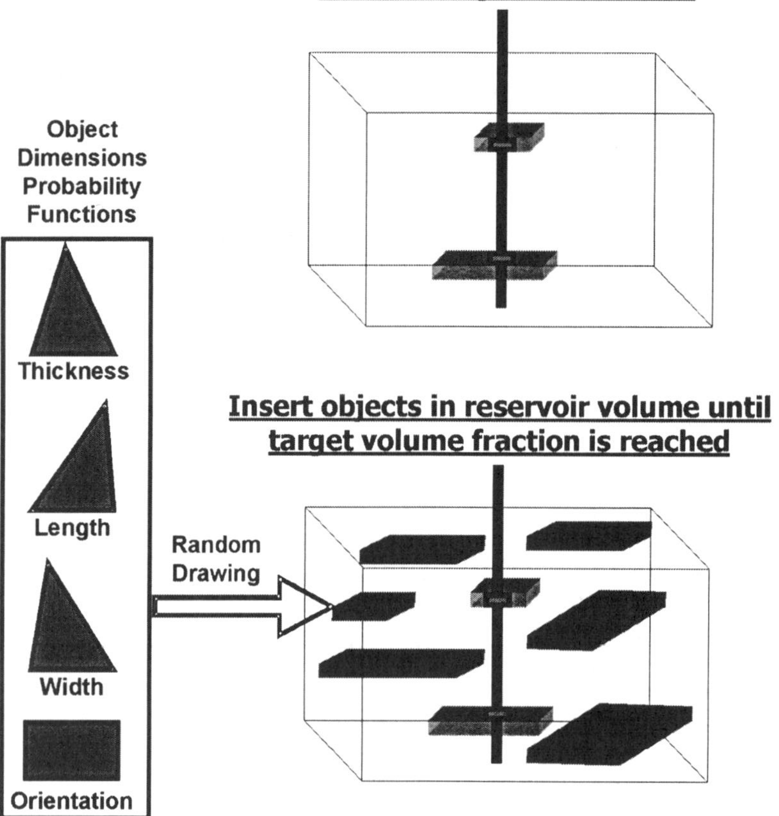

Fig. 7.1—Major steps for the marked point simulation process.

6. Calculate the volume fraction of the objects. Return to step four if the volume fraction is below the target value. Stop the simulation when the volume fraction reaches the target volume fraction.

This simulation process can be repeated for other types of geologic facies that are represented with objects of different properties and volume fractions.

7.1.2 Properties of Objects. Several types of objects can be used to describe different geologic facies. The most common objects used include parallelepipeds, ribbon-shaped half-cylinders and ellipsoids. **Fig. 7.2** shows several types of objects commonly used in object modeling applications. A common application of object modeling describes shale and sands with box-shaped parallelepipeds and ribbon-shaped half-cylinders to represent shales and sand channels, respectively. Reservoir models can include more than one type of object. Objects with other shapes can be used for specific applications, as long as their dimensions are described with a finite number of parameters. For example, spherical objects have been used to describe the microscopic pores of rocks, and line segments have been used to describe fractures.

Parallelepipeds have a convenient shape to represent a wide range of shale types. **Fig. 7.3** shows a realization of shale objects generated with the marked point process. Thickness, length, width and orientation describe the geometry of parallelepipeds. Different types of shales are described with a different range of values for these parameters. To illustrate the range of shale dimensions, consider floodplain and drape shales. Floodplain shales, in a shallow marine environment, are represented with large widths and lengths of several thousand feet or even miles, and uniform orientations. Drape shales, in a deep marine environment, are represented with small widths and lengths of a few feet to just inches, and orientations along the paleocurrent direction.

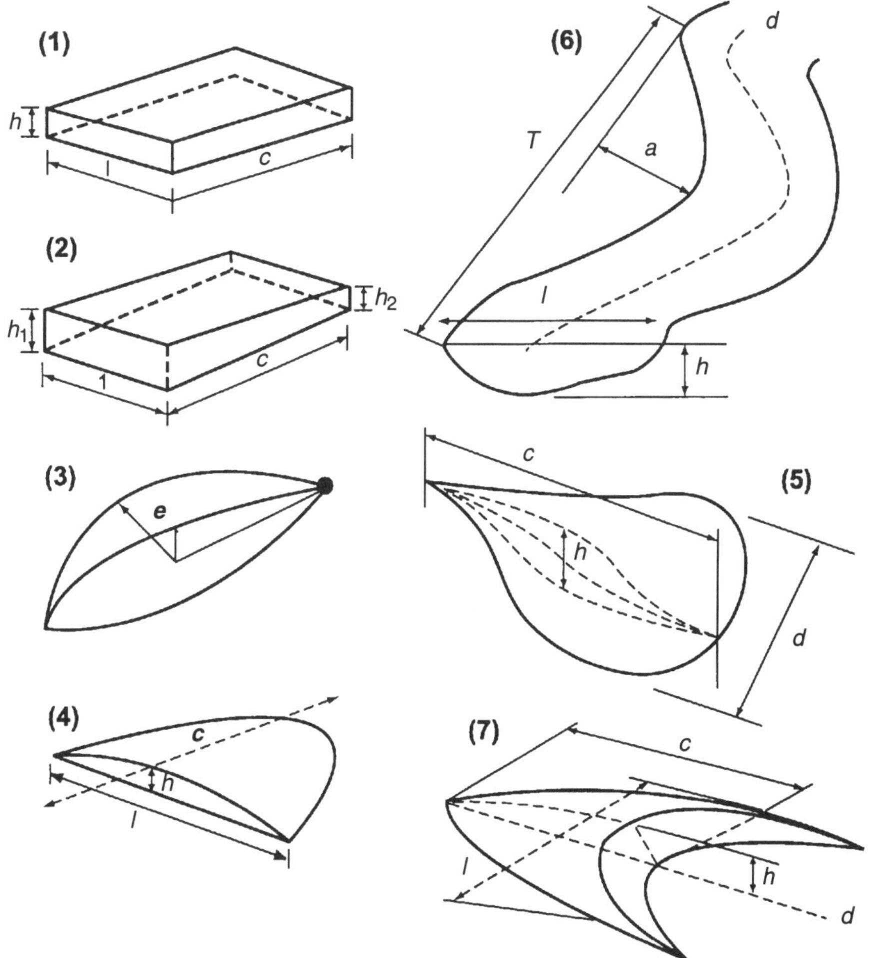

Fig. 7.2—Types of geologic shapes for object modeling: 1) parallelepiped, 2) wedge, 3) ellipsoid, 4) lobe, 5) sigmoid, 6) channel and 7) dune (from Ref. 1).

In the marked point process, the parameters of the objects are given as probability distribution functions. **Fig. 7.4** shows the different types of probability functions commonly used to describe the properties of geologic objects. The selection of a probability distribution function depends on how much is known about the object's dimensions. In practice, plenty of information is available about the thickness from well data, but only a little direct information about the lateral dimensions is known. Often, field and modern analogs and outcrops are used to obtain approximate lateral dimensions of objects. The simplest probability distribution function is enough to approximate this type of information. A uniform distribution is the simplest type of probability distribution, requiring only the minimum and maximum values. Triangular probability functions are a popular choice when all three values (minimum, most likely and maximum) are available. Multipoint distribution functions can take any form and provide the most flexible way to approximate any probability function. Numerical Example 7.1 illustrates how probability functions are used to draw object dimensions in the marked point process.

Ribbon-shaped half-cylinders are used to describe channels in several environments, including meandering, braided and deep marine. **Fig. 7.5** shows horizontal sections of channel objects with ribbon shapes. **Fig. 7.6** shows an example of fluvial channels represented by half-cylinder shapes in a vertical section. The parameters that describe channels include thickness, width, wavelength, amplitude and orientation. These are specified with uniform, triangular or multipoint probability functions, as previously described for parallelepiped objects. Sometimes the ratio of parameters, such as channel width-to-thickness ratio or width-to-length ratio, is specified in order to honor the relations among parameters observed in the data.

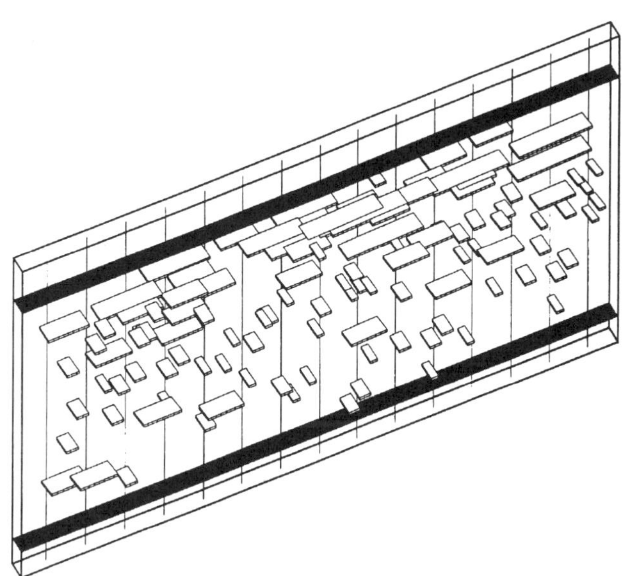

Fig. 7.3—Object model for shale, generated by the marked point process with parallelepipeds in a vertical cross section (from Ref. 2).

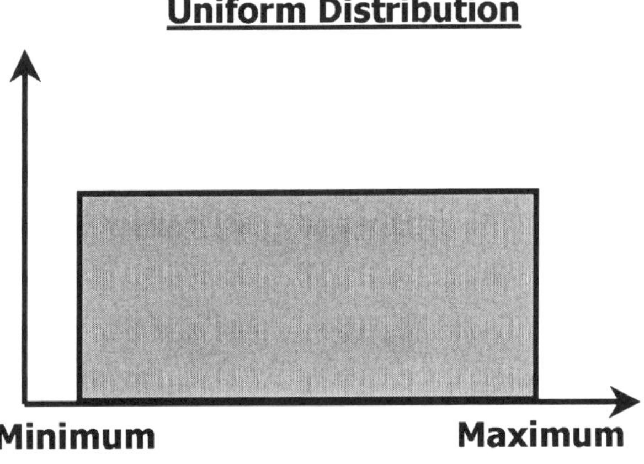

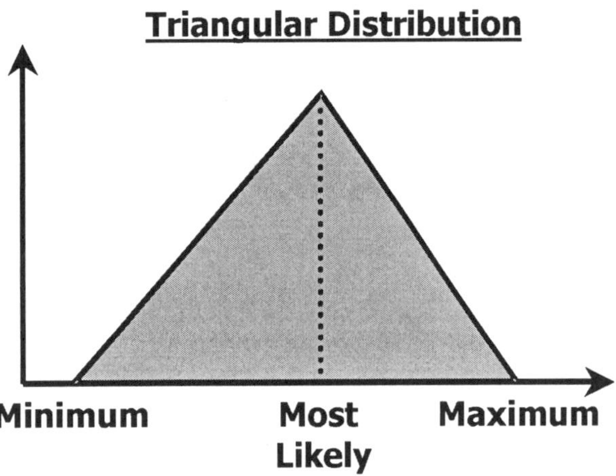

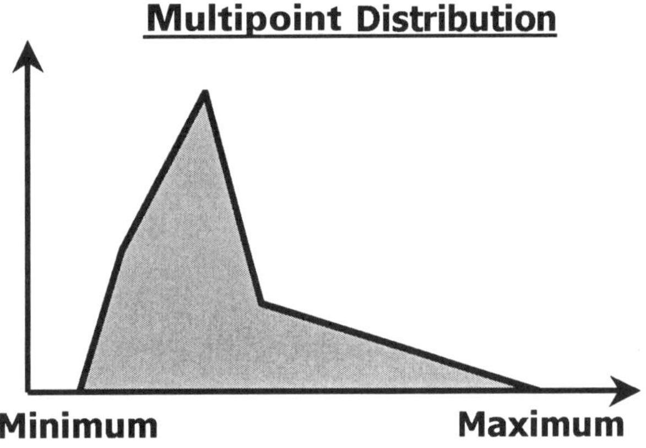

Fig. 7.4—Common types of probability distribution functions for description of object dimensions.

The volume fraction is another important parameter because it controls the amount of objects in a reservoir model. The volume fraction of each object type is defined as the bulk volume of an object type divided by the bulk volume of the reservoir. The marked point process inserts objects in a reservoir model until the volume fraction of each object type is reached.

7.1.3 Conditioning Data. Similar to the conditional simulation techniques described in previous chapters, conditioning data for object modeling techniques consist of geologic facies information at a limited number of locations inside the reservoir volume. Conditioning data are derived from geologic descriptions of the facies based on cores and logs from wells. Fig. 7.1 portrays how conditioning data are used in object modeling. The first steps of the marked point process ensure that objects of the specified types are placed in every conditioning data location.

Conditioning data points are represented as either indicators or discrete intervals. Indicators are data at regular intervals along wells that indicate the presence or absence of facies with a value of one or zero, respectively. More often, the geologic descriptions of the facies are recorded as interval data, and they include the top and base of each interval containing a facies type along a well. **Fig. 7.7** gives an example of conditioning data from a well showing the shale intervals identified from geological and petrophysical interpretations. Both indicator and interval data convey the same information.

7.1.4 Interactions Among Objects. The interaction among objects in the marked point process assumes that objects erode into each other. However, in some depositional environments, facies interact among themselves in different fashions. Some facies tend to cluster close to each other, while other facies tend to be isolated.

One way to account for known interactions among facies in the marked point process is the attraction and repulsion mechanisms developed by Clemetsen *et al.*[6] Instead of drawing the locations of the object at random, the attraction and repulsion mechanisms draw the locations of objects from user-defined probability functions that quantify the tendency of objects to be close or far from each other. The different types of interactions among objects are illustrated in **Fig. 7.8.**

Numerical Example 7.1—Shale Object-Modeling in Fluvial Reservoir. This example illustrates the major steps to generate a description of shale in a fluvial reservoir with the marked point process. Shales are represented by objects of parallelepiped shapes and given probability functions for their dimensions.

Given:

• The reservoir volume of interest covers an area of 2,000 ft by 2,000 ft and has a thickness of 50-ft (**Fig. 7.9**).

• Only one type of shale is present, and it is represented by objects with parallelepiped shapes. See **Fig. 7.10** and **Table 7.1** for the probability functions for dimensions and orientation.

• Assume that shales erode into each other.

• The volume fraction of shales is 5%.

• One conditioning data point is available with x and y coordinates both equal to 1,000 ft and with a shale thickness of 5 feet (Figure 7.9).

Calculate:

• Generate one realization of the shale objects with the marked point process.

Solution. Step 1—Select the location of the conditioning data point to insert the first object (Fig. 7.9). There is only one conditioning data point with x and y coordinates, both at 1,000 ft, and it is located at the center of the reservoir volume. For simplicity, the centroid of the object is placed at this location. In general, the conditioning data can intercept the object at any point within the object.

Step 2—Calculate the cumulative probability functions required in the marked point process. As shown in **Fig. 7.11,** cumulative probability distribution functions are calculated by integrating the probability density functions. The thickness is assigned a value of 5 ft based on the conditioning data; therefore, it is not necessary to draw it from the probability functions. Sample the cumulative distribution functions for the

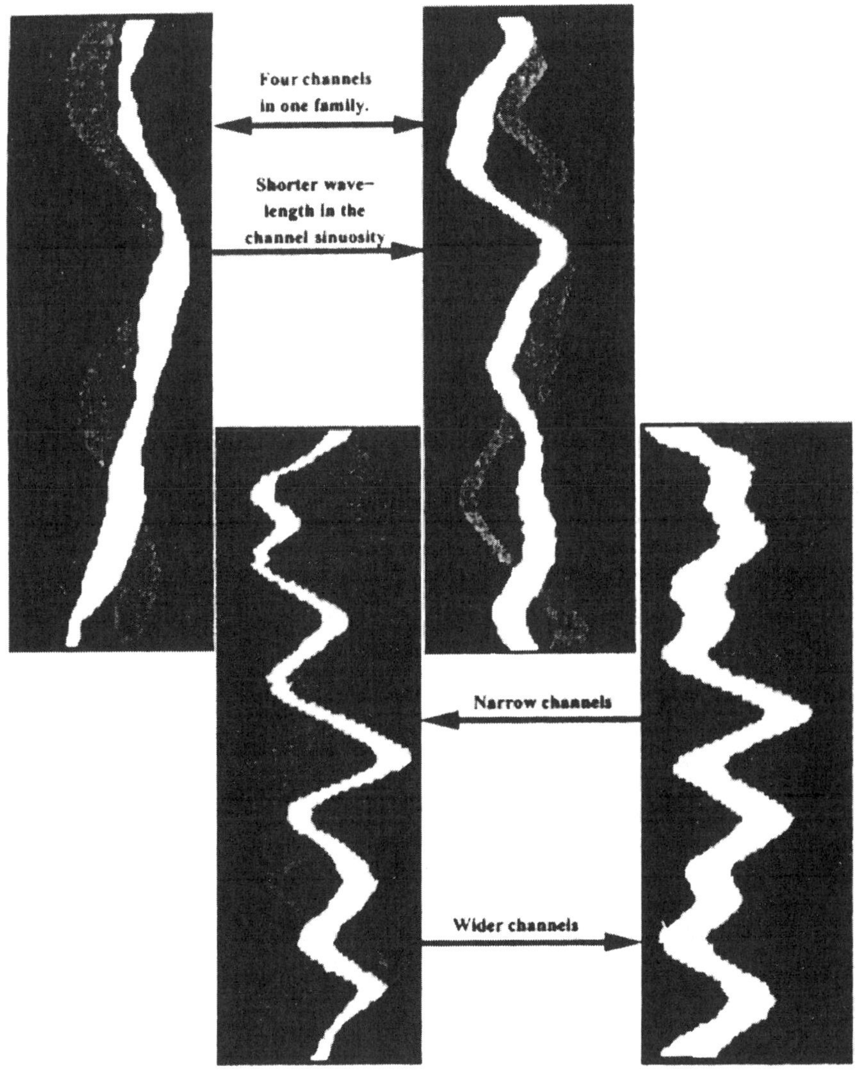

Fig. 7.5—Object models for channels generated by the marked point process with ribbon-shaped objects (from Ref. 3).

other object dimensions and orientation with a uniform random number generator that gives values between zero and one. Draw the object's length for a random number equal to 0.7352 by interpolating the points from the cumulative distribution function in Figure 7.11.

$$\text{Length} = 50 + (400 - 50)(0.7352 - 0)/(1 - 0) = 307.3 \text{ ft.}$$

Draw the object's width from the cumulative distribution function (Figure 7.11) for a random number equal to 0.3793.

$$\text{Width} = 50 + (200 - 50)(0.3793 - 0)/(1 - 0) = 106.9 \text{ ft.}$$

Draw the object's orientation from the cumulative distribution function (Figure 7.11) for a random number equal to 0.2856.

$$\text{Angle} = 80 + (80 - 90)(0.2856 - 0)/(0.5 - 0) = 85.7°.$$

The angles are measured clockwise from the North so that the y axis in the positive direction is 0°, and the x axis in the positive direction corresponds to 90°.

The shale object at the conditioning data location is shown in **Fig. 7.12.** The bulk volume for this shale is:

$$\text{Shale Volume} = 5 \times 307.3 \times 106.9 = 164{,}254.9 \text{ ft}^3.$$

The bulk volume for the reservoir region is:

$$\text{Reservoir Volume} = 2{,}000 \times 2{,}000 \times 50 = 200 \times 10^6 \text{ ft}^3.$$

The shale volume fraction for this first shale object is:

$$\text{Shale Fraction} = 164{,}254.9/200 \times 10^6 \times 100 = 0.08\%.$$

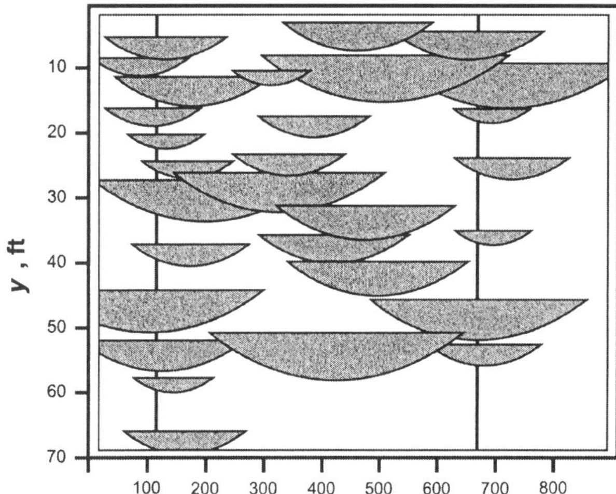

Fig. 7.6—Object models for channels generated by the marked point process with half-cylinder shaped objects in a vertical section (from Ref. 4).

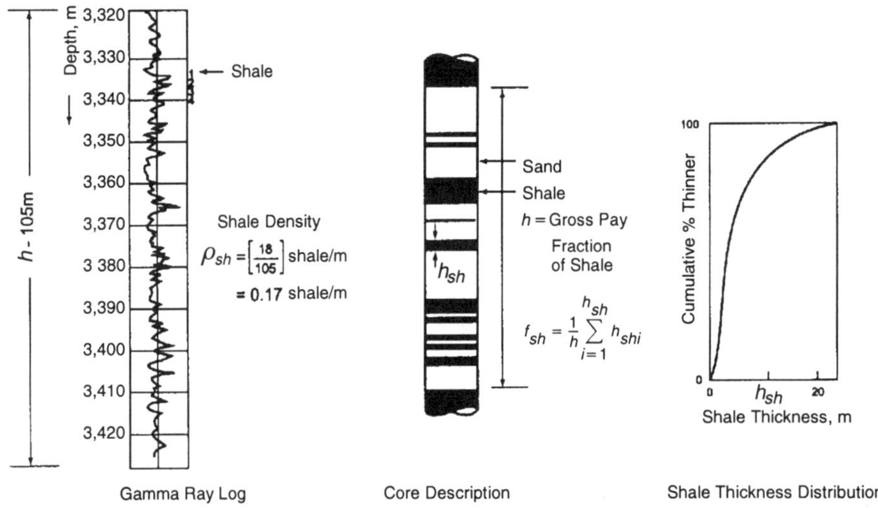

Fig. 7.7—Conditioning data showing shale interval along a well (from Ref. 5).

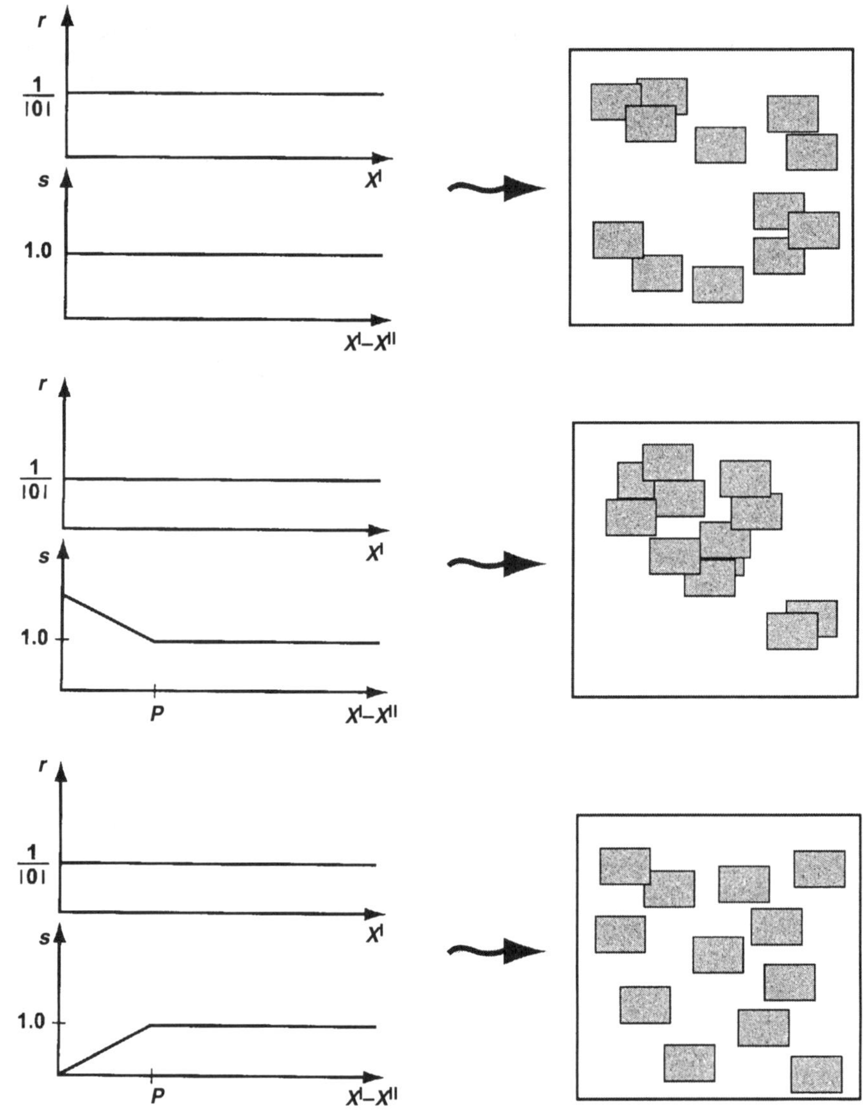

Fig. 7.8—Repulsion and attraction interactions among objects for marked point process (from Ref. 6).

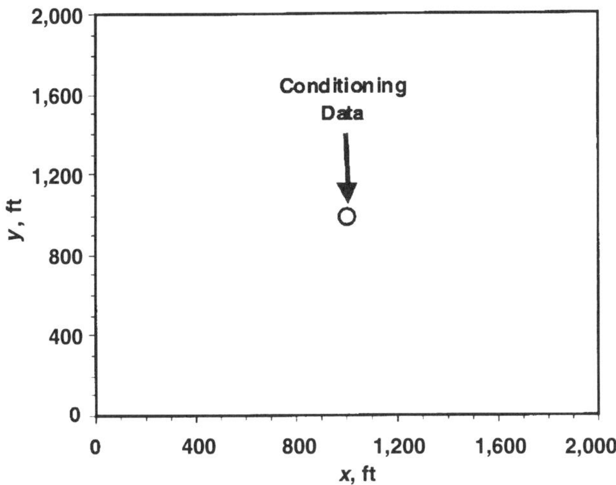

Fig. 7.9—Reservoir region for object modeling in Numerical Example 7.1.

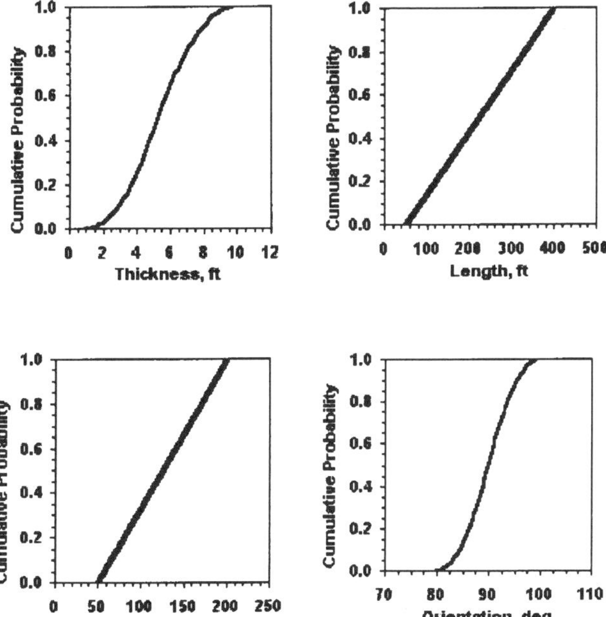

Fig. 7.11—Cumulative probability functions for shale dimensions for Numerical Example 7.1.

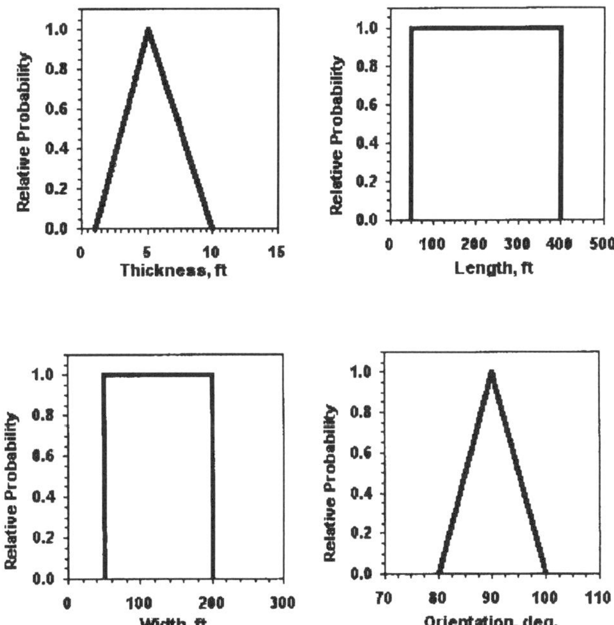

Fig. 7.10—Probability functions for shale dimensions for Numerical Example 7.1.

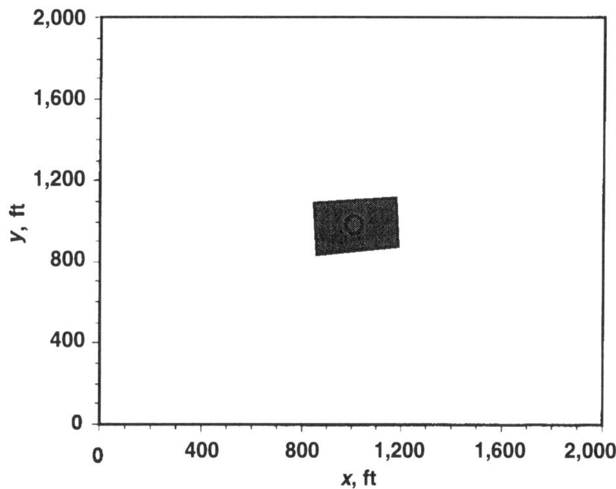

Fig. 7.12—First shale object at the conditioning data location in Numerical Example 7.1.

Step 3—Because an object has been inserted at the only conditioning data location, proceed to the next step.

Step 4—Select a location for the centroid of the next object. Draw at random the x and y coordinates with a uniform distribution between the minimum and maximum limits of the reservoir area. The random numbers used to sample the distributions are obtained from a uniform random generator. For a random number equal to 0.7199, the x coordinate is:

$$x = 0 + (2{,}000 - 0)(0.7199 - 0)/(1 - 0) = 1{,}439.8 \text{ ft}.$$

For a random number equal to 0.4172, the y coordinate is:

$$y = 0 + (2{,}000 - 0)(0.4172 - 0)/(1 - 0) = 834.4 \text{ ft}.$$

Step 5—Similar to step two, draw the dimensions and orientation of the object by sampling the cumulative distribution functions (Fig. 7.11) with a uniform random number generator. Draw the object's thickness using a random number equal to 0.2765.

$$\text{Thickness} = 1 + (5 - 1)(0.2765 - 0)/(0.4444 - 0) = 3.5 \text{ ft}.$$

TABLE 7.1—DATA FOR NUMERICAL EXAMPLE 7.1

Parameter	Probability Function	Minimum	Most Likely	Maximum
Thickness, ft	Triangular	1	5	10
Length, ft	Uniform	50	—	400
Width, ft	Uniform	50	—	200
Orientation, degree	Triangular	80	90	100

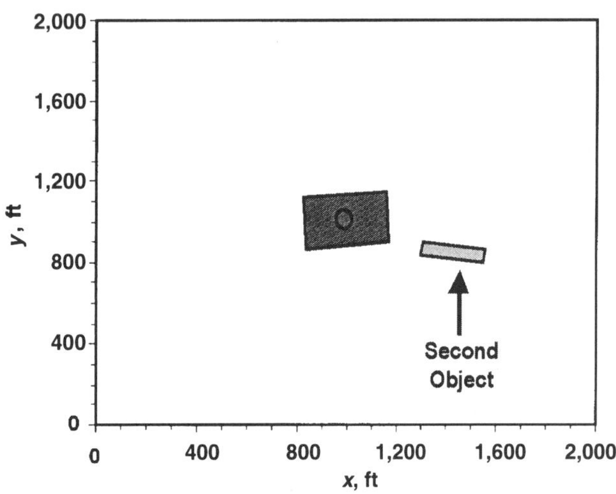

Fig. 7.13—First two shale objects in Numerical Example 7.1.

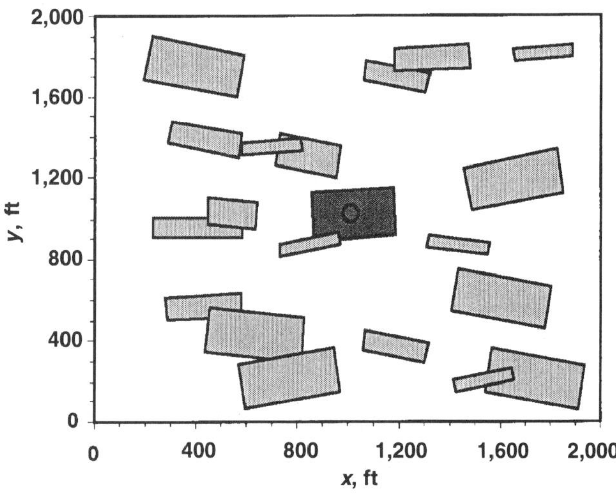

Fig. 7.14—Shale realization with a few more objects in Numerical Example 7.1.

Draw the object's length using a random number equal to 0.5399.

$$\text{Length} = 50 + (400 - 50)(0.5399 - 0)/(1 - 0) = 239.0 \text{ ft.}$$

Draw the object's width using a random number equal to 0.0611.

$$\text{Width} = 50 + (200 - 50)(0.0611 - 0)/(1 - 0) = 59.2 \text{ ft.}$$

Draw the object's orientation using a random number equal to 0.8325.

$$\text{Angle} = 90 + (100 - 90)(0.8325 - 0.5)/(1 - 0.5) = 96.6°.$$

The two objects generated so far are shown in **Fig. 7.13.**

Step 6—The shale volume of this object and the current volume fraction of all shale objects are:

$$\text{Shale Volume} = 3.5 \times 239.0 \times 59.2 = 49{,}318.4 \text{ ft}^3.$$

$$\text{Shale Volume Fraction} =$$

$$(164{,}254.9 + 49{,}318.4) / 200 \times 10^6 \times 100 = 0.11\%.$$

Because the volume fraction is below the target of 5%, continue iterating between steps four and six until the volume fraction reaches the specified target.

Fig. 7.14 shows the realization with a few more shale objects. To complete the simulation, repeat the previous steps, and continue inserting objects until the shale volume fraction reaches 5%.

7.2 Criteria for Choosing Object Modeling

The most accepted conditional simulation techniques to described geologic facies are object modeling and Sequential Indicator Simulation (SIS) technique for categorical variables described in Chap. 6. Two groups of geostatistics practitioners that advocate the use of either object modeling or SIS have emerged in recently published field cases; therefore, there is some degree of confusion when selecting the right technique. Although, there are no general rules for choosing one technique over the other, a few guidelines, for deciding the conditions that make object modeling more appropriate, can be derived, based on the type of information available.

7.2.1 Depositional Environment. Object modeling is recommended for fluvial, deltaic and deep marine depositional environments. The shape of most facies in these environments, like channels, mouth bars, levees and different types of shale, are often reasonably preserved in the subsurface and can be represented by discrete objects with unique shapes. Usually, geologists can estimate dimensions and volume fractions of facies based on modern river analogs, outcrops, ground-penetrating radar and seismic data. Object modeling is less applicable to carbonate environments because facies have gone through post-depositional processes, such as dolomatization. Therefore, it is difficult to approximate the shape of facies and estimate their dimensions in carbonate environments. Still, a few applications of object modeling, such as the field case presented by Petit *et al.*[7] in a mixed carbonate platform, have been found attractive for quantifying the distribution of laterally continuous, but thin, sedimentary bodies in a carbonate environment.

7.2.2 Conditioning Data. Object modeling is recommended during early appraisal stages of a reservoir and when limited wells with conditioning data are available. It is difficult for object modeling to honor conditioning data in several wells that are closely spaced and when facies extend across multiple wells. In addition, conditioning data at wells does not have as much influence in the outcome of object modeling as it does for SIS.

7.2.3 Proportion of Objects. Object modeling works better when volume fractions of facies represented by objects are small and volume fractions of the background facies are large. Examples include a reservoir that has 5% volume fraction of shale objects within a sand background or a reservoir that has about 30% volume fraction of channel objects within a shale background. Object modeling can fail to reproduce the specified dimensions, in cases where volume fractions of facies are large and of about the same magnitude. Excessive eroding among objects can make reproduction of target volume fractions difficult and can result in abnormal shapes.

7.2.4 Object Dimensions. Object modeling can encounter difficulties, while generating descriptions when object dimensions are large, relative to the well spacing and simulation area. The problems arise when the marked point simulation process attempts to place an object at a conditioning well, and because of its large dimensions, the object extends over to a nearby well, which does not have conditioning data at this location. The simulation process attempts to iterate around these conflicts with nearby wells; however, because the ob-

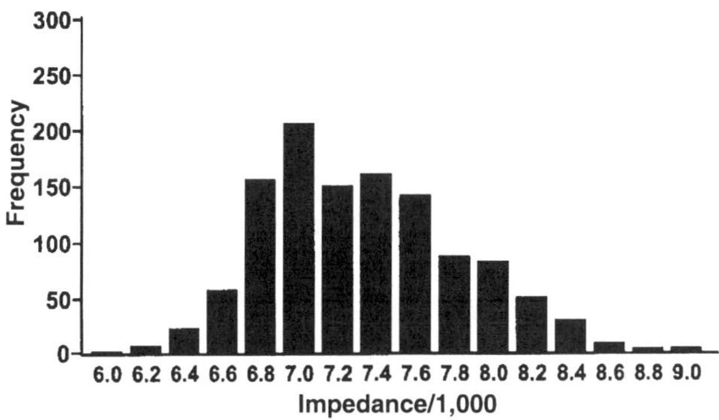

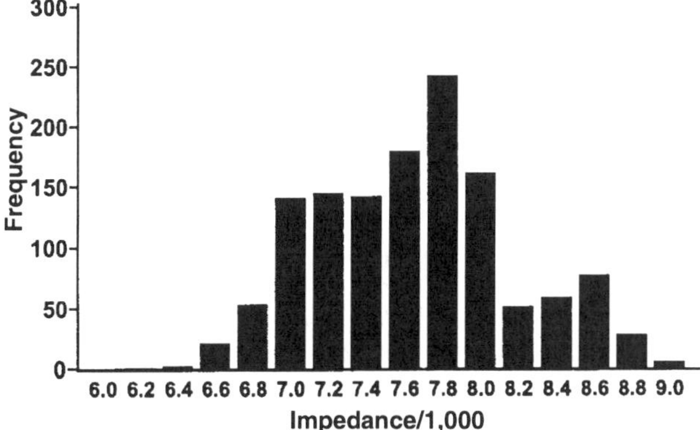

Fig. 7.15—Histograms of seismic impedance for channel and nonreservoir facies (from Ref. 8).

ject's dimensions are large, it will not converge, even after several iterations.

7.2.5 Seismic Data. Object modeling is not well suited to condition descriptions with seismic data. The marked point process does not have a mechanism that can easily use the volume of reservoir attributes, generated with recently developed seismic inversion techniques. Seismic inversion provides reservoir attributes, which can be correlated to geologic facies or convey direct information about trends and lateral variations of facies. The relationships between seismic impedance and channel facies and nonreservoir facies are shown in **Fig. 7.15.** One approach proposed to incorporate the seismic attribute data into the marked point process (Tjølsen *et al.*[8]), based on the relationships observed in Fig. 7.15, which indicate that channels tend to have a lower seismic impedance than nonreservoir facies. However, only limited results were presented, and more evaluations of this technique are needed to prove the effectiveness of incorporating seismic attribute data.

7.2.6 Available Data. Object modeling requires information external to the data available from well logs and cores. The data about dimensions, shapes and orientations of various objects can be obtained from outcrop analogues, modern analogues, and literature data. However, if such information is not readily available, object modeling could render results that are meaningless and far removed from reality.

7.3 Hybrid Conditional Simulation Approaches

This chapter, as well as previous ones, have shown that individual conditional simulation techniques suffer from some shortcomings because these techniques are not applicable for all scenarios, even within the same depositional environment. To overcome these deficiencies, geostaticians have proposed

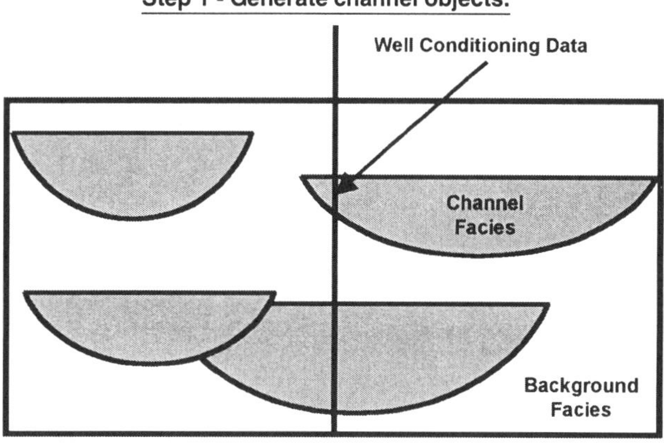

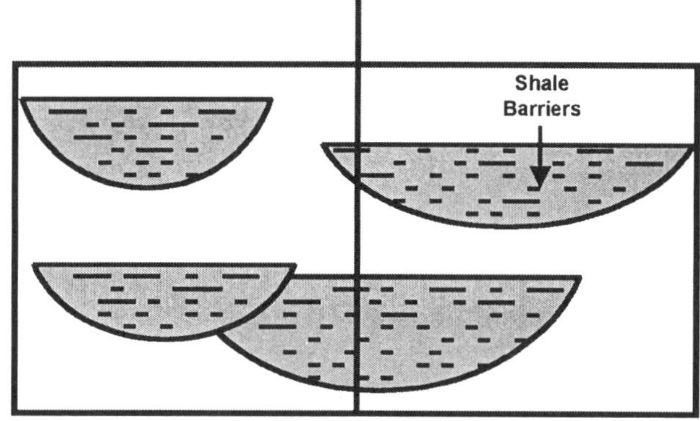

Fig. 7.16—Two-step object modeling approach.

different hybrid techniques that are more flexible and generate more realistic reservoir descriptions. Hybrid approaches combine two or more conditional simulation techniques to take advantage of the best features of each technique. The field cases that illustrate the application of these hybrid approaches are given in Sec. 7.3.1.

7.3.1 Multistep Approach Using Object Modeling. The multistep object modeling approach applies the marked point process multiple times in several stages. At each stage, the simulation process generates geologic objects at different scales. In the first stage, large-scale objects are simulated to fill the reservoir volume with major geologic facies or facies associations that extend throughout the simulation area. In subsequent stages, facies at smaller scales or lithotypes, which occur within major facies, are simulated within each of the objects from the previous stage. Several stages of this process can be repeated to describe the necessary level of geologic detail.

The following example considers the description of channels and shale in a meandering river environment. **Fig. 7.16** illustrates the major steps of the two-step approach. In the first stage, channels with 30% volume fraction are distributed with object modeling within the floodplain shale background. In the second stage, drape shale with 5% volume fraction are described within each of the channel objects.

7.3.2 Multistep Hybrid Approaches Using Object Modeling and Pixel-Based Techniques. It is impractical for a single geostatistical technique to describe reservoir heterogeneities at all scales. Therefore, multistep approaches, using more than one technique, were introduced to characterize dis-

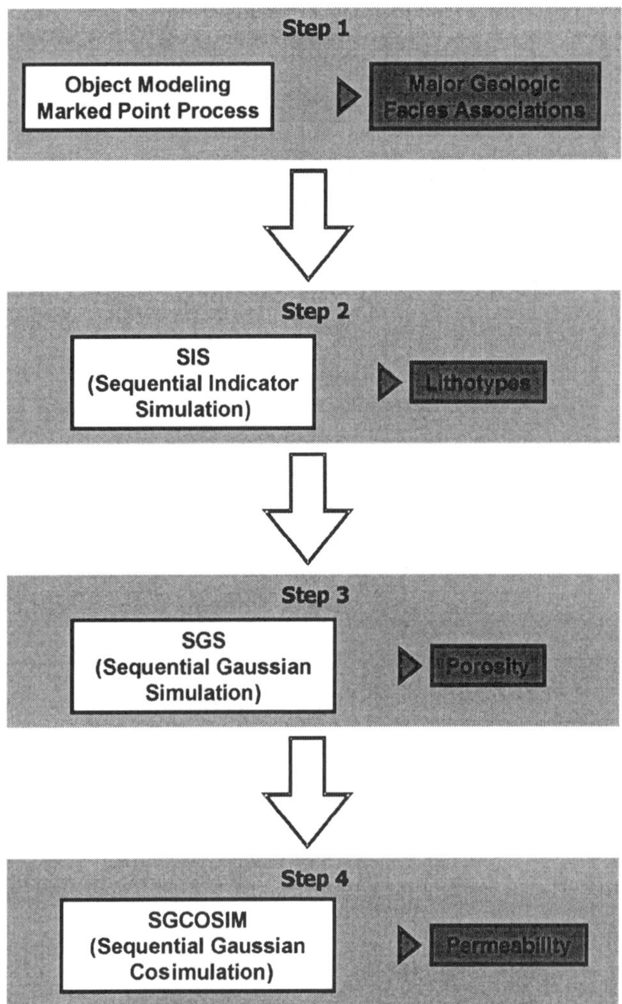

Fig. 7.17—Steps in a hybrid four-step approach.

tinct styles of heterogeneity at different reservoir scales. For example, in a fluvial environment, the character and degree of correlation are different for major geologic facies, such as channels, and for beds of sand grain within each channel, which control porosity and permeability. Channels are correlated across the field along the paleoflow direction, while, at smaller scales within individual channels, the arrangement of sand grain is highly influenced by the distance from the source of the river. For some reservoir characterization applications, it is important to properly describe the heterogeneities at the scale of major facies and at the smaller scales within the facies.

The number of steps required in a hybrid approach depends on the variation of correlation at different scales and the type of reservoir properties being described. A hybrid approach, consisting of four steps that describe major facies and reservoir properties in a fluvial depositional environment, is depicted in **Fig. 7.17.** Other approaches with different steps can be designed to fit other modeling purposes. Steps are executed sequentially, so the descriptions from previous steps can be used in subsequent steps. The four steps are:

Step 1—Describe major geologic facies associations with object modeling. This includes facies that extend across a significant region of the field, such as channels, mouth bars, levees and floodplain shale. The dimensions, type of objects, volume fractions, and conditioning data are specified for each facies.

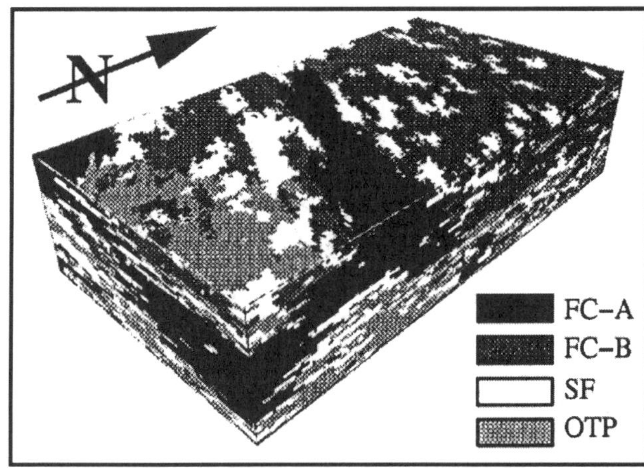

Fig. 7.18—SIS facies description in braided fluvial reservoir (from Ref. 9).

Step 2—Describe lithotypes with Sequential Indicator Simulation (SIS). Lithotypes include beds of different lithogies, such as fine and coarse grain sandstones, mudstones and calcite concretions, which are present in different proportions within each geologic facies described in step one. The variograms, volume fraction and conditioning data are specified for each lithotype within each facies.

Step 3—Describe porosity with Sequential Gaussian Simulation (SGS). The variograms, porosity mean value and conditioning data are specified for each lithotype described in step two. This approach allows us to assign different properties to the same lithotype present in more than one facies.

Step 4—Describe permeability with Sequential Gaussian Cosimulation (SGCOSIM). The variograms, permeability mean value, conditioning data, and crosscorrelation between porosity and permeability are specified for each lithotype described in step two.

7.4 Field Case Studies

The three field case studies described below illustrate the application of object modeling techniques to reservoir in different depositional environments.

7.4.1 Comparison of Object Modeling and SIS for a Braided Fluvial Reservoir. Seifert and Jensen[9] presented a quantitative comparison of object modeling and SIS models for a reservoir deposited in a braided fluvial environment. The comparison is based on ten realizations generated with each technique. The four facies or "genetic units" considered are low angle cross-bedded channels (FC-A), planar or cross-laminated channels (FC-B), good reservoir quality sheetfloods (SF), and other than pay (OTP), which represents poor reservoir quality rocks. The width, length and thickness of reservoir volume described in the models are 5,270 ft, 9,700 ft and 80 ft, respectively, and these were filled using cells that were 50 × 50 × 2 ft.

For this depositional environment, Seifert and Jensen[9] found that a two-step SIS works better than the conventional SIS where all facies are described in a single step. The SIS parameters (variogram ranges and orientations) were obtained by sensitivity analyses and conditioning data from two wells. The first step described the FC-A channel facies and the background with SIS. In the second step, the FC-B, SF and OTP facies were modeled with SIS within the background of the first step. **Fig. 7.18** shows a realization of the SIS description.

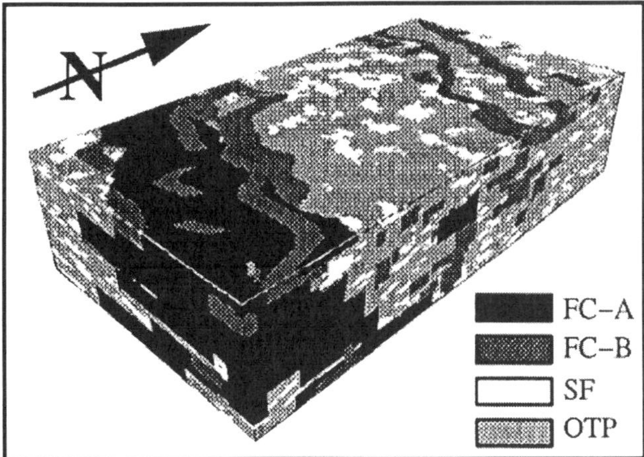

Fig. 7.19—Object modeling facies description in braided fluvial reservoir (from Ref. 9).

The input for object modeling, such as channel dimensions, aspect ratios and sinuosity, were estimated from a detailed quantitative sedimentological study and published outcrop analogs. The object modeling approach consisted of four steps. In each step, each channel type was described and then merged, based on specified erosion rules. **Fig. 7.19** shows one object modeling realization.

Visual inspection of the object modeling and SIS realizations (Figs. 7.18 and 7.19) indicate that the channel architecture, generated by SIS, is less confined than the object models. Qualitatively, the techniques were compared based on their effective permeabilities in the *x* and *y* directions that were calculated by scaling-up the models into one cell that covered reservoir volume. Along the channel direction (approximately oriented in the East/West direction and aligned with the *x* axis), the effective permeabilities of the object models are only 20% lower than the SIS models. While, in the across-channel orientation (approximately oriented in the North/South direction and aligned with the *y* axis), the effective permeabilities of the object models are 103% lower than the SIS models. The permeability anisotropy ratio for the *x* and *y* directions is 0.46 for the object models and 0.93 for the SIS models. These differences in permeabilities are mainly caused by the poorer hydraulic communication in the across-channel direction in the object models. Seifert and Jensen[9] concluded that the optimum approach for this depositional environment could be a hybrid approach, which uses object modeling to describe channel facies (FC-A and FC-B) and SIS to describe the background facies (SF and OTP).

7.4.2 Two-Step Hybrid Simulation Approach in an Upper Shoreface Environment. The depositional environment described in this field case study is an upper shoreface consisting of several stacked and laterally adjacent lobe-shaped mouth bars. Damsleth *et al.*[10] proposed a two-step hybrid conditional simulation, based on object modeling and Gaussian simulation, to develop a reservoir simulation model that investigates the effect of different levels of heterogeneities in recovery of a waterflood process.

The first stage describes geologic facies using the marked point process. The following six facies types were described with objects: mouthbar sands (M1A), mouthbar fines (M1B), calcite (K), channel deposits (P1), distal deltafront fines (D1) and delta plain fines (P3). The geometries of the facies were derived from extensive studies of analog outcrop sequences. Conditioning data at two wells were included in the models.

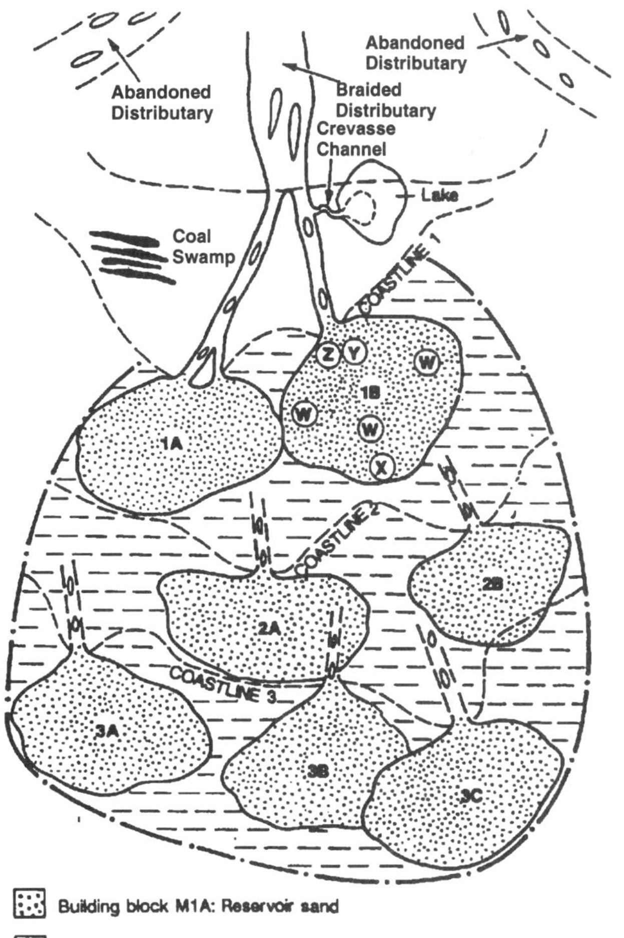

Fig. 7.20—Schematic of geologic setting delta mouth bar setting (from Ref. 10).

Fig. 7.20 shows a conceptual description of facies in a delta mouthbar setting. The shapes of the objects used to represent the facies were rectangular boxes and ellipsoid. The reservoir model consists of nine layers, and each layer contains between one and four facies types. Horizontal and vertical sections for five object modeling realizations are shown in **Figs. 7.21** and **7.22**. The differences among the five realizations illustrate the degree of uncertainty that exists in spatial arrangement of facies.

The second stage uses Gaussian simulation to describe horizontal and vertical permeabilities within the facies framework constructed in the first step. The parameters used to describe the permeabilities, such as mean, standard deviation and variograms, were assigned different values according to the facies. Because the lateral variogram ranges could not be estimated from the data, these were defined based on geologic experience. The original paper shows five realizations of horizontal permeability corresponding to the facies realizations shown in Fig. 7.21. The realizations exhibit a significant degree of heterogeneity, and according to Damsleth *et al.*,[10] they appear realistic from a geological viewpoint. For one facies realization, several permeability realizations can also be generated, but these will exhibit less variation than those based on different facies realizations.

Several realizations of the two-step descriptions were used in a flow simulation model. The permeability realizations were scaled-up from a 1.2 million cell grid to the simulation grid with 5,500 cells. The waterflood simulations included

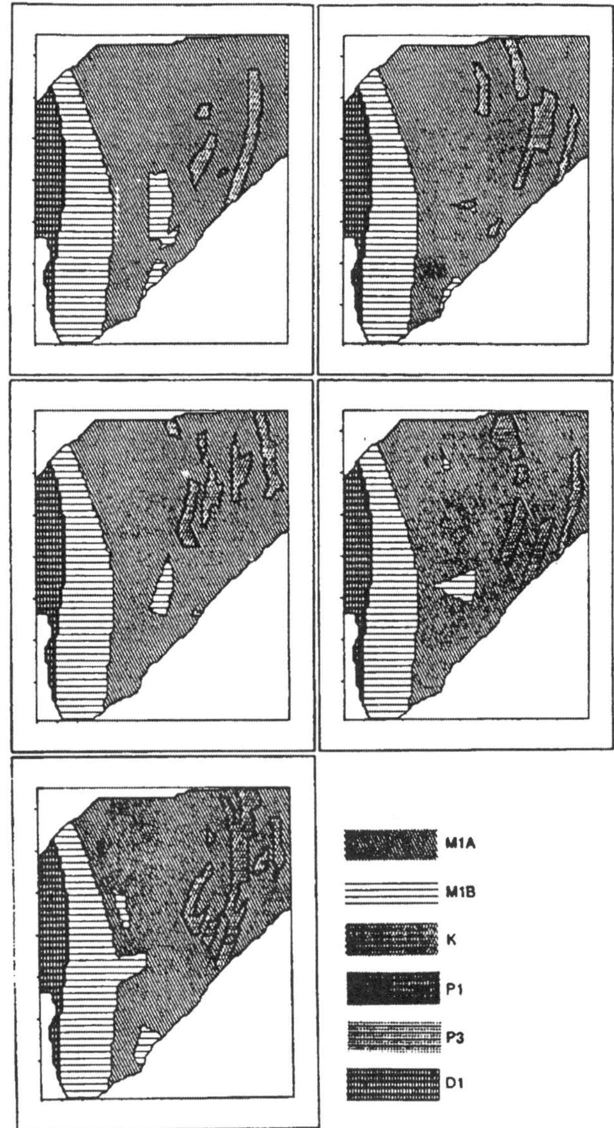

Fig. 7.21—Horizontal sections showing five realizations of facies objects (from Ref. 10).

Fig. 7.22—Simulated water cut for five permeability realizations based on the same facies model (from Ref. 10).

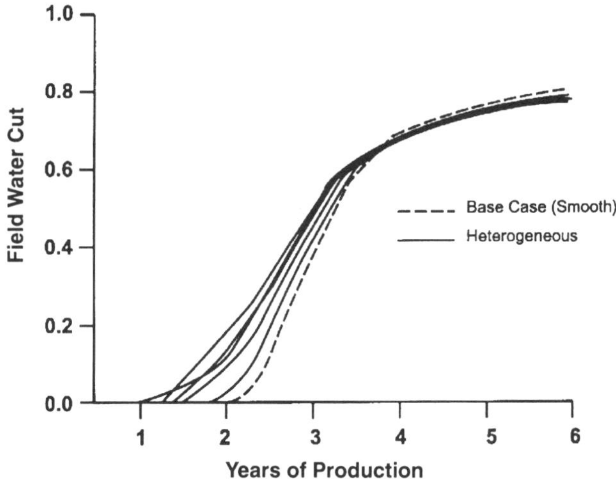

Fig. 7.23—Simulated water cut for five permeability realizations based on five facies models (from Ref. 10).

four injection wells and five oil production wells. The simulation results illustrate the effect of different degrees of heterogeneities caused by facies and permeability variations on field watercut. **Fig. 7.22** shows simulations for five permeability realizations based on the same facies model, while **Fig. 7.23** shows simulations for five permeability realizations based on different facies models. The spread in the simulated watercut is greater for the five realizations based on different facies models. Figs. 7.22 and 7.23 also show a smooth case that was generated based on a smooth interpolation of permeability. For both sets of simulations, the watercuts for the smooth case are lower than those obtained with the stochastic models. These simulations indicate, by ignoring the reservoir heterogeneities and using a traditional smooth description, the oil recovery will be overestimated.

7.4.3 Description of Shale in Submarine Fan Environments. Object modeling is used in this field case to describe shale in a submarine fan environment (Haldorsen[5]). The descriptions were designed so single-well simulation models could investigate the effect of shale barriers on gas and water conning from the gas cap and aquifer.

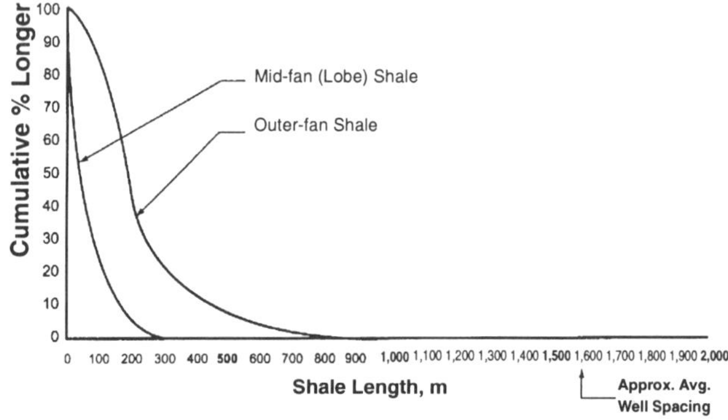

Fig. 7.24—Shale lengths cumulative probability distributions for shale in mid-fan and outer-fan facies (from Ref. 5).

The two major geologic facies in this reservoir are an outer fan and a mid fan (lobe). The shale dimensions, within each facies, are different, as shown in **Fig. 7.24** by the probability distribution of the shale lengths. Shale is smaller in the mid fan (lobe) than in the outer fan. Based on studies of submarine-fan depositional environments and paleohydrologic reconstructions, shale object shapes were represented with parallelepipeds of width equal to one half the length. The conditioning data for the well located in the center of the mod-

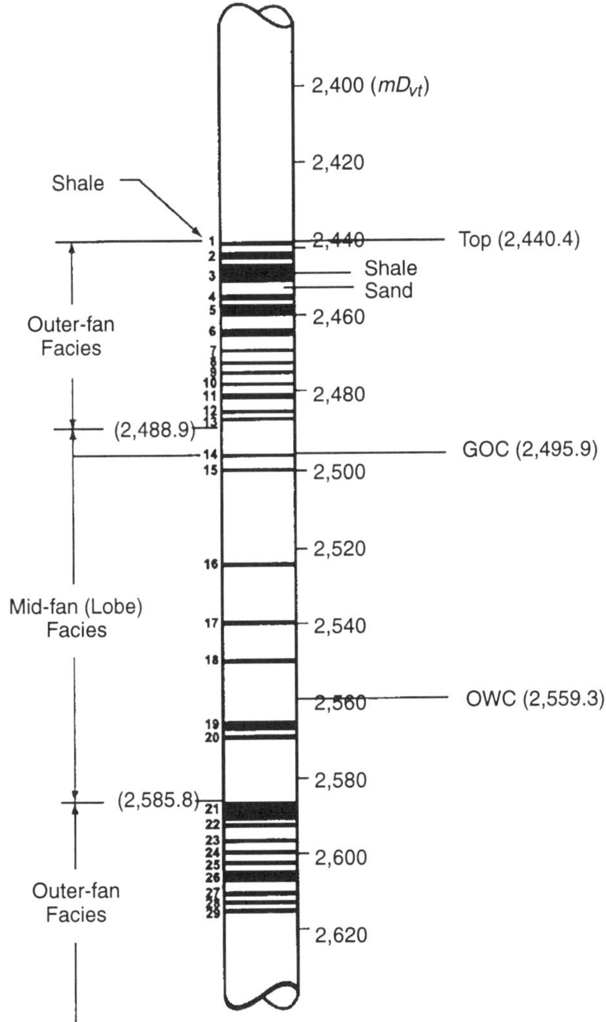

Fig. 7.25—Shale conditioning data (from Ref. 5).

el is shown in **Fig. 7.25.** In addition to conditioning, the shale descriptions at wells are the main source to estimate the distribution of thickness and the volume fraction of shale. Only Shale 11 through 20, which fall within, just above and below the oil column, were included explicitly in the simulation model.

Descriptions of the shale objects were generated by the marked point process. Because of the limitations in the number of cells that the simulation model can handle, types of shale were described only in an area of about 240 meters around the well. **Fig. 7.26** shows the shale objects that intercept the conditioning well. Shale objects are placed only in the locations indicated by the conditioning data (Fig. 7.25). Shale are scaled-up into vertical transmissibility modifiers, which can be used in the flow simulation model.

7.5 Other Applications of Object Modeling

In addition to geologic facies descriptions, object-based modeling techniques have been used to characterize other reservoir properties, such as faults, fractures and pore structure networks. Sec. 7.5.1 discusses the use of object modeling to describe subseismic faults.

7.5.1 Subseismic Fault Modeling. Subseismic faults are faults with displacements of less than 50 ft and cannot be interpreted even in high-resolution, 3D, seismic data. Despite their small displacements and sizes, subseismic faults can have a significant effect on permeabilities and connectivity of the reservoir.

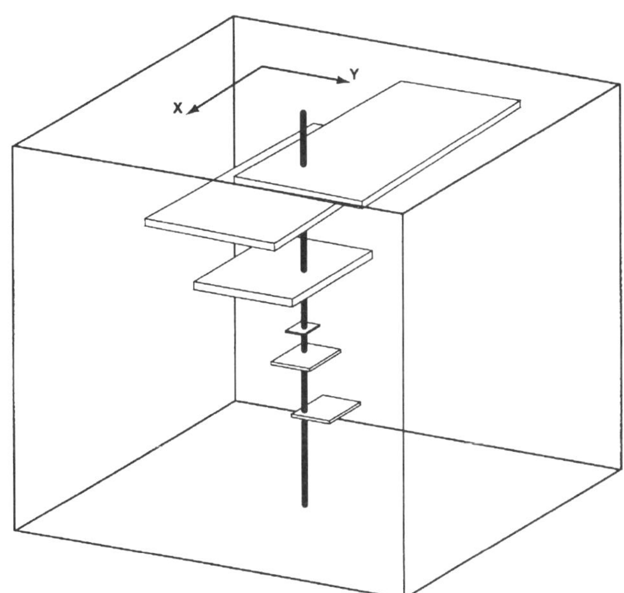

Fig. 7.26—Shale objects that intercept the conditioning well (from Ref. 5).

A technique to describe subseismic faults using the marked point process has been proposed by Munthe *et al.*[11] Faults are assumed to be vertical planes represented by line traces in horizontal planes and displacements. The parameters that describe faults are the center point of the fault line, length, width, orientation, vertical displacement and number of faults. Because subseismic faults are assumed to be a local phenomena, ellipsoids are defined around fault objects to bind the reservoir volume that is displaced by the faults.

The subseismic fault simulation process is similar to the process previously described for geologic facies. Fault parameters are represented by conditional probability distribution functions, which account for the conditional relationships among parameters. The conditional relationships among fault parameters are observed in field data, and these are used to develop models for the simulations. For example, fault displacement can be a function of location, and fault length can be a function of the displacement. Because subseismic faults are not seen in the field data, there are uncertainties when defining the number of faults and probability functions for the fault dimensions. A solution is to extrapolate the relationships observed at the field to the subseismic scale. Some studies have shown that the number of faults and their dimensions can be described by fractal scaling laws. In the simulation process, fault parameters are sampled sequentially in the order that is required to account for the prescribed conditional relationships.

Summary

In Chap. 7, we briefly describe object-based modeling of reservoir heterogeneities. Compared to grid-based simulations, object-based modeling is superior in representing geologically consistent, reservoir architecture. Object-based modeling also has an advantage of generating crisp images of geological bodies. The physical relationships among various geological objects can be explicitly accounted for, thereby, creating realistic images.

Object-based modeling falls short when diagenetic overprint is significant so that the original geological object shapes are significantly altered. It is also difficult to apply when the reservoir is mature, and a significant number of conditioning well data exists. In general, we conclude that object-based methods work best in clastic environments with a limited number of well data and when significant external information, regarding geological object shapes and sizes, is available.

Ultimately, the method that works best depends on the type of information available, as well as the geological environment. As discussed in this chapter, in some cases, a combination of object-based and grid-based methods may provide the best solution.

Nomenclature

f = fraction
h = thickness
mD_{vt} = measured vertical depth
r = repulsion function
s = attraction function
p = probability
x = axis
y = axis
ρ = density

Subscripts

i = dummy variable
sh = shale

References

1. Lanzarini, W.L. *et al.*: "Stochastic Modeling of Geometric Objects and Reservoir Heterogeneties," paper SPE 38953 presented at the 1997 Latin American and Caribbean Petroleum Engineering Conference and Exhibition, Rio de Janeiro (30 August–3 September).
2. Haldorsen, H.H. and Damsleth, E.: "Stochastic Modeling," *JPT* (April 1990) 404.
3. Bratvold, R.B. *et al.*: "STORM: Integrated 3D Stochastic Reservoir Modeling Tool for Geologists and Reservoir Engineers," paper SPE 27563 presented at the 1994 European Petroleum Computer Conference, Aberdeen, 15–17 March.
4. Taheri, M.: "Stochastic Modeling of Reservoir Architecture for Field Management," *SPERE* (November 1992) 433.
5. Haldorsen, H.H.: "On the Modeling of Vertical Permeability Barriers in Single-Well Simulation Models," *SPEFE* (September 1989) 349.
6. Clemetsen, R. *et al.*: "A Computer Program for Evaluation of Fluvial Reservoirs," Paper 23, North Sea Oil and Gas Reservoirs-II, The Norwegian Inst. of Technology, Graham & Trotman (1990).
7. Petit, F.M. *et al.*: "Multistep Geostatistical Modeling of a Complex Reservoir Based on Quantified Sedimentary Models and Characterization of Core Petrophysical Groups," paper SPE 30618 presented at the 1995 SPE Annual Technical Conference and Exhibition, Dallas, 22–25 October.
8. Tjølsen, C.B. *et al.*: "Seismic Data Can Improve Stochastic Facies Modeling," *SPEFE* (September 1996) 141.
9. Seifert, D. and Jensen, J.L.: "Object and Pixel-Based Reservoir Modeling of a Braided Fluvial Reservoir," paper presented at the 1997 Intl. Assn. of Mathematical Geology Annual Conference, Barcelona, Spain, 22–27 September.
10. Damsleth, E. *et al.*: "A Two-Stage Stochastic Model Applied to a North Sea Reservoir," *JPT* (April 1992) 402.
11. Munthe, K.L. *et al.*: "Subseismic Faults in Reservoir Description and Simulation," paper SPE 26500 presented at the 1993 SPE Annual Technical Conference and Exhibition, Houston, 3–6 October.

SI Metric Conversion Factors

ft	× 3.048*	E − 01 = m
ft^3	× 2.831 685	E − 02 = m^3

*Conversion factor is exact.

Chapter 8
Scaleup

This chapter presents scaleup techniques used to calculate effective reservoir properties for a coarse scale grid that usually has blocks several times larger than the fine scale in which the geologic and petrophysical descriptions are generated. Often, reservoir properties are described at small scales in order to capture variations observed in cores, logs, geologic models and seismic data. However, the resolution of models needed for other applications, such as flow simulations, is much coarser because the maximum number of gridblocks that can be used is limited by computer resources. **Fig. 8.1** shows an example of fine- and coarse-scale grids. In this example, one coarse block contains 25 fine-scale blocks. Recently, with the advent of more sophisticated gridding algorithms in flow simulators, such as corner point geometry and local grid refinement, it is possible to tailor grids to account for complexities, such as faults and high fluid velocity profiles near wells. **Fig. 8.2** shows an example of local grid refinement around a well. Scaleup techniques provide effective petrophysical properties for coarse gridblocks, such as the ones shown in Figs. 8.1 and 8.2, based on a fine-scale grid.

Sec. 8.1 covers scaleup techniques for static reservoir properties, including gross thickness, porosity and initial water saturation, and Secs. 8.2 and 8.3 cover the techniques used to scaleup dynamic properties, including permeability and relative permeabilities. Scaleup techniques described here are formulated, so the outcome of using effective properties in the coarse scale should yield the same result as the fine scale. For example, the scaleup technique for porosity ensures that the pore volume, at either scale, is the same. In general, the scale-up process for dynamic properties is more challenging than the process for static properties because several factors, such as boundary conditions, flow directions and interactions among the phases, affect the effective dynamic properties. Therefore, several approaches, which account for different conditions, have been proposed to the scaleup process for dynamic properties.

8.1 Scaleup of Static Properties

This section describes the techniques used to scaleup static properties. These properties include gross thickness, porosity and saturation. Scaleup of these properties is relatively straightforward and requires simple averaging. The details are discussed in the following subsections.

8.1.1 Gross Thickness. The scaleup technique for gross thickness calculates an effective thickness for a coarse block, based on the thicknesses of the small-scale blocks that fall inside the coarse block. The scaleup calculation is based on the principle that the bulk volume of the coarse-scale block must be equal to the sum of the bulk volumes of the fine-scale blocks that fall inside the coarse block, as described in the following equations. The bulk volume of a coarse block, V_b, is given by

$$V_b = A_b\tilde{h}, \quad \text{(8.1)}$$

where A_b is the surface area of the block, and $\tilde{h}$ is the effective gross thickness of the coarse block. Using the properties of the fine-scale blocks that fall inside the coarse block, the bulk volume of the coarse block is given by

$$V_b = \sum_{i=1}^{n} A_i h_i, \quad \text{(8.2)}$$

and the surface area is given by

$$A_b = \sum_{i=1}^{n} A_i, \quad \text{(8.3)}$$

where n is the number of fine-scale blocks that fall inside the coarse-scale block; A_i is the surface area, and h_i is the gross thickness of fine-scale block i. The effective gross thickness of the coarse block is obtained by combining Eqs. 8.1–8.3, and it is given by

$$\tilde{h} = \frac{\sum_{i=1}^{n} A_i h_i}{\sum_{i=1}^{n} A_i}. \quad \text{(8.4)}$$

In many cases, the surface area of the fine-scale blocks is the same, such as the example shown in Fig. 8.1. For these cases, the effective gross thickness, given by Eq. 8.4 becomes

$$\tilde{h} = \frac{1}{n}\sum_{i=1}^{n} h_i. \quad \text{(8.5)}$$

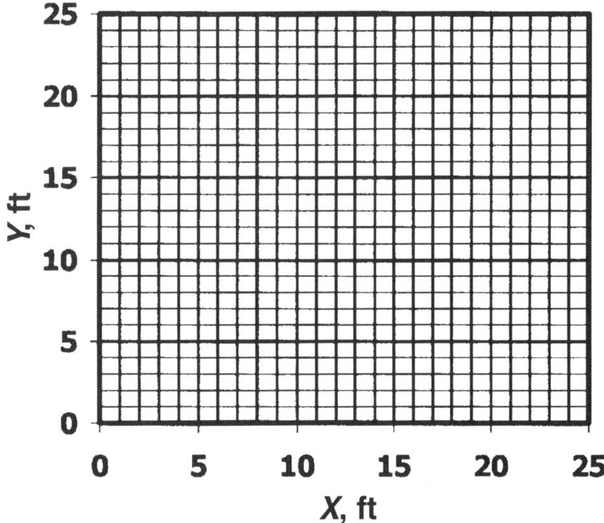

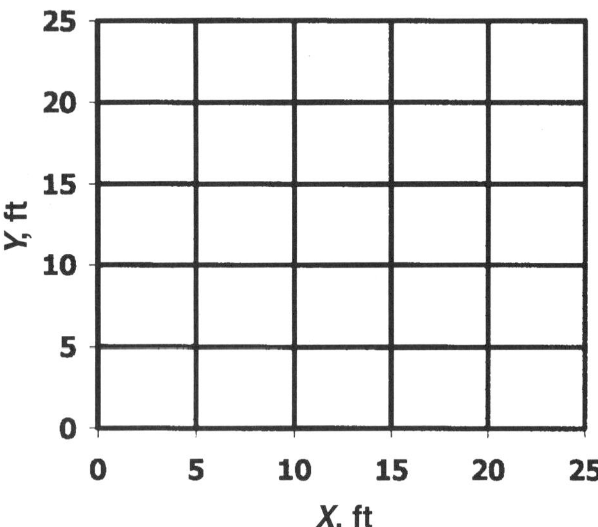

Fig. 8.1—Fine- and coarse-scale grids.

For this special case, the effective gross thickness is equal to the arithmetic average of the thicknesses of the fine-scale blocks.

8.1.2 Porosity. The scaleup calculation for porosity is based on the principle that the pore volume for the coarse-scale block equals the sum of the pore volumes of the fine-scale blocks that fall inside the coarse block. The pore volume of the coarse-scale block, pv_b, is given by

$$pv_b = V_b\tilde{\phi}, \quad \text{(8.6)}$$

where V_b is the bulk volume, and $\tilde{\phi}$ is the effective porosity of the coarse-scale block. The pore volume can be calculated by adding the pore volumes of the fine-scale blocks that fall inside the coarse block. This is given by

$$pv_b = \sum_{i=1}^{n} A_i h_i \phi_i, \quad \text{(8.7)}$$

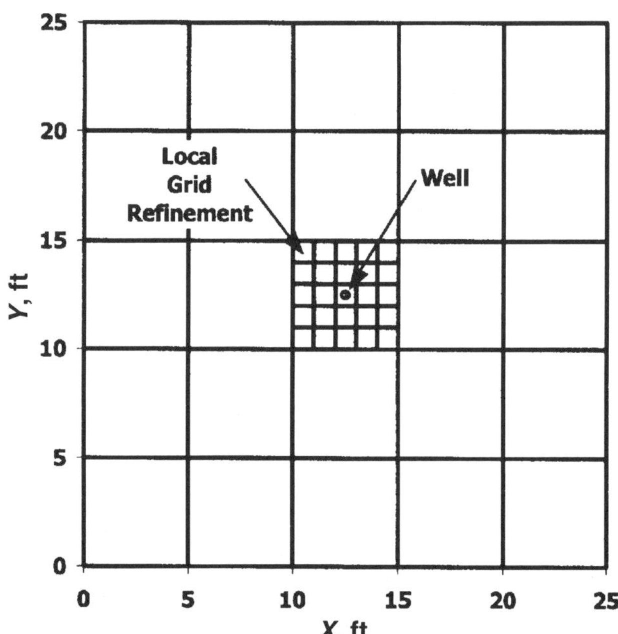

Fig. 8.2—Coarse grid with local grid refinement around a well.

where n is the number of fine-scale blocks that fall inside the coarse-scale block; A is the surface area; h is the gross thickness, and ϕ is the fine-scale block porosity. Substituting Eqs. 8.7 and 8.2 into Eq. 8.6 yields the equation for effective porosity,

$$\tilde{\phi} = \frac{\sum_{i=1}^{n} A_i h_i \phi_i}{\sum_{i=1}^{n} A_i h_i}. \quad \text{(8.8)}$$

There are cases when the surface area and gross thickness of the fine-scale blocks are common. For these cases, Eq. 8.8 can be simplified. For the case in which the surface area of the fine-scale gridblock is constant, the effective porosity is given by the gross thickness weighted average,

$$\tilde{\phi} = \frac{\sum_{i=1}^{n} h_i \phi_i}{\sum_{i=1}^{n} h_i}. \quad \text{(8.9)}$$

For the case in which the gross thickness and surface area of the fine-scale blocks are constant, the effective porosity is given by

$$\tilde{\phi} = \frac{1}{n}\sum_{i=1}^{n} \phi_i, \quad \text{(8.10)}$$

which is the arithmetic average of the porosities at the fine scale.

8.1.3 Water Saturation. The scaleup technique for initial water saturation is based on the conservation principle that the initial volume of water at the fine and coarse scales must be equal. The total initial volume of water, Wpv_b, at the coarse scale is given by

$$Wpv_b = pv_b\tilde{S}_{wi}, \quad \text{(8.11)}$$

Gross Thickness, ft

20	25	14	7	22
10	15	6	9	7
13	21	23	10	12
6	24	15	8	19
22	22	9	8	9

Legend
>21
17-21
13-17
9-13
<9

Porosity

0.18	0.25	0.06	0.09	0.05
0.18	0.05	0.08	0.18	0.28
0.23	0.11	0.15	0.25	0.12
0.27	0.19	0.09	0.22	0.22
0.11	0.27	0.17	0.09	0.13

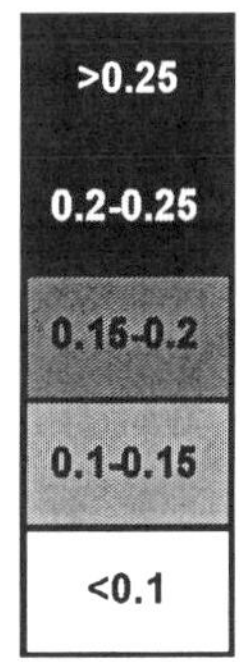

Water Saturation

0.09	0.31	0.13	0.25	0.26
0.10	0.37	0.26	0.15	0.18
0.22	0.37	0.39	0.37	0.26
0.12	0.10	0.25	0.10	0.29
0.15	0.22	0.15	0.37	0.40

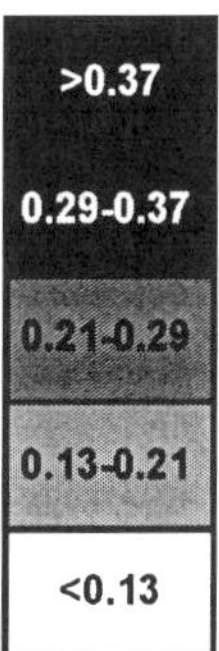

Fig. 8.3—Static reservoir properties for a fine-scale grid (1 × 1 ft).

where pv_b is the pore volume, and $\tilde{S}_{wi}$ is the effective initial water saturation of the coarse-scale block. The volume of water is given by the sum of the volumes of the fine-scale blocks that fall inside the coarse block.

$$Wpv_b = \sum_{j=1}^{n} A_j h_j \phi_j S_{wi_j}, \quad \ldots\ldots (8.12)$$

where n is the number of fine-scale blocks that fall inside the coarse-scale block; A_j is the surface area; h_j is the gross thickness; ϕ_j is the porosity, and S_{wi_j} is the initial water saturation of the fine-scale blocks. The effective initial water saturation is obtained by substituting Eqs. 8.11 and 8.7 into Eq. 8.12, and it is given by

$$\tilde{S}_{wi} = \frac{\sum_{j=1}^{n} A_j h_j \phi_j S_{wi_j}}{\sum_{j=1}^{n} A_j h_j \phi_j}. \quad \ldots\ldots (8.13)$$

The left side of this equation is often referred to as the pore volume weighted average.

For some cases commonly encountered, Eq. 8.13 can be simplified to calculate the effective initial water saturation. For the case in which the surface area of the fine-scale blocks is constant, the effective initial water saturation is

$$\tilde{S}_{wi} = \frac{\sum_{j=1}^{n} h_j \phi_j S_{wi_j}}{\sum_{j=1}^{n} h_j \phi_j}. \quad \ldots\ldots (8.14)$$

For the case in which the surface area and gross thickness of the fine-scale blocks are constant, the effective initial water saturation is

$$\tilde{S}_{wi} = \frac{\sum_{j=1}^{n} \phi_j S_{wi_j}}{\sum_{j=1}^{n} \phi_j}. \quad \ldots\ldots (8.15)$$

For the case in which the surface area, gross thickness and porosity of the fine-scale blocks are constant, the effective water saturation is

$$\tilde{S}_{wi} = \frac{1}{n}\sum_{j=1}^{n} S_{wi_i}, \quad \ldots\ldots (8.16)$$

which corresponds to the arithmetic average.

Numerical Example 8.1—Scaleup of Static Properties. This example illustrates the scaleup calculations for gross thickness, porosity and initial water saturation. The fine-scale grid, consisting of 1 × 1-ft blocks, and the reservoir properties for each block are shown in **Fig. 8.3.** Calculate the effective properties for a coarse block with the dimensions 5 × 5 ft, which includes all fine-scale blocks.

Solution. Eq. 8.5 gives the effective gross thickness of the coarse block.

$$\sum_{i=1}^{n} h_i = 20 + 25 + 14 + \ldots + 9 = 356,$$

and

$$\tilde{h} = \frac{1}{n}\sum_{i=1}^{n} h_i = \frac{1}{25}(356) = 14.24 \text{ ft}.$$

Eq. 8.9 gives effective porosity of the coarse block.

$$\sum_{i=1}^{n} h_i \phi_i = 20 \times 0.18 + 25 \times 0.25$$

$$+ 14 \times 0.06 + \ldots + 9 \times .013 = 56.97,$$

and

$$\tilde{\phi} = \frac{\sum_{i=1}^{n} h_i \phi_i}{\sum_{i=1}^{n} h_i} = \frac{56.97}{356} = 0.160.$$

Eq. 8.14 gives effective water saturation of the coarse block.

$$\sum_{j=1}^{n} h_j \phi_j S_{wi_j} = 20 \times 0.18 \times 0.09 + 25 \times 0.25 \times 0.31$$

$$+ 14 \times 0.06 \times 0.13 + \ldots + 9 \times 0.13 \times 0.4 = 13.15,$$

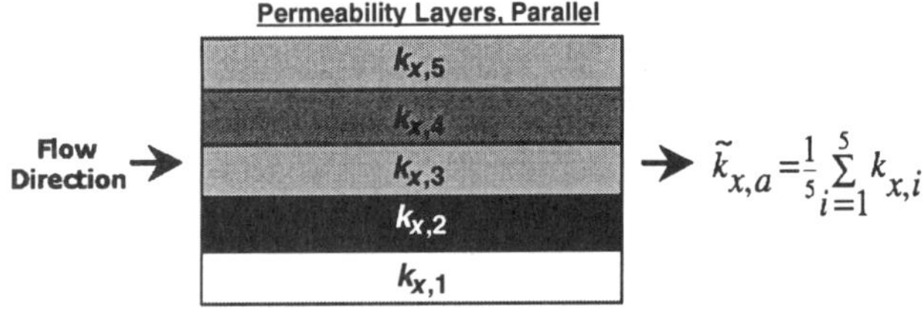

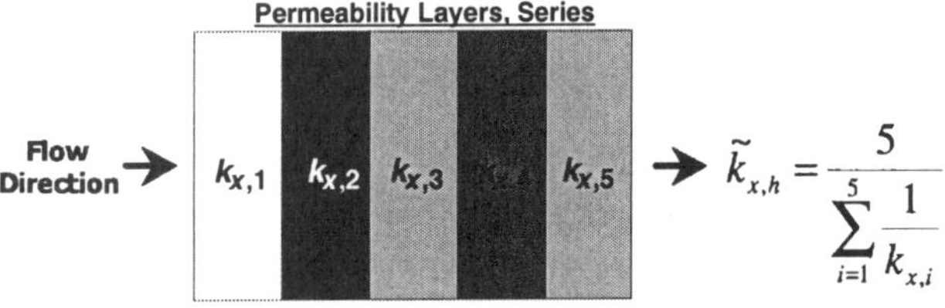

Fig. 8.4—Effective permeabilities for ideal cases where beds are arranged parallel and in a series, with respect to flow direction.

and

$$\tilde{S}_{wi} = \frac{\sum_{j=1}^{n} h_j \phi_j S_{wi_j}}{\sum_{j=1}^{n} h_j \phi_j} = \frac{13.15}{56.97} = 0.231.$$

8.2 Scaleup for Dynamic Properties—Single-Phase Flow

This section discusses scaleup techniques for dynamic properties for single-phase flow. The most important dynamic property of a reservoir is permeability. Calculation of the effective permeability for a coarse block from fine-scale permeabilities is more challenging than for static properties. Because permeability is a dynamic property, scaleup depends on the relative spatial arrangement of the fine-scale permeabilities, boundary conditions and anisotropy. For some ideal spatial arrangements of permeabilities and boundary conditions, analytical equations for scaleup are available. For more general cases, permeability scaleup is based on numerical solutions of the partial differential equations that govern single-phase flow.

8.2.1 Effective Permeabilities for Ideal Cases. The effective permeabilities for three ideal cases are given by arithmetic, harmonic and geometric averages.[1] Although ideal cases seldomly represent the real heterogeneity of rocks, these provide useful measures of upper and lower bounds of effective permeability.

The arithmetic average provides an effective permeability for the case, when flow occurs along layers of constant permeability (**Fig. 8.4**). The permeability of individual layers might change, but it is constant along the flow direction. The arithmetic average permeability for the x direction, $\tilde{k}_{x,a}$, is given by

$$\tilde{k}_{x,a} = \frac{1}{n}\sum_{i=1}^{n} k_{x,i}, \qquad (8.17)$$

where n is the number of fine-scale blocks that fall inside the coarse-scale block, and k_x is the permeability along the x-direction for the fine-scale blocks. This equation assumes that the blocks have constant dimensions along the x, y, and z directions. If the thickness of the layers varies, the effective permeability is given by

$$\tilde{k}_{x,a} = \frac{\sum_{i=1}^{n} k_{x,i} h_i}{\sum_{i=1}^{n} h_i}. \qquad (8.18)$$

Similarly, effective permeabilities for the y and z directions, $\tilde{k}_{y,a}$ and $\tilde{k}_{z,a}$, can be derived.

The harmonic average provides the effective permeability for the case in which the permeability is arranged in series, and it changes only perpendicular to the direction of flow (Fig. 8.4). The effective permeability along the x direction, given by the harmonic average, is

$$\tilde{k}_{x,h} = \frac{n}{\sum_{i=1}^{n} \frac{1}{k_{x,i}}}. \qquad (8.19)$$

Similar equations can be derived for $\tilde{k}_{y,h}$ and $\tilde{k}_{z,h}$.

For the case in which the block length varies along the flow direction, the effective permeability is given by

$$\tilde{k}_{x,h} = \frac{\sum_{i=1}^{n} L_i}{\sum_{i=1}^{n} \frac{L_i}{k_{x,i}}}.$$

The geometric average provides the effective permeability for the case in which the distribution of permeability is completely uncorrelated or when the correlation range of the variogram is zero. The geometric mean effective permeability, $\tilde{k}_{x,g}$, for blocks with the same dimensions, is given by

$$\tilde{k}_{x,g} = \left(\prod_{i=1}^{n} k_{x,i}\right)^{\frac{1}{n}}. \qquad (8.20)$$

Similar equations can be derived for $\tilde{k}_{y,g}$ and $\tilde{k}_{z,g}$.

For the same fine-scale values, the arithmetic average gives the highest effective permeability; whereas, the harmonic average gives the lowest effective permeability, and the geometric average provides an effective permeability that is between these two.

The power average is a more general averaging technique, which includes the above averages as limiting cases. To calculate the effective permeability, $\tilde{k}_{x,p}$, with the power average, use the equation,[2]

$$\tilde{k}_{x,p} = \left[\frac{1}{n}\sum_{i=1}^{n}\left(k_{x,i}\right)^{p}\right]^{\frac{1}{p}}, \qquad (8.21)$$

where p is the power that ranges from -1 to $+1$, except for zero. $p=1$ represents the arithmetic average; $p=-1$ represents the harmonic average, and $p=0$, as a limiting case, represents the geometric average.

8.2.2 Incomplete Layer Method. The incomplete layer method provides a scaleup approximation when permeabilities vary in more than one direction. Cardwell and Parsons[3] showed that the effective permeability lies between the upper and lower bounds, which are given by combinations of the

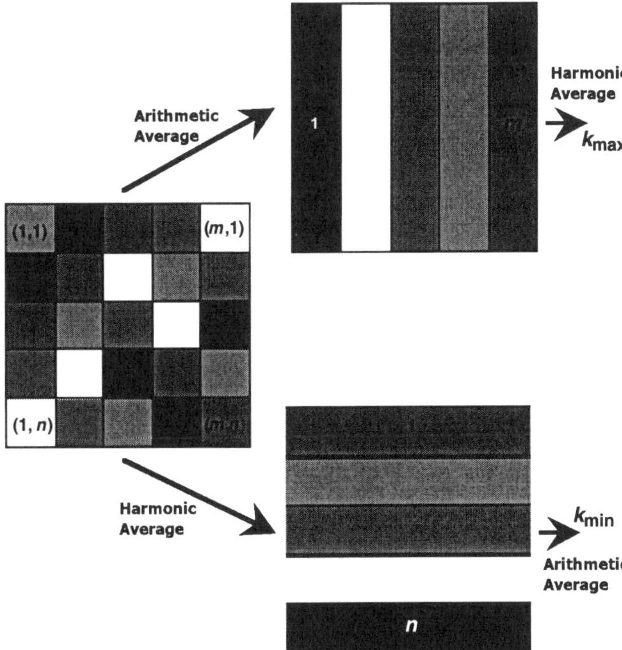

Fig. 8.5—Upper and lower permeability bounds for an incomplete layer method.

arithmetic and harmonic averages. **Fig. 8.5** illustrates the calculation of the upper and lower bounds for the effective permeability. The details are provided in Appendix F. The upper bound, k_{max}, consists of the harmonic average of the arithmetic average values.

$$k_{max} = \frac{m}{\sum_{i=1}^{m}\left[\frac{1}{n}\sum_{j=1}^{n} k_{i,j}\right]^{-1}}. \qquad (8.22)$$

The lower bound, k_{min}, consists of the arithmetic average of the harmonic average values.

$$k_{min} = \frac{1}{n}\sum_{j=1}^{n}\left(\frac{m}{\sum_{i=1}^{m}\frac{1}{k_{i,j}}}\right). \qquad (8.23)$$

An approximation for the effective permeability for a coarse block is given by the geometric average of the bounds,[4]

$$\tilde{k} = \sqrt{k_{min}k_{max}}. \qquad (8.24)$$

8.2.3 Renormalization Method. The renormalization method uses the analogy between flow in porous media and electric circuits to calculate the effective permeability of a coarse block. King[5] introduced the method. The effective permeability is calculated recursively in several steps. First, the fine-scale grid is divided into groups of four two-dimensional (2D) blocks or eight three-dimensional (3D) blocks. The effective permeability is calculated for each group with the equations derived from circuit analogs. In the subsequent steps, new groups of four or eight blocks are created with the effective permeabilities from the previous step. The procedure continues until the blocks reach the required scale. The equation, which calculates the effective permeability in 2D for the configuration, is shown in **Fig. 8.6.** (See Appendix F for details.)

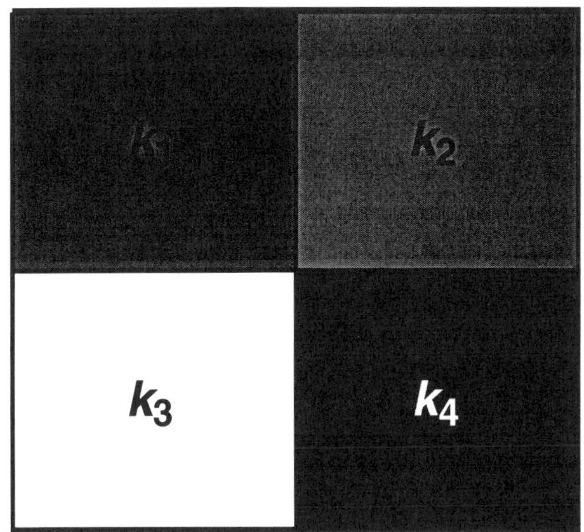

Fig. 8.6—Permeability arrangement for renormalization and tensor methods.

$$\bar{k}_x = \frac{4(k_1 + k_3)(k_2 + k_4)C_1}{(k_1 + k_2 + k_3 + k_4)C_1 + 3C_2}, \qquad (8.25)$$

$$C_1 = k_2k_4(k_1 + k_3) + k_1k_3(k_2 + k_4), \qquad (8.26)$$

and

$$C_2 = (k_1 + k_2)(k_3 + k_4)(k_1 + k_3)(k_2 + k_4). \qquad (8.27)$$

This effective permeability, $\bar{k}$, must be calculated for the y and z directions. The equation assumes closed boundaries in the directions perpendicular to flow. King notes that the derivation for 3D does not yield a closed form, as Eq. 8.25; therefore, it must be solved numerically.

8.2.4 Permeability Tensor. Permeability tensors allow a more rigorous representation of permeability. They correctly account for crossflow, created by the heterogeneity within the gridblock. Full permeability tensors are required to describe the effective permeability of gridblocks when[6-8] the layered structure of the permeability is not aligned with the coordinate axes of the gridblocks, and heterogeneity is not centered within the gridblocks.

In two dimensions, permeability tensors have four components. Two components account for flow along the principal directions because of pressure gradients in those directions (k_{xx} and k_{yy}), and two components account for crossflow because of pressure gradients in the perpendicular directions (k_{xy} and k_{yx}). The Darcy's law equations, which describe the velocity field for the permeability tensor, are

$$(v_x)_x = -\frac{k_{xx}}{\mu}\left(\frac{\partial p}{\partial x}\right)_x - \frac{k_{xy}}{\mu}\left(\frac{\partial p}{\partial y}\right)_x, \qquad (8.28)$$

$$(v_y)_x = -\frac{k_{yx}}{\mu}\left(\frac{\partial p}{\partial x}\right)_x - \frac{k_{yy}}{\mu}\left(\frac{\partial p}{\partial y}\right)_x, \qquad (8.29)$$

$$(v_x)_y = -\frac{k_{xx}}{\mu}\left(\frac{\partial p}{\partial x}\right)_y - \frac{k_{xy}}{\mu}\left(\frac{\partial p}{\partial y}\right)_y, \qquad (8.30)$$

and

$$(v_y)_y = -\frac{k_{yx}}{\mu}\left(\frac{\partial p}{\partial x}\right)_y - \frac{k_{yy}}{\mu}\left(\frac{\partial p}{\partial y}\right)_y. \qquad (8.31)$$

Here, $(v_x)_x$ and $(v_x)_y$ are the velocity components in the x direction for flow along the x and y directions, respectively. $(v_y)_x$ and $(v_y)_y$ are the corresponding velocity components in the y direction. $(\partial p/\partial x)_x$ and $(\partial p/\partial x)_y$ are the pressure gradients in the x direction flow along the x and y directions, respectively. $(\partial p/\partial y)_x$ and $(\partial p/\partial y)_y$ are the corresponding pressure gradients in the y direction.

The method proposed by Aasum *et al.*[7] calculates the effective permeability tensor of a coarse block, recursively, in groups with a two by two arrangement with the finer-scale blocks, as shown in Fig. 8.6. The equations that calculate the effective permeability tensor for a two by two arrangement of blocks are

$$\tilde{k}_{xx} = \frac{\tilde{k}_{xapp}}{1 - \left\{\left[\left(\frac{\partial \tilde{p}}{\partial x}\right)_y \Big/ \left(\frac{\partial \tilde{p}}{\partial y}\right)_y\right] \times \left[\left(\frac{\partial \tilde{p}}{\partial y}\right)_x \Big/ \left(\frac{\partial \tilde{p}}{\partial x}\right)_x\right]\right\}}, \quad \ldots (8.32)$$

$$\tilde{k}_{yy} = \frac{\tilde{k}_{yapp}}{1 - \left\{\left[\left(\frac{\partial \tilde{p}}{\partial x}\right)_y \Big/ \left(\frac{\partial \tilde{p}}{\partial y}\right)_y\right] \times \left[\left(\frac{\partial \tilde{p}}{\partial y}\right)_x \Big/ \left(\frac{\partial \tilde{p}}{\partial x}\right)_x\right]\right\}}, \quad \ldots (8.33)$$

$$\tilde{k}_{xy} = -\tilde{k}_{xx}\left(\frac{\partial \tilde{p}}{\partial x}\right)_y \Big/ \left(\frac{\partial \tilde{p}}{\partial y}\right)_y, \quad \ldots (8.34)$$

and

$$\tilde{k}_{yx} = -\tilde{k}_{yy}\left(\frac{\partial \tilde{p}}{\partial y}\right)_x \Big/ \left(\frac{\partial \tilde{p}}{\partial x}\right)_x. \quad \ldots (8.35)$$

$\tilde{k}_{xapp}$ and $\tilde{k}_{yapp}$ are the x and y apparent effective permeabilities and are used as effective permeabilities, if the simulator is not equipped to handle a tensor formulation for permeability. $\tilde{k}_{xapp}$ will reduce to $\tilde{k}_{xx}$, and $\tilde{k}_{yapp}$ will reduce to $\tilde{k}_{yy}$, if the potential gradient ratios in Eqs. 8.32 and 8.33 are zero. The potential gradient ratio is defined as the ratio of the transverse to the longitudinal pressure gradients. $(\partial \tilde{p}/\partial x)_y/(\partial \tilde{p}/\partial y)_y$ is the potential gradient ratio induced by the flow along the y axis; whereas, $(\partial \tilde{p}/\partial y)_x/(\partial \tilde{p}/\partial x)_x$ is the potential gradient ratio, induced by the flow along the x axis. For details of the derivation, see Appendix F.

The method of permeability tensors can be extended to 3D flow[9]; however, the equations become more complex.

In addition to the methods discussed here, other analytical methods have also been developed. These include methods that account for shale/sand distributions[10-11] and methods that use wavelet theory.[12-13] The methods for sand/shale distributions are useful for only specific applications. The methods using wavelets seem to predict the upscaling very well; however, they are not grounded in physical flow behavior.

All the methods discussed so far are applicable for linear flow in gridblocks. However, near the wellbore, the flow is radial rather than linear. Recently, an analytical method was developed to account for such flow.[14]

8.2.5 Steady-State Flow—Numerical Solution. This method uses the numerical solution of the partial differential equation that governs single-phase steady-state flow conditions, to calculate the effective permeability of a coarse-scale block. The equation for the system consisting of fine-scale blocks that fall inside the coarse block is

$$\frac{\partial}{\partial x}\left(k_x\frac{\partial p}{\partial x}\right) + \frac{\partial}{\partial y}\left(k_y\frac{\partial p}{\partial y}\right) + \frac{\partial}{\partial z}\left(k_z\frac{\partial p}{\partial z}\right) = 0. \quad \ldots (8.36)$$

This equation is solved with constant pressure conditions at inlet and outlet faces, which are normal to the direction in which the effective permeability is being calculated, with closed boundaries in the other directions. Boundary conditions that calculate the effective permeability for the x direction are

$$p(x = 0) = p_0, \quad \ldots (8.37)$$

$$p(x = L) = p_L, \quad \ldots (8.38)$$

$$\left.\frac{\partial p}{\partial y}\right|_{y=0,L_y} = 0, \quad \ldots (8.39)$$

and

$$\left.\frac{\partial p}{\partial z}\right|_{z=0,L} = 0. \quad \ldots (8.40)$$

The numerical solution for Eqs. 8.36–8.40 provides the pressure distribution and flow rates in the fine-scale system. The sum of the flow rates at the outlet face calculate the total flow rate, Q_x. Then, the effective permeability for the coarse block for the x direction is calculated with Darcy's Law, written as

$$\tilde{k}_x = \frac{Q_x \mu L_x}{A(P_0 - P_L)}. \quad \ldots (8.41)$$

Similar equations can be derived for the y and z directions by using the appropriate boundary conditions in Eqs. 8.37–8.40.

Numerical Example 8.2—Permeability Scaleup. In this example several scaleup methods are used to calculate the effective permeability of a gridblock. The fine-scale permeability distribution consisting of 5×5 ft blocks, is shown in **Fig. 8.7.** With the scaleup methods discussed in this chapter, calculate the effective permeabilities for a coarse block that contains all the fine-scale blocks.

Solution. **Fig. 8.8** summarizes the effective permeabilities of the coarse block estimated with the different methods. The arithmetic and harmonic averages are the highest and lowest estimates of all methods. For this example, the numerical solution for effective permeability in the x direction is closer to the arithmetic average, while the numerical solution for the permeability in the y direction is closer to the harmonic average. The power average varies from values close to the arith-

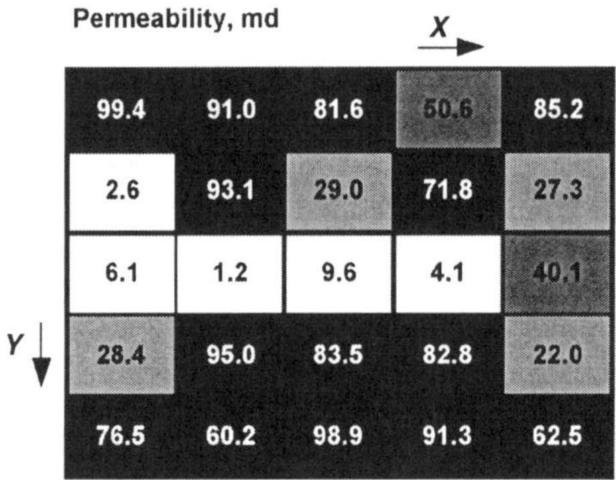

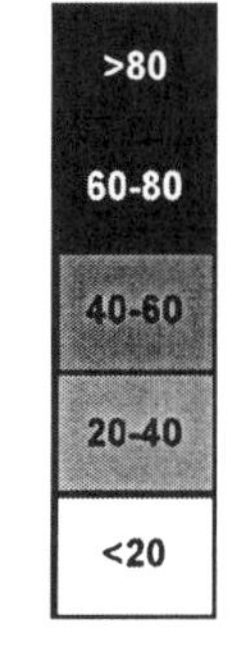

Fig. 8.7—Fine-scale permeability description.

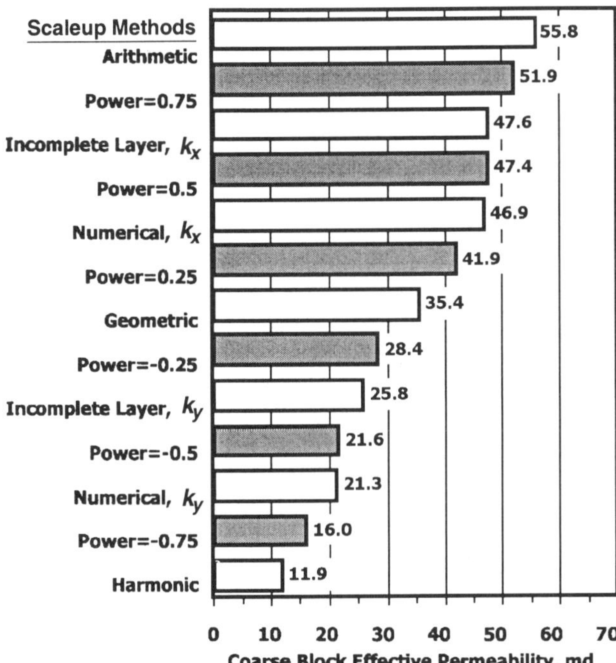

Fig. 8.8—Effective permeability for a coarse block, calculated with different scaleup methods.

metic average, for exponents close to one, to values close to the harmonic average, for exponents close to -1. This example shows that the exponents needed to estimate the x and y permeabilities are significantly different. The incomplete layer method provides estimates for x and y permeabilities, which are closer to the numerical solution than the simple averages. The following are the details of the calculations.

The arithmetic average (Eq. 8.17) is

$$\tilde{k}_{x,a} = \frac{1}{25}(99.4 + 91.0 + 81.6 + \ . \ . \ . + 62.5)$$
$$= 55.8 \text{ md}.$$

The harmonic average (Eq. 8.19) is

$$\tilde{k}_{x,h} = \frac{25}{\frac{1}{99.4} + \frac{1}{91.0} + \frac{1}{81.6} + \ . \ . \ . + \frac{1}{62.5}} = 11.9 \text{ md}.$$

The geometric average (Eq. 8.20) is

$$\tilde{k}_{x,g} = (99.4 \times 91.0 \times 81.6 \times \ . \ . \ . \times 62.5)^{\frac{1}{25}}$$
$$= 35.4 \text{ md}.$$

The power average (Eq. 8.21) for a power equal to 0.5 is

$$\tilde{k}_{x,p} = \frac{1}{25}\left(99.4^{0.5} + 91.0^{0.5} + 81.6^{0.5} + \ . \ . \ . + 62.5^{0.5}\right)^{\frac{1}{0.5}}$$
$$= 47.4 \text{ md}.$$

The incomplete layer method (Eqs. 8.22 through 8.24) provides an estimate for each direction. The maximum effective permeability for the x direction is

$$\frac{1}{n}\sum_{j=1}^{n} k_{1,j} = \frac{1}{5}(99.4 + 2.6 + 6.1 + 28.4 + 76.5) = 42.6,$$

$$\frac{1}{n}\sum_{j=1}^{n} k_{2,j} = \frac{1}{5}(91.0 + 93.1 + 1.2 + 95.0 + 60.2) = 68.1,$$

and

.

.

.

$$\tilde{k}_{x,\max} = \frac{5}{\sum_{i=1}^{5}\left(\frac{1}{42.6} + \frac{1}{68.1} + \ . \ . \ .\right)} = 54.1 \text{ md}.$$

The minimum effective permeability for the x direction is

$$\frac{m}{\sum_{i=1}^{m}\frac{1}{k_{i,1}}} = \frac{5}{\frac{1}{99.4} + \frac{1}{91.0} + \frac{1}{81.6} + \frac{1}{50.6} + \frac{1}{85.2}} = 77.2,$$

$$\frac{m}{\sum_{i=1}^{m}\frac{1}{k_{i,2}}} = \frac{5}{\frac{1}{2.6} + \frac{1}{93.1} + \frac{1}{29.0} + \frac{1}{71.8} + \frac{1}{27.3}} = 10.4,$$

and

$$\tilde{k}_{x,\min} = \frac{1}{5}(77.2 + 10.4 + \ . \ . \ .) = 41.9 \text{ md}.$$

The effective permeability for the x direction is obtained by combining the maximum and minimum effective permeabilities, written as

$$\tilde{k}_x = \sqrt{54.1 \times 41.9} = 47.6 \text{ md}.$$

The effective permeability calculations for the y direction are similar, but it is assumed that flow moves from the top towards the base of the grid.

Field Example 8.1—Scaleup of Static Properties and Permeability. The scaleup techniques are used in this field to calculate effective reservoir properties for a grid with large blocks. The fine-scale models were generated with conditional simulation techniques that capture variations observed in the well data and directional trends inferred from the geologic model. The number of blocks required to fully describe the reservoir heterogeneities is too large for flow simulators; therefore, a model with larger blocks must be generated. Scaleup techniques transform the reservoir properties from the fine-scale grid into effective properties for a grid with coarser blocks, such that the simulation results will be the same at both scales.

The following are the objectives of this field case:

• Calculate the effective gross thickness, porosity and permeabilities for a coarse-scale grid using scaleup techniques.

• Compare the statistics of the reservoir properties at the fine and coarse scales.

The following information is given:

• The fine-scale grid consists of 100×100 blocks, and block sizes are equal to 100×100 ft for the x and y directions, respectively. Gross thickness, porosity and permeability descriptions were generated in this grid.

• The gross thickness description (**Fig. 8.9**) was generated with the sequential Gaussian simulation technique described in Chap. 6. The uncorrelated appearance of this description is a result of the large nugget. The contribution of the nugget is approximately 60% of the total variogram sill.

• The porosity description (**Fig. 8.10**) was generated with sequential Gaussian simulation. This description has strong anisotropy. The principal direction of correlation is 22.5° southeast, and the anisotropy ratio is equal to 2.9.

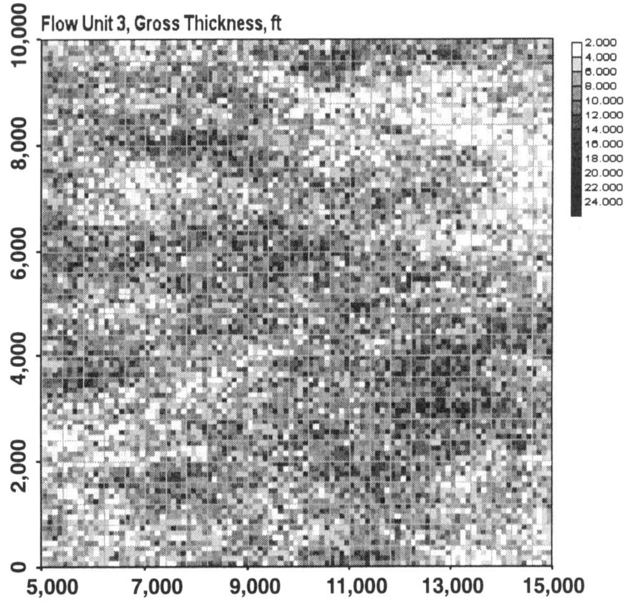

Fig. 8.9—Flow Unit 3 gross thickness map with 100 × 100 ft blocks.

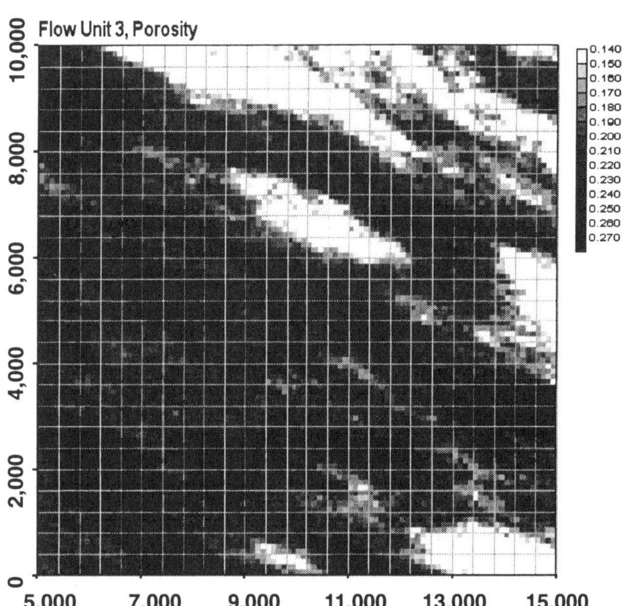

Fig. 8.10—Flow Unit 3 porosity map with 100 × 100 ft blocks.

• The permeability description (**Fig. 8.11**) was generated with linear equations between porosity and the logarithm of permeability. The permeabilities for the x and y directions are assumed to be same at the fine scale.

• The coarse-scale grid consists of 25 × 25 blocks with block sizes equal to 400 × 400 ft in the x and y directions. The mesh of the coarse grid has been superimposed on the fine-scale descriptions in Figs. 8.9–8.11, which shows the fine-scale properties used to calculate the effective properties of coarse blocks.

The scaleup results for gross thickness, porosity and permeability are shown in **Figs. 8.12–8.15,** respectively. The scaled-up properties at the coarse grid have less variability than the fine-scale properties because of the averaging nature of the techniques that removes the high frequency noise from the maps. However, the major scale directional trends and anisotropies are preserved because the ratio of the coarse to fine scale is only four to one. Permeability was scaled-up using various methods. Figs. 8.14a–8.14c show the permeability scaleup using arithmetic, harmonic and geometric averaging, respectively. Figs. 8.15a and 8.15b represent scaled-up permeabilities in the x and y directions using the incomplete layer method, and Figs. 8.15c and 8.15d represent the scaled-up permeabilities in the x and y directions using the renormalization method. The x and y permeabilities at the coarse scale for the last two methods are different, even though these are equal at the fine scale. In general, the directional permeabilities of the coarse blocks will depend on the spatial arrangement of the permeabilities in the fine-scale grid. **Fig. 8.16** shows that most blocks have a k_y/k_x ratio close to one. However, k_y/k_x is significantly less than one for a few blocks in the areas where the contrast of the fine-scale permeabilities is great. The changes of the k_y/k_x ratios result from the directional correlation introduced by the variogram at the fine scale. k_x tends to be larger than k_y because the permeabilities at the fine scale have a longer correlation range along the prin-

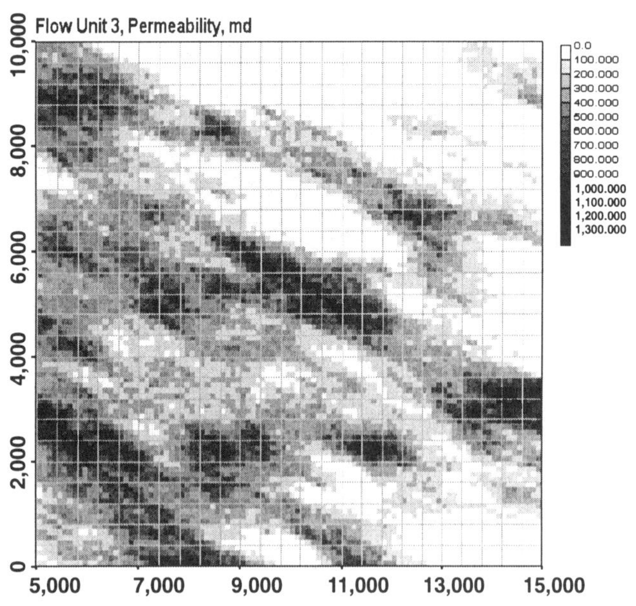

Fig. 8.11—Flow Unit 3 permeability map with 100 × 100 ft blocks.

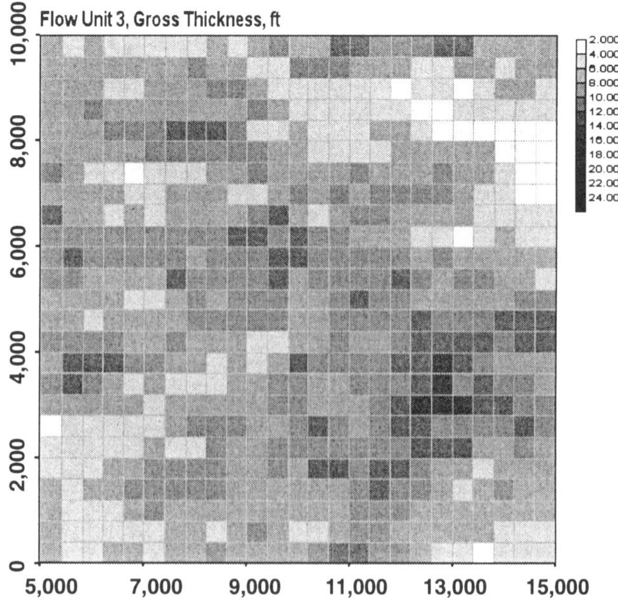

Fig. 8.12—Flow Unit 3 scaleup gross thickness map with 400 × 400 ft blocks.

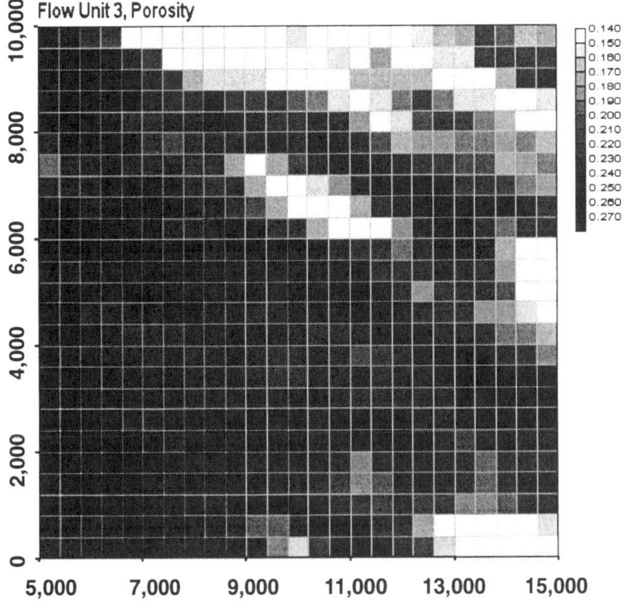

Fig. 8.13—Flow Unit 3 scaleup porosity map with 400 × 400 ft blocks.

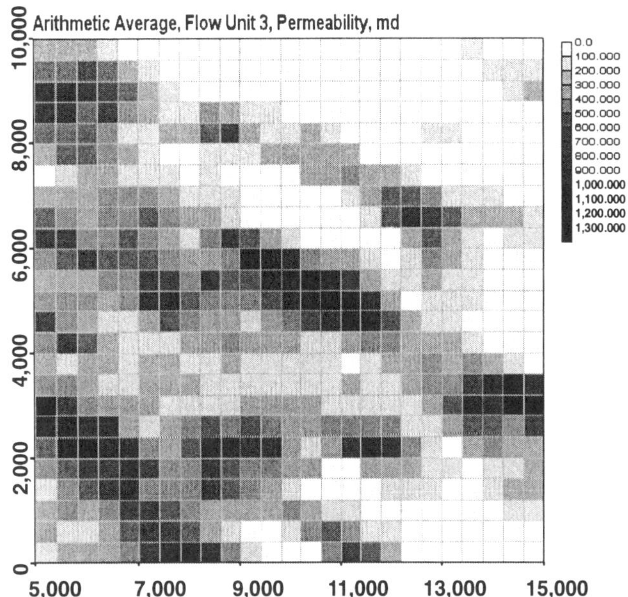

Fig. 8.14a—Flow Unit 3 scaleup permeability map for 400 × 400 ft blocks using arithmetic average.

cipal direction (only 22.5° off from the x direction), which is more closely aligned to the x direction rather than the y direction. Also, scaleup has an effect on the correlation between porosity and permeability. **Fig. 8.17** illustrates that as the scale increases, the correlation between porosity and permeability deviates from the linear relations observed in the fine scale. Therefore, it is important to note that usually the correlations observed at small scales (e.g., from core data) cannot be used to describe the relations between porosity and permeability for the coarse scales (e.g., simulation grid blocks).

In general, the statistics of the effective properties become narrower as the scale increases because the high frequency variations are smoothed in the scaleup process (**Figs. 8.18–8.20**). The standard deviation and range of the coarse scale reservoir properties are smaller than those for the fine scale. Still, the distributions at the coarse-scale preserve the overall character of the fine-scale distributions. In this example, the fine- and coarse-scale distributions of gross thickness are close to normal (Gaussian); those for porosity are close to bimodal, and those for permeability are close to log-normal.

8.3 Scaleup for Dynamic Properties—Multiphase Flow

Under multiphase flow conditions, we must consider two additional rock properties. They are capillary pressure and relative permeability. In this section, we concentrate on the upscaling of relative permeabilities with only passing references to the upscaling of capillary pressure. The upscaling of relative permeabilities has received a great amount of attention recently, as the ability to generate multimillion gridblock reservoir descriptions has improved.

In general, the upscaling procedures can be divided into two groups: analytical and numerical. Analytical procedures are applicable under limiting cases and under limited descriptions of heterogeneities. These procedures, however, serve a

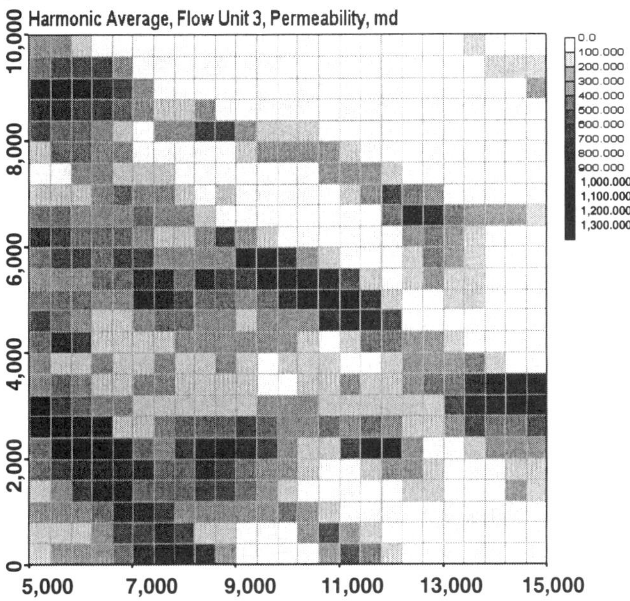

Fig. 8.14b—Flow Unit 3 scaleup permeability map for 400 × 400 ft blocks using harmonic average.

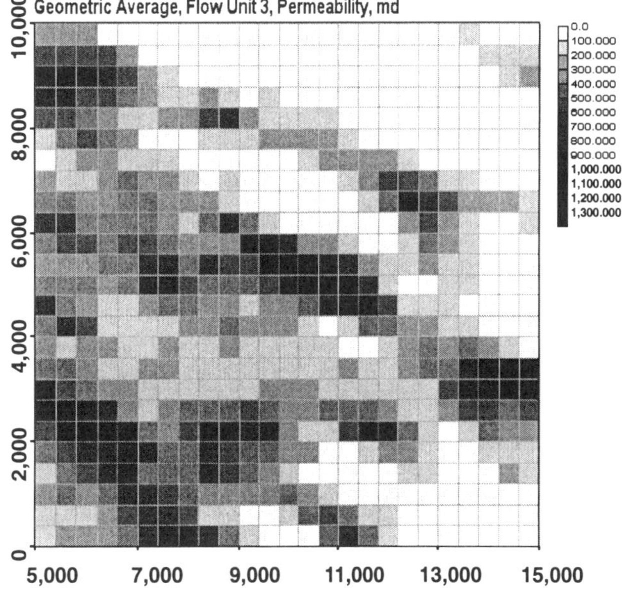

Fig. 8.14c—Flow Unit 3 scaleup permeability map for 400 × 400 ft blocks using geometric average.

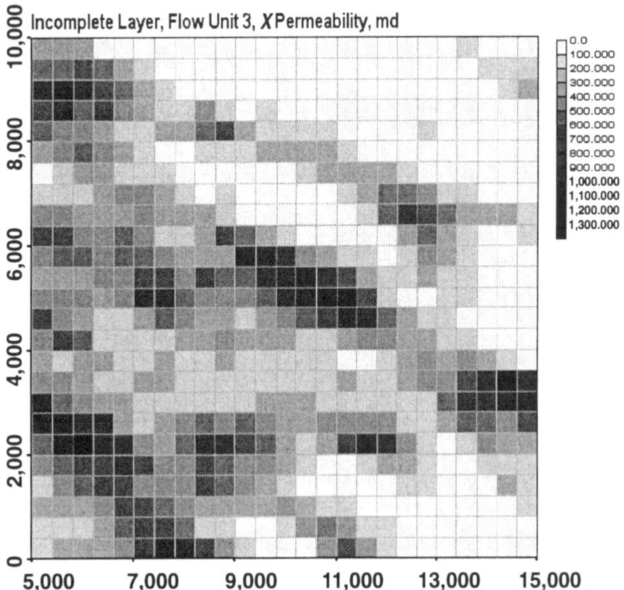

Fig. 8.15a—Flow Unit 3 scaleup *x* direction permeability map for 400 × 400 ft blocks using incomplete layer method.

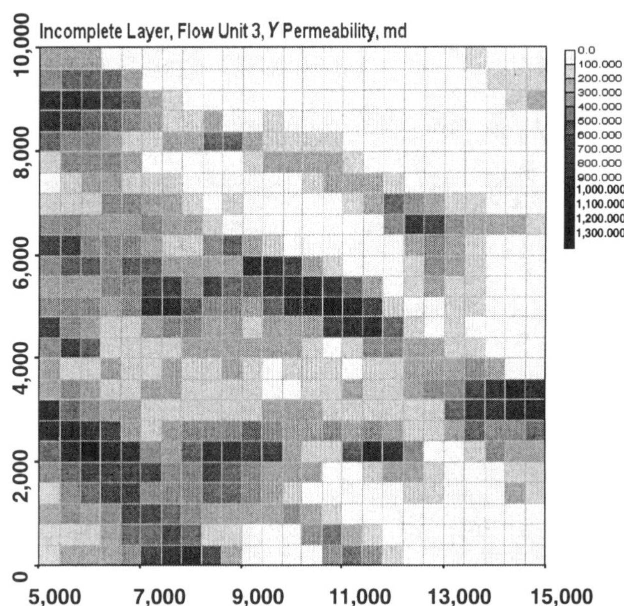

Fig. 8.15b—Flow Unit 3 scaleup *y* direction permeability map for 400 × 400 ft blocks using incomplete layer method.

useful purpose, and they have applications in many practical situations. Numerical methods, on the other hand, are more general; however, they are more computationally demanding.

8.3.1 Analytical Methods. When considering multiphase flow in a porous medium, we can think of three important forces that affect flow behavior. They are viscous, gravity and capillary forces. The relative magnitude of the individual forces determines the manner in which the flow behavior is affected. In limiting cases, one of these forces dominates the overall flow behavior; in this situation an analytical upscaling procedure may be applied.

In general, the two principal limiting cases we can consider are the no crossflow and vertical equilibrium cases. Under a no crossflow condition, no communication in the vertical direction exists. Each layer is isolated from the others, and the only communication is at the wellbore. This is possible if the vertical permeability is very small, or a lower permeability barrier exists between the two layers. Under vertical equilibrium conditions, the potential gradient in the vertical direction is zero. The flow in the vertical direction is so rapid that the pressure in the vertical direction reaches equilibrium quickly. Zapata and Lake[15] provided a dimensionless number to predict the presence of a particular limiting case

$$R_L = \frac{L}{H}\sqrt{\frac{k_v}{k_h}}, \qquad (8.42)$$

where k_v is the thickness-weighted average harmonic vertical permeability; k_h is the thickness-weighted average arithmetic horizontal permeability; L is the length in the horizontal direction, and H is the thickness. Zapata and Lake[15] observed that if $R_L > 10$, the assumption of vertical equilibrium holds, and if $R_L < 0.1$, the assumption of no crossflow holds. For values between 0.1 and 10, the limiting cases will not hold.

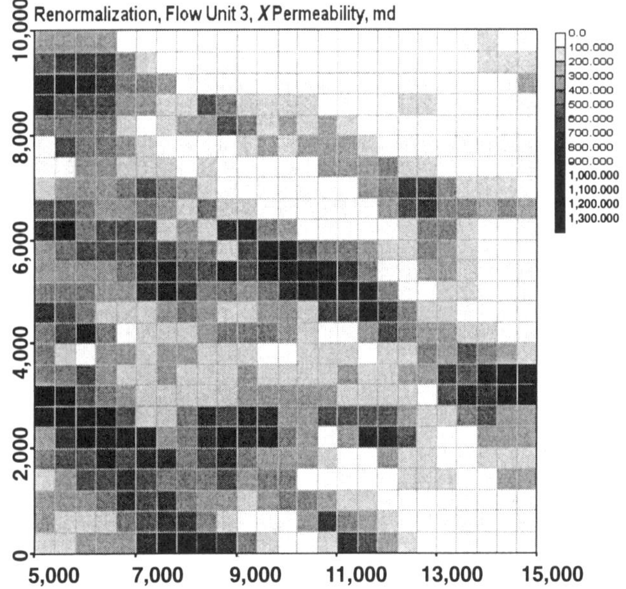

Fig. 8.15c—Flow Unit 3 scaleup *x* direction permeability map for 400 × 400 ft blocks using renormalization method.

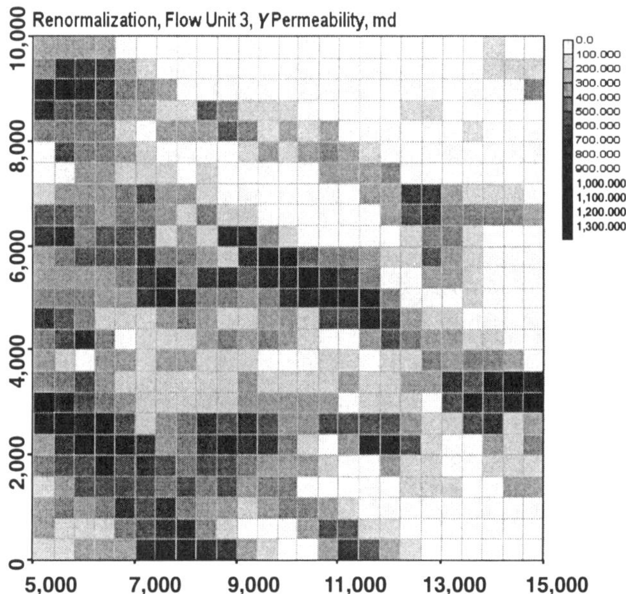

Fig. 8.15d—Flow Unit 3 scaleup *y* direction permeability map for 400 × 400 ft blocks using renormalization method.

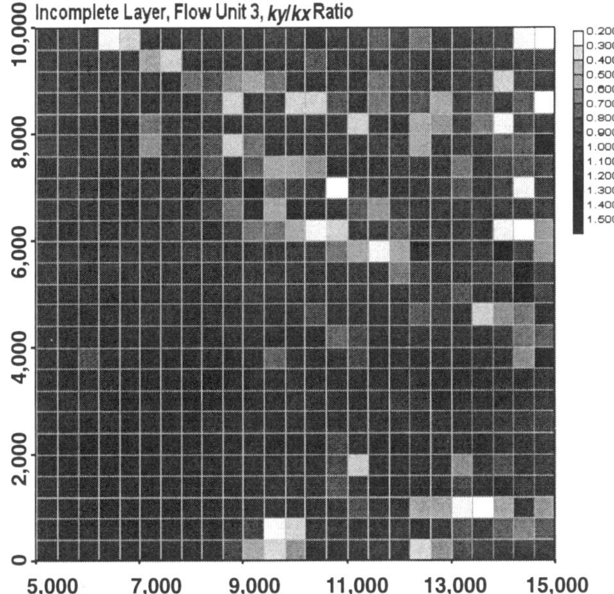

Fig. 8.16a—Flow Unit 3 scaleup *ky/kx* permeability ratio map for 400 × 400 ft blocks using incomplete layer method.

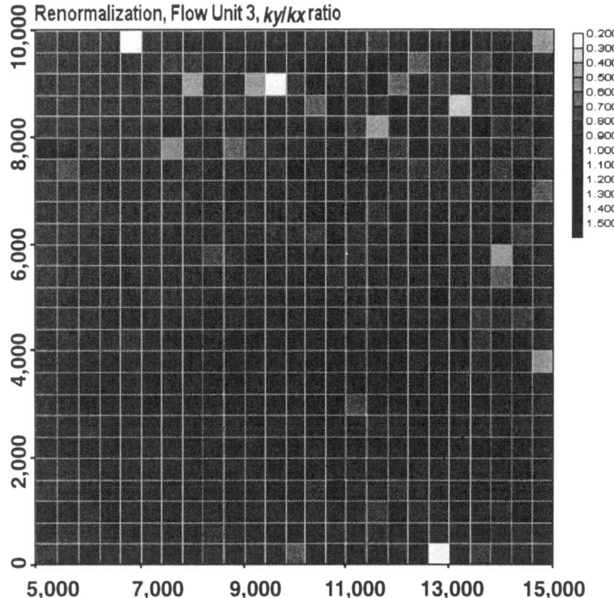

Fig. 8.16b—Flow Unit 3 scaleup *ky/kx* permeability ratio map for 400 × 400 ft blocks using renormalization method.

Under a vertical equilibrium assumption, the following four possibilities exist:[16]

• Viscous crossflow equilibrium—under this condition, gravity and capillary forces are negligible, and only viscous forces establish vertical equilibrium.

• Capillary/gravity equilibrium—both capillary and gravity forces dominate to establish the saturation distribution and, hence, the flow behavior.

• Gravity equilibrium—gravity forces dominate.

• Capillary equilibrium—only capillary forces determine the saturation distribution in the reservoir.

No specific guidelines exist to determine the forces that dominate under certain conditions. However, we can examine two dimensionless numbers with

$$N_c = \frac{1.127 \times 10^{-3} k_h A p_c}{\mu q L}, \quad \text{(8.43)}$$

where k_h is the horizontal permeability in md; p_c is the capillary pressure in psia; μ is the viscosity in cp; q is the flow in the horizontal direction; L is the horizontal length in feet, and

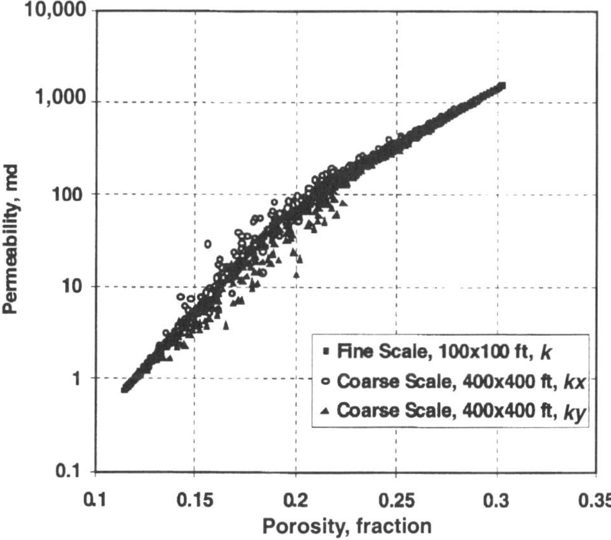

Fig. 8.17—Flow Unit 3 porosity vs. permeability crossplots for fine and coarse scales (incomplete layer method).

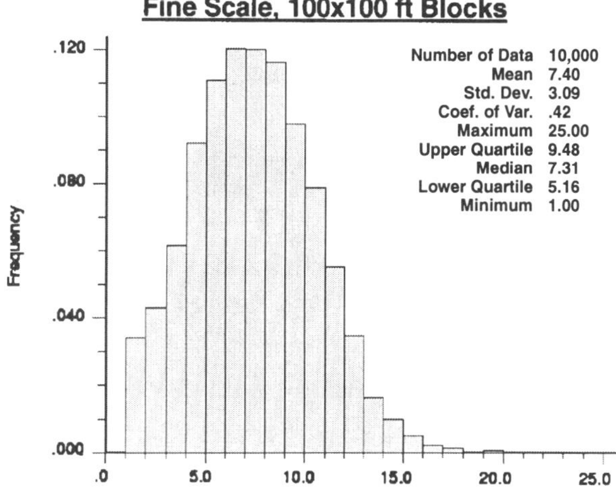

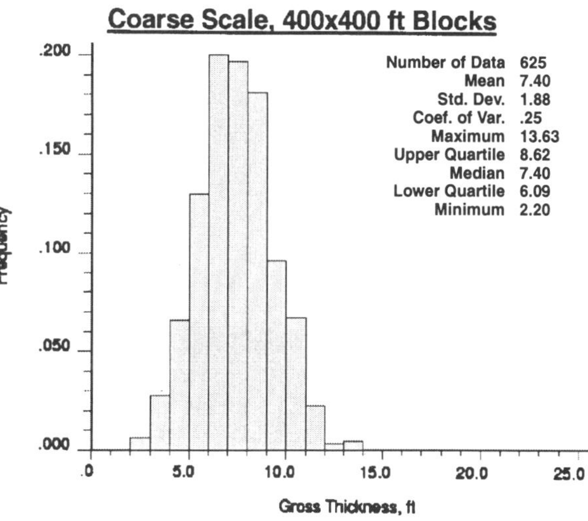

Fig. 8.18—Histograms of Flow Unit 3 gross thickness for fine and coarse scales.

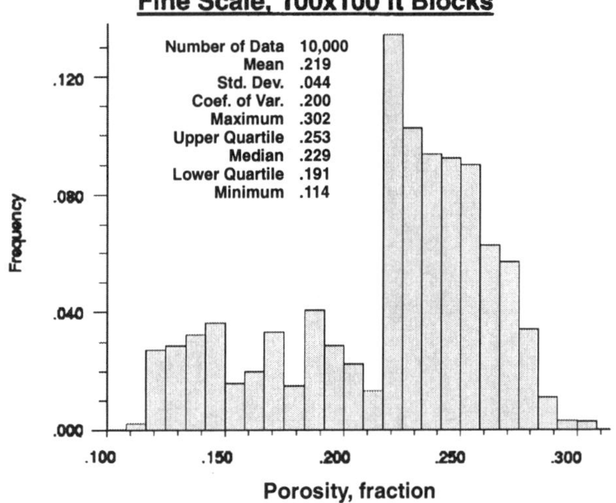

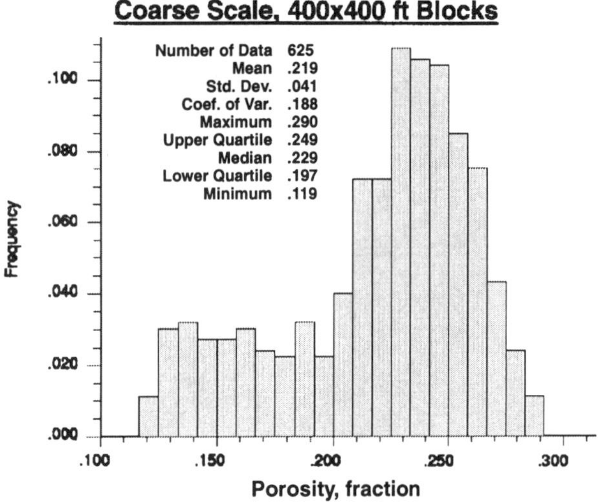

Fig. 8.19—Histograms of Flow Unit 3 porosity for fine and coarse scales.

A is the horizontal cross-sectional area in feet. N_c represents the ratio of capillary to viscous forces.

$$N_g = \frac{1.127 \times 10^{-3} \sqrt{k_h k_v}\,(0.4335)\Delta\gamma A}{q\mu}, \quad \ldots\ldots (8.44)$$

where $\Delta\gamma$ is the specific gravity difference; A is the cross-sectional horizontal area in ft^2; q is the flow rate in B/D; μ is the viscosity in cp, and k_h and h_v are the thickness-averaged horizontal and vertical permeabilities, respectively. N_g represents the ratio of the gravity to viscous forces. Some evidence exists to indicate that for values of N_g greater than 0.1, the flow can be considered gravity-dominated.[17] No such criterion exists for determining the dominance of capillary forces.

Analytical solutions exist for all the limiting cases. These solutions are restricted to layered reservoirs with different properties for each layer. However, we assume that each layer is homogeneous. We also have restricted the development of these equations for a waterflooding process, although the equations can be generalized for other displacements. The de-

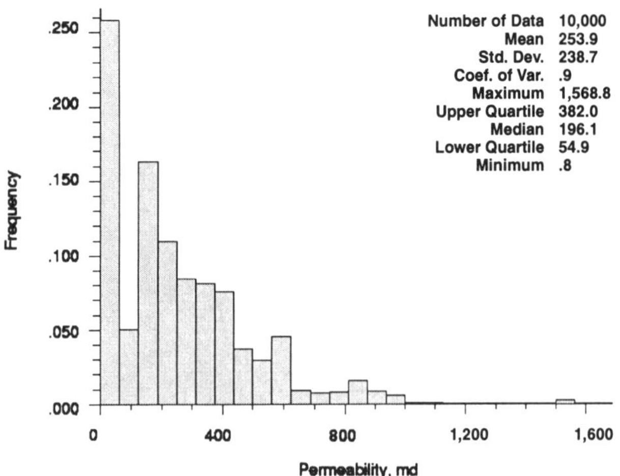

Fig. 8.20a—Histograms of Flow Unit 3 permeability for a fine scale.

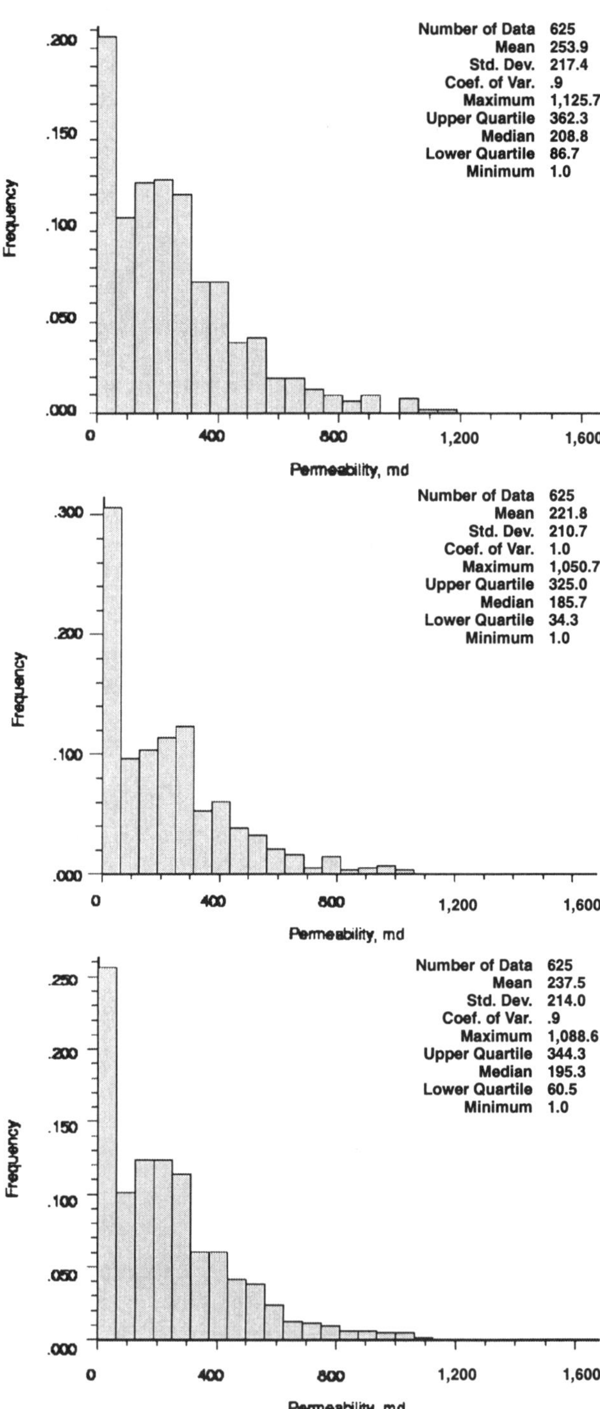

Fig. 8.20b—Histograms of Flow Unit 3 permeability for a coarse scale generated with arithmetic, harmonic and geometric averages.

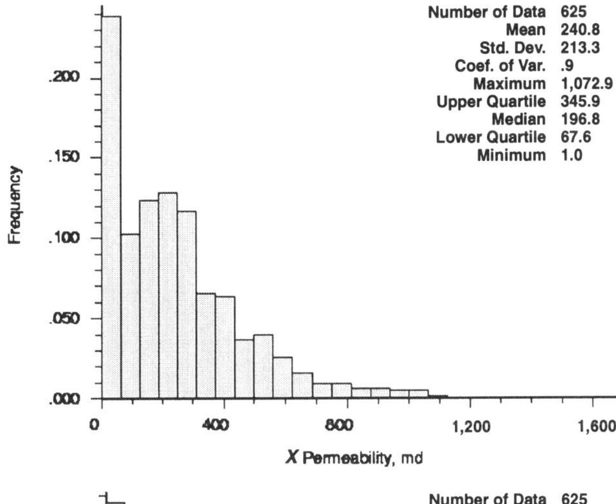

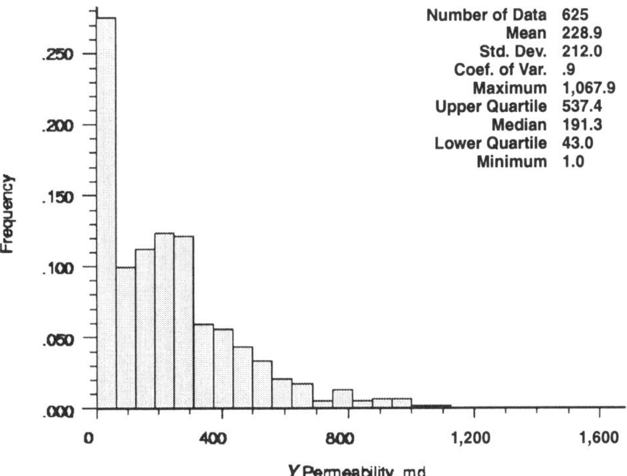

Fig. 8.20c—Histograms of Flow Unit 3 permeability for a coarse scale, generated with an incomplete layer method.

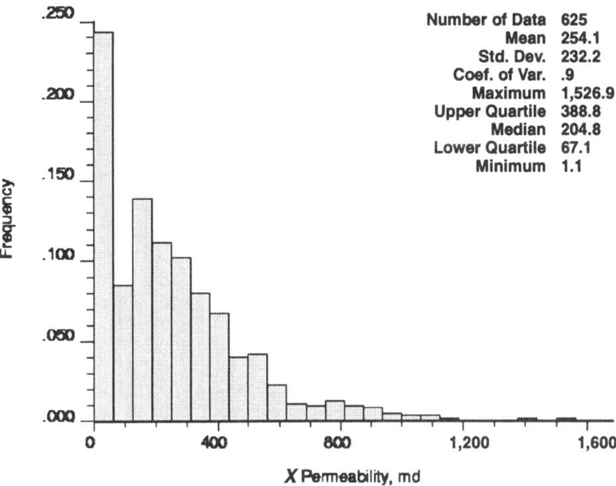

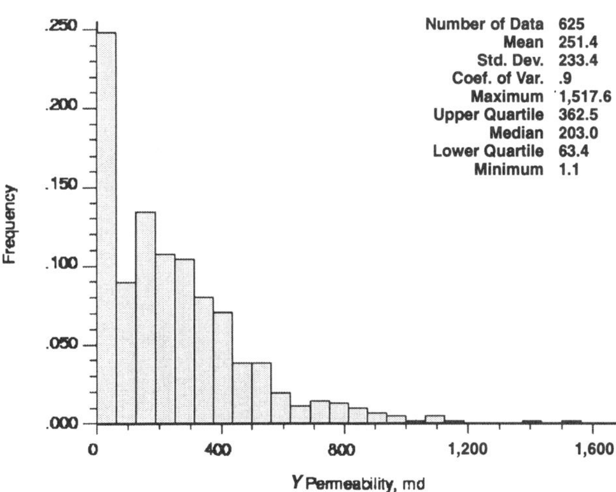

Fig. 8.20d—Histograms of Flow Unit 3 permeability for a coarse scale, generated with the renormalization method.

tails of the derivations are presented in Appendix F. In the text, only the important steps are described.

No Crossflow. We define the reservoir as a layered reservoir, as shown in **Fig. 8.21.** The layers are arranged according to the velocity at which fluid flows through it. The fastest layer is at the top and will breakthrough first, and the slowest layer is at the bottom and will breakthrough last. The in-situ velocity in a given layer is related to the quantity $k/\phi\Delta S_w$, where k is the permeability; ϕ is the porosity, and ΔS_w is $(1 - S_{w_i} - S_{o_r})$. S_{w_i} is the connate water saturation, and S_{o_r} is the residual oil saturation. We assume a piston-like displacement in each layer, and further assume that the following mobility ratio is the same for all the layers.

$$M = \frac{k'_{rw}/\mu_w}{k'_{ro}/\mu_o}. \qquad (8.45)$$

In Equation 8.44, k'_{rw} and k'_{ro} are the end point relative permeabilities of water and oil, respectively, and μ_o and μ_w are the viscosities of oil and water, respectively.

Assuming that layer l has just broken through, we can define

$$R_{jl} = \frac{\left(\dfrac{\phi\Delta S_w}{kk'_{rw}}\right)_l}{\left(\dfrac{\phi\Delta S_w}{kk'_{rw}}\right)_j}, \qquad (8.46)$$

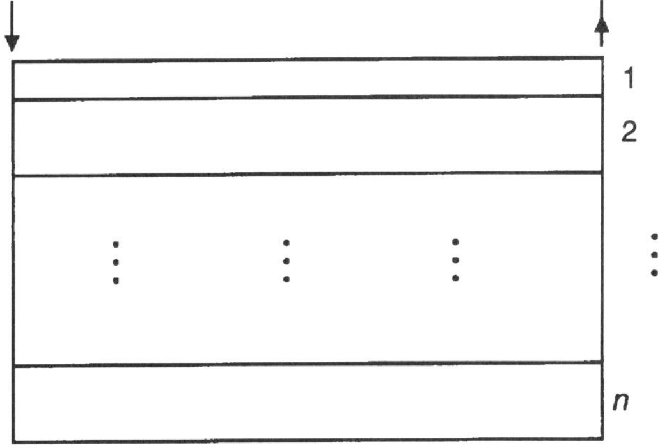

Fig. 8.21—Layered reservoir with *n* layers.

where j is any bed below l and has yet to breakthrough. The distance traveled by the flood front, in bed j, at the time the lth bed has broken through, is calculated as

$$x_{Dj} = \frac{-M + \sqrt{M^2 + R_{jl}(1 - M^2)}}{1 - M}, \text{ if } M \neq 1$$

$$= R_{jl}, \text{ if } M = 1, \qquad (8.47)$$

where x_D is the dimensionless distance in the x direction and is equal to x/L. We define the weighted arithmetic average for permeability as

TABLE 8.1—DATA FOR NUMERICAL EXAMPLE 8.3					
Layer	h_i	k_i	ϕ_i	S_{w_i}	S_{o_r}
1	15	700	0.24	0.2	0.15
2	2	450	0.21	0.26	0.2
3	9	80	0.18	0.24	0.2
4	22	200	0.2	0.3	0.16
5	11	50	0.16	0.25	0.2

TABLE 8.2—RANKING OF LAYERS		
Layer	$\dfrac{k_i}{\phi_i \Delta S_{wi}}$	Ranking
1	4,487	1
2	3,968	2
3	794	4
4	1,852	3
5	568	5

$$\tilde{k}H = \frac{\sum_{i=1}^{n} k_i h_i}{n}, \quad \text{(8.48)}$$

where H is the total thickness. The relative permeabilities are calculated as

$$\tilde{k}_{rw_l} = \frac{\sum_{i=1}^{n} (kk'_{rw}h)_i}{\tilde{k}H}, \quad \text{(8.49)}$$

and

$$\tilde{k}_{ro_l} = \frac{\sum_{i=l+1}^{n} \dfrac{(kk'_{ro}h)_i}{\frac{1}{M}x_{Di} + (1 - x_{Di})}}{\tilde{k}H}. \quad \text{(8.50)}$$

Notice that Eq. 8.50 will be simplified, if $M = 1$. The average saturation is

$$\tilde{S}_{w_l} = \frac{\sum_{i=1}^{l}\left[h\phi(1 - S_{o_r})\right]_i \sum_{i=l+1}^{n}\left\{\left[h\phi(1 - S_{o_r})\right]_i x_{Di} + \left(h\phi S_{w_i}\right)(1 - x_{Di})\right\}}{\sum_{i=1}^{l}(h\phi)_i}. \quad \text{(8.51)}$$

By repeating the calculations for l, varying between 1 and n, the range of the relative permeabilities can be calculated.

Numerical Example 8.3. The following data for a given layered reservoir is provided in **Table 8.1.** Assuming no crossflow, generate the effective relative permeability curves.

$$k'_{rw} = 0.4, \; k'_{ro} = 1.0, \; \mu_o = 0.6 \text{ cp}, \; \mu_w = 0.8 \text{ cp}.$$

Solution. The mobility ratio for a piston-like displacement is

$$M = \frac{k'_{rw}/\mu_w}{k'_{ro}/\mu_o} = \frac{0.4/0.8}{1.0/0.6} = 0.3.$$

The first step is to arrange the ranking in which the layers will break through. **Table 8.2** shows the ranking.

With Eqs. 8.46 and 8.47, we can calculate the distance traveled by the flood front, when each bed breaks through. (See **Table 8.3.**)

For example, at the time when layer one broke through, layer two traveled 0.92 of the total distance between the injector and producer.

$$\tilde{k}H = \sum_{i=1}^{n} k_1 h_i = 17,070,$$

and

$$\sum_{i=1}^{n} h_i \phi_i = 11.8$$

Once x_D is known, we can calculate the effective relative permeabilities with Eqs. 8.49 and 8.50 and can calculate the associated saturation with Eq. 8.51.

For example, when layer four breaks through,

$$\tilde{k}_{rw} = \frac{0.4(700 \times 15 + 450 \times 2 + 80 \times 9 + 200 \times 22)}{17,070}$$

$$= 0.387,$$

$$\tilde{k}_{ro} = \frac{\dfrac{1.0 \times 11 \times 50}{\frac{1}{0.3}(0.8) + (1 - 0.8)}}{17,070} = 0.0112,$$

and

$$\tilde{S}_w = \frac{\begin{array}{l} 15 \times 0.24 \times (1 - 0.15) + 2 \times 0.21 \times (1 - 0.2) \\ + \ldots + 22 \times 0.2 \times (1 - 0.16) \\ + \{11 \times 0.16[(1 - 0.2) \times 0.8 + 0.25 \times 0.2]\} \end{array}}{11.8}$$

$$= 0.814.$$

Fig. 8.22 shows the effective relative permeability curves over the entire range of saturation. For comparison, the effective permeabilities for $M = 1$ and $M = 10$ are also shown on the same graph. As expected, for an unfavorable mobility ratio of ten, the relative permeabilities are also unfavorable—higher relative permeability for water and lower relative permeability for oil.

TABLE 8.3—DISTANCE TRAVELLED BY THE FLOOD FRONT										
	Breakthrough Layer									
	1		2		3		4		5	
Layer	R_{jl}	x_D	R_{jl}	x_D	R_{jl}	x_D	R_{jl}	x_D	R_{jl}	x_D
1	1.00	1.0		1.00		1.00		1.00		1.00
2	0.88	0.92	1.00	1.00		1.00		1.00		1.00
3	0.41	0.55	0.47	0.60	1.00	1.00		1.00		1.00
4	0.18	0.29	0.20	0.32	0.43	0.56	1.00	1.00		1.00
5	0.13	0.22	0.14	0.24	0.31	0.44	0.72	0.80	1.00	1.00

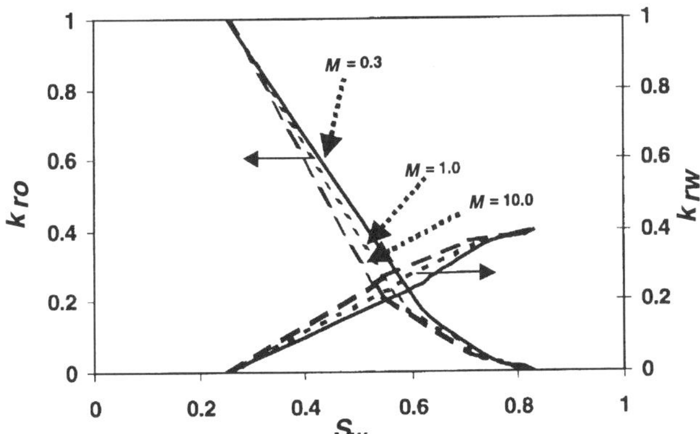

Fig. 8.22—Effective relative permeability for a no crossflow case.

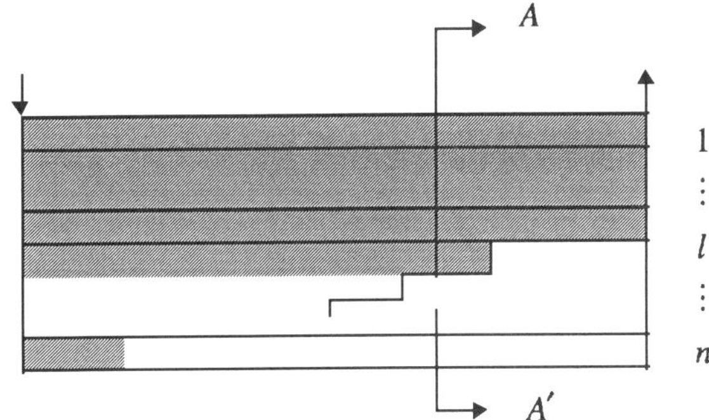

Fig. 8.23—Viscous-dominated crossflow.

Vertical Equilibrium. Under vertical equilibrium conditions, four situations are possible.

- Vertical crossflow by viscous forces
- Displacement dominated by gravity
- Displacement dominated by gravity and capillary forces
- Displacement dominated by capillary pressure

In this section, we discuss only the first three cases. The solution for the fourth case is provided in Appendix F; however, it does not have many practical applications.

Crossflow Dominated by Viscous Force. This situation can exist in a high permeability, high capacity reservoir with a high degree of vertical communication. The flow rates are high enough that gravity is not important. The density difference between the two phases is small; and for a high permeability reservoir, capillary forces are negligible.

Fig. 8.23 exhibits the schematics of a layered reservoir. Similar to no crossflow, the ordering of layers is dependent on the value of $k/\phi\Delta S_w$—the higher its value, the quicker is the breakthrough.

The characteristic feature of vertical equilibrium is the no pressure gradient in the vertical direction. That is, the horizontal pressure gradient across AA' (see Fig. 8.23) is the same for all the layers.

Assume that the lth bed has broken through. Below l, there are $(n-l)$ flood fronts. Depending on the relative location of the flood front, the pressure across the cross section can be calculated with equation

$$\left(\frac{dp}{dx}\right)_{n-l+1} = \frac{q_t}{\left(\sum_{i=1}^{l}\frac{k_{w_i}h_i}{\mu_w} + \sum_{i=l+1}^{n}\frac{k_{o_i}h_i}{\mu_o}\right)}, \quad \ldots\ldots (8.52)$$

where q_t is the total flow rate, and $\left(\frac{dp}{dx}\right)_{n-l+1}$ is the pressure gradient across a section drawn through a flood front, where the water front is in the $(l+1)^{\text{th}}$ bed, and oil is below that bed. The flood front location below the bed can be calculated as

$$\frac{x_j}{L} = \frac{u_j}{u_l} = \frac{\left(\frac{k}{\phi\Delta S_w}\right)_j}{\left(\frac{k}{\phi\Delta S_w}\right)_l}, \quad \ldots\ldots (8.53)$$

where x_j is the flood front location in bed j. By knowing the pressure gradient across each segment, we can calculate the total pressure drop across the length, L, with equation

$$\left(\frac{\Delta p}{L}\right)_t = \frac{1}{L}\sum_{j=1}^{n-l+1}\left(\frac{dp}{dx}\right)_j \Delta x_j, \quad \ldots\ldots (8.54)$$

where

$$\Delta x_j = x_{j+1} - x_j. \quad \ldots\ldots (8.55)$$

The effective relative permeabilities are calculated as

$$\tilde{k}_{rw_l} = \frac{\sum_{i=1}^{l}(kk'_{rw})_i h_i \left(\frac{dp}{dx}\right)_{n-l+1}}{\tilde{k}H\left(\frac{\Delta p}{L}\right)_T}, \quad \ldots\ldots (8.56)$$

where $\tilde{k}H$ is the weighted arithmetic average,

$$\tilde{k}_{ro_l} = \frac{\sum_{i=l+1}^{n}(kk'_{ro})_i h_i \left(\frac{dp}{dx}\right)_{n-l+1}}{\tilde{k}H\left(\frac{\Delta p}{L}\right)_T}. \quad \ldots\ldots (8.57)$$

The average water saturation is

$$\tilde{S}_{w_l} = \frac{\sum_{i=1}^{l}[h\phi(1-S_{or})]_i + \sum_{i=l+1}^{n}[h\phi(1-S_{or})]_i x_{D_i} + (h\phi S_{wi})(1-x_{D_i})}{\sum_{i=1}^{n}(h\phi)_i}. \quad \ldots\ldots (8.58)$$

Numerical Example 8.4. Use the same data given in Numerical Example 8.3. Calculate the effective relative permeabilities using the assumption of vertical equilibrium with viscous flow domination.

Solution. We can assume the unit flow rate per width. The relative locations of the flood fronts can be calculated with Eq. 8.53. (See **Table 8.4.**)

By assuming the unit total flow rate, we can calculate $\left(\frac{dp}{dx}\right)$ across each front (Eq. 8.52). By knowing $\left(\frac{dp}{dx}\right)$ and Δx_j, we can calculate $\left(\frac{\Delta p}{L}\right)_t$ for each bed's breakthrough.

TABLE 8.4—RELATIVE LOCATIONS OF FLOOD FRONT

	Breakthrough Layer									
	1		2		3		4		5	
Layer	x_D	Δx_D	x_D	Δx_D	x_D	Δx_D	x_D	Δx_D	x_D	Δx_D
1	1.00	0.12	1.00		1.00		1.00		1.00	
2	0.88	0.47	1.00	0.53	1.00		1.00		1.00	
3	0.41	0.24	0.47	0.27	1.00	0.57	1.00		1.00	
4	0.18	0.05	0.20	0.05	0.43	0.12	1.00	0.28	1.00	
5	0.13	0.13	0.14	0.14	0.31	0.31	0.72	0.72	1.00	

TABLE 8.5—RELATIVE PERMEABILITY FOR NUMERICAL EXAMPLE 8.4

$\tilde{S}_w$	$\tilde{k}_{rw}$	$\tilde{k}_{ro}$
0.25	0.00	1.00
0.57	0.18	0.29
0.59	0.21	0.26
0.73	0.35	0.07
0.81	0.37	0.03
0.83	0.40	0.00

For example, when the third bed breaks through,

$$\left(\frac{\Delta p}{L}\right)_t = \left(0.57 \times 9.98 \times 10^{-5} + 0.12 \times 1.09 \times 10^{-4} + 0.31 \times 1.17 \times 10^{-4}\right) = 1.06 \times 10^{-4},$$

and

$$\tilde{k}H = 17,070.$$

The effective relative permeabilities, when the third bed breaks through, are

$$\tilde{k}_{rw_3} = \frac{0.4\left(15 \times 700 \times 6.17 \times 10^{-5} + 2 \times 450 \times 6.6 \times 10^{-5} + 22 \times 200 \times 9.98 \times 10^{-5}\right)}{17,070 \times 1.06 \times 10^{-4}} = 0.35,$$

and

$$\tilde{k}_{ro_3} = \frac{1.09(9 \times 80 \times 1.09 \times 10^{-4} + 11 \times 50 \times 1.17 \times 10^{-4})}{17,070 \times 1.06 \times 10^{-4}} = 0.07.$$

The calculation of saturation is straightforward. **Table 8.5** and **Fig. 8.24** show the results of the entire range of relative permeabilities.

Crossflow Dominated by Gravity Force. When the injection rate is low and capillary effects are negligible, we can assume crossflow dominated by gravity force. In some instances, it is possible to assume gravity-dominated flow in the presence of capillary forces, if the

$$\text{capillary transition zone} >> H. \quad \ldots\ldots\ldots\ldots (8.59)$$

That is, the transition zone is so large, compared to the total thickness, that the saturation changes across the thickness can be neglected.

In gravity-dominated displacement, we assume that water segregates to the bottom, and the lowest bed breaks through first. We, therefore, do not rearrange the beds. Instead, starting with the lowest bed, the effective relative permeabilities are calculated as

$$\tilde{k}_{rw_l} = \frac{\sum_{i=1}^{l}(kk'_{rw})_i h_i}{\tilde{k}H}, \quad \ldots\ldots (8.60)$$

$$\tilde{k}_{ro_l} = \frac{\sum_{i=1}^{n}(kk'_{ro})_i h_i}{\tilde{k}H}, \quad \ldots\ldots (8.61)$$

and

$$\tilde{S}_w = \frac{\sum_{i=1}^{l}\left[h\phi(1 - S_{wi})_i\right] + \sum_{i=l+1}^{n}(h\phi S_{wi})_i}{\sum_{i=1}^{n}(h\phi)_i}. \quad \ldots\ldots (8.62)$$

Numerical Example 8.5. Assume the same data given in Numerical Example 8.3. Estimate the effective relative permeabilities using gravity-dominated displacement.

Solution. The lowest layer will break through first.

For example, when the second layer from the bottom breaks through,

$$\tilde{k}_{rw} = \frac{0.4(11 \times 50 + 22 \times 200)}{17,070} = 0.12,$$

$$\tilde{k}_{ro} = \frac{1.0(9 \times 80 + 2 \times 450 + 15 \times 700)}{17,070} = 0.71,$$

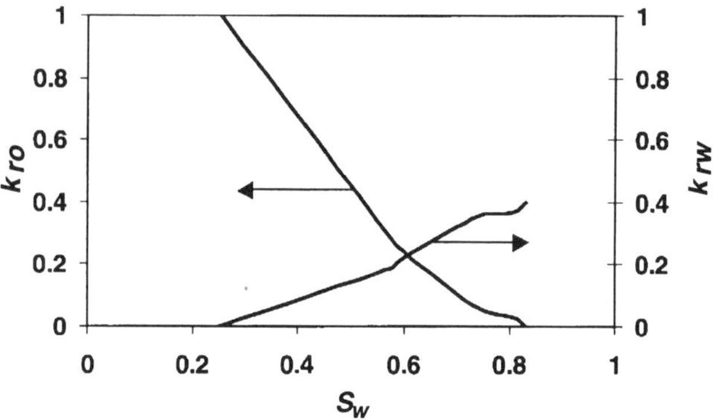

Fig. 8.24—Effective relative permeabilities for viscous-dominated crossflow.

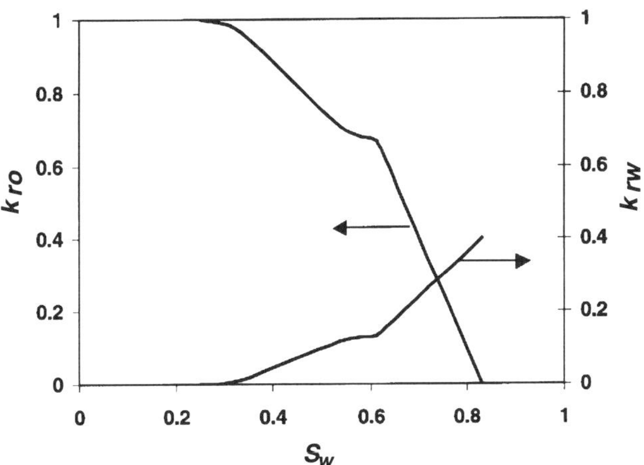

Fig. 8.25—Effective relative permeabilities for gravity-dominated crossflow.

and

$$\tilde{S}_w = \frac{(11 \times 50 + 0.8 + 22 \times 200 \times 0.84 + 9 \times 80 \times 0.24 + 2 \times 450 \times 0.26 + 15 \times 700 \times 0.2)}{11.8}$$

$$= 0.54.$$

Other values are calculated in a similar fashion. The overall relative permeability curves are shown in **Fig. 8.25.** As observed, when the lowest permeability layer is at the bottom, it creates a favorable displacement scenario because gravity compensates for the low permeability. In contrast, if a high permeability layer is at the bottom, it creates an extremely unfavorable displacement profile because the effect of the high permeability contrast is exacerbated by gravity.

Crossflow Dominated by Gravity and Capillary Forces. When gravity and capillary forces dominate the flow, the saturation distribution in each layer is governed by the transition zone. By knowing the equation,

$$p_c = \Delta\rho g h, \quad \text{(8.63)}$$

we can calculate the saturation as a function of depth. If we assume a unique capillary pressure curve for all the layers, then we can generate the saturation distribution across all the layers, as shown in **Fig. 8.26.**

If we assume a saturation at the bottom of the lowest layer, then with Eq. 8.64, we can calculate the saturation at any height. A similar procedure can be used, if each layer has its own unique capillary pressure curve. When changing from one capillary pressure curve to another, we always assume pressure continuity and saturation discontinuity. See Appendix F.

Once the saturation distribution is known, we can calculate the average saturation in each layer as

$$\tilde{S}_{w_l} = \frac{1}{(h\phi)_l} \int_o^{(h\phi)_l} S_w d(h\phi)_l, \quad \text{(8.64)}$$

where we can assume that porosity varies as a function of depth.

In the absence of a simplified analytical function for capillary pressure, Eq. 8.64 has to be evaluated numerically. Once

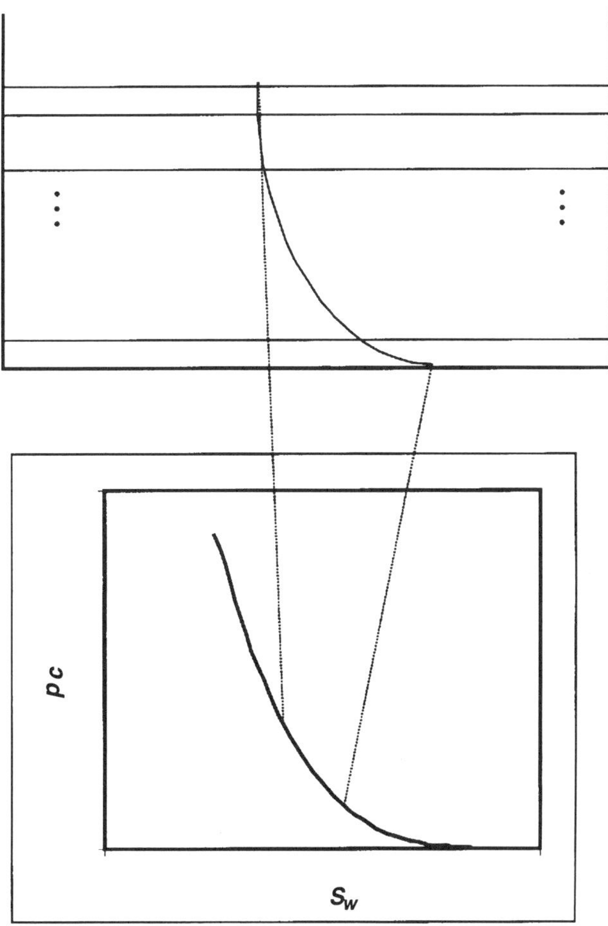

Fig. 8.26—Saturation distribution in a layered reservoir.

the saturation in each layer is known, the relative permeabilities are calculated as

$$\tilde{k}_{rw} = \frac{\sum_{i=1}^{n} [kk'_{rw}(\bar{S}_w)]_i h_i}{\tilde{k}H}, \quad \text{(8.65)}$$

$$\tilde{k}_{ro} = \frac{\sum_{i=1}^{n} [kk'_{ro}(\bar{S}_w)]_i h_i}{\tilde{k}H}, \quad \text{(8.66)}$$

and

$$\tilde{S}_w = \frac{\sum_{i=1}^{n} h_i \phi_i \bar{S}_{wi}}{\sum_{i=1}^{n} h_i \phi_i}, \quad \text{(8.67)}$$

where the relative permeabilities for each layer are calculated as a function of saturation. The procedure is repeated by assuming different saturation values at the bottom of the lowest layer so that the entire range can be constructed.[18-19]

Crossflow Dominated by Capillary Force. In the absence of gravity and viscous crossflow, the flow is dominated by capillary force. This limiting case has very little practical value. The analytical derivation for calculating the effective relative permeabilities is provided in Appendix F and is not repeated here.

To summarize, among the analytical expressions for effective relative permeabilities, the two most practical expressions are the no crossflow case and gravity-dominated displacement. Dake[20] has shown that in many field examples, these two limiting cases are able to describe the waterflooding history.

In addition to the methods described here, other analytical solutions are also available.[21-22] These equations are based on more general principles; therefore, they are applicable over a broader range of conditions. However, the mathematical details are beyond the scope of this text.

Another force, transverse dispersion, which is especially important for miscible displacements, has been neglected in the development of the previous equations. Dake[20] provides an excellent discussion on the effect of transverse dispersion on miscible displacements.

8.3.2 Numerical Methods. Analytical methods have the advantages of being simple to use and beneficial implements. The methods, however, cannot be applied under all conditions.

- Analytical methods are mostly restricted to layered reservoirs. In a typical scaleup process, we might be interested in scaling up relative permeabilities from multiple grids, in either two or three dimensions, to a single equivalent grid. The reservoir properties such as permeability, porosity and saturation can vary in all three directions.
- Analytical methods are useful under limiting conditions. If the combination of forces is such that all three forces play some role in determining the fluid flow distribution, no single analytical solution can be applied.
- Analytical methods assume that the effective relative permeabilities are independent of the boundary conditions. That is, these properties do not depend on the flow rate or whether fluid is injected under variable or constant pressure. This assumption holds true, as long as the saturations of the displacing fluid move at constant velocities. In other words, the Buckley-Leverett analytical solution for the one-dimensional (1D) displacement case should be applicable. The fractional flow curve typically should exhibit a flood front and gradual change with the saturation after breakthrough. The velocity of a given saturation is shown by

$$\frac{dx}{dt} \alpha \left(\frac{df_w}{dS_w}\right)_{S_w}, \quad \text{(8.68)}$$

where $\left(\frac{df_w}{dS_w}\right)$ is the tangent at a given saturation on a fractional flow curve. Based on the fractional flow curve, the slope, after breakthrough, decreases as saturation increases; therefore, high saturations always move at lower velocities and never overrun low saturation values. Unfortunately, in the presence of reservoir heterogeneities, this well-behaved phenomenon might not be applicable. Saturations might not move at constant velocities, thus violating the basic Buckley-Leverett theory.[23-24] Under these conditions the effective relative permeabilities might have to be defined as functions of initial and boundary conditions.

When the previous mentioned conditions arise, numerical procedures can provide an avenue in which the effective relative permeabilities can be estimated. In many papers, the effective relative permeabilities are referred to as pseudofunctions. The distinction between these two terminologies is important. The effective relative permeabilities are still assumed to be rock properties and independent of boundary conditions; whereas, pseudofunctions not only depend on the rock properties but also depend on the boundary conditions.

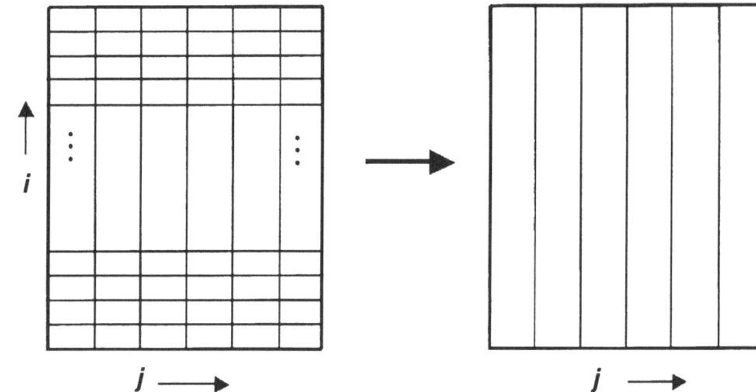

Fig. 8.27—Upscaling from 2D cross section to 1D flow.

Thus, pseudofunctions are assumed to account for nonideal flow behavior (deviations from Buckley-Leverett theory) and must be customized to the appropriate flux conditions under which the actual flow simulation is conducted.

The first attempt at estimating pseudofunctions was by Jacks *et al.*[25] The primary motivation for that work was to reduce a 3D model to a 2D model and, hence, reduce the computational costs. By simulating vertical cross section runs, they defined pseudofunctions for the entire cross section. By using these pseudofunctions, they could then run a 2D areal model.

Conceptually, the upscaling is shown in **Fig. 8.27.** If we have n gridblocks in the vertical direction, we can simulate the vertical cross section, and by knowing the flow rates, we can define pseudofunctions, so the n gridblocks, in the vertical direction, are reduced to one gridblock.

To calculate pseudofunctions, Jacks *et al.*[25] used the equations,

$$q_{o_t} = \sum_{i=1}^{n} q_{o_i}; q_{w_t} = \sum_{i=1}^{n} q_{w_i}, \quad \text{(8.69)}$$

$$q_{o_i} = \frac{C(kk_{ro})_i h_i \Delta p_i}{\mu_o \Delta Y}, \quad \text{(8.70)}$$

$$q_{w_i} = \frac{C(kk_{rw})_i h_i \Delta p_i}{\mu_w \Delta Y}, \quad \text{(8.71)}$$

$$q_{o_t} = \frac{C(\tilde{k}H)\tilde{k}_{ro}}{\mu_o}\frac{\Delta p}{\Delta Y}, \quad \text{(8.72)}$$

and

$$q_{w_t} = \frac{C(\tilde{k}H)\tilde{k}_{rw}}{\mu_w}\frac{\Delta p}{\Delta Y}, \quad \text{(8.73)}$$

where $\tilde{k}H$ is the thickness-weighted permeability; C is a units conversion constant, and $\frac{\Delta p}{\Delta Y}$ is the pressure gradient in the horizontal direction. We can substitute Eqs. 8.70–8.73 in Eq. 8.69 and calculate pseudofunctions, if we assume Δp_i is the same as Δp. This assumption holds for vertical equilibrium but not for the other cases. However, Jacks *et al.*[25] made this assumption and solved for dynamic pseudorelative permeabilities.

Kyte and Berry[26] improved on the Jacks *et al.*[25] procedure with similar equations, except for assuming that the Δp terms in Eqs. 8.73 and 8.74 are the differences in the volumetrically-averaged pressure values of the fine-scale gridblocks. That is, they defined

$$\tilde{p}_{oj} = \frac{\sum_{i=1}^{n}\left[\left(k_i k_{r_i} h_i\right) p_{o_i}\right]_j}{\sum_{i=1}^{n}\left[\left(k_i k_{ro_i} h_i\right)\right]_j}. \qquad (8.74)$$

This is the weighted average for the oil pressure for block j. Similar values can be calculated for other blocks and, hence, the Δp values. A similar equation can also be used for the water phase. Although this method did improve the Jacks *et al.*[25] method, no theoretical explanation is provided for such an averaging scheme. Although the method should work in principle, it is possible that, because of gravity, the net flow could be in a negative direction, with reference to the direction of the average pressure drop, thus resulting in nonphysical values for the pseudofunctions.[27]

Stone[28] improved on the drawbacks of these methods and proposed that pseudofunctions can be calculated, based on total mobility functions. The fractional flow, which can be measured from the simulation, can be written as

$$f_w = \frac{1}{1 + \dfrac{\tilde{k}_{ro}}{\tilde{k}_w}\cdot\dfrac{\mu_w}{\mu_o}}. \qquad (8.75)$$

We write this equation in the absence of capillary and gravity effects.

By knowing the total flow rate coming out of the fine-scale gridblocks, we can calculate the total mobility as

$$M_t = \left(\frac{\tilde{k}_{rw}}{\mu_w} + \frac{\tilde{k}_{ro}}{\mu_o}\right) = \frac{\sum_{i=1}^{n} k_i h_i\left(\dfrac{k_{ro_i}}{\mu_{o_i}} + \dfrac{k_{rw_i}}{\mu_{w_i}}\right)}{\sum_{i=1}^{n} k_i h_i}. \qquad (8.76)$$

By knowing the fractional flow and total mobility, the pseudorelative permeabilities are calculated as

$$\tilde{k}_{rw} = M_t f_w \mu_w, \qquad (8.77)$$

and

$$\tilde{k}_{ro} = M_t(1 - f_w)\mu_o. \qquad (8.78)$$

This method is rigorous; however, in the presence of significant gravity or capillary forces, the method can give poor results. Further, the method might not work well if significant variations in mobility exist.[27]

Stone's method has been improved, by suggesting better ways to calculate the total mobility (Eq. 8.77).[29] However, these methods, to some extent, also neglect gravity and might not exactly reproduce the fine-scale solution on a coarse grid.

In addition to the methods discussed here, an approach of quasisteady state has been used to estimate the pseudorelative permeabilities. In this approach, for a given saturation[30] or for saturations[31] on a fine-scale level, the effective phase permeabilities, k_o and k_w, are calculated on a fine scale. By using the methods discussed in the previous section, these fine-scale phase permeabilities are upscaled to an effective value with the single-phase permeability upscaling equation. The main limitation of this technique is the assumption of steady-state conditions. The pseudofunctions are not independent of boundary conditions, and, hence, the general applicability under a wide range of conditions is questionable.

It is clear from the previous discussion that no single method exists to adequately calculate dynamic pseudofunctions. Ideally, each coarse gridblock, depending on the fine-scale permeability distribution, should be assigned an array of pseudorelative permeabilities, which can vary depending on the boundary conditions. This is obviously impractical. One possible compromise might be to divide the reservoir into sufficiently small coarse blocks so that each coarse block can be assigned only one rock type and, hence, one set of fine-scale relative permeabilities. Generate a sufficient number of representative, fine-scale descriptions, and generate pseudofunctions for each rock type. Each rock type can be assigned a set of pseudofunctions, which are dependent on the well rates or other boundary conditions. Verify that the coarse gridblocks are able to generate the fine-scale simulation results, using a sector model for representative boundary conditions. Once verified, the pseudofunctions can be used for the simulation exercise.[27] This approach appears practical; however, it has not been tested for a field case study.

8.3.3 Effective Capillary Pressure. So far, we have concentrated on effective relative permeabilities. We did not discuss the upscaling of capillary pressures. If capillary forces are significant, the generation of upscaled capillary pressures is relatively straightforward. On a fine scale, if the saturation distribution is known, assign capillary pressure values for each fine-scale gridblock. This value might depend on the saturation, as well as the rock type and permeability. Then, calculate the volumetrically-averaged capillary pressure, as shown.

$$\tilde{p}_c = \frac{\sum_{i=1}^{n} \phi_i h_i p_{c_i}}{\sum_{i=1}^{n} \phi_i h_i}. \qquad (8.79)$$

By knowing the volumetrically-averaged water saturation, an effective capillary pressure can be generated.

Summary

To summarize the findings of this chapter, the upscaling of static properties is relatively straightforward and well-established. Methods exist for upscaling permeability under single-phase flow. These methods work well and are able to capture important features of the fine-scale model. An argument could be made that once the absolute permeabilities are correctly upscaled, the fine-scale relative permeabilities can be used on a coarse scale as well. By correctly identifying fine-scale features, which are important, and coarsening the other grids by using adaptive gridding techniques, only the absolute permeability must be upscaled, without upscaling the multiphase properties.[32] Although a very attractive proposition, rigorous methods for properly defining adaptive grids do not exist; therefore, the appropriateness of the method must be validated.

The upscaling of dynamic properties under multiphase flow conditions is not very easy. In the limiting cases, simplified solutions can be established, which work well in practice if the physical conditions are satisfied. The two conditions, in which analytical solutions are available, include no crossflow and vertical equilibrium conditions. If these conditions are not satisfied, numerical methods can be used to upscale the relative permeabilities. However, as discussed in the chapter, these methods have limitations and might not work under all conditions. Practical solutions to upscale these properties are still not available in the literature and are the subject of ongoing research.

Nomenclature

A = area
C = values defined in Eqs. 8.26 and 8.27
f = fractional flow
g = gravitational acceleration
h = thickness of layer
H = total thickness
j = dummy variable
k = permeability
L = length or distance
m,n = number of gridblocks
M = mobility ratio
N_c = capillary number, Eq. 8.43
N_g = gravity number, Eq. 8.44
p = pressure
p_c = capillary pressure
pv = pore volume
q = flow rate
Q = flow rate
R_{jl} = velocity ratio between layers j and l
R_L = dimensionless number, Eq. 8.42
S = saturation
t = time
V = volume
Wpv = water pore volume
x = distance of the flood front
x,y,z = coordinate axes
Y = direction of flow, Eq. 8.70
Δ = incremental distance or pressure
ϕ = porosity
∂ = partial derivative
γ = specific gravity
μ = viscosity
π = a constant, 3.142 . . .
ρ = density
Σ = summation

Subscripts

a = arithmetic average
app = apparent
b = bulk
D = dimensionless
g = geometric average
h = harmonic average, horizontal
i = gridblock i
j = dummy variable representing layer
l = layer which has broken through
o = oil
or = residual oil
p = power average
r = relative
t = total
v = vertical
w = water
x,y,z = coordinate axes
xx,yy = principal directions
xy,yx = off-diagonal elements in permeability tensor

Superscripts

$\sim$ = average
p = power average

References

1. Amyx, J.W., Bass, D.M. Jr., and Whiting, R.L.: *Petroleum Reservoir Engineering Physical Properties*, McGraw-Hill Book Co. Inc., New York City (1960) 71–77.
2. Deutsch, C.: "Calculating Effective Absolute Permeability in Sandstone/Shale Sequences," *SPEFE* (September 1989) 343.
3. Cardwell, W.T. Jr. and Parsons, R.L.: "Average Permeabilities of Heterogeneous Oil Sands," *Trans.*, AIME (March 1945) **1852.**
4. Abbaszadeh, M. and Koide, N.: "Evaluation of Permeability Upscaling Techniques and a New Algorithm for Interblock Transmissibilities," paper SPE 36179 presented at the 1996 Abu Dhabi Intl. Petroleum Exhibition and Conference, Abu Dhabi, UAE, 13–16 October.
5. King, P.R.: "Renormalization Calculations of Immiscible Flow," *Transport in Porous Media* (1993) **12,** 237–260.
6. Kasap, E. and Lake, L.W.: "Calculating the Effective Permeability Tensor of a Gridblock," *SPEFE* (June 1990) 192.
7. Aasum, Y., Kasap, E. and Kelkar, M.: "Analytical Up-Scaling of Small-Scale Permeability Using a Full Tensor," *Petroleum Geoscience* (1995) **1,** 365–377.
8. Pickup, G.E. *et al.*: "Geology, Geometry and Effective Flow," *Petroleum Geoscience* (1995) **1,** 37–42.
9. Lee, J., Kasap, E., and Kelkar, M.G.: "Analytical Upscaling of Permeability for 3D Gridblocks," *SPEJ* (March 1996) 59.
10. Haldorsen, H.H. and Lake, L.W.: "A New Approach to Shale Management in Field-Scale Models," *SPEJ* (August 1984) 447.
11. Beggs, S.H. and King, P.R.: "Modeling the Effects of Shales on Reservoir Performance: Calculation of Effective Vertical Permeability," paper SPE 13529, presented at the 1985 SPE Symposium on Reservoir Simulation, Dallas, 10–13 February.
12. Chu, L., Schatzinger, R.A., and Tham, M.K.: "Application of Wavelet Analysis to Upscaling of Rock Properties," paper SPE 36517 presented at the 1996 SPE Annual Technical Conference and Exhibition, Denver, 6–9 October.
13. Jensen, F.E. and Kelkar, M.G.: "Upscaling of Reservoir Properties Using Wavelets," paper SPE 39495 presented at the 1998 India Oil and Gas Conference and Exhibition, New Delhi, India, 17–19 February.
14. Soeriawinata, T., Kasap, E., and Kelkar, M.G.: "Permeability Upscaling for Near-Wellbore Heterogeneities," *SPEFE* (December 1997) 255.
15. Zapata, V. J. and Lake, L.W.: "A Theoretical Analysis of Viscous Crossflow," paper SPE 10111 presented at the 1981 SPE Annual Technical Conference and Exhibition, San Antonio, Texas, 5–7 October.
16. Ahmadi, A., Labastie, A., and Quintard, M.: "Large-Scale Properties for Flow Through a Stratified Medium: Various Approaches," *SPERE* (August 1993) 214.
17. Tompang, R.: "Relative Effects of Gravitational and Viscous Forces on Water Flooding Performance in a Linear Stratified Reservoir," MS thesis, U. of Tulsa, Tulsa (1987).
18. Lake, L.W.: *Enhanced Oil Recovery*, Prentice Hall, Englewood Cliffs, New Jersey (1989) Chap. 6.
19. Coats, K.H., Dempsey, J.R., and Henderson, J.H.: "The Use of Vertical Equilibrium in Two-Dimensional Simulation of Three-Dimensional Reservoir Performance," *SPEJ* (March 1971) 63.
20. Dake, L.P.: *The Practice of Reservoir Engineering*, Elsevier, Amsterdam, The Netherlands (1994) Chap. 5.
21. Quintard, M. and Whitaker, S.: "Two-Phase Flow in Heterogeneous Porous Media, the Methods of Large Scale Averaging," *Transport in Porous Media* (1988) **5,** 257.
22. Mohanty, S. and Sharma, M.M.: "A Recursive Method for Estimating Single and Multiphase Permeabilities," paper SPE 20477 presented at the 1990 SPE Annual Technical Conference and Exhibition, New Orleans, 23–26 September.

23. Hewett, T.A. and Behrens, R.A.: "Conditional Simulation of Reservoir Heterogeneity With Fractals," *SPEFE* (September 1990) 217.
24. Hewett, T.A. and Behrens, R.A.: "Considerations Affecting the Scaling of Displacements in Heterogeneous Permeability Distributions," paper SPE 20739 presented at the 1993 SPE Annual Technical Conference and Exhibition, New Orleans, 23–26 September.
25. Jacks, H.H., Smith, O.J.E., and Mattax, C.C.: "The Modeling of a Three-Dimensional Reservoir With a Two-Dimensional Reservoir Simulator—The Use of Dynamic Pseudo Functions," *SPEJ* (June 1973) 175.
26. Kyte, J.R. and Berry, D.W.: "New Pseudo Functions To Control Numerical Dispersion," *SPEJ* (August 1975) 269.
27. Barker, J.W. and Thibeau, S.: "A Critical Review of the Use of Pseudo Relative Permeabilities for Upscaling," paper SPE 35491 presented at the 1996 European 3D Reservoir Modelling Conference, Stavanger, 16–17 April.
28. Stone, H.L.: "Rigorous Black Oil Pseudo Functions," paper SPE 21207 presented at the SPE 1991 Symposium on Reservoir Simulation, Anaheim, California, 17–20 February.
29. Christie, M.A. *et al.*: "A Renormalization-Based Up-scaling Technique for WAG Floods in Heterogeneous Reservoirs," paper SPE 29127 presented at the 1995 SPE Symposium on Reservoir Simulation, San Antonio, Texas, 12–15 February.
30. Saad, N., Cullick, A.S., and Honarpour, M.M.: "Effective Relative Permeability in Scale-Up and Simulation," paper SPE 29592 presented at the 1995 SPE Rocky Mountain Regional/Low-Permeability Reservoir Symposium, Denver, 20–22 March.
31. Tambe, D.E. and Chopra, A.K.: "Scaleup of Geological Facies for Multiphase Flow Simulation," paper SPE 39758 presented at the 1998 SPE Asia Pacific Conference on Integrated Modeling for Asset Management, Kuala Lumpur, 23–24 March.
32. Durlofsky, L.J. *et al.*: "Scale-Up of Heterogeneous Three Dimensional Reservoir Descriptions," paper SPE 30709 presented at the 1995 SPE Annual Technical Conference and Exhibition, Dallas, 22–25 October.

SI Metric Conversion Factors

cp	× 1.0	E − 03 = Pa·s
ft	× 3.048*	E − 01 = m

*Conversion factor is exact.

Chapter 9
Looking Ahead

In previous chapters, we discussed various strides that have been made in the area of geostatistics in describing the reservoir properties. In this chapter, we discuss some of the future challenges for geostatistics and the solutions that seem promising. We believe that the two main challenges facing geostatistics are the integration of data from various disciplines and effective solutions for solving inverse problems. Although various solutions have been proposed, and field applications have been demonstrated, the applications still remain largely in the research domain. We expect, within the next 10 years, we will see more solutions to these vexing problems.

9.1 Integration of Data

As we discussed in the chapters on conditional simulation techniques, it is critical that as much information as possible is integrated as part of the reservoir description so that uncertainties in the prediction of future performance are reduced. The type of data we must include to describe reservoirs comes mainly from three disciplines: geology, geophysics, and engineering. Some integration of data occurs as we describe reservoirs; however, it is not done effectively. Several reasons exist for such a lack of integration. Here, we offer some of them.

9.1.1 Series Approach. We briefly discussed this problematic approach in Chap. 1, when we explained the conventional history-matching problem. We still have not found an effective solution. Typically, in a history-matching process, geologists provide information on the correlation between wells, and geophysicists provide information on structure. This information is carried forward to the engineer who populates the interwell areas with petrophysical properties to describe the reservoir, eventually for flow simulation purposes. During the history-matching phase, the engineer, will most likely adjust the petrophysical properties of the reservoir to match the historical data. Rarely, the geophysical or geological model will be altered to match the historical data. History-matching might eventually be achieved, but the changes in the petrophysical properties might not be consistent with the engineering data.[1] For example, the engineer might be required to change the value for the permeability of layers to a value greater than or less than that observed in the core or well test data to match the water breakthrough, or he might be required to reduce the porosity, obtained from log values, to match the decline in pressures. We must ask why it is necessary to make changes in reservoir properties that are inconsistent with the observed data to be able to match the historical production data. There could be several reasons, including uncertainties in the measurements; however, selecting the series approach is one reason.

Similar to the uncertainty in engineering properties, uncertainty also exists in the geological and geophysical interpretations.[2] Typically, the scale in which geological and geophysical models are constructed is much bigger than the scale in which engineering properties are estimated. When only one geological and geophysical model is provided to the engineer, we have assumed that the uncertainty in the geological and geophysical model is not that critical; although in reality, the scale in which uncertainties are defined has a much bigger impact on the reservoir performance. As a result, when the engineer tries to match the historical performance, the solution domain in which he or she can explore is already reduced. Note that the geological and geophysical models are largely based on static information. With that static model, the engineer attempts to match the historical performance (dynamic data) with reduced degrees of freedom (**Fig. 9.1**). Because the geological and geophysical models might be far away from reality, the engineer is forced to adjust the properties that deviate from the observed engineering information. Therefore, although history-matching is achieved, it might be rendered meaningless because it only matches the production data and not necessarily the static information that was collected.

This problem can be avoided if we can make the loop bigger (Fig. 9.1). That is, we understand the importance of uncertainties in the geological and geophysical models and encourage both the geologists and engineers to develop alternate models that match the observed data. To be meaningful, these models should capture the uncertainties at the appropriate scale. We know that interpretation of the data, whether geophysical or geological, is subjective, and different persons can develop alternate descriptions. The goal here is to develop sufficiently different alternate models so that reality can be bounded. For

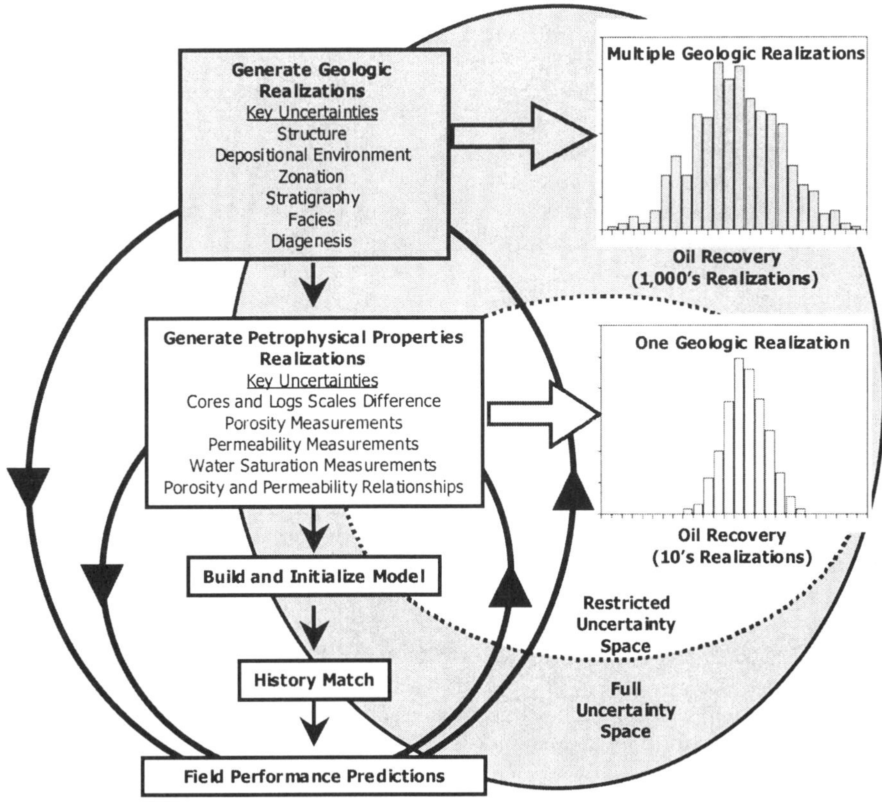

Fig. 9.1—Summary of reservoir modeling process showing impact of geologic and rock property uncertainties on field performance predictions.

example, if we build alternate geological and geophysical models for a reservoir, then if a new well is drilled, the structural information and the geological facies information from that well should fall within the alternate models. What commonly happens is that if a new well is drilled and the observed data do not match the predicted results, instead of re-examining the entire data set, we only make local adjustments to match the results. The large-scale model essentially remains intact. Instead of making these types of local refinements, our goal should be to develop alternate geological and geophysical models that can truly capture the uncertainties. Armed with those models, while going through the history-matching process, if the engineer does not match the historical data, he or she can loop all the way back to the geological and geophysical models and sample a new model to iterate over the history-matching process. If such integration is achieved, the engineering properties will not need to be arbitrarily adjusted to match the historical production performance. Instead, with a larger solution domain, the engineer is able to describe a reservoir that is consistent with geological, geophysical, and engineering information.

9.1.2 Scale Consistency. In achieving the integration of the data, we might be aware that the scales of reservoir properties are different. We already discussed the problem in Chap. 1. It is, however, worthwhile to revisit the issue while discussing the proposed solutions. We can discuss several examples in which the scale consistency becomes an issue. It is not only related to the fact that different reservoir parameters are measured over different scales but, for example, to well test permeability vs. core permeability, or core porosity vs. log porosity, or core porosity vs. seismic-derived porosity (**Fig. 9.2**).[3,4] The problem is also related to the fact that cross correlations established on one scale might not be valid on other scales. This is because attributes scale differently on different scales. For example, if we establish a relationship between permeability and porosity on a core level, how do we know that such a relationship will also hold for a gridblock scale? If we establish a relationship between impedance and porosity based on sonic logs, how do we know that such a relationship will also hold when we invert the amplitude data that does not have the same resolution as the log data? The relationship between the two attributes can be perfect on one scale and totally different on another scale. The randomness in a relationship, which seems to be based on a physical relationship, can be solely caused from the fact that the scale on which we are establishing the relationship is not appropriate.

We must correctly address both issues of relating multiple attributes to each other and generating reservoir descriptions. The issue of integrating properties measured on different scales has been addressed in recent literature.[3,4] However, the

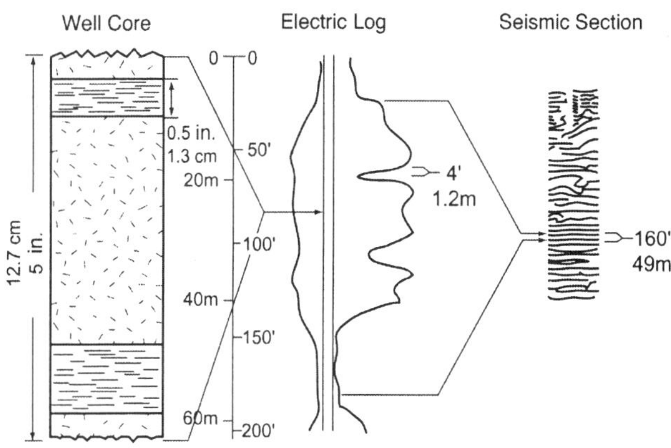

Fig. 9.2—Different sources of reservoir data from different scales (from Ref. 5).

solutions offered so far are largely restricted to static data. For example, one proposed solution is the integration of log porosity data with seismic-derived porosity data. We still do not have methods for effectively integrating dynamic measurements of the same attribute. For example, the integration of core-derived permeability and well test permeability is largely done on an ad-hoc basis. We need more robust techniques for generating reservoir descriptions that can properly include the core permeability and well test permeability in a consistent manner.

The second issue of the relationship between multiple attributes and their dependence on scale remains largely unresolved. Practically, we simply cannot generate gridblocks that are small enough to be consistent with the scale in which reservoir properties are sometimes measured and correlated. However, with the increasing emphasis on integrating data, we are required to generate reservoir descriptions that allow for relationships among various attributes at the local level. The main problem is that we do not know the relationship at that local level. Cokriging and cosimulation are becoming increasingly popular to account for such relationships, but in most instances, we routinely use relationships that are based on a different scale—whether it is a relationship between permeability and porosity, a relationship between geological facies and porosity, or a relationship between seismic attributes and geological facies. We must develop a better understanding of scale-consistent relationships before we can truly integrate information on various scales.

9.1.3 Scale Resolution. One of the issues, which must be addressed in the integration process, is choosing the most appropriate scale, which must be resolved to properly describe the reservoir. As we discussed in Chap. 1, a more detailed reservoir description is required as the complexity in the flow process increases. However, the issue is even more relevant when integrating information. With improved tools and computer speed, there is an increasing tendency to keep describing the reservoir on a finer and finer scale. For example, because of the use of three-dimensional (3D) seismic data and the increased frequency with which samples are collected, seismic resolution has increased. With microresistivity logs, geologists can examine detailed microfacies on a millimeter scale, which would not have been previously possible without taking core samples. Engineers are prepared to generate millions of gridblock models to simulate the reservoir performance. Is this search for detail really contributing to an improved resolution? The answer is a guarded yes. Without a question, today, we are able to describe the reservoir in greater detail than what was possible just a few years ago. However, it is not necessary that we are improving the reservoir description.

Let us first understand an important principle. As the scale becomes smaller, the uncertainty in describing the properties increases rapidly. See **Fig. 9.3,** which shows a plot of error variance as a function of scale. As the scale gets bigger, the uncertainty is reduced. For example, it is much easier to say that a particular part of the reservoir has a better quality than the other. On a larger scale, this is indeed true because that part of the reservoir has performed better than other parts of the reservoir. On the other hand, it is much harder to say that the permeability of a gridblock, precisely 100 ft. away from a well in a particular direction, is within the range of 100 to 1,000 md. As the scale becomes smaller and smaller, it becomes increasingly difficult to quantify uncertainty. If we could truly quantify uncertainty on an extremely small scale, it would be frightening. Fortunately, although the uncertainty increases as the scale gets smaller, its relative impact on the performance might be small enough to be ignored. What we say about the spatial scale, we can also say about the time scale. The changes in daily production rates can be much more severe and random, compared to monthly or even yearly production rates. To match the daily production rates, because the time scale has become smaller, we are required to resolve the reservoir on a much finer scale. To match the well production data, instead of overall reservoir production, we are required to resolve the reservoir on a much finer scale.

We must ask ourselves what scale is the most appropriate to describe the reservoir. Because, as we decide to describe the reservoir on a finer scale, we also need to be able to increase the associated uncertainty. Many times, in a history-matching process, the engineer gets carried away with matching the individual well performance. The engineer tends to forget that if the individual well data are matched, the local properties must be adjusted, and the uncertainty, with respect to local properties, is much higher than large-scale properties. This means that if alternate pictures of the reservoir are developed that match the well performance, much more variability should exist on a local level than on a larger scale. If the geologist decides to describe a fine scale reservoir description, based on microresistivity or other logs, and decides to correlate information between wells, based on that information, he or she must understand that as the scale gets smaller, the uncertainty, with respect to correlation, will increase much more rapidly. Important markers can be easily identified, even with relatively less sophisticated logs, and can be reasonably correlated. There will also be uncertainty with respect to those at interwell scales, but not as much. As the scale gets smaller, and the geologist is forced to correlate information on a finer scale, the uncertainty increases rapidly. The same can be said about geophysical data. As the sampling rate increases, the geophysicist is increasingly eager to generate reservoir properties at the gridblock level. However, as the scale in which the reservoir properties are resolved based on geophysical attributes gets smaller, geophysical uncertainty increases quickly. It is much easier to identify a large-scale anomaly with geophysical data. Although uncertainty exists, with respect to such an anomaly, it is less than when geophysical data are used to quantify reservoir porosity on a figure scale. With increasing emphasis on generating more detailed, integrated,

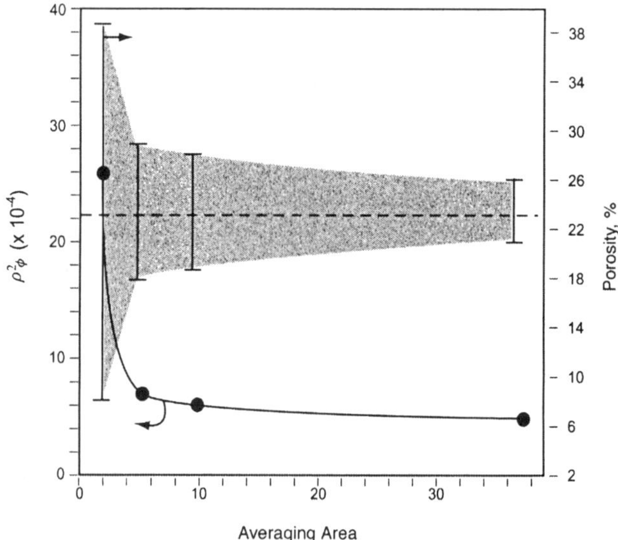

Fig. 9.3—Porosity variance and spread vs. averaging scale.

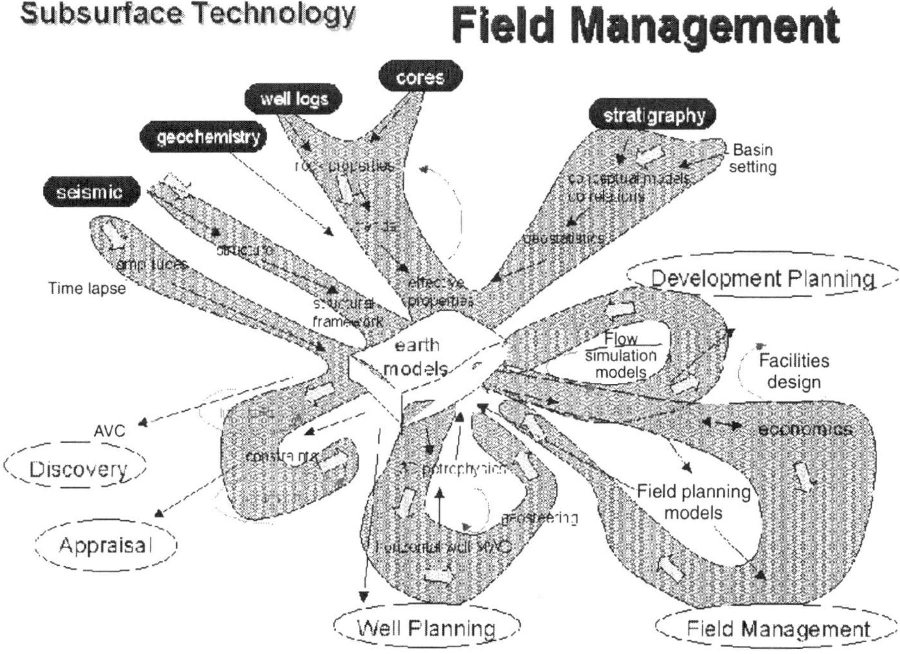

Fig. 9.4—Shared Earth model proposed by POSC to integrate reservoir data.[6]

reservoir descriptions on a finer scale, how do we ensure that the relationship between the resolution and uncertainty is properly honored? We still have not solved this problem effectively. The only cautionary note we can offer is that we should tolerate a greater uncertainty as the scale gets smaller. For example, the engineer should be willing to accept the fact that individual well rates might not match as well as the overall reservoir performance. That way, he or she is implicitly accepting the fact that we have not resolved the uncertainty for small scale features; instead, we are concentrating on large-scale features only. Rather than generating extremely fine-scale models, which are unable to correctly quantify the uncertainty at that level, we should concentrate on generating alternate models that integrate information from various disciplines and quantify uncertainty on a coarser scale. Once we master the integration on a coarser scale, we might be able to move to a finer scale. At this point, however, we are still struggling with quantification of uncertainty on a coarser scale.

9.1.4 Data Management. The management of data coming from various disciplines is not a truly technical issue; however, in practice, the lack of data management can hamper the integration of data. Therefore, we briefly mention it here. Geophysicists, geologists, and engineers typically use different types of software to process raw data. Despite some change in recent years, the use of different software is still more the norm than the exception. **Figs. 9.4 and 9.5** show the schematic of a shared Earth model, which can incorporate and integrate many types of information in a seamless fashion. This is a concept advanced by the Petroleum Open Software Corporation (POSC)[6] and supported by many oil companies.[7] In many instances, when different software packages are used to generate information, the formats from these packages can be very specific, and it may become very difficult to use another type of software to read the data from the first. (For example, one software program might use 99,999 as missing data, whereas others use 1.0e31 for missing data). If the exercise of integrating data becomes frustrating and time consuming because of the incompatibility of software, there is a lesser tendency to integrate information from various disciplines. As our ability to store large amount of information increases, it is more critical than ever to establish data transparency procedures, so data used by one discipline can easily be used by other disciplines in a meaningful way.

9.2 Inverse Problem

In any reservoir description process, we must solve an inverse problem. That is, knowing an output, we must determine the input. An important characteristic of inverse problems is nonunique solutions. That is, we can obtain multiple solutions that will satisfy the output. The inverse problem is faced by geologists, geophysicists, and engineers. For example, knowing the well information, geologists are required to construct a depositional model that can describe the interwell depositional environment. This requires solving an inverse problem. If the depositional environment is known, it is easy to describe the facies, if a well is drilled at a particular location. It is much more difficult to construct the depositional model based on the well data only. Obviously, we can develop multiple solutions that satisfy the well information. Geophysicists also routinely deal with inverse problems when they develop the velocity model to match seismic response. Alternate velocity models are possible that will provide the same seismic response. Engineers deal with inverse problems when they try to match the pressure response from the well test to establish permeability data, or when they try to describe the reservoir so that production data can be matched. Although inverse problems are important in all these disciplines, here we consider only the inverse problem faced by engineers.

9.2.1 Incorporation of Dynamic Data. The main problem faced by engineers is how to incorporate reservoir performance information as part of the reservoir description. Traditional history matching also solves an inverse problem; however, the procedure is much more cumbersome. Changes in the reservoir properties are made, largely based on experience and the subjective judgement of the simulation engineer, until

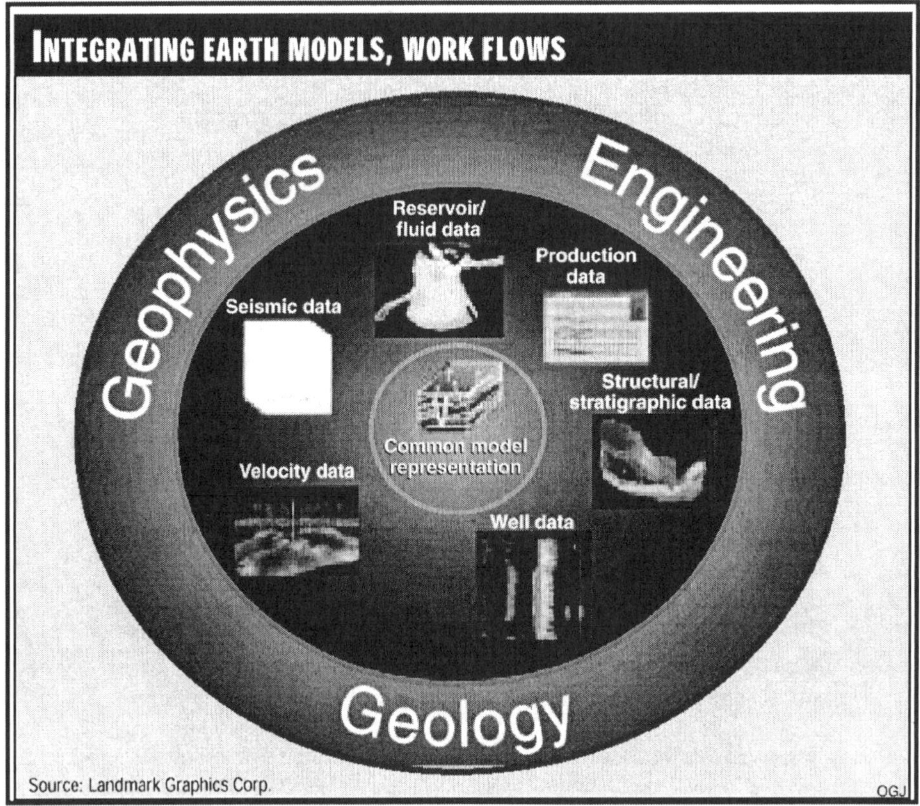

Fig. 9.5—Integration concepts for shared Earth model proposed by POSC showing interaction among different disciplines and sources of reservoir data.

a reasonably satisfactory match is obtained. A reasonably satisfactory match is also very subjective. However, once a match is obtained, the same reservoir description is used to predict the future performance.

The difficulty in using this traditional approach is that the process takes a long time to match the performance. As a result, multiple descriptions are very rarely constructed to match the historical performance and, at the same time, predict uncertainty in the future performance. If we have a desire to match the historical performance and predict uncertainty in the future performance, we must automate the procedure so that multiple descriptions can be easily constructed. This will allow us to match the historical performance and predict the uncertainty in the future performance as well.

Over the last 10 years, much progress has been made in automating the history matching procedure. Although the early work on automatic history-matching was done in the middle and late '70s,[8-10] the work was largely abandoned because of computational limitations. Computers were not powerful enough to match the historical data in a reasonable amount of time; therefore, the work was considered impractical and was ignored until recent years.

As geostatistics became more commonplace, the need to incorporate production data as part of the reservoir description became increasingly important. The first applications of the incorporation of production data as part of the reservoir description involved developing static parameters that could represent the dynamic characteristics.[11] This procedure allowed the establishment of static parameters, which could constrain the reservoir description, to match the production data. These parameters, for example, include well test-derived permeability, well productivity, or connectivity between wells. The technique, however, had limitations, in terms of generalizing, to other systems and environments.

Simulated annealing or similar procedures were also extensively used to match the historical performance of the reservoir. Because of the inherent flexibility of the annealing procedure, constraints including production data can be incorporated as part of the objective function. The reservoir description can then be perturbed until a reasonable match is obtained.[12-15] Some field applications demonstrated the utility of the procedure; however, the procedure is computationally demanding and is not able to generate multiple descriptions very efficiently. Several perturbations are needed to match the historical performance, and to generate multiple descriptions, the same procedure might have to be repeated several times. This makes the procedure very difficult to use in practice.

The current trend in the incorporation of production data is to use gradient-based methods.[16] In simple terms, gradient-based methods work on the principle that the changes in the input parameters affect the observable parameters. If we can calculate the sensitivity of the observable parameter with respect to the input parameter, by knowing the difference between the observed and predicted values, we can make an appropriate adjustment in the input parameter to match the observable value. For example, if we consider the observable parameter to be the production rate of a well, then, by calculating the sensitivity of production rate with respect to permeability of a gridblock, we can adjust the permeability of the gridblock so that we can match the production rate. The amount of alteration needed is also dependent on the difference between the observed production rate and the predicted production rate. The procedure is iterative. We start with an initial reservoir model, which might be based on static information only. We simulate the performance and compare the simulated performance with the observed values, and if the match is not acceptable, we adjust our input model, based on sensitivity calculations, and repeat the procedure until a satisfactory match is obtained. In general, the match is obtained much more rapidly than what is possible using the simulated annealing procedure. At present, most of the applications are restricted to single-phase flow;[17-20] however, multiphase flow applications are slowly emerging.[21-22]

Although the procedure based on sensitivity coefficients is a significant improvement over other methods, several issues remain unresolved. Next we discuss some of these issues.

Input Parameters. In using the sensitivity coefficient method, we must be able to determine the sensitivity of the observable parameter with respect to every input parameter. As the number of grids increases, the number of input parameters increases very rapidly. For example, even for single-phase flow, if we have 1,000 gridblocks, we can have at least two parameters for each of the gridblocks, namely permeability and porosity. If we move to multiphase flow, then in addition to porosity and permeability, we also must include saturation and relative permeabilities. One can imagine the number of input parameters as we start solving problems with 10,000 or more gridblocks. To use that many input parameters and to use the sensitivity coefficients, with respect to all of them, becomes progressively difficult, and it makes the procedure impractical.

Several solutions have been suggested to overcome this problem. The pilot points technique uses only a few gridblocks as the pilot points and calculates sensitivity coefficients only at those locations, and once the input values are adjusted at those locations, it uses a kriging-type procedure to estimate the input parameter values at other locations.[23-25] This technique reduces the number of input parameters for computing sensitivity coefficients; however, the issue of the selection of the correct pilot points and the efficacy of convergence, in using only selected points, still remains unresolved. Another solution is to use streamline simulators rather than conventional flow simulators.[26] By using a simplified flow process, streamline simulators run much faster than conventional simulators. This way, we can complete more iterations more efficiently. Although streamline simulators are more efficient than conventional simulators, they are limited in their applications. For example, they can not be used for a primary depletion process. They also become less efficient as operational constraints are added (such as variations in production and injection rates, shut-in of wells, drilling of in-fill wells, etc.).

We must address the issue of the selection of appropriate input parameters to make the procedure more practical. We know, for example, that the change in the permeability of a gridblock value, far away from a well, does not significantly affect the well performance. However, the gridblock closest to that well has a much bigger impact. How can we prioritize these parameters based on our understanding of these relationships? Another, more relevant issue related to input parameters is the question of progressively developing a reservoir description with increasing details. Based on our discussion, we know the importance of a large-scale description. Is it possible that we can first describe the reservoir on a large scale, thus reducing the number of input parameters, and then progressively increase the resolution as we match the information on the coarser scale? Currently, most of the efforts are restricted to changing the traditional history-matching parameters such as the permeability and porosity of the individual gridblocks. No effort is made in incorporating geological or geophysical uncertainties. In other words, more emphasis is placed on small-scale heterogeneities. If we can generate a large-scale description with fewer parameters, we might be better able to quantify uncertainties in future performance.

Output Parameters. The issue of output parameters is as relevant as that of the input parameters. The sensitivity coefficient calculations require the calculation of sensitivity with respect to every observable (output) parameter. The larger the number of observable parameters, the greater is the number of sensitivity coefficients. For example, if we use monthly production data instead of yearly production data, we will significantly increase the number of observable parameters. Computational effort increases very rapidly as the number of observable parameters increases. This happens as the data are collected over a longer period of time from several wells. For multiphase production data, it becomes even more complicated as we keep track of multiphase rates from individual wells.

Similar to input parameters, we must address the issue of the scale of the output parameters. For example, if we can first solve the inverse problem on a coarse scale by matching the overall field performance, we might be able to successfully describe the reservoir on a finer scale, as detailed, individual well production data are incorporated progressively. The scale issue is not only relevant in terms of space (overall vs. individual wells) but is also relevant in terms of time (yearly vs. monthly production rates).

Spatial Relationships. Part of the requirement for inverse modeling is the spatial relationships of various attributes. In the absence of such a relationship, the incorporation of production data results in a highly heterogeneous and discontinuous reservoir description. To create a more continuous and smoother description, a variogram model based on static parameters is typically included as part of the constraint in solving the inverse problem.

We question if such a spatial relationship is adequate for matching the historical performance data. Should we use more flexible spatial relationships, which might be based on the production data themselves? Field experience tells us that the reservoir is highly heterogeneous in nature and highly nonstationary. Injection in a well will significantly improve the performance of a producer 2,000 ft away, but it might not affect the performance of a producer on the other side, which is only 600 ft away. This type of dynamic connectivity is quite common in the oil fields and yet is difficult to capture using conventional spatial relationships. We need more flexible, nonstationary approaches that better model dynamic connectivity in the reservoir so that such connectivities can be incorporated as part of the reservoir description during the inverse modeling process.

Summary

In this chapter, we presented some of the challenges we will face in the future as the discipline of reservoir description matures. The two main problems we presented are the integration of data from various disciplines and the ability to effectively solve inverse modeling. Solutions to these problems are emerging very rapidly, and we expect that within the next 10 years, books will be written offering solutions to these problems, while discussing new challenges.

References

1. Shiralkar, G.S. *et al.*: "Parallel Computer Alters Approaches, Raises Integration Challenges in Reservoir Modeling," *Oil and Gas J.* (20 May 1996) 48.
2. Richardson, J.G., Sangree, J.B., and Sneider, R.M.: "Applications of Geophysics to Geological Models and Reservoir Description," *JPT* (July 1987) 753.
3. Behrens, R.A. *et al.*: "Incorporating Seismic Attribute Maps in 3D Reservoir Models," paper SPE 36499 presented at the 1996 SPE Annual Technical Conference and Exhibition, Denver, 6–9 October.
4. Doyen, P.M. *et al.*: "Reconciling Data at Seismic and Well Log Scales in 3D Earth Modeling," paper SPE 38698 presented at the 1997 SPE Annual Technical Conference and Exhibition, San Antonio, Texas, 5–8 October.
5. Haldorsen, H.H., Brand, P.J., and MacDonald, C.J.: "Review of Stochastic Nature of Reservoirs," *Mathematics in Oil Production,* S.S. Edwards and P.R. King (eds.), Clarendon Press, Oxford (1988) 109.
6. POSC, www.posc.org.
7. Tippee, B.: "Collaborative Effort Seeks Computing Platform for Shared Earth Model," *Oil and Gas J.* (30 November 1998) 58.
8. Gavalas, G.R., Shah, P.C., and Seinfeld, J.H.: "Reservoir History Matching by Bayesian Estimation," *SPEJ* (December 1976) 337.
9. Carter, R.D. *et al.*: "Performance Matching With Constraints," *SPEJ* (April 1974) 187.
10. Jacquard, P. and Jain, C.: "Permeability Distribution from Field Pressure Data," *SPEJ* (December 1965) 281.
11. Hird, K.B. and Kelkar, M.G.: "Conditional Simulation Method for Reservoir Description Using Spatial and Well-Performance Constraints," *SPERE* (May 1994) 145.
12. Deutsch, C.V.: "Annealing Techniques Applied to Reservoir Modeling and the Integration of Geological and Engineering (Well Test) Data," PhD dissertation, Stanford U., Stanford, California (1992).
13. Sagar, R.K.: "Reservoir Description by Integrating Well Test Data and Spatial Statistics," PhD dissertation, U. of Tulsa, Tulsa (1993).
14. Ouenes, A. *et al.*: "A New Algorithm for Automatic History Matching: Application of Simulated Annealing Method (SAM) to Reservoir Inverse Modeling," paper SPE 26297 available from SPE, Richardson, Texas (1993).
15. Huang, X., Gajraj, A., and Kelkar, M.: "The Impact of Integrating Static and Dynamic Data in Quantifying Uncertainties in the Future Prediction of Multiphase Systems," *SPEFE* (December 1997) 263.
16. Tarantola, A.: *Inverse Problem Theory, Methods for Data Fitting and Model Parameter Estimation,* Elsevier Science Publishers, Amsterdam (1987).
17. Oliver, D.S.: "Incorporation of Transient Pressure Data into Reservoir Characterization," *In Situ* (1994) **18,** 243.
18. Reynolds, A.C. *et al.*: "Reparameterization Techniques for Generating Reservoir Descriptions Conditioned to Variograms and Well-Test Pressure Data," *SPEJ* (December 1996) 413.
19. Rahon, D., Edoa, P.F.G., and Masmoudi, M.: "Inversion of Geological Shapes in Reservoir Engineering Using Well Tests and History Matching of Production Data," paper SPE 38656 presented at the 1997 SPE Annual Technical Conference and Exhibition, San Antonio, Texas, 5–8 October.
20. Landa, J.L. and Horne, R.N.: "A Procedure to Integrate Well Test Data, Reservoir Performance History, and 4D Seismic Information Into a Reservoir Description," paper SPE 38653 presented at the 1997 SPE Annual Technical Conference and Exhibition, San Antonio, Texas, 5–8 October.
21. Tang, Y.N. *et al.*: "Generalized Pulse Spectrum Technique for 2-D and 2-Phase History Matching," *Applied Numerical Mathematics* (1989) **5,** 529.
22. Vasco, D.W., Datta-Gupta, A., and Long, J.C.S.: "Integrating Field Production History in Stochastic Reservoir Characterization," paper SPE 36567 presented at the 1996 SPE Annual Technical Conference and Exhibition, Denver, 6–9 October.
23. de-Marsily, G. *et al.*: "Interpretation of Interference Tests in a Well Field Using Geostatistical Techniques to Fit the Permeability Distribution in a Reservoir Model," *Geostatistics for Natural Resources Characterization,* G. Verly *et al.* (eds.), D. Reidel, Norwell, Massachusetts (1984) Part 2, 831.
24. Xue, G. and Datta-Gupta, A.: "Structure Preserving Inversion: An Efficient Approach to Conditioning Stochastic Reservoir Models to Dynamic Data," paper SPE 38727 presented at the 1997 SPE Annual Technical Conference and Exhibition, San Antonio, Texas, 5–8 October.
25. Wen, X., Deutsch, C.V., and Cullick, A.S.: "High Resolution Reservoir Models Integrating Multiple-Well Production Data," paper SPE 38728 paper presented at the 1997 SPE Annual Technical Conference and Exhibition, San Antonio, Texas, 5–8 October.
26. Datta-Gupta, A., Vasco, D.W., and Long, J.C.S.: "On the Sensitivity and Spatial Resolution of Transient Pressure and Tracer Data for Heterogeneity Characterization," *SPEFE* (June 1997) 137.

SI Metric Conversion Factor

ft × 3.048*   E − 01 = m

*Conversion factor is exact.

Appendix A
Field Data

The data used for field examples in this book comes from the Burbank Oil Field in Oklahoma. The field produces from a sandstone reservoir that consists of several flow units. Hird[1] describes the data for this field in detail. This appendix lists and describes the data for wells in the area of the field considered for the examples. This appendix is useful for reviewing field example calculations in detail and for designing further exercises. Data include reservoir top depths, flow unit thickness, porosity and permeability. This data set was selected for field examples because it is the type of data commonly available to geoscientists and engineers in reservoir description studies. The disk enclosed with this book includes computer files with the data described in this appendix.

A.1 Wells

Fig. A.1 shows the locations of the 79 wells in the field. All the wells are vertical with constant coordinates through the flow units. The *x* and *y* coordinates of the wells are listed in **Table A.1.** These coordinates are used to describe the locations for all properties that are needed for kriging and conditional simulation exercises.

A.2 Reservoir Top

Subsea depths for the tops of reservoirs are listed in Table A.1 for several wells. Note that tops are not available for all the wells in the field. The top of the reservoir gradually dips to the west.[1] In three-dimensional (3D) models, the top of a structure map is the reference point where flow units hang. Reservoir description applications require the top of a structure map for simulation models, estimation of height from fluid contacts in reserves calculations, and 3D visualization.

A.3 Flow Unit Thickness

The reservoir consists of ten flow units that capture changes in geologic description and variation of petrophysical properties. Gross thickness for many wells is included in Table A.1. Flow units were described at wells by analyzing several geologic cross sections throughout the field, as well as the porosity and permeability from cores and logs. Flow units are designated by numbers from one at the top to ten at the bottom. Flow units three through seven, in the middle of the reservoir, are the most productive units and have the best petrophysical properties. Flow units one, two, and eight through ten are less productive. The overall gross thickness of all flow units ranges from 46–85 ft, and the mean is 65 ft. Flow unit ten is the thickest, while flow unit nine is the thinnest interval.

A.4 Porosity

Porosity data was calculated as the arithmetic average values from cores and logs within each flow unit. Flow units one and two have the lowest porosity with an average less than 12%. Flow units three through seven, which are the most productive, have the highest porosity with average values greater than 19%. Flow units eight through ten have an average porosity less than 17%, and these are considered to have minor contributions to production.

A.5 Permeability and Porosity Core Data

Permeability and porosity core plug measurements were available for several wells. The highest average permeabilities are in flow units three through seven. Flow units two, nine, and ten have low average permeabilities of less than

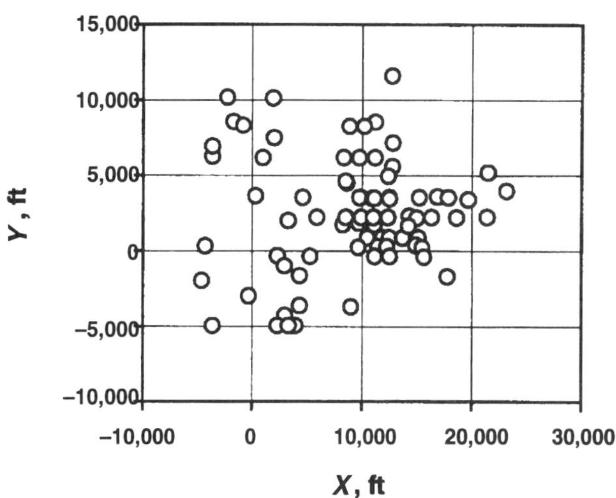

Fig. A.1—Well locations.

TABLE A.1—WELL LOCATIONS, NAMES, THICKNESS AND TOPS									
Well	X, ft	Y, ft	Gross Thickness ft	Top, ft	Well	X, ft	Y, ft	Gross Thickness ft	Top, ft
GRAHM281	–4,620	–1,980	67.		254A	10,207	8,273	69.	1,824.5
GRAHAM1A	–4,300	330	57.		50W24	10,485	890	66.	1,838.0
LEOLA1	–3,630	–4,950	59.			10,860	9,580		1,801.0
WHITLK1A	–3,630	6,260	76.		254A	10,207	8,273		1,824.5
WHITLK2	–3,630	6,930	65.		50W24	10,485	890		1,838.0
CONK92	–2,280	10,200	63.			10,860	9,580		1,801.0
CONK97	–1,700	8,580	71.		42D32	10,985	3,477	78.	1,817.0
CONK98	–800	8,320	81.		51W33	11,050	2,220	69.	1,828.0
GRAHAM1	–300	–2,990	58.		51W25	11,125	2,220	73.	1,824.0
388	330	3,630	80.		51W29	11,126	1,750	66.	1,837.0
30W22	1,005	6,170	50.		42W29	11,128	3,477	73.	1,817.0
30W27	2,005	7,495	61.		59W25	11,160	–360	55.	
CALVIN1	2,310	–4,950	62.		34W28	11,200	6,170	66.	1,826.0
WARE1X	2,310	–330	56.		26D29	11,225	8,547	78.	1,803.0
5614	2,940	–990	58.		5131	11,465	300	57.	1,850.0
64W1	2,970	–4,290	65.		51W22	11,880	890	55.	1,836.0
48W25	3,280	2,000	59.	1,860.0	5132	12,295	300	65.	1,842.0
6410	3,300	–4,950	66.		51W28	12,410	2,220	65.	1,814.0
648	3,890	–4,950	66.		42W28	12,432	4,944	80.	1,814.0
6411	4,290	–3,630	65.		51W30	12,510	890	63.	1,827.0
5613	4,290	–1,650	61.		42W30	12,535	3,455	73.	1,818.0
	4,300	8,220		1,858.0	42W23	12,535	3,530	61.	1,822.0
39W23	4,565	3,530	79.	1,832.0	59W27	12,540	–360	60.	
56W28	5,240	–350	58.		18W24	12,760	11,610	58.	
	5,580	8,220		1,850.5	3429	12,825	5,580	65.	1,811.0
49W25	5,845	2,220	70.	1,861.0	34D30	12,836	7,150	69.	1,833.0
	6,260	9,580		1,863.0	52W33	13,736	856	78.	1,818.0
	7,205	3,530		1,845.0	5228	14,255	1,660	67.	1,808.0
505A	8,220	1,745	72.	1,854.0	52D31	14,337	2,318	89.	1,799.0
	8,220	9,580		1,842.0	5229	14,935	375	63.	1,832.0
33W28	8,320	6,170	49.	1,864.0	52W32	15,050	2,195	72.	1,803.0
50W25	8,485	2,220	69.	1,867.0	52W23	15,170	890	68.	1,814.0
41W31S	8,485	4,605	75.	1,838.0	43W24	15,275	3,530	59.	1,795.0
50W29	8,560	2,220	71.	1,868.0	5230	15,535	225	60.	1,848.0
M41	8,615	4,505	78.	1,841.0	60W28	15,700	–400	46.	
252A	8,847	8,273	65.	1,818.0	53W25	16,405	2,220	59.	1,802.0
6613	8,970	–3,690	75.		44D29	16,960	3,585	50.	1,800.0
	9,017	9,512		1,797.0	61W23	17,820	–1,700	50.	
	9,580	9,580		1,841.0	44W24	17,915	3,530	58.	1,785.0
5030	9,633	247	58.	1,879.0		18,180	9,580		1,793.0
50D31	9,685	1,838	55.	1,845.0	54W25	18,640	2,200	57.	
41W28	9,770	3,530	82.	1,818.0	45W22	19,680	3,390	51.	
33W29	9,770	6,170	69.	1,846.0	55W25	2,170	2,250	57.	
41W23	9,845	3,530	79.	1,819.0	BEAGLE12	21,450	5,180	56.	
50W28	9,920	2,220	61.	1,841.0	BEAGLE9 W	23,100	3,960	74.	

20 md. Each table includes a collection of core plug measurements for wells that have data in a flow unit. The correlation between permeability and porosity, which can be obtained from these tables, is used in permeability description techniques.

A.6 Distribution Disk Computer Files

Data described in Appendix A and used for the field examples are included in the disk distributed with this book. The data files are in ASCII format for PCs and can be imported into most editor, word processor and spreadsheet programs. Files include data for well names and locations; top of reservoir, gross thickness and porosity for flow units; and core plug permeability and porosity for individual wells and permeability. Data are stored using GSLIB convention.[2] The first few lines of the files include header titles used to describe the data, and the rest of the file contains columns of numbers with the data.

The file **wellnam.dat** contains well names, coordinates and total gross thickness for all wells. The headers of this GSLIB file and first few lines of data are:

```
Gross Thickness for Burbank Unit
4
xcoord
ycoord
gross thickness
Burbank Field well name
21450.000    5180.000    56.000    BEAGLE12
-2280.000    10200.000   63.000    CONK92
-1700.000    8580.000    71.000    CONK97
```

The first line is the title. The second line is the number of columns of data. Lines three through six are the labels of each column. The remainder of the file consists of four columns of data. Columns are separated by one or more blank spaces. In this file, each line of data corresponds to a different well.

The file **tops.dat** includes the top of reservoir subsea depth for several wells in the field. The tops for all wells in the previous file are not available. The well names can be obtained by matching the *X* and *Y* coordinates with those in the file **wellnam.dat**.

The header and the first few lines of the file are:

tops (subsea depth)
3

x-coord	ft	F12.3
y-coord	ft	F12.3
top	ft	F12.3
12900.000	9580.000	1792.000
10860.000	9580.000	1836.000
11805.000	8810.000	1832.000

Flow unit gross thickness data are in the files **fu*h.dat** (* represents the number of the flow unit). The header and first few lines of data for flow unit one are

Flow Unit 1 gross thickness data
3
X, ft
Y, ft
Gross thickness, ft

–4620	–1980	7
–4300	330	3
–3630	–4950	2

Flow unit average porosity data are in the files **fu*por.dat**. The header and the first few lines of data for flow unit one are

Flow Unit 1 Porosity
4
xcoord
ycoord
zcoord
Porosity

21450.000	5180.000	0.0	6.340
–2280.000	10200.000	0.0	8.940
–800.000	8320.000	0.0	8.550

Flow units core plug permeability and porosity data are in the files **fu*prm.dat**. The header and the first few lines of data for flow unit one are

Flow Unit 1 Core Por vs Perm
4

depth	ft		
perm	md		
porosity	%		
log(perm)			
2944.500	0.300	10.500	-0.523
2945.500	4.300	14.400	0.633
2947.500	43.000	18.900	1.633

Note that the data in this file comes from different wells, and therefore the *X* and *Y* coordinates (not included) are different.

The files **"well_name".dat** contain permeability and porosity data for wells with core plug measurements. The header and the first few lines of data for one well are:

Well No. 23-W23 Core Data
3

Depth	(ft)	
Perm	(md)	
Porosity	(–)	
3039.500	53.000	0.209
3040.500	0.570	0.106
3041.500	1560.000	0.284

References

1. Hird, K.B.: "A Conditional Simulation Method for Reservoir Description Using Geological and Well Performance Constraints," PhD dissertation, U. of Tulsa, Tulsa (1993).
2. Deutsch, C.V. and Journel, A.G.: *GSLIB: Geostatistical Software Library and User's Guide*, Oxford U. Press, New York City (1995).

SI Metric Conversion Factor

$$\text{ft} \times 3.048^* \qquad \text{E}-01 = \text{m}$$

*Conversion factor is exact.

Appendix B
Derivations for Principles of Statistics

In Appendix B, we expand and derive many of the equations discussed in Chap. 2. For additional explanation, please refer to the references provided at the end of Chap. 2.

B.1 Expected Value

The generalized expression for expected value is given by

$$E[u(X)] = \sum_{i=1}^{k} u(x_i)P(X = x_i), \quad \text{(B.1)}$$

for a discrete random variable, and

$$E[u(X)] = \int_{-\infty}^{\infty} u(x)f(x)dx, \quad \text{(B.2)}$$

for a continuous random variable. $P(X=x_i)$ is a probability mass function, and $f(x)$ is the probability density function. In the following, we develop several of the expected value characteristics for a continuous variable. Similar arguments follow for a discrete variable.

B.1.1 Characteristics of Expected Value. If K is a constant,

$$E(K) = \int_{-\infty}^{\infty} Kf(x)dx = K\int_{-\infty}^{\infty} f(x)dx;$$

however,

$$\int_{-\infty}^{\infty} f(x)dx = 1 \rightarrow \text{property of probability density function.}$$

Therefore,

$$E(K) = K$$

$$E[Ku(X)] = \int_{-\infty}^{\infty} Ku(x)f(x)dx = K\int_{-\infty}^{\infty} u(x)f(x)dx. \quad \text{(B.3)}$$

Therefore,

$$E[Ku(X)] = KE[u(X)]. \quad \text{(B.4)}$$

Similarly,

$$E[u_1(X) + u_2(X)] = \int_{-\infty}^{\infty} [u_1(x) + u_2(x)]f(x)dx$$

$$= \int_{-\infty}^{\infty} u_1(x)f(x)dx + \int_{-\infty}^{\infty} u_2(x)f(x)dx.$$

Therefore,

$$E[u_1(X) + u_2(X)] = E[u_1(X)] + E[u_2(X)]. \quad \text{(B.5)}$$

B.1.2 Important Parameters of Univariate Distribution.

$$E(X) = \mu = \int_{-\infty}^{\infty} xf(x)dx, \quad \text{(B.6)}$$

and

$$E\left[(X-\mu)^2\right] = \sigma^2 = \int_{-\infty}^{\infty} (x-\mu)^2 f(x)dx. \quad \text{(B.7)}$$

For variance,

$$\sigma^2 = Var(X) = E\left[(X-\mu)^2\right] = E(X^2 - 2\mu X + \mu^2)$$

$$= E(X^2) - E(2\mu X) + E(\mu^2),$$

but μ is constant; therefore, by using Eqs. B.3 and B.4,

$$\sigma^2 = E(X^2) - 2\mu E(X) + \mu^2,$$

but $E(X)=\mu$ by definition.
Therefore,

$$\sigma^2 = E(X^2) - 2\mu^2 + \mu^2 = E(X^2) - \mu^2.$$

Therefore,

$$\sigma^2 = E(X^2) - [E(X)]^2. \quad \ldots\ldots (B.8)$$

This is the same as Eq. 2.65.

Properties of Variance. If K is constant,

$$Var(KX) = E\left[(KX)^2\right] - [E(KX)]^2$$

with Eq. B.8,

$$= KE(K^2X^2) - K^2[E(X)]^2$$

with Eq. B.4.

Therefore,

$$Var(KX) = K^2\left\{E(X^2) - [E(X)]^2\right\} = K^2Var(X). \quad \ldots\ldots (B.9)$$

If K and b are constant,

$$Var(KX + b) = E\left[(KX + b)^2\right] - [E(KX + b)]^2$$

$$= E(K^2X^2) + 2E(KXb) + E(b^2) - [E(KX) + E(b)]^2$$

$$= K^2E(X^2) + 2KbE(X) + b^2 - K^2[E(X)]^2 - b^2$$

$$- 2KbE(X)$$

$$= K^2E(X^2) - K^2[E(X)]^2.$$

Therefore,

$$Var(KX + b) = K^2Var(X). \quad \ldots\ldots (B.10)$$

B.1.3 Important Properties of Bivariate Distribution. The expected value for a bivariate continuous distribution is given by

$$E[u(X, Y)] = \int_{-\infty}^{\infty}\int_{-\infty}^{\infty} u(x, y)f(x, y)dxdy. \quad \ldots\ldots (B.11)$$

Covariance is defined as

$$C(X, Y) = E[(X - \mu_X)(Y - \mu_Y)] = E(XY) - E(Y\mu_X)$$

$$- E(X\mu_Y) + E(\mu_X\mu_Y).$$

However, μ_X and μ_Y are constant.

$$= E(XY) - \mu_XE(Y) - \mu_YE(X) + \mu_X\mu_Y$$

$$= E(XY) - \mu_X\mu_Y - \mu_X\mu_Y + \mu_X\mu_Y.$$

Therefore,

$$C(X, Y) = E(XY) - E(X)E(Y). \quad \ldots\ldots (B.12)$$

Similarly,

$$C(X + Z, Y) = E[(X + Z - \mu_X - \mu_Z)(Y - \mu_Y)]$$

$$= E(XY) + E(ZY) - \mu_YE(X) - \mu_YE(Y) - \mu_ZE(Y)$$

$$+ \mu_X\mu_Y + \mu_Z\mu_Y$$

$$= E(XY) - \mu_Y\mu_X - \mu_X\mu_Y + \mu_X\mu_Y$$

$$+ E(ZY) - \mu_Y\mu_Z - \mu_Y\mu_Z + \mu_Z\mu_Y$$

$$= [E(XY) - \mu_X\mu_Y] + [E(ZY) - \mu_Z\mu_Y].$$

Therefore,

$$C(X + Z, Y) = C(X, Y) + C(Z, Y). \quad \ldots\ldots (B.13)$$

Similarly,

$$C\left(\sum_{i=1}^{n} X_i, Y\right) = E\left[\left(\sum_{i=1}^{n} X_i - \sum_{i=1}^{n}\mu_{X_i}\right)(Y - \mu_Y)\right]$$

$$= E\left(\sum_{i=1}^{n} X_i \cdot Y\right) - \sum_{i=1}^{n}\mu_{X_i}E(Y) - \mu_Y E\sum_{i=1}^{n} X_i$$

$$+ \sum_{i=1}^{n}\mu_{X_i}\mu_Y$$

$$= \sum_{i=1}^{n} E(X_iY) - \mu_Y\sum_{i=1}^{n}\mu_{X_i} - \mu_Y\sum_{i=1}^{n}\mu_{X_i} + \sum_{i=1}^{n}\mu_{X_i}\mu_Y$$

$$= \sum_{i=1}^{n}\left[E(X_iY) - \mu_{X_i}\mu_Y\right].$$

Therefore,

$$C\left(\sum_{i=1}^{n} X_i, Y\right) = \sum_{i=1}^{n} C(X_i, Y). \quad \ldots\ldots (B.14)$$

This equation can be further generalized by writing

$$C\left(\sum_{i=1}^{n} X_i, \sum_{j=1}^{m} Y_j\right) = \sum_{i=1}^{n}\sum_{j=1}^{m} C(X_i, Y_j). \quad \ldots\ldots (B.15)$$

The proof of Eq. B.15 is very similar to Eq. B.14 and is not provided here.

If X and Y are independent, we can state that

$$f(x, y) = f_X(x) \cdot f_Y(y), \quad \ldots\ldots (B.16)$$

where $f(x,y)$ is a bivariate probability density function, and $f_X(x)$ and $f_Y(y)$ are marginal distribution functions of random variable X and Y, respectively.

Using the definition of covariance,

$$C(X, Y) = E\left[(X - \mu_x)(Y - \mu_y)\right] = E(XY) - E(X)E(Y). \quad \ldots\ldots (B.12)$$

The first term on the right side of Eq. B.12 can be written as

$$E(X, Y) = \int_{-\infty}^{\infty}\int_{-\infty}^{\infty} xyf(x, y)dxdy.$$

If X and Y are independent, substituting Eq. B.16 in Eq. B.12 yields

$$= \int_{-\infty}^{\infty}\int_{-\infty}^{\infty} xyf_X(x)f_Y(y)dxdy.$$

Because $f_X(x)$ and $f_Y(y)$ are only functions of variables X and Y respectively, the integral can be written as

$$= \int_{-\infty}^{\infty} x f_X(x)dx \cdot \int_{-\infty}^{\infty} y f_Y(y)dy.$$

Using the definition of expected value,

$$= E(X)E(Y).$$

Because

$$E(XY) = E(X)E(Y), \ C(X, Y) = 0. \quad \text{(B.17)}$$

If the two variables are independent, the covariance is zero. It should be noted that if the covariance is zero, the two variables may be uncorrelated but not necessarily independent.

B.2 Important Distribution Functions

B.2.1 Uniform Distribution Function. The probability density function is given by

$$f(x) = \frac{1}{b-a} \text{ for } a \le X \le b$$

$$= 0 \text{ otherwise.} \quad \text{(B.18)}$$

The cumulative distribution function is given by

$$F(u) = \int_{-\infty}^{u} f(x)dx = \int_{-\infty}^{a} 0dx + \int_{a}^{u} \frac{1}{b-a}dx.$$

Therefore,

$$F(u) = \frac{u-a}{b-a}. \quad \text{(B.19)}$$

Mean.

$$E(X) = \int_{a}^{b} xf(x)dx = \int_{a}^{a} \frac{x}{b-a}dx = \frac{1}{b-a}\left(\frac{x^2}{2}\right)_a^b$$

$$= \frac{b^2 - a^2}{2(b-a)} = \frac{b+a}{2}. \quad \text{(B.20)}$$

Variance.

$$Var(X) = E(X^2) - [E(X)]^2. \quad \text{(B.21)}$$

$$E(X^2) = \int_{a}^{b} x^2 f(x)dx = \frac{1}{b-a}\int_{a}^{b} x^2 dx,$$

$$\frac{1}{b-a}\left(\frac{x^3}{3}\right)_a^b = \frac{b^2 + ab + a^2}{3},$$

and

$$[E(X)]^2 = \frac{(b+a)^2}{4} \text{ (from Eq. B.20).}$$

Substituting in Eq. B.21,

$$Var(X) = \frac{b^2 + ab + a^2}{3} - \frac{b^2 + a^2 + 2ab}{4} = \frac{(b-a)^2}{12}. \quad \text{(B.22)}$$

B.2.2 Normal Distribution Function. The probability density function is given by

$$f(x) = \frac{1}{\sqrt{2\Pi\sigma^2}}\exp\left[-\frac{1}{2}\frac{(X-\mu)^2}{\sigma^2}\right]$$

$$\text{for } -\infty < x < \infty. \quad \text{(B.23)}$$

As we discussed in Chap. 2, we cannot get an analytical expression for the cumulative distribution function. It must be solved with a numerical integration.

Mean.

$$E(X) = \int_{-\infty}^{\infty} x \cdot \frac{1}{\sqrt{2\Pi\sigma^2}}\exp\left[-\frac{1}{2}\frac{(x-\mu)^2}{\sigma^2}\right]dx. \quad \text{(B.24)}$$

Substituting,

$$t = \frac{x-\mu}{\sigma}; \ x = t\sigma + \mu,$$

and

$$E(X) = \frac{1}{\sqrt{2\Pi}}\int_{-\infty}^{\infty} (t\sigma + \mu)\exp\left(\frac{-t^2}{2}\right)dt$$

$$= \frac{\sigma}{\sqrt{2\Pi}}\int_{-\infty}^{\infty} t\exp\left(\frac{-t^2}{2}\right)dt + \mu\left[\frac{1}{\sqrt{2\Pi}}\int_{-\infty}^{\infty}\exp\left(\frac{-t^2}{2}\right)dt\right]$$

$$= \frac{\sigma}{\sqrt{2\Pi}}\left[-\exp\left(\frac{-t^2}{2}\right)\right]_{-\infty}^{\infty} + \mu. \quad \text{(B.25)}$$

The first term is zero, and the term in the bracket, in the second term (Eq. B.25), represents an integral of probability density function if the mean is zero and the variance is one. Therefore, that term should be equal to one. Hence,

$$E(X) = \mu. \quad \text{(B.26)}$$

Variance.

$$Var(X) = E\left[(X-\mu)^2\right]$$

$$= \int_{-\infty}^{\infty} (X-\mu)^2 \frac{1}{\sqrt{2\Pi\sigma^2}}\exp\left[\frac{1}{2}\left(\frac{x-\mu}{\sigma}\right)^2\right]dx. \quad \text{(B.27)}$$

Substituting, $t = \frac{x-\mu}{\sigma}; x - \mu = t\sigma.$

$$Var(X) = \sigma^2\left[\frac{1}{\sqrt{2\Pi}}\int_{-\infty}^{\infty} t^2 \cdot \exp\left(\frac{-t^2}{2}\right)dt\right].$$

Integrating by parts,

$$= \sigma^2\left[\frac{1}{\sqrt{2\Pi}}\left(-te^{\frac{-t^2}{2}}\right)_{-\infty}^{\infty} + \frac{1}{\sqrt{2\Pi}}\int_{-\infty}^{\infty} e^{\frac{-t^2}{2}}dt\right].$$

The second term represents the normal distribution with a mean of zero and a variance of one.

$$= \sigma^2(0 + 1).$$

Therefore,

$$Var(X) = \sigma^2. \quad \text{(B.28)}$$

Standard Normal Distribution. If we define,

$$Z = \frac{X - \mu}{\sigma}. \quad \text{(B.29)}$$

The mean of standardized variable is

$$E(Z) = E\left(\frac{X - \mu}{\sigma}\right) = \frac{1}{\sigma}[E(X) - \mu] = 0. \quad \text{(B.30)}$$

The variance of standardized variable is

$$Var(Z) = E(Z^2) - [E(Z)]^2 = E\left[\left(\frac{X - \mu}{\sigma}\right)^2\right] - 0$$

$$= \frac{1}{\sigma^2} E\left[(X - \mu)^2\right],$$

but

$$E\left[(X - \mu)^2\right] = \sigma^2.$$

Therefore,

$$Var(Z) = 1. \quad \text{(B.31)}$$

B.2.3 Log Normal Distribution. For log normal distribution, if X is log normal distributed, then $Y = \ln X$ is log normal distributed. If the mean of Y is α and the variance is β^2, we can write the distribution function for Y as

$$f(y) = \frac{1}{\sqrt{2\Pi}\beta} \exp\left[-\frac{1}{2}\left(\frac{y - \alpha}{\beta}\right)\right], \quad -\infty < y < \infty. \quad \text{(B.32)}$$

Mean. We can write

$$E(X) = E(e^Y) \text{ because } \ln X = Y. \quad \text{(B.33)}$$

$$E(e^Y) = \frac{1}{\sqrt{2\Pi}} \int_{-\infty}^{\infty} \frac{1}{\beta} e^y e^{-\frac{1}{2}\left(\frac{y-\alpha}{\beta}\right)^2} dy$$

$$= \frac{1}{\beta\sqrt{2\Pi}} \int_{-\infty}^{\infty} e^{-\frac{1}{2\beta^2}\left(y^2 + \alpha^2 - 2y\alpha - 2\beta^2 y\right)} dy$$

$$= \frac{1}{\beta\sqrt{2\Pi}} \int_{-\infty}^{\infty} e^{-\frac{1}{2\beta^2}\left\{\left[y - \left(\alpha + \beta^2\right)\right]^2 + \alpha^2 - \left(\alpha + \beta^2\right)^2\right\}} dy$$

$$= e^{-\frac{1}{2\beta^2}\left[\alpha^2 - \left(\alpha + \beta^2\right)^2\right]} \left[\frac{1}{\beta\sqrt{2\Pi}} \int_{-\infty}^{\infty} e^{\left\{-\left[\frac{y - \left(\alpha + \beta^2\right)}{\beta}\right]^{\frac{1}{2}}\right\}} dy\right]. \quad \text{(B.34)}$$

However, the term on the right side, in the big bracket, is equal to one because it represents an integration of normal probability distribution function with a mean of $(\alpha + \beta^2)$ and a variance of β^2.

$$= e^{-\frac{1}{2\beta^2}\left(\alpha^2 - \alpha^2 - 2\alpha\beta^2 - \beta^4\right)}.$$

Therefore,

$$\mu = e^{\alpha + \frac{\beta^2}{2}}. \quad \text{(B.35)}$$

Variance.

$$Var(X) = E(X^2) - [E(X)]^2. \quad \text{(B.36)}$$

We already calculated $E[X]$, so we must calculate $E[X^2]$, which can be written as $E[e^{2Y}]$.

We can write

$$E(e^{2Y}) = \frac{1}{\beta\sqrt{2\Pi}} \int_{-\infty}^{\infty} e^{2y} e^{-\frac{1}{2}\left(\frac{y-\alpha}{\beta}\right)^2} dy$$

$$= \frac{1}{\beta\sqrt{2\Pi}} \int_{-\infty}^{\infty} e^{-\frac{1}{2\beta^2}\left(y^2 + \alpha^2 - 2\alpha y - 4\beta^2 y\right)} dy$$

$$= \frac{1}{\beta\sqrt{2\Pi}} \int_{-\infty}^{\infty} e^{-\frac{1}{2\beta^2}\left\{\left[y - \left(\alpha + 2\beta^2\right)\right]^2 + \alpha^2 - \left(\alpha + \beta^2\right)^2\right\}} dy$$

$$= e^{-\frac{1}{2\beta^2}\left[\alpha^2 - \left(\alpha + 2\beta^2\right)^2\right]} \left[\frac{1}{\beta\sqrt{2\Pi}} \int_{-\infty}^{\infty} e^{-\frac{1}{2}\left\{\left[\frac{y - \left(\alpha + 2\beta^2\right)}{\beta}\right]^2\right\}} dy\right]. \quad \text{(B.37)}$$

The term in the big bracket is equal to one,

$$= e^{-\frac{1}{2\beta^2}\left(\alpha^2 - \alpha^2 - 4\beta^2 - 4\alpha\beta^2\right)}.$$

Therefore,

$$E(X^2) = e^{2(\beta^2 + \alpha)}. \quad \text{(B.38)}$$

By substituting Eqs. B.38 and B.35 in Eq. B.36,

$$Var(X) = e^{2(\beta^2 + \alpha)} - e^{2\left(\alpha + \frac{\beta^2}{2}\right)} = e^{2\left(\alpha + \frac{\beta^2}{2}\right)}\left(e^{\beta^2} - 1\right).$$

Therefore,

$$Var(X) = \mu^2\left(e^{\beta^2} - 1\right). \quad \text{(B.39)}$$

B.3 Linear Regression Using MVUE Technique

In this appendix, we provide a simple example of a linear regression using the minimum variance unbiased estimate (MVUE) technique.

We can write an expression for linear regression as

$$y = mx + b. \quad \text{(B.40)}$$

In terms of random variables, where Y is the true value,

$$Y = mX + b + \varepsilon. \quad \text{(B.41)}$$

If we write an equation in terms of estimated value, we can write

$$\hat{Y} = mX + b, \quad \text{(B.42)}$$

where ε represents an error associated with the estimate. We apply two requirements—unbiasedness and variance minimization.

B.3.1 Unbiasedness. An unbiasedness condition requires that

$$E\left(Y - \hat{Y}\right) = 0. \quad \text{(B.43)}$$

In other words, $E(\varepsilon)=0$.

The expected value of error should be zero.

B.3.2 Minimum Variance. For minimum variance,

$$Var\left(Y - \hat{Y}\right) = \sigma_E^2. \quad \text{(B.44)}$$

We can write the equation for variance as

$$\sigma_E^2 = Var\left(Y - \hat{Y}\right) = E\left[\left(Y - \hat{Y}\right)^2\right] - \left[E\left(Y - \hat{Y}\right)\right]^2. \quad \text{(B.45)}$$

However $E\left(Y - \hat{Y}\right) = 0$, based on an unbiased condition.

By substituting Eq. B.42 in Eq. B.45,

$$\sigma_E^2 = E\left[(Y - mX - b)^2\right]. \quad \text{(B.46)}$$

To minimize the error variance,

$$\frac{\partial\sigma_E^2}{\partial m} = E[-2X(Y - mX - b)] = 0 = E(YX)$$

$$- mE\left(X^2\right) - bE(X) = 0. \quad \text{(B.47)}$$

Also,

$$\frac{\partial\sigma_E^2}{\partial b} = E[-2(Y - mX - b)] = 0 = E(Y)$$

$$- mE(X) - b = 0. \quad \text{(B.48)}$$

To solve for m and b, we can use Eqs. B.47 and B.48.

By multiplying Eq. B.48 with $E(X)$ and subtracting from Eq. B.47, we obtain

$$E(YX) - E(Y)E(X) - m\left\{\left(X^2\right) - [E(X)]^2\right\} = 0,$$

but

$$E(YX) - E(Y)\cdot E(X) = C(X, Y),$$

and

$$E\left(X^2\right) - [E(X)]^2 = Var(X).$$

Therefore,

$$m = \frac{C(X, Y)}{\sigma_X^2}. \quad \text{(B.49)}$$

By knowing the value of m, using Eq. B.48, we can write,

$$b = E(Y) - mE(X). \quad \text{(B.50)}$$

B.4 Generalized Linear Regression

If we write a generalized expression as

$$Y = a_0 + \sum_{i=1}^{n} a_i X_i + \varepsilon, \quad \text{(B.51)}$$

by applying an unbiased condition, we obtain

$$E(Y) = a_0 + \sum_{i=1}^{n} a_i E(X_i),$$

or

$$a_0 = E(Y) - \sum_{i=1}^{n} a_i E(X_i). \quad \text{(B.52)}$$

By applying the minimum variance condition, we obtain

$$\sigma_E^2 = Var\left(Y - a_o - \sum_{i=1}^{n} a_i X_i\right)$$

$$= Var(Y) + Var\left(\sum_{i=1}^{n} a_i X_i\right) - 2\left[\sum_{i=1}^{n} a_i C(Y, X_i)\right]$$

$$= Var(Y) + \sum_{i=1}^{n}\sum_{j=1}^{n} a_i a_j C\left(X_i, X_j\right) - 2\sum_{i=1}^{n} a_i C(Y, X_i).$$

To minimize variance, we can equate the partial derivative of variance with respect to individual coefficients equal to zero, which is written as

$$\frac{\partial\sigma_E^2}{\partial a_i} = 2\sum_{i=1}^{n} a_i C\left(X_i, X_j\right) - 2C\left(Y, X_j\right) = 0 \text{ for } j=1, \ldots, n. \quad \text{(B.53)}$$

Eq. B.53 can be written as

$$\sum_{j=1}^{n} a_j C\left(X_i, X_j\right) = C(Y, X_i) \text{ for } i=1, \ldots, n.$$

In matrix form, it is written as

$$\begin{bmatrix} C(X_1, X_1) & \ldots & C(X_1, X_n) \\ C(X_n, X_1) & \ldots & C(X_n, X_n) \end{bmatrix} \begin{bmatrix} a_1 \\ \vdots \\ a_n \end{bmatrix} \begin{bmatrix} C(Y, X_1) \\ \vdots \\ C(Y, X_n) \end{bmatrix}. \quad \text{(B.54)}$$

To solve for coefficients $a_1, \ldots, a_n$, we must invert the matrix on the left side and multiply it by the vector on the right side. Once a_i is obtained, by substituting the values in Eq. B.52, we can obtain the value of a_0.

Nomenclature

a_i = coefficients in a linear regression
b = constant
$C(X,Y)$ = covariance between variables X and Y
$E[u(X)]$ = expected value of function $u(x)$
$f(x)$ = probability density function
$f_X(x)$ = marginal distribution function for variable X
$f_Y(y)$ = marginal distribution function for variable Y
$f(x,y)$ = bivariate probability density function
$F(u)$ = cumulative distribution function
i = dummy variable

$j =$ dummy variable
$k =$ dummy variable
$K =$ constant
$m =$ slope in a linear regression
$n =$ dummy variable
$P(X = x_i) =$ probability mass function for variable X
$t =$ dummy variable
$u(x) =$ function of variable X
$u(y) =$ function of variable Y
$Var(X) =$ variance of variable X
$X =$ variable
$Y =$ variable
$Z =$ standardized normal variable
$\alpha =$ arithmetic mean of natural log of the variable
$\beta^2 =$ variance of natural log of the variable
$\varepsilon =$ error of estimation
$\mu =$ arithmetic mean of variable
$\sigma^2 =$ variance of variable
$\sigma_E^2 =$ error variance

Subscripts

$x =$ variable
$y =$ variable

Appendix C
Derivations for Spatial Relationships

C.1 Relationships Between Variograms and Covariance

To understand the relationship between variogram and covariance, we must start with a definition of variance. Recall that

$$Var(X) = E(X^2) - [E(X)]^2. \quad \text{(C.1)}$$

If we define the two random variables, X and Y, we can write

$$Var(X - Y) = E\left[(X - Y)^2\right] - [E(X - Y)]^2. \quad \text{(C.2)}$$

Expanding,

$$Var(X - Y) = E(X^2) + E(Y^2) - 2E(XY) - [E(X)]^2$$

$$[E(Y)]^2 + 2E(X)E(Y). \quad \text{(C.3)}$$

However,

$$E(X^2) - [E(X)]^2 = Var(X), \quad \text{(C.1)}$$

$$E(Y^2) - [E(Y)]^2 = Var(Y), \quad \text{(C.4)}$$

and

$$E(XY) - E(X)E(Y) = C(X, Y). \quad \text{(C.5)}$$

By substituting Eqs. C.1, C.4, and C.5 in Eq. C.3,

$$Var(X - Y) = Var(X) + Var(Y) - 2C(X, Y). \quad \text{(C.6)}$$

If we assume that variable $X = X(\vec{u})$ and variable $Y = X(\vec{u} + \vec{L})$, we can rewrite Eq. C.6 as

$$Var\left[X(\vec{u}) - X(\vec{u} + \vec{L})\right] = Var[X(\vec{u})] + Var\left[X(\vec{u} + \vec{L})\right]$$

$$- 2C\left[X(\vec{u}), X(\vec{u} + \vec{L})\right]. \quad \text{(C.7)}$$

Recall that

$$\gamma(\vec{L}) = \frac{1}{2} Var\left[X(\vec{u}) - X(\vec{u} + \vec{L})\right]. \quad \text{(C.8)}$$

Also, the second order stationarity requires that

$$Var[X(\vec{u})] = Var\left[X(\vec{u} + \vec{L})\right] = C(0), \quad \text{(C.9)}$$

and

$$C\left[X(\vec{u}), X(\vec{u} + \vec{L})\right] = C(\vec{L}). \quad \text{(C.10)}$$

By substituting Eqs. C.8, C.9, and C.10 in Eq. C.7,

$$2\gamma(\vec{L}) = 2C(0) - 2C(\vec{L}),$$

or

$$\gamma(\vec{L}) = C(0) - C(\vec{L}). \quad \text{(C.11)}$$

This is the same equation as Eq. 3.19 in Chap. 3.

In practice, when we estimate the variogram, we write the equation as[1]

$$\hat{\gamma}(\vec{L}) = \frac{1}{2n(\vec{L})} \sum_{i=1}^{n(\vec{L})} \left[x(\vec{u}_i) - x(\vec{u}_i + \vec{L})\right]^2. \quad \text{(C.12)}$$

Expanding,

$$\hat{\gamma}(\vec{L}) = \frac{1}{2n(\vec{L})} \sum_{i=1}^{n(\vec{L})} x^2(\vec{u}_i) + \frac{1}{2n(\vec{L})} \sum_{i=1}^{n(\vec{L})} x^2(\vec{u}_i + \vec{L})$$

$$- \frac{1}{n(\vec{L})} \sum_{i=1}^{n(\vec{L})} x(\vec{u}_i) x(\vec{u}_i + \vec{L}).$$

By adding and subtracting same terms, we can write

$$\hat{\gamma}(\vec{L}) = \frac{1}{2n(\vec{L})}\sum_{i=1}^{n(\vec{L})} x^2(\vec{u}_i) - \frac{1}{2}\left[\frac{1}{n(\vec{L})}\sum_{i=1}^{n(\vec{L})} x(\vec{u}_i)\right]^2$$

$$+ \frac{1}{2n(\vec{L})}\sum_{i=1}^{n(\vec{L})} x^2(\vec{u}_i + \vec{L}) - \frac{1}{2}\left[\frac{1}{n(\vec{L})}\sum_{i=1}^{n(\vec{L})} x(\vec{u}_i + \vec{L})\right]^2$$

$$- \frac{1}{n(\vec{L})}\sum_{i=1}^{n(\vec{L})} x(\vec{u}_i)x(\vec{u}_i + \vec{L}) + \frac{1}{n(\vec{L})}\sum_{i=1}^{n(\vec{L})} x(\vec{u}_i)$$

$$\cdot\frac{1}{n(\vec{L})}\sum_{i=1}^{n(\vec{L})} x(\vec{u}_i + \vec{L})$$

$$+ \frac{1}{2}\left\{\left[\frac{1}{n(\vec{L})}\sum_{i=1}^{n(\vec{L})} x(\vec{u}_i)\right]^2 + \left[\frac{1}{n(\vec{L})}\sum_{i=1}^{n(\vec{L})} x(\vec{u}_i + \vec{L})\right]^2\right.$$

$$\left. - \frac{2}{n(\vec{L})}\sum_{i=1}^{n(\vec{L})} x(\vec{u}_i)\cdot\frac{1}{n(\vec{L})}\sum_{i=1}^{n(\vec{L})} x(\vec{u}_i + \vec{L})\right\}. \quad \text{(C.13)}$$

If we define the following terms as

$$s^2_{-L} = \frac{1}{n(\vec{L})}\sum_{i=1}^{n(\vec{L})} x^2(\vec{u}_i) - \left[\frac{1}{n(\vec{L})}\sum_{i=1}^{n(\vec{L})} x(\vec{u}_i)\right]^2, \quad \text{(C.14)}$$

$$s^2_{+L} = \frac{1}{n(\vec{L})}\sum_{i=1}^{n(\vec{L})} x^2(\vec{u}_i + \vec{L}) - \left[\frac{1}{n(\vec{L})}\sum_{i=1}^{n(\vec{L})} x(\vec{u}_i + \vec{L})\right]^2, \quad \text{(C.15)}$$

$$\bar{x}_{-L} = \frac{1}{n(\vec{L})}\sum_{i=1}^{n(\vec{L})} x(\vec{u}_i), \quad \text{(C.16)}$$

and

$$\bar{x}_{+L} = \frac{1}{n(\vec{L})}\sum_{i=1}^{n(\vec{L})} x(\vec{u}_i + \vec{L}), \quad \text{(C.17)}$$

where s^2_{-L} and s^2_{+L}, respectively, represent the variance of the first and the second data points in sample pairs located a lag distance $\vec{L}$ apart, and $\bar{x}_{-L}$ and $\bar{x}_{+L}$, respectively, represent the arithmetic means of the first and the second data points in the sample pairs located at a lag distance $\vec{L}$ apart.

By substituting Eqs. C.14, C.15, C.16, and C.17 in Eq. C.13, we obtain

$$\hat{\gamma}(\vec{L}) = \frac{1}{2}(s^2_{-L} + s^2_{+L}) - c(\vec{L}) + \frac{1}{2}(\bar{x}_{-L} - \bar{x}_{+L})^2, \quad \text{(C.18)}$$

where Eq. C.18 represents the covariance estimation, which is defined as

$$c(\vec{L}) = \frac{1}{n(\vec{L})}\sum_{i=1}^{n(\vec{L})} x(\vec{u}_i)x(\vec{u}_i + \vec{L})$$

$$- \frac{2}{n(\vec{L})}\sum_{i=1}^{n(\vec{L})} x(\vec{u}_i)\cdot\frac{1}{n(\vec{L})}\sum_{i=1}^{n(\vec{L})} x(\vec{u}_i + \vec{L}). \quad \text{(C.19)}$$

By examining Eq. C.18, one realizes that this equation is not the same as

$$\hat{\gamma}(\vec{L}) = c(0) - c(\vec{L}). \quad \text{(C.20)}$$

Eqs. C.18 and C.20 can be identical only when

$$s^2_{-L} = s^2_{+L} = c(0) \text{ and } \bar{x}_{-L} = \bar{x}_{+L}. \quad \text{(C.21)}$$

That is, when the sample variance is the same as the lag variance, and the sample mean is the same as the lag mean. If we have sufficient samples represented at each lag distance, Eq. C.21 can be satisfied, and we can assume Eq. C.20 to be valid. Unfortunately, we might not have representative pairs for each lag distance. As a result, the sample mean and the lag mean can be different, and the sample variance and the lag variance might also be different.

To account for these differences, we use the modifications of the variogram discussed in Sec. 3.3.4. For example, the nonergodic variogram is defined as

$$\hat{\gamma}_{NE}(\vec{L}) = c(0) - c(\vec{L}), \quad \text{(C.22)}$$

where $c(0)$ represents the sample variance. If we substitute Eq. C.22 in Eq. C.18, we obtain

$$\hat{\gamma}(\vec{L}) = \frac{1}{2}(s^2_{-L} + s^2_{+L}) - c(0) + \frac{1}{2}(\bar{x}_{-L} - \bar{x}_{+L})^2 + \hat{\gamma}_{NE}(\vec{L}). \quad \text{(C.23)}$$

Eq. C.23 establishes the relationship between nonergodic and conventional variogram estimations. The nonergodic equation, through the definition of covariance, accounts for the variations in lag mean. As explained before, if the lag mean and the sample mean are the same, and the lag variance is identical to the sample variance, we can state

$$\hat{\gamma}(\vec{L}) = \hat{\gamma}_{NE}(\vec{L}). \quad \text{(C.24)}$$

Similarly, the nonergodic correlogram is defined as

$$r_{NE}(\vec{L}) = \frac{c(\vec{L})}{s_{-L}s_{+L}}. \quad \text{(C.25)}$$

Eq. C.25 accounts for the variation in the sample and lag variances.

C.2 Indicator Variable

Indicator variables allow transformation of continuous variables into discrete variables. It is defined as

$$I(\vec{L}, x_t) = 1, \text{ if } X(u) \le x_t$$
$$= 0, \text{ if } X(u) > x_t. \quad \text{(C.26)}$$

With this basic definition, we can derive some important properties of the indicator function.[2]

C.2.1 Expected Value.

$$E[I(\vec{u},x_t)] = 1 \times p[X(\vec{u}) \le x_t] + 0 \times p[x(\vec{u}) > x_t]. \quad \text{(C.27)}$$

The indicator function can only take two values, with the probabilities for each outcome defined by the distribution function of the original variable. However, we know that

$$p[X(\vec{u}) \le x_t] = F(x_t). \quad \text{(C.28)}$$

This is the definition of the cumulative distribution function. Therefore, we can write

$$E[I(\vec{u},x_t)] = p[X(\vec{u}) \le x_t] = F(x_t). \quad \text{(C.29)}$$

C.2.2 Variance.

$$Var[I(\vec{u},x_t)] = E[I^2(\vec{u},x_t)] - \{E[I(\vec{u},x_t)]\}^2. \quad \text{(C.30)}$$

Because the indicator variable can only take values of zero or one, we can write

$$I(\vec{u},x_t) = [I(\vec{u},x_t)]^2. \quad \text{(C.31)}$$

Note that $0^2 = 0$, and $1^2 = 1$.

By substituting Eqs. C.31 and C.28 in Eq. C.30,

$$Var[I(\vec{u},x_t)] = F(x_t) - [F(x_t)]^2$$

$$= F(x_t)[1 - F(x_t)]. \quad \text{(C.32)}$$

C.2.3 Noncentered Covariance. One of the unique properties of an indicator variable is its ability to define the connectivity between two points for a given threshold. This information cannot be obtained by conventional variograms. For example, we can write

$$H_I(\vec{L},x_t) = E\left[I(\vec{u},x_t)I(\vec{u}+\vec{L},x_t)\right]$$

$$= 1 \times \left\{p\left[X(\vec{u}) \le x_t, X(\vec{u}+\vec{L}) \le x_t\right]\right\}$$

$$+ 0 \times \left\{p\left[X(\vec{u}) \le x_t, X(\vec{u}+\vec{L}) > x_t\right]\right\}$$

$$+ 0 \times \left\{p\left[X(\vec{u}) > x_t, X(\vec{u}+\vec{L}) \le x_t\right]\right\}$$

$$+ 0 \times \left\{p\left[X(\vec{u}) > x_t, X(\vec{u}+\vec{L}) > x_t\right]\right\}. \quad \text{(C.33)}$$

In calculating the expected value of a product of indicator variables, we must consider four possibilities; of these four, however, only one will result in a value of one. The other three possibilities will result in a value of zero. The left side of Eq. C.33 is called the noncentered covariance. It can be written as

$$H_I(\vec{L},x_t) = E\left[I(\vec{u},x_t)\cdot I(\vec{u}+\vec{L},x_t)\right]$$

$$= p\left[X(\vec{u}) \le x_t, X(\vec{u}+\vec{L}) \le x_t\right]. \quad \text{(C.34)}$$

This represents a joint probability that two values, located at a distance $\vec{L}$ apart, are less than a given threshold. In reservoir description, such information might be very important. For example, the connectivity of porosities, both below a threshold (productive) level, can be used to represent a nonproductive zone.

If we define an indicator variable as

$$J(\vec{u},x_t) = 0, \text{if } X(u) \le x_t$$

$$= 1, \text{if } X(u) > x_t, \quad \text{(C.35)}$$

we can easily see that

$$J(\vec{u},x_t) = 1 - I(\vec{u},x_t). \quad \text{(C.36)}$$

Similar to $I(\vec{u},x_t)$, we can also use $J(\vec{u},x_t)$ to represent the joint probability that two values located at a distance $\vec{L}$ apart are both greater than a certain threshold. An example is the connectivity of high permeability values representing a thief zone. We can write

$$H_J(\vec{L},x_t) = E\left[J(\vec{u},x_t)\cdot J(\vec{u}+\vec{L},x_t)\right]$$

$$= p\left[X(\vec{u}) > x_t, X(\vec{u}+\vec{L}) > x_t\right]. \quad \text{(C.37)}$$

By substituting Eq. C.36 in Eq. C.37, we can easily show that

$$H_J(\vec{L},x_t) = 1 + H_I(\vec{L},x_t) - 2F(x_t). \quad \text{(C.38)}$$

As we discussed in the multipoint histogram section, this type of relationship can be extended to multiple point connectivity, which allows a joint probability determination of how multiple points are connected to each other.

Indicator variables are also useful for describing categorical or discrete variables. A good example is geological facies. We can define an indicator variable as

$$I(\vec{u},K_t) = 1, \text{if } K(u) = K_t$$

$$= 0, \text{if } K(u) \ne K_t, \quad \text{(C.39)}$$

where K_t is a particular categorical facies, and $K(\vec{u})$ is the facies present at location $\vec{u}$. Similar to Eq. C.33, we can easily define the joint probability distribution as

$$H_I(\vec{L},K_t) = E\left[I(\vec{u},K_t)\cdot I(\vec{u}+\vec{L},K_t)\right]$$

$$= p\left[K(\vec{u}) = K_t, K(\vec{u}+\vec{L}) = K_t\right]. \quad \text{(C.40)}$$

Instead of defining a joint connectivity of two locations with respect to the same facies, we can also define the joint connectivity of two locations with respect to different facies as

$$H_I(\vec{L},K_t,K_t') = E\left[I(\vec{u},K_t)\cdot I(\vec{u}+\vec{L},K_t)\right]$$

$$= p\left[K(\vec{u}) = K_t, K(\vec{u}+\vec{L}) = K_t'\right]. \quad \text{(C.41)}$$

This type of relationship can be easily extended to multipoint histograms for the categorical variables.

C.3 Cross Relationships

C.3.1 Theoretical Equations. As explained in Chap. 3, an equation for the cross variogram can be written as

$$\gamma_c(\vec{L}) = \frac{1}{2}E\left\{\left[X(\vec{u}) - X(\vec{u} + \vec{L})\right]\left[Y(\vec{u}) - Y(\vec{u} + \vec{L})\right]\right\}. \quad \text{(C.42)}$$

Similarly, we can write, for a lag of $-\vec{L}$, an equation as

$$\gamma_c(-\vec{L}) = \frac{1}{2}E\left\{\left[X(\vec{u} + \vec{L}) - X(\vec{u})\right]\left[Y(\vec{u} + \vec{L}) - Y(\vec{u})\right]\right\}, \quad \text{(C.43)}$$

which can be written as

$$= \frac{1}{2}E\left\{\left[X(\vec{u}) - X(\vec{u} + \vec{L})\right]\left[Y(\vec{u}) - Y(\vec{u} + \vec{L})\right]\right\} = \gamma_c(\vec{L}). \quad \text{(C.44)}$$

That is, the cross variogram is symmetric.

The cross covariance equation can be written as

$$C_c(\vec{L}) = E\left[X(\vec{u})Y(\vec{u} + \vec{L})\right] - E[X(\vec{u})]E\left[Y(\vec{u} + \vec{L})\right]. \quad \text{(C.45)}$$

For a lag of $-\vec{L}$, we can write the equation as

$$C_c(-\vec{L}) = E\left[X(\vec{u} + \vec{L})Y(\vec{u})\right] - E\left[X(\vec{u} + \vec{L})\right]E[Y(\vec{u})]. \quad \text{(C.46)}$$

If we assume that first order stationarity holds, we can write

$$E[X(\vec{u})] = E\left[X(\vec{u} + \vec{L})\right], \text{and } E[Y(\vec{u})] = E\left[Y(\vec{u} + \vec{L})\right]. \quad \text{(C.47)}$$

Therefore, the second term in Eqs. C.45 and C.46 can be assumed to be equal. However, $E\left[X(\vec{u})Y(\vec{u} + \vec{L})\right]$ is not necessarily equal to $E\left[X(\vec{u} + \vec{L})Y(\vec{u})\right]$. In some mining applications, they are observed to be significantly different. This phenomenon is called a lag effect. In the presence of a lag effect, we can state that

$$C_c(\vec{L}) \neq C_c(-\vec{L}). \quad \text{(C.48)}$$

In other words, the cross covariance is not symmetric. In most reservoir characterization applications, however, we assume that it is symmetric.

Expanding Eq. C.42, we can write

$$2\gamma_c(\vec{L}) = E[X(\vec{u})Y(\vec{u})] + E\left[X(\vec{u} + \vec{L})Y(\vec{u} + \vec{L})\right]$$
$$E\left[X(\vec{u})Y(\vec{u} + \vec{L})\right] - E\left[Y(\vec{u})X(\vec{u} + \vec{L})\right]. \quad \text{(C.49)}$$

We can rewrite it as,

$$2\gamma_c(\vec{L}) = E[X(\vec{u})Y(\vec{u})] - E[X(\vec{u})Y(\vec{u})]$$
$$+ E\left[X(\vec{u} + \vec{L})Y(\vec{u} + \vec{L})\right] - E\left[X(\vec{u} + \vec{L})Y(\vec{u} + \vec{L})\right]$$
$$- E\left[X(\vec{u})Y(\vec{u} + \vec{L})\right] + E[X(\vec{u})]E\left[Y(\vec{u} + \vec{L})\right]$$
$$- E\left[X(\vec{u} + \vec{L})Y(\vec{u})\right] + E\left[X(\vec{u} + \vec{L})\right]E[Y(\vec{u})]. \quad \text{(C.50)}$$

In Eq. C.50, we are assuming first order stationarity. See Eq. C.47. Knowing that,

$$C_c(0) = E[X(\vec{u})Y(\vec{u})] - E[X(\vec{u})]E[Y(\vec{u})]$$
$$= E\left[X(\vec{u} + \vec{L})Y(\vec{u} + \vec{L})\right] - E\left[X(\vec{u} + \vec{L})\right]E\left[Y(\vec{u} + \vec{L})\right]. \quad \text{(C.51)}$$

We can write

$$2\gamma_c(\vec{L}) = 2C_c(0) - C_c(\vec{L}) - C_c(-\vec{L}). \quad \text{(C.52)}$$

If we assume the cross covariance to be symmetric,

$$\gamma_c(\vec{L}) = C_c(0) - C_c(\vec{L}). \quad \text{(C.53)}$$

C.3.2 Practical Considerations. In practice, the cross variogram is calculated as

$$\hat{\gamma}_c(\vec{L}) = \frac{1}{2n(\vec{L})}\sum_{i=1}^{n(\vec{L})}\left[x(\vec{u}_i) - x(\vec{u}_i + \vec{L})\right]\left[y(\vec{u}_i) - y(\vec{u}_i + \vec{L})\right]. \quad \text{(C.54)}$$

Expanding,

$$\hat{\gamma}_c(\vec{L}) = \frac{1}{2n(\vec{L})}\sum_{i=1}^{n(\vec{L})} x(\vec{u}_i)y(\vec{u}_i)$$
$$+ \frac{1}{2n(\vec{L})}\sum_{i=1}^{n(\vec{L})} x(\vec{u}_i + \vec{L})y(\vec{u}_i + \vec{L})$$
$$- \frac{1}{2n(\vec{L})}\sum_{i=1}^{n(\vec{L})} x(\vec{u}_i)y(\vec{u}_i + \vec{L})$$
$$- \frac{1}{2n(\vec{L})}\sum_{i=1}^{n(\vec{L})} x(\vec{u}_i + \vec{L})y(\vec{u}_i). \quad \text{(C.55)}$$

If we define terms similar to the conventional variogram formulation,

$$\bar{x}_{-L} = \frac{1}{n(\vec{L})}\sum_{i=1}^{n(\vec{L})} x(\vec{u}_i), \bar{x}_{+L} = \frac{1}{n(\vec{L})}\sum_{i=1}^{n(\vec{L})} x(\vec{u}_i + \vec{L}),$$

and

$$\bar{y}_{-L} = \frac{1}{n(\vec{L})}\sum_{i=1}^{n(\vec{L})} y(\vec{u}_i), \bar{y}_{+L} = \frac{1}{n(\vec{L})}\sum_{i=1}^{n(\vec{L})} y(\vec{u}_i + \vec{L}). \quad \text{(C.56)}$$

We can rewrite Eq. C.55 as

$$\hat{\gamma}_c(\vec{L}) = \frac{1}{2n(\vec{L})}\sum_{i=1}^{n(\vec{L})} x(\vec{u}_i)y(\vec{u}_i) - \frac{1}{2}\bar{x}_{-L}\bar{y}_{-L}$$

$$+ \frac{1}{2n(\vec{L})}\sum_{i=1}^{n(\vec{L})} x(\vec{u}_i + \vec{L})y(\vec{u}_i + \vec{L}) - \frac{1}{2}\bar{x}_{+L}\bar{y}_{+L}$$

$$- \frac{1}{2n(\vec{L})}\sum_{i=1}^{n(\vec{L})} x(\vec{u}_i)y(\vec{u}_i + \vec{L}) + \frac{1}{2}\bar{x}_{-L}\bar{y}_{+L}$$

$$- \frac{1}{2n(\vec{L})}\sum_{i=1}^{n(\vec{L})} x(\vec{u}_i + \vec{L})y(\vec{u}_i) + \frac{1}{2}\bar{x}_{+L}\bar{y}_{-L}$$

$$+ \frac{1}{2}(\bar{x}_{-L}\bar{y}_{-L} + \bar{x}_{+L}\bar{y}_{+L} - \bar{x}_{-L}\bar{y}_{+L} - \bar{x}_{+L}\bar{y}_{-L}). \quad \text{(C.57)}$$

Using a similar notation as for the conventional variogram, we can define

$$c_{c_{-L}}(0) = \frac{1}{n(\vec{L})}\sum_{i=1}^{n(\vec{L})} x(\vec{u}_i)y(\vec{u}_i) - \bar{x}_{-L}\bar{y}_{-L},$$

and

$$c_{c_{+L}}(0) = \frac{1}{n(\vec{L})}\sum_{i=1}^{n(\vec{L})} x(\vec{u}_i + \vec{L})y(\vec{u}_i + \vec{L}) - \bar{x}_{+L}\bar{y}_{+L}, \quad \text{(C.58)}$$

where the left sides in Eq. C.58 represent local cross covariance. Substituting Eq. C.58 and the standard definition of cross covariance, we can write Eq. C.57 as

$$\hat{\gamma}_c(\vec{L}) = \frac{1}{2}\left[c_{c_{-L}}(0) + c_{c_{+L}}(0)\right] - \frac{1}{2}\left[c_c(-\vec{L}) + c_c(+\vec{L})\right]$$

$$+ \frac{1}{2}\left[(\bar{x}_{-L} - \bar{x}_{-L})(\bar{y}_{-L} - \bar{y}_{+L})\right]. \quad \text{(C.59)}$$

In most instances, we assume that $c_c(-\vec{L}) = c_c(\vec{L})$. The nonergodic cross variogram is defined as

$$\hat{\gamma}_{c_{NE}}(\vec{L}) = c_c(0) - c_c(\vec{L}). \quad \text{(C.60)}$$

By substituting Eq. C.60 in Eq. C.59, we observe,

$$\hat{\gamma}_c(\vec{L}) = \frac{1}{2}\left[c_{c_{-L}}(0) + c_{c_{+L}}(0)\right]$$

$$+ \frac{1}{2}\left[(\bar{x}_{-L} - \bar{x}_{-L})(\bar{y}_{-L} - \bar{y}_{+L})\right] + \hat{\gamma}_{c_{NE}}(\vec{L}). \quad \text{(C.61)}$$

For $\hat{\gamma}_c(\vec{L})$ to be equal to $\hat{\gamma}_{c_{NE}}(\vec{L})$, the local cross covariances must be equal to the global cross covariance, and the local mean of the first data point in a pair should be equal to the local mean of the second data point in a pair.

C.3.3 Markov-Bayes Approximation for Cross Correlation.[3] The Markov-Bayes approximation minimizes the modeling requirements. If we assume that X is the principal variable, a variable we are interested in estimating, and Y is the covariable, we can write the covariance equation for the X variable as

$$C_X(\vec{L}) = E\left[X(\vec{u})X(\vec{u} + \vec{L})\right] - E[X(\vec{u})]E\left[X(\vec{u} + \vec{L})\right]. \quad \text{(C.62)}$$

The equation for the cross covariance between the two variables is written as

$$C_{XY}(\vec{L}) = E\left[Y(\vec{u})X(\vec{u} + \vec{L})\right] - E[Y(\vec{u})]E\left[X(\vec{u} + \vec{L})\right]. \quad \text{(C.63)}$$

If we further assume a linear relationship between the two variables as

$$Y(\vec{u}) = \beta X(\vec{u}), \quad \text{(C.64)}$$

the slope, β, can be calculated as

$$\beta = \frac{C_{XY}(0)}{C_X(0)}, \quad \text{(C.65)}$$

where $C_X(0)$ represents the variance of variable X. Further, if we assume that the value of variable Y at a given location $\vec{u}$ is only influenced by the variable, X, at the same location and no other X value, we can write

$$E\left[Y(\vec{u})|X(\vec{u}) = x_u, X(\vec{u} + \vec{L}) = x_{u+L}\right],$$

which is the same as

$$E[Y(\vec{u})|X(\vec{u}) = x_u]. \quad \text{(C.66)}$$

This is similar to the Markov chain assumption because the variable value depends on another variable value at the same location only. Transition probability calculations require a similar assumption. With Eq. C.46, we can substitute Eq. C.64 in Eq. C.62. We obtain

$$C_{XY}(\vec{L}) = \beta\left\{E\left[X(\vec{u})X(\vec{u} + \vec{L})\right] - E[X(\vec{u})]E\left[X(\vec{u} + \vec{L})\right]\right\}. \quad \text{(C.67)}$$

By substituting Eqs. C.62 and C.65 in Eq. C.67,

$$C_{XY}(\vec{L}) = C_{XY}(0)\frac{C_X(\vec{L})}{C_X(0)}. \quad \text{(C.68)}$$

Eq. C.68 allows us to calculate the cross covariance based on the covariance of the X variable. Once the covariance of the X variable is modeled, we can infer cross covariance without modeling.

Nomenclature

$C(0)$ = variance
$C(\vec{L})$ = covariance at lag distance $\vec{L}$
$C(X,Y)$ = covariance between variables and
$E(X)$ = expected value
F = cumulative distribution function
$H_I(\vec{L}, x_t)$ = noncentered covariance at lag distance $\vec{L}$
i = dummy variable
$I(\vec{u}, x_t)$ = indicator variable for threshold x_t
$J(\vec{u}, x_t)$ = indicator variable
K = categorical variable
$\vec{L}$ = lag distance

$n(\vec{L})=$ number of pairs at a distance $\vec{L}$
$p=$ probability
$r(\vec{L})=$ correlation coefficient for data points $\vec{L}$ distance apart
$s^2(\vec{L})=$ variance of samples at lag distance $\vec{L}$
$Var(X)=$ variance of variable
$x(\vec{u})=$ sample at location $\vec{u}$
$X=$ variable
$Y=$ variable
$\beta=$ slope as defined in Eq. C.65
$\gamma(\vec{L})=$ variogram at lag distance $\vec{L}$
$\hat{\gamma}(\vec{L})=$ estimated variogram at lag distance $\vec{L}$

Subscripts

$c=$ cross covariance or cross variogram
$+\vec{L}=$ second data point in pairs located $\vec{L}$ distance apart
$-\vec{L}=$ first data point in pairs located $\vec{h}$ distance apart
$NE=$ nonergodic
$t=$ threshold value
$x=$ variable X
$xy=$ cross variogram/covariance between variables X and Y

References

1. Isaaks, E.H. and Srivastava, R.M.: "Spatial Continuity Measures for Probabilistic and Deterministic Geostatistics," *Math Geology* (1988) **20,** No. 4, 313.
2. Alabert, F.: "Stochastic Imaging of Spatial Distributions Using Hard and Soft Information," MS thesis, Stanford U., Stanford, California (1987).
3. Xu, W. *et al.*: "Integrating Seismic Data in Reservoir Modeling: The Collocated Cokriging Alternative," paper SPE 24742 presented at the 1992 SPE Annual Technical Conference and Exhibition, Washington, DC, 4–7 October.

Appendix D
Derivations of Conventional Estimation Methods

D.1 Simple Kriging

We start with the equation,

$$X^*(\vec{u}_0) = \lambda_0 + \sum_{i=1}^{n} \lambda_i X(\vec{u}_i). \quad \text{(D.1)}$$

D.1.1 Unbiased Condition.

$$E[X(\vec{u}_0) - X^*(\vec{u}_0)] = 0.$$

That is,

$$E\left[X(\vec{u}_0) - \lambda_0 - \sum_{i=1}^{n} \lambda_i X(\vec{u}_i)\right] = 0.$$

Expanding,

$$E[X(\vec{u}_0)] = \lambda_0 + \sum_{i=1}^{n} \lambda_i E[X(\vec{u}_i)].$$

In principle, the expected value at each location can be different from each other. We can write

$$m(\vec{u}_0) = \lambda_0 + \sum_{i=1}^{n} \lambda_i m(\vec{u}_i).$$

Therefore,

$$\lambda_0 = m(\vec{u}_0) - \sum_{i=1}^{n} \lambda_i m(\vec{u}_i). \quad \text{(D.2)}$$

If we assume first order stationarity,

$$m(\vec{u}_0) = m(\vec{u}_i) = m. \quad \text{(D.3)}$$

Substituting,

$$\lambda_0 = m\left(1 - \sum_{i=1}^{n} \lambda_i\right). \quad \text{(D.4)}$$

D.1.2 Minimum Variance Condition.

$$\hat{\sigma}_E^2 = Var[X(\vec{u}_0) - X^*(\vec{u}_0)]$$

$$= Var\left[X(\vec{u}_0) - \sum_{i=1}^{n} \lambda_i X(\vec{u}_i) - \lambda_0\right]. \quad \text{(D.5)}$$

Note that λ_0 is a constant. Also,

$$Var(X - Y) = Var(X) + Var(Y) - 2C(X, Y). \quad \text{(D.6)}$$

With Eq. D.6, we can expand Eq. D.5 as

$$\hat{\sigma}_E^2 = Var[X(\vec{u}_0)] + Var\left[\sum_{i=1}^{n} \lambda_i X(\vec{u}_i)\right]$$

$$- 2C\left[X(\vec{u}_0), \sum_{i=1}^{n} \lambda_i X(\vec{u}_i)\right]$$

$$= C(\vec{u}_0, \vec{u}_0) + \sum_{i=1}^{n} \sum_{j=1}^{n} \lambda_i \lambda_j C(\vec{u}_i, \vec{u}_j) - 2\sum_{i=1}^{n} \lambda_i C(\vec{u}_i, \vec{u}_0), \quad \text{(D.7)}$$

where $c(\vec{u}_0, \vec{u}_0)$ is the variance; $c(\vec{u}_i, \vec{u}_j)$ is the covariance between $X(\vec{u}_i)$ and $X(\vec{u}_j)$, and $c(\vec{u}_i, \vec{u}_0)$ is the covariance between $X(\vec{u}_i)$ and $X(\vec{u}_0)$.

To minimize the variance, we can write

$$\frac{\partial \hat{\sigma}_E^2}{\partial \lambda_i} = 0 = 2\sum_{j=1}^{n} \lambda_j C(\vec{u}_i, \vec{u}_j) - 2C(\vec{u}_i, \vec{u}_0) \text{ for } i = 1, \ldots, n. \quad \text{(D.8)}$$

Eq. D.8 can be written as

$$\sum_{j=1}^{n} \lambda_j C(\vec{u}_i, \vec{u}_j) = C(\vec{u}_i, \vec{u}_0) \text{ for } i = 1, \ldots, n. \quad \ldots \text{(D.9)}$$

The covariance values are estimated based on the spatial model. Eq. D.9 in matrix form can be solved to obtain λ_i values. Eq. D.9 can also be substituted in Eq. D.7.

$$\hat{\sigma}_E^2 = C(\vec{u}_0, \vec{u}_0) + \sum_{i=1}^{n} \lambda_i C(\vec{u}_i, \vec{u}_0) - 2\sum_{i=1}^{n} \lambda_i C(\vec{u}_i, \vec{u}_0),$$

and

$$\hat{\sigma}_E^2 = C(\vec{u}_0, \vec{u}_0) - \sum_{i=1}^{n} \lambda_i C(\vec{u}_i, \vec{u}_0). \quad \ldots \text{(D.10)}$$

Eq. D.10 allows the estimation of error variance.

Similar equations can also be written in terms of the variogram. The equation for an unbiased condition remains the same. However, Eq. D.9 can be written as

$$\sum_{j=1}^{n} \lambda_j \left[C(0) - \gamma(\vec{u}_i, \vec{u}_j)\right] = C(0) - \gamma(\vec{u}_i, \vec{u}_0) \text{ for } i = 2, \ldots, n.$$

$$\ldots \text{(D.11)}$$

In matrix form,

$$\begin{bmatrix} C(0) - \gamma(\vec{u}_1, \vec{u}_1) & \ldots & C(0) - \gamma(\vec{u}_1, \vec{u}_n) \\ \vdots & & \\ C(0) - \gamma(\vec{u}_n, \vec{u}_1) & \ldots & C(0) - \gamma(\vec{u}_n, \vec{u}_n) \end{bmatrix} \begin{pmatrix} \lambda_1 \\ \vdots \\ \lambda_n \end{pmatrix} =$$

$$\begin{bmatrix} C(0) - \gamma(\vec{u}_1, \vec{u}_0) \\ \vdots \\ C(0) - \gamma(\vec{u}_n, \vec{u}_0) \end{bmatrix}. \quad \ldots \text{(D.12)}$$

Eq. D.12 is very similar to Eq. 4.10, except that the covariance terms are replaced by variogram terms. Solving Eq. D.12 requires that the value of $C(0)$ be available. If we model a variogram without a sill, we can still arbitrarily define a large variogram value at a distance, which is beyond the region of interest, as $C(0)$. This will not affect the weight calculations, as shown in Eq. D.12. However, the choice of $C(0)$ would affect the calculation of error variance, as shown in Eq. D.13. Depending on the choice of $C(0)$, the error variance value will be significantly affected.

Similar to Eq. D.10, the error variance is calculated as

$$\hat{\sigma}_E^2 = \left(1 - \sum_{j=1}^{n} \lambda_i\right) C(0) + \sum_{i=1}^{n} \lambda_i \gamma(\vec{u}_i, \vec{u}_0). \quad \ldots \text{(D.13)}$$

Although Eqs. D.1 through D.13 are written in terms of a random variable X, in practice, we only have one realization of the sample. Therefore, we can write Eq. D.1 as

$$x^*(\vec{u}_0) = \lambda_0 + \sum_{i=1}^{n} \lambda_i x(\vec{u}_i), \quad \ldots \text{(D.14)}$$

where $x(\vec{u}_i)$ is the sample value at location $\vec{u}_i$.

D.2 Simple Block Kriging

Similar to point kriging, we start with the equation,

$$X_v^*(\vec{u}_0) = \lambda_0 + \sum_{i=1}^{n} \lambda_i X(\vec{u}_i). \quad \ldots \text{(D.15)}$$

D.2.1 Unbiased Condition. An unbiased condition requires that

$$E\left[X_v(\vec{u}_0) - X_v^*(\vec{u}_0)\right] = 0. \quad \ldots \text{(D.16)}$$

Similar to simple point kriging, to satisfy the unbiased condition, we write

$$\lambda_0 = m\left(1 - \sum_{i=1}^{n} \lambda_i\right). \quad \ldots \text{(D.17)}$$

D.2.2 Minimum Variance.

$$\hat{\sigma}_E^2 = Var\left[X_v(\vec{u}_0) - X_v^*(\vec{u}_0)\right]$$

$$= Var\left[X_v(\vec{u}_0) - \sum_{i=1}^{n} \lambda_i X(\vec{u}_i) - \lambda_0\right]. \quad \ldots \text{(D.18)}$$

Expanding,

$$= Var[X_v(\vec{u}_0)] + Var\left[\sum_{i=1}^{n} \lambda_i X(\vec{u}_i)\right]$$

$$- 2C\left[X_v(\vec{u}_0), \sum_{i=1}^{n} \lambda_i X(\vec{u}_i)\right]$$

$$= C_{vv}(\vec{u}_0, \vec{u}_0) + \sum_{i=1}^{n}\sum_{j=1}^{n} \lambda_i \lambda_j C(\vec{u}_i, \vec{u}_j) - 2\sum_{i=1}^{n} \lambda_i C_v(\vec{u}_i, \vec{u}_0),$$

$$\ldots \text{(D.19)}$$

where $C_{vv}(\vec{u}_o, \vec{u}_0)$ represents block variance, and $C_v(\vec{u}_i, \vec{u}_0)$ represents covariance between a block at $\vec{u}_0$ and a point at $\vec{u}_i$.

Similar to simple point kriging, taking the derivative with respect to λ_i and equating it to zero, we obtain

$$\sum_{j=1}^{n} \lambda_j C(\vec{u}_i, \vec{u}_j) = C_v(\vec{u}_i, \vec{u}_0) \text{ for } i = 1, \ldots, n. \quad \ldots \text{(D.20)}$$

Substituting Eq. D.20 in Eq. D.19, we obtain the expression for error variance as

$$\hat{\sigma}_E^2 = C_{vv}(\vec{u}_0, \vec{u}_0) - \sum_{i=1}^{n} \lambda_i C_v(\vec{u}_i, \vec{u}_0). \quad \ldots \text{(D.21)}$$

Notice the similarity between the point and the block kriging expressions except for the covariance values.

D.3 Ordinary Kriging

Similar to simple kriging, we can start with

$$X^*(\vec{u}_0) = \sum_{i=1}^{n} \lambda_i X(\vec{u}_i) + \lambda_0. \quad \ldots \text{(D.22)}$$

D.3.1 Unbiased Condition.

$$E\left[X(\vec{u}_0) - X^*(\vec{u}_0)\right] = 0.$$

By substituting Eq. D.22 in the above equation,

$$E[X(\vec{u}_0)] = \lambda_0 + \sum_{i=1}^{n} \lambda_i E[X(\vec{u}_i)]. \quad \text{(D.23)}$$

If we assume that $E[X(\vec{u}_0)] = E[X(\vec{u}_i)] = m(\vec{u}_0)$, where $m(\vec{u}_0)$ is the local mean within the search neighborhood, we can write Eq. D.23 as

$$\lambda_0 = m(\vec{u}_0)\left(1 - \sum_{i=1}^{n} \lambda_i\right). \quad \text{(D.24)}$$

Because we do not know the value of $m(\vec{u}_0)$, we can force λ_0 to be zero. This results in

$$\sum_{i=1}^{n} \lambda_i = 1. \quad \text{(D.25)}$$

This is the unbiased condition.

D.3.2 Minimum Variance Condition.

$$\hat{\sigma}_E^2 = Var[X(\vec{u}_0) - X^*(\vec{u}_0)]$$

$$= Var\left[X(\vec{u}_0) - \sum_{i=1}^{n} \lambda_i X(\vec{u}_i)\right]. \quad \text{(D.26)}$$

Expanding,

$$\hat{\sigma}_E^2 = C(\vec{u}_0, \vec{u}_0) + \sum_{i=1}^{n}\sum_{j=1}^{n} \lambda_i \lambda_j C(\vec{u}_i, \vec{u}_j) - 2\sum_{i=1}^{n} \lambda_i C(\vec{u}_i, \vec{u}_0). \quad \text{(D.27)}$$

We must minimize the error variance with a constraint defined in Eq. D.25. To achieve this, we use the Lagrange multiplier method. We define the function, F, as

$$F = \hat{\sigma}_E^2 + \left(\sum_{i=1}^{n} \lambda_i - 1\right)$$

$$= C(\vec{u}_0, \vec{u}_0) + \sum_{i=1}^{n}\sum_{j=1}^{n} \lambda_i \lambda_j C(\vec{u}_i, \vec{u}_j) - 2\sum_{i=1}^{n} \lambda_i C(\vec{u}_i, \vec{u}_0)$$

$$+ 2\mu\left(\sum_{i=1}^{n} \lambda_i - 1\right), \quad \text{(D.28)}$$

where μ is a Lagrange parameter.

To minimize the error variance, we can write

$$\frac{\partial F}{\partial \lambda_i} = 0 = 2\sum_{j=1}^{n} \lambda_j C(\vec{u}_i, \vec{u}_j) + 2\mu - 2C(\vec{u}_i, \vec{u}_0),$$

$$\text{for } i = 1, \ldots, n, \quad \text{(D.29)}$$

and

$$\frac{\partial F}{\partial \mu} = 0 = \sum_{i=1}^{n} \lambda_i - 1. \quad \text{(D.30)}$$

Eq. D.29 can be written as

$$\sum_{j=1}^{n} \lambda_j C(\vec{u}_i, \vec{u}_j) + \mu = C(\vec{u}_i, \vec{u}_0) \text{ for } i = 1, \ldots, n. \quad \text{(D.31)}$$

By solving Eqs. D.30 and D.31 simultaneously, we can obtain the values of λ_i and μ. The matrix form is provided in Eq. 4.21, which results in the $(n+1) \times (n+1)$ matrix.

Once the weights are determined, we can estimate the value at the unsampled location by

$$x^*(\vec{u}_0) = \sum_{i=1}^{n} \lambda_i x(\vec{u}_i). \quad \text{(D.32)}$$

Further, we can simplify the expression for $\hat{\sigma}_E^2$ by substituting Eq. D.31 in Eq. D.27. We then obtain

$$\hat{\sigma}_E^2 = C(\vec{u}_0, \vec{u}_0) + \sum_{i=1}^{n} \lambda_i [C(\vec{u}_i, \vec{u}_o) - \mu] - 2\sum_{i=1}^{n} \lambda_i C(\vec{u}_i, \vec{u}_0)$$

$$= C(\vec{u}_0, \vec{u}_0) - \sum_{i=1}^{n} \lambda_i C(\vec{u}_i, \vec{u}_0) - \mu. \quad \text{(D.33)}$$

The error variance will always be greater than or equal to zero.

D.4 Cokriging

We first develop the set of equations for two variables. We assume they are the principal variable and the covariable.

The estimation equation can be written as

$$X^*(\vec{u}_0) = \lambda_0 + \sum_{i=1}^{n} \lambda_{X_i} X(\vec{u}_{X_i}) + \sum_{k=1}^{m} \lambda_{Y_k} Y(\vec{u}_{Y_k}). \quad \text{(D.34)}$$

D.4.1 Unbiased Condition.

$$E[X(\vec{u}_0) - X^*(\vec{u}_0)] = 0.$$

By substituting Eq. D.34 in the above equation, we obtain

$$E[X(\vec{u}_0)] = E\left[\lambda_0 + \sum_{i=1}^{n} \lambda_{X_i} X(\vec{u}_{X_i}) + \sum_{k=1}^{m} \lambda_{Y_k} Y(\vec{u}_{Y_k})\right]. \quad \text{(D.35)}$$

Invoking the first order stationarity,

$$m_X = \lambda_0 + m_X \sum_{i=1}^{n} \lambda_{X_i} + m_Y \sum_{k=1}^{m} \lambda_{Y_k},$$

where m_X and m_Y are the means of the X and Y variables, respectively. Therefore,

$$\lambda_0 = m_X\left(1 - \sum_{i=1}^{n} \lambda_{X_i}\right) - m_Y \sum_{k=1}^{m} \lambda_{Y_k}. \quad \text{(D.36)}$$

For a simple cokriging system, Eq. D.36 constitutes an unbiased condition. If the means are not known, we can force λ_0 to be zero by the equation,

$$\sum_{i=1}^{n} \lambda_{X_i} = 1, \text{and} \sum_{k=1}^{m} \lambda_{Y_k} = 0. \quad \text{(D.37)}$$

These two constraints in Eq. D.37 will result in a traditional ordinary cokriging system. One of the drawbacks of Eq. D.37 is that it forces some of the weight values, assigned to the covariable, to be negative because all weight values must add to zero. This can result in a negative estimate at a particular unsampled location, which might not have physical meaning. To avoid forcing the negative weight values, one option is to define

$$\sum_{i=1}^{n} \lambda_{X_i} + \sum_{k=1}^{m} \lambda_{Y_k} = 1,$$

and

$$\lambda_0 = (-m_Y + m_X) \sum_{k=1}^{m} \lambda_{Y_k}. \quad \text{(D.38)}$$

Eq. D.38 provides the two constraints, such that all the weight values add to one, and λ_0 requires the knowledge of the means. If we substitute Eq. D.37 in Eq. D.34, we obtain

$$X^*(\vec{u}_o) = \sum_{i=1}^{n} \lambda_{X_i} X(\vec{u}_{X_i}) + \sum_{k=1}^{m} \lambda_{Y_k} Y(\vec{u}_{Y_k}). \quad \text{(D.39)}$$

On the other hand, if we substitute Eq. D.38 in Eq. D.34, we obtain

$$X^*(\vec{u}_o) = \sum_{i=1}^{n} \lambda_{X_i} X(\vec{u}_{X_i}) + \sum_{k=1}^{m} \lambda_{Y_k} \left[Y(\vec{u}_{Y_k}) - m_Y + m_X \right]. \quad \text{(D.40)}$$

Eq. D.38 is another way in which the unbiased condition is satisfied. It provides a constraint in an ordinary cokriging system. Depending on whether Eq. D.37 or Eq. D.38 is chosen as an unbiased constraint, the estimation at the unsampled location is done with either Eq. D.39 or Eq. D.40, respectively.

D.4.2 Minimum Variance. Depending on the unbiased condition, a different expression for minimum variance can be established. We develop the expression based on the unbiased condition provided in Eq. D.37.

To minimize variance, we must minimize

$$\hat{\sigma}_E^2 = Var\left[X^*(\vec{u}_0) - \sum_{i=1}^{n} \lambda_{X_i} X(\vec{u}_{X_i}) - \sum_{k=1}^{m} \lambda_{Y_k} Y(\vec{u}_{Y_k}) \right]. \quad \text{(D.41)}$$

Noting that

$$Var(X - Y - Z) = Var(X) + Var(Y) + Var(Z) - 2C(X,Y) + 2C(Y,Z) - 2C(X,Z)$$

we can expand Eq. D.41 as

$$\hat{\sigma}_E^2 = C_X(\vec{u}_0, \vec{u}_0) + \sum_{j=1}^{n} \sum_{i=1}^{n} \lambda_{X_i} \lambda_{X_j} C_X(\vec{u}_{X_i}, \vec{u}_{X_j})$$
$$+ \sum_{k=1}^{m} \sum_{l=1}^{m} \lambda_{Y_k} \lambda_{Y_l} C_Y(\vec{u}_{Y_k}, \vec{u}_{Y_l})$$
$$- 2\sum_{i=1}^{n} \lambda_{X_i} C_X(\vec{u}_0, \vec{u}_{X_i}) - 2\sum_{k=1}^{m} \lambda_{Y_k} C_c(\vec{u}_0, \vec{u}_{Y_k})$$
$$+ 2\sum_{i=1}^{n} \sum_{k=1}^{m} \lambda_{X_i} \lambda_{Y_k} C_c(\vec{u}_{X_i}, \vec{u}_{Y_k}). \quad \text{(D.42)}$$

The error variance equation defined in Eq. D.42 must be minimized with two constraints, as described in Eq. D.37. With the Lagrange multiplier method, we can define the function, F, as

$$F = \hat{\sigma}_E^2 + 2\mu_X \left(\sum_{i=1}^{n} \lambda_{X_i} - 1 \right) + 2\mu_Y \left(\sum_{k=1}^{m} \lambda_{Y_k} \right). \quad \text{(D.43)}$$

Taking the derivative with respect to λ and μ, we obtain

$$\sum_{j=1}^{n} \lambda_{X_j} C_X(\vec{u}_{X_i}, \vec{u}_{X_j}) + \sum_{k=1}^{m} \lambda_{Y_k} C_c(\vec{u}_{X_i}, \vec{u}_{Y_k}) + \mu_X = C_X(\vec{u}_{X_i}, \vec{u}_o),$$

for $i = 1, \ldots, n$;

$$\sum_{i=1}^{n} \lambda_{X_i} C_C(\vec{u}_{X_i}, \vec{u}_{Y_k}) + \sum_{l=1}^{m} \lambda_{Y_l} C_Y(\vec{u}_{Y_k}, \vec{u}_{Y_l}) + \mu_Y = C_C(\vec{u}_{Y_k}, \vec{u}_o),$$

for $k = 1, \ldots, m$;

$$\sum_{i=1}^{n} \lambda_{X_i} = 1;$$

and

$$\sum_{k=1}^{m} \lambda_{Y_k} = 0. \quad \text{(D.44)}$$

In Eq. D.44, C_X represents the covariance for variable X; C_Y represents the covariance for variable Y, and C_C represents the cross covariance between X and Y. In matrix form, Eq. D.44 can be written as

$$\begin{bmatrix} C_X(\vec{u}_{X_1}, \vec{u}_{X_1}) & \cdots & C_X(\vec{u}_{X_1}, \vec{u}_{X_n}) & C_C(\vec{u}_{X_1}, \vec{u}_{Y_1}) & \cdots & C_C(\vec{u}_{X_1}, \vec{u}_{Y_m}) & 1 & 0 \\ \vdots & & \vdots & \vdots & & \vdots & \vdots & \vdots \\ C_X(\vec{u}_{X_n}, \vec{u}_{X_1}) & \cdots & C_X(\vec{u}_{X_n}, \vec{u}_{X_n}) & C_C(\vec{u}_{X_n}, \vec{u}_{Y_1}) & \cdots & C_C(\vec{u}_{X_n}, \vec{u}_{Y_m}) & 1 & 0 \\ C_C(\vec{u}_{Y_1}, \vec{u}_{X_1}) & \cdots & C_C(\vec{u}_{Y_1}, \vec{u}_{X_n}) & C_Y(\vec{u}_{Y_1}, \vec{u}_{Y_1}) & \cdots & C_Y(\vec{u}_{Y_1}, \vec{u}_{Y_{1m}}) & 0 & 1 \\ \vdots & & \vdots & \vdots & & \vdots & \vdots & \vdots \\ C_C(\vec{u}_{Y_m}, \vec{u}_{X_1}) & \cdots & C_C(\vec{u}_{Y_m}, \vec{u}_{X_n}) & C_Y(\vec{u}_{Y_m}, \vec{u}_{Y_1}) & \cdots & C_Y(\vec{u}_{Y_m}, \vec{u}_{Y_m}) & 0 & 1 \\ 1 & \cdots & 1 & 0 & \cdots & 0 & 0 & 0 \\ 0 & \cdots & 0 & 1 & \cdots & 1 & 0 & 0 \end{bmatrix}$$

$$\begin{bmatrix} \lambda_{X_1} \\ \vdots \\ \lambda_{X_n} \\ \lambda_{Y_1} \\ \vdots \\ \lambda_{X_m} \\ \mu_X \\ \mu_Y \end{bmatrix} = \begin{bmatrix} C_X(\vec{u}_{X_1}, \vec{u}_0) \\ \vdots \\ C_X(\vec{u}_{X_n}, \vec{u}_0) \\ C_C(\vec{u}_{Y_1}, \vec{u}_0) \\ \vdots \\ C_C(\vec{u}_{Y_m}, \vec{u}_0) \\ 1 \\ 0 \end{bmatrix}. \quad \text{(D.45)}$$

Eq. D.45 can be solved for λ and μ. Further, by substituting Eq. D.44 in Eq. D.42, we obtain

$$\hat{\sigma}_E^2 = C_X(\vec{u}_o, \vec{u}_o) + \sum_{i=1}^{n} \lambda_{X_i} C_X(\vec{u}_{X_i}, \vec{u}_o) - \sum_{k=1}^{m} \lambda_{Y_k} C_c(\vec{u}_{Y_k}, \vec{u}_o) - \mu_X. \quad \text{(D.46)}$$

As before, the error variance is always greater than or equal to zero.

If we choose Eq. D.38 as our unbiased condition, the overall development of equations remains the same. The final expression, in the matrix form, can be written as

$$\begin{bmatrix} C_X(\vec{u}_{X_1},\vec{u}_{X_1}) & \cdots & C_X(\vec{u}_{X_1},\vec{u}_{X_n}) & C_C(\vec{u}_{X_1},\vec{u}_{Y_1}) & \cdots & C_C(\vec{u}_{X_1},\vec{u}_{Y_m}) & 1 \\ \vdots & & \vdots & \vdots & & \vdots & \vdots \\ C_X(\vec{u}_{X_n},\vec{u}_{X_1}) & \cdots & C_X(\vec{u}_{X_n},\vec{u}_{X_n}) & C_C(\vec{u}_{X_n},\vec{u}_{Y_1}) & \cdots & C_C(\vec{u}_{X_n},\vec{u}_{Y_m}) & 1 \\ C_C(\vec{u}_{Y_1},\vec{u}_{X_1}) & \cdots & C_C(\vec{u}_{Y_1},\vec{u}_{X_n}) & C_Y(\vec{u}_{Y_1},\vec{u}_{Y_1}) & \cdots & C_Y(\vec{u}_{Y_1},\vec{u}_{Y_{1m}}) & 1 \\ \vdots & & \vdots & \vdots & & \vdots & \vdots \\ C_C(\vec{u}_{Y_m},\vec{u}_{X_1}) & \cdots & C_C(\vec{u}_{Y_m},\vec{u}_{X_n}) & C_Y(\vec{u}_{Y_m},\vec{u}_{Y_1}) & \cdots & C_Y(\vec{u}_{Y_m},\vec{u}_{Y_m}) & 1 \\ 1 & \cdots & 1 & 0 & \cdots & 0 & 0 \end{bmatrix}$$

$$\begin{bmatrix} \lambda_{X_1} \\ \vdots \\ \lambda_{X_n} \\ \lambda_{Y_1} \\ \vdots \\ \lambda_{X_m} \\ \mu \end{bmatrix} = \begin{bmatrix} C_X(\vec{u}_{X_1},\vec{u}_0) \\ \vdots \\ C_X(\vec{u}_{X_n},\vec{u}_0) \\ C_C(\vec{u}_{Y_1},\vec{u}_0) \\ \vdots \\ C_C(\vec{u}_{Y_m},\vec{u}_0) \\ 1 \end{bmatrix}. \qquad \text{(D.47)}$$

This equation is similar to Eq. D.45, except the matrix size is reduced by one because one constraint, with respect to weight, must be incorporated. The error variance is calculated as

$$\hat{\sigma}_E^2 = C_X(\vec{u}_0,\vec{u}_0) + \sum_{i=1}^{n}\lambda_{X_i}C_X(\vec{u}_{X_i},\vec{u}_0) - \sum_{k=1}^{m}\lambda_{Y_k}C_c(\vec{u}_{Y_k},\vec{u}_0) - \mu. \qquad \text{(D.48)}$$

Note that to estimate a value under Eqs. D.47 and D.48, we must use Eq. D.40. In practice, this translates into

$$X^*(\vec{u}_0) = \sum_{i=1}^{n}\lambda_{X_i}X(\vec{u}_{X_i}) + \sum_{k=1}^{m}\lambda_{Y_k}\left[Y(\vec{u}_{Y_k}) - m_Y + m_X\right]. \qquad \text{(D.49)}$$

As stated before, the means must be known to use Eq. D.49.

We can also develop a similar set of equations for a simple cokriging algorithm. This requires the use of Eq. D.36 as an unbiased condition. There are no constraints on the assigned weights.

D.4.3 Collocated Cokriging. In this method, we assume that to estimate $X^*(\vec{u}_0)$, we must only use the covariable value, $Y^*(\vec{u}_0)$. Because $m = 1$, Eq. D.40 reduces to

$$X^*(\vec{u}_0) = \sum_{i=1}^{n}\lambda_{X_i}X(\vec{u}_{X_i}) + \lambda_{Y_0}\left[Y(\vec{u}_0) - m_y + m_X\right], \qquad \text{(D.50)}$$

where m_X and m_Y are assumed to be known. The unbiased condition requires that

$$\sum_{i=1}^{n}\lambda_{X_i} + \lambda_{Y_0} = 1. \qquad \text{(D.51)}$$

Applying the minimum variance constraint, and following a similar development like conventional cokriging, we obtain

$$\begin{bmatrix} C_X(\vec{u}_{X_1},\vec{u}_{X_1}) & \cdots & C_X(\vec{u}_{X_1},\vec{u}_{X_n}) & C_C(\vec{u}_{X_1},\vec{u}_0) & 1 \\ \vdots & & \vdots & \vdots & \vdots \\ C_X(\vec{u}_{X_n},\vec{u}_{X_1}) & \cdots & C_X(\vec{u}_{X_n},\vec{u}_{X_n}) & C_C(\vec{u}_{X_n},\vec{u}_0) & 1 \\ C_C(\vec{u}_0,\vec{u}_{X_1}) & \cdots & C_C(\vec{u}_0,\vec{u}_{X_n}) & C_Y(\vec{u}_0,\vec{u}_0) & 1 \\ 1 & \cdots & 1 & 1 & 0 \end{bmatrix}$$

$$\begin{bmatrix} \lambda_{X_1} \\ \vdots \\ \lambda_{X_n} \\ \lambda_{Y_0} \\ \mu \end{bmatrix} = \begin{bmatrix} C_X(\vec{u}_{X_1},\vec{u}_0) \\ \vdots \\ C_X(\vec{u}_{X_n},\vec{u}_0) \\ C_C(\vec{u}_0,\vec{u}_0) \\ 1 \end{bmatrix}. \qquad \text{(D.52)}$$

For n principal variable samples, Eq. D.52 results in the $(n+2)\times(n+2)$ matrix. This is not significantly different from the ordinary kriging matrix. Further, we only need the variance of variable Y, $C_Y(\vec{u}_0,\vec{u}_0)$. No spatial model for variable Y is needed. We do need the cross covariance model to solve Eq. D.52. However, as explained in Appendix C, we can infer the cross covariance model as

$$C_C(\vec{h}) = C_C(0)\frac{C_X(\vec{h})}{C_X(0)}. \qquad \text{(D.53)}$$

That is, if we know the covariance model for the principal variable, $C_X(\vec{h})$, we can calculate cross covariance at the same lag distance with Eq. D.53. We only need to estimate the cross covariance at lag distance zero, which is easy to estimate. As a result, no independent modeling of cross covariance is necessary.

By combining Eqs. D.53 and D.52, both the modeling and the computational effort is effectively reduced to the corresponding ordinary kriging method. At the same time, information gathered from the covariable is used in the estimation process.

D.4.4 Generalized Cokriging Procedure. In the previous section, we only considered a cokriging procedure with one principal and covariable. The procedure can be generalized when the value of the principal variable is estimated with multiple covariables.

If we assume that we are interested in estimating $X_0^*(\vec{u}_0)$ at an unsampled location using L covariables, we can write the equation for estimation as

$$X_0^*(\vec{u}_0) = \sum_{i=1}^{n_0}\lambda_i^0 X_0(\vec{u}_i) + \sum_{l=1}^{L}\sum_{j=1}^{n_l}\lambda_j^l X_l(\vec{u}_j), \qquad \text{(D.54)}$$

where the superscript for λ represents the weight assigned to a particular variable. For example, λ_i^0 represents the weight assigned to a sample of a principal variable located at $\vec{u}_i$. $X_l(\vec{u}_j)$ represents a covariable sample of variable l located at $\vec{u}_j$. n_l is the number of samples of variable l within the search neighborhood. X_0 represents the principal variable.

Eq. D.54 can also be written as

$$X_0^*(\vec{u}_0) = \sum_{l=1}^{L}\sum_{j=1}^{n_l}\lambda_j^l X_l(\vec{u}_j). \quad \text{(D.55)}$$

To estimate the weights assigned to the individual sample points, we impose the Minimum Variance Unbiased Estimation (MVUE) technique.

Unbiasedness. Imposing the unbiased requirement, we observe

$$E[X(\vec{u}_0) - X_0^*(\vec{u}_0)] = 0. \quad \text{(D.56)}$$

By substituting Eq. D.55 in Eq. D.56,

$$m_0 = \sum_{i=1}^{n_0}\lambda_i^0 m_0 - \sum_{l=1}^{L} m_l \sum_{j=1}^{n_l}\lambda_j^l = 0. \quad \text{(D.57)}$$

One way to satisfy the unbiased condition is to require that

$$\sum_{i=1}^{n_0}\lambda_i^0 = 1, \text{and} \sum_{i=1}^{n_l}\lambda_i^l = 0, \text{ for } l = 1,2,\ldots L. \quad \text{(D.58)}$$

Other conditions are also possible; however, we assume Eq. D.58 to be used for further analysis.

Minimum Variance. The variance between the true and the estimated value is defined as

$$\hat{\sigma}_E^2 = Var\left[X_0(\vec{u}_0) - \sum_{l=0}^{L}\sum_{i=1}^{n_l}\lambda_i^l X_l(\vec{u}_i)\right]. \quad \text{(D.59)}$$

This is the variance that must be minimized with the constraints imposed by Eq. D.58.

By expanding Eq. D.59, we obtain

$$\hat{\sigma}_E^2 = C_{00}(\vec{u}_0, \vec{u}_0) + Var\left[\sum_{l=0}^{L}\sum_{i=1}^{n_l}\lambda_i^l X_l(\vec{u}_i)\right]$$

$$- 2C\left[X_o(\vec{u}_0), \sum_{l=0}^{L}\sum_{i=1}^{n_l}\lambda_i^l X_l(\vec{u}_i)\right]$$

$$= C_{00}(\vec{u}_0, \vec{u}_0) + \sum_{l=0}^{L}\sum_{m=0}^{L}\sum_{i=1}^{n_l}\sum_{j=1}^{n_m}\lambda_i^l\lambda_j^m C_{lm}(\vec{u}_i, \vec{u}_j)$$

$$- 2\sum_{l=0}^{L}\sum_{i=1}^{n_l}\lambda_i^l C_{l0}(\vec{u}_i, \vec{u}_0). \quad \text{(D.60)}$$

We must minimize Eq. D.60 with constraints defined by Eq. D.58. Therefore, we can define function F as

$$F = \hat{\sigma}_E^2 + 2\mu_0\left(\sum_{i=1}^{n_0}\lambda_i^0 - 1\right) + \sum_{l=1}^{L}2\mu_l\sum_{i=1}^{n_l}\lambda_i^l. \quad \text{(D.61)}$$

Differentiating F with respect to λ_i^l, we obtain

$$\frac{\partial F}{\partial \lambda_i^l} = \sum_{m=0}^{L}\sum_{j=1}^{n_m}\lambda_j^m C_{lm}(\vec{u}_i, \vec{u}_j) - C_{l0}(\vec{u}_i, \vec{u}_0) + \mu_l = 0.$$

Simplifying,

$$\sum_{m=0}^{L}\sum_{j=1}^{n_m}\lambda_j^m C_{lm}(\vec{u}_i, \vec{u}_j) + \mu_l = C_{l0}(\vec{u}_i, \vec{u}_0),$$

$$\text{for } i = 1, \ldots, n_l$$

and

$$l = 0, 1, \ldots, L. \quad \text{(D.62)}$$

Differentiating F with respect to μ_l, we obtain

$$\sum_{i=1}^{n_0}\lambda_i^0 = 1, \quad \text{(D.63)}$$

and

$$\sum_{i=1}^{n_l}\lambda_i^l = 0, \text{ for } l = 1, \ldots, L. \quad \text{(D.64)}$$

In matrix form, the equation can be written as

$$\begin{bmatrix} \vec{C}_{00} & \vec{C}_{01} & \ldots & \vec{C}_{0L} & I_0^T \\ \vec{C}_{10} & \vec{C}_{11} & \ldots & \vec{C}_{1L} & I_1^T \\ \vdots & \vdots & & & \vdots \\ \vec{C}_{L0} & \vec{C}_{L1} & \ldots & \vec{C}_{LL} & I_L^T \\ I_0 & I_1 & \ldots & I_L & 0 \end{bmatrix} \begin{bmatrix} \wedge^0 \\ \wedge^1 \\ \vdots \\ \wedge^L \\ \vec{\mu} \end{bmatrix} = \begin{bmatrix} \vec{C}_{00}(\vec{u}_0) \\ \vec{C}_{01}(\vec{u}_0) \\ \vdots \\ \vec{C}_{0L}(\vec{u}_0) \\ \Delta \end{bmatrix}. \quad \text{(D.65)}$$

In Eq. D.65, each term on the left side of the matrix itself is a matrix. For example,

$$\vec{C}_{lm} = \begin{bmatrix} C_{lm}(\vec{u}_1, \vec{u}_1) & \ldots & C_{lm}(\vec{u}_1, \vec{u}_{n_l}) \\ & \vdots & \\ C_{lm}(\vec{u}_{n_m}, \vec{u}_1) & \ldots & C_{lm}(\vec{u}_{n_m}, \vec{u}_{n_l}) \end{bmatrix}, \quad \text{(D.66)}$$

where $\vec{C}_{lm}$ is a matrix with the dimensions of $n_l \times n_m$.

$$I_l = \begin{bmatrix} 00 & \ldots & 0 \\ \vdots & & \vdots \\ 11 & \ldots & 1 \\ \vdots & & \vdots \\ 0 & \ldots & 0 \end{bmatrix} \leftarrow l^{th} \text{ row}, \quad \text{(D.67)}$$

where I_l is the $n_l \times n_m$ matrix, and I_l^T is the transpose of I_l, which can be written as

$$I_l^T = \begin{bmatrix} 0 & \ldots & 1 & \ldots & 0 \\ 0 & \ldots & 1 & \ldots & 0 \\ \vdots & & & & \vdots \\ 0 & \ldots & 1 & \ldots & 0 \end{bmatrix}. \quad \text{(D.68)}$$

$\uparrow l^{th}$ column

Each term in the vector on the left side is a vector itself. For example,

$$\wedge^l = \begin{bmatrix} \lambda_1^l \\ \vdots \\ \lambda_{n_l}^l \end{bmatrix} \text{and } \vec{\mu} = \begin{bmatrix} \mu_0 \\ \mu_1 \\ \vdots \\ \mu_L \end{bmatrix}. \quad \text{(D.69)}$$

The individual terms on the right side of the matrix are represented by

$$\vec{C}_{0l}(\vec{u}_0) = \begin{bmatrix} \vec{C}_{0l}(\vec{u}_0, \vec{u}_1) \\ \vdots \\ \vec{C}_{0l}(\vec{u}_0, \vec{u}_{n_l}) \end{bmatrix}, \text{and } \Delta = \begin{pmatrix} 1 \\ 0 \\ \vdots \\ 0 \end{pmatrix}. \quad \text{(D.70)}$$

Each $\wedge^l$ is a vector of size $n_l \times 1$; $\vec{\mu}$ is a size of $(L+1)\times 1$, and Δ is also a vector of size $(L+1)\times 1$. To generalize, the size of the matrix on the left is

$$\left(\sum_{l=0}^{L} n_l + L + 1\right) \times \left(\sum_{l=0}^{L} n_l + L + 1\right),$$

and the size of the vector on the right is $\left(\sum_{l=0}^{L} n_l + L + 1\right) \times 1.$

Obviously, as more covariables are used, in addition to an extra modeling effort, the computational effort also increases because of the bigger matrix size.

In addition to estimating the value of the variable, we can also estimate the error variance by substituting Eq. D.62 in Eq. D.60. The final expression can be written as

$$\hat{\sigma}_E^2 = C_{00}(\vec{u}_0, \vec{u}_0) - \sum_{l=0}^{L} \sum_{i=1}^{n_l} \lambda_i^l C_{l0}(\vec{u}_0, \vec{u}_i) - \mu_0. \quad \text{(D.71)}$$

A further generalization of the equations, in which all the variables at the unsampled locations are estimated simultaneously is possible.[1] The procedure is powerful if all the variables are sampled at each sampled location, and all the variables are unsampled at each unsampled location. If some of the variables are undersampled, additional constraints might be necessary to solve the system of equations.[1] Further, it might not be possible to estimate only a few of the variables at the unsampled locations (e.g., collocated co-kriging). Considering that cokriging is mainly useful when a principal variable is undersampled, the utility of this generalized procedure remains uncertain.

D.5 Universal Kriging

Universal kriging is applicable when estimation is required in the presence of a trend. We assume that the value at the unsampled location can be estimated as

$$X^*(\vec{u}_0) = \sum_{i=1}^{n} \lambda_i X(\vec{u}_i). \quad \text{(D.72)}$$

The trend in the data can be defined by

$$m(\vec{u}) = \sum_{l=0}^{L} a_l f_l(\vec{u}), \quad \text{(D.73)}$$

where a_l is a coefficient, and $f_l(\vec{u})$ is a function of location $\vec{u}$. By definition, $f_0(\vec{u})$ is assumed to be one. For example, if we assume a quadratic trend, we can write

$$m(\vec{u}) = a_0 + a_1\vec{u} + a_2\vec{u}^2, \quad \text{(D.74)}$$

where $f_1(\vec{u}) = \vec{u}$ and $f_2(\vec{u}) = \vec{u}^2$.

D.5.1 Unbiased Condition. The unbiased condition requires that

$$E\left[X^*(\vec{u}_0) - \sum_{i=1}^{n} \lambda_i X(\vec{u}_i)\right] = 0, \quad \text{(D.75)}$$

and

$$m(\vec{u}_0) - \sum_{i=1}^{n} \lambda_i m(\vec{u}_i) = 0. \quad \text{(D.76)}$$

By substituting Eq. D.73 in Eq. D.76, we obtain

$$\sum_{l=0}^{L} a_l \left[f_l(\vec{u}_0) - \sum_{i=1}^{n} \lambda_i f_l(\vec{u}_i)\right] = 0. \quad \text{(D.77)}$$

To satisfy Eq. D.77, we can impose the condition,

$$f_l(\vec{u}_0) = \sum_{i=1}^{n} \lambda_i f_l(\vec{u}_i) \text{ for } l=0,1,\ldots,L. \quad \text{(D.78)}$$

The $L+1$ constraints defined in Eq. D.78 must be used to minimize variance.

D.5.2 Minimum Variance. Minimizing variance requires $Var\left[X^*(\vec{u}_0) - \sum_{i=1}^{n} \lambda_i X(\vec{u}_i)\right]$ to be minimized. Expanding,

$$\hat{\sigma}_E^2 = C(\vec{u}_0, \vec{u}_0) + \sum_{i=1}^{n} \sum_{j=1}^{n} \lambda_i \lambda_j C(\vec{u}_i, \vec{u}_j) - 2\sum_{i=1}^{n} \lambda_i C(\vec{u}_i, \vec{u}_0). \quad \text{(D.79)}$$

Eq. D.79 must be minimized with the constraint, Eq. D.78. We can write the function, F, as

$$F = \hat{\sigma}_E^2 - \sum_{l=0}^{L} 2\mu_l \left[f_l(\vec{u}_0) - \sum_{i=1}^{n} \lambda_i f_l(\vec{u}_i)\right]. \quad \text{(D.80)}$$

Taking the derivative, with respect to λ_i and μ_l, we obtain

$$\frac{\partial F}{\partial \lambda_i} = 2\sum_{j=1}^{n} \lambda_j C(\vec{u}_i, \vec{u}_j) + 2\sum_{l=0}^{L} 2\mu_l f_l(\vec{u}_i) - 2C(\vec{u}_i, \vec{u}_0) = 0, \text{ for } i=1,\ldots,n;$$

and

$$\frac{\partial F}{\partial \mu_l} = \sum_{i=1}^{n} \lambda_i f_l(\vec{u}_i) - f_l(\vec{u}_0) = 0, \text{ for } l=0,\ldots,L. \quad \text{(D.81)}$$

Eq. D.81 can be simplified as

$$\sum_{j=1}^{n} \lambda_i C(\vec{u}_i, \vec{u}_j) + \sum_{l=0}^{L} \mu_l f_l(\vec{u}_i) = C(\vec{u}_i, \vec{u}_0), \text{ for } i=1,\ldots,n; \quad \text{(D.82)}$$

and

$$\sum_{i=1}^{n} \lambda_i f_l(\vec{u}_i) = f_l(\vec{u}_0), \text{ for } l=0,1,\ldots,L. \quad \text{(D.83)}$$

In matrix form, Eqs. D.82 and D.83 can be written as

$$\begin{bmatrix} C(\vec{u}_1,\vec{u}_1) & \ldots & C(\vec{u}_1,\vec{u}_n) & f_0(\vec{u}_1) & \ldots & f_L(\vec{u}_1) \\ \vdots & & & \vdots & & \\ C(\vec{u}_n,\vec{u}_1) & \ldots & C(\vec{u}_n,\vec{u}_n) & f_0(\vec{u}_n) & \ldots & f_L(\vec{u}_n) \\ f_0(\vec{u}_1) & \ldots & f_0(\vec{u}_n) & 0 & \ldots & 0 \\ \vdots & & & \vdots & & \\ f_L(\vec{u}_1) & \ldots & f_L(\vec{u}_n) & 0 & \ldots & 0 \end{bmatrix} \begin{bmatrix} \lambda_1 \\ \vdots \\ \lambda_n \\ \mu_0 \\ \vdots \\ \mu_L \end{bmatrix}$$

$$= \begin{bmatrix} C(\vec{u}_1,\vec{u}_0) \\ \vdots \\ C(\vec{u}_n,\vec{u}_0) \\ f_0(\vec{u}_0) \\ \vdots \\ f_L(\vec{u}_0) \end{bmatrix}. \qquad \text{(D.84)}$$

Eq. D.84 can be solved to obtain the values of λ_i and μ_l. Notice that Eq. D.84 reduces to an ordinary kriging equation, if $L=0$, and $f_0(\vec{u})$ is one. The error variance can be estimated as

$$\hat{\sigma}_E^2 = C(\vec{u}_0,\vec{u}_0) - \sum_{i=1}^{n}\lambda_i C(\vec{u}_i,\vec{u}_0) - \sum_{l=0}^{L}\mu_l f_l(\vec{u}_0). \qquad \text{(D.85)}$$

D.5.3 Generalized Covariance Function. Application of the universal kriging method requires the knowledge of a covariance model for residuals, as well as the type of trend present in the original data set. Determining both can be difficult.

Matheron,[2] and later Delfiner,[3] provided an alternative to overcome this problem. Instead of using a variogram as a spatial function, they proposed a more generalized intrinsic random function of order *k*. The principle behind this random function is the idea that any trend in the data can be eliminated by proposing an appropriate function. A variogram is a special case of this generalized function. We provide a brief outline of this approach.

The function, $K(\vec{h})$, is a generalized covariance function of order *k*, if the variance of any linear combination of value $X(\vec{u}_i)$ with weight value λ_i is

$$\left[\sum_{i=0}^{n}\lambda_i X(\vec{u}_i)\right] = \sum_{i=0}^{n}\sum_{j=0}^{n}\lambda_i\lambda_j K(\vec{u}_i,\vec{u}_j), \qquad \text{(D.86)}$$

for any set of weights constrained to

$$\sum_{i=0}^{n}\lambda_i f_l(\vec{u}_i) = 0, \text{ for } l=0,1,\ldots,L. \qquad \text{(D.87)}$$

In the universal kriging equation, we showed that

$$\sum_{i=1}^{n}\lambda_i f_l(\vec{u}_i) = f_l(\vec{u}_0), \text{ for } l=0,1,\ldots,L. \qquad \text{(D.83)}$$

Eq. D.87 can be written by simply assigning $\lambda_0=-1$, and combining the left and right sides of Eq. D.83.

We define variable *Y* as

$$Y = X^*(\vec{u}_0) - \sum_{i=1}^{n}\lambda_i X(\vec{u}_i) = -\sum_{i=0}^{n}\lambda_i X(\vec{u}_i), \qquad \text{(D.88)}$$

with λ_i subjected to the same constraints, as explained in Eq. D.87.

For example, if $L=0$, we can define $Y = X(\vec{u}) - X(\vec{u}+\vec{h})$, then the coefficients of $X(\vec{u})$ and $X(\vec{u}+\vec{h})$ satisfy the requirement,

$$\sum_{i=0}^{1}\lambda_i = 0 = -1+1. \qquad \text{(D.89)}$$

Similarly, if $L=1$, we can define *Y* as

$$Y = X(\vec{u}) - 2X(\vec{u}+\vec{h}) + X(u+2\vec{h}). \qquad \text{(D.90)}$$

This equation satisfies the requirement,

$$\sum_{i=0}^{2}\lambda_i = 0 = 1-2+1, \qquad \text{(D.91)}$$

and, for a linear trend where $f_1(\vec{u}) = \vec{u}$,

$$\sum_{i=0}^{n}\lambda_i f_1(\vec{u}) = 0 = 1(\vec{u}) - 2(\vec{u}+\vec{h}) + (u+2\vec{h}) = 0. \qquad \text{(D.92)}$$

A similar equation can be written for a quadratic trend, where $f_1(\vec{u}) = \vec{u}$ and $f_2(\vec{u}) = \vec{u}^2$. The *Y* function can be written as

$$Y = X(\vec{u}) - 3X(\vec{u}+\vec{h}) + 3X(u+2\vec{h}) - X(u+3\vec{h}). \qquad \text{(D.93)}$$

One can easily check that

$$\sum_{i=0}^{3}\lambda_i = 0; \text{ and } \sum_{i=0}^{3}\lambda_i\vec{u} \text{ and } \sum_{i=0}^{3}\lambda_i\vec{u}^2 = 0. \qquad \text{(D.94)}$$

This provides the framework for defining an appropriate function to calculate a generalized covariance.

If the data have no trends, we can define function *Y* as

$$Y = X(\vec{u}) - X(\vec{u}+\vec{h}). \qquad \text{(D.95)}$$

Therefore, the variance of *Y*, in practice, can be calculated as

$$K(\vec{h}) = Var(Y) = \hat{\sigma}_Y^2 = \frac{1}{n(\vec{h})}\sum_{i=1}^{n(\vec{h})}\left[x(\vec{u}_i) - x(\vec{u}_i+\vec{h})\right]^2. \qquad \text{(D.96)}$$

This is a conventional variogram equation multiplied by two. If a trend is assumed to be linear, we can calculate the variance as

$$K(\vec{h}) = Var(Y) = \hat{\sigma}_Y^2 = \frac{1}{n(\vec{h})}\sum_{i=1}^{n(\vec{h})}$$

$$\left[x(\vec{u}_i) - 2x(\vec{u}_i+\vec{h}) + x(\vec{u}_i+2\vec{h})\right]^2. \qquad \text{(D.97)}$$

A similar equation can be written for a quadratic model. The second term in $Var(Y)=[E(Y)]^2$ drops out.

For example, for a linear trend, variance is defined as

$Var(Y)=E(Y)^2-[E(Y)]^2$

$$= E\left\{\left[X(\vec{u}) - 2X\left(\vec{u} + \vec{h}\right) + X\left(\vec{u} + 2\vec{h}\right)\right]^2\right\}$$

$$- \left\{E\left[X(\vec{u}) - 2X\left(\vec{u} + \vec{h}\right) + X\left(\vec{u} + 2\vec{h}\right)\right]\right\}^2. \quad \text{(D.98)}$$

The second term on the right side is now zero;

$$E\left[X(\vec{u}) - 2X\left(\vec{u} + \vec{h}\right) + X\left(\vec{u} + 2\vec{h}\right)\right]$$

$$= m(\vec{u}) - 2m\left(\vec{u} + \vec{h}\right) + m\left(\vec{u} + 2\vec{h}\right)$$

$$= a_0 + a_1\vec{u} - 2a_0 - 2a_1\left(\vec{u} + \vec{h}\right)$$

$$+ a_0 + a_1\left(\vec{u} + 2\vec{h}\right) = 0. \quad \text{(D.99)}$$

Therefore, the estimation of variance in Eq. D.98 reduces to Eq. D.97.

In practice, assuming a particular trend, experimental values of $K\left(\vec{h}\right)$ are calculated and fitted with the model,

$$K_k\left(\vec{h}\right) = \sum_{j=0}^{k} b_j(-1)^{j+1}|\vec{h}|^{2j+1}, \quad \text{(D.100)}$$

where k is the order of covariance.

For example, for a linear model, we can write

$$K_1\left(\vec{h}\right) = b_0 - b_1|\vec{h}|^3. \quad \text{(D.101)}$$

Whichever trend that minimizes, the difference between experimental and model $K\left(\vec{h}\right)$ is accepted.

Once the generalized covariance is modeled, equations similar to universal kriging can be used. The final equation, in matrix form, can be written as

$$\begin{bmatrix} K(\vec{u}_1,\vec{u}_1) & \cdots & K(\vec{u}_1,\vec{u}_n) & f_0(\vec{u}_1) & \cdots & f_L(\vec{u}_1) \\ \vdots & & & \vdots & & \\ K(\vec{u}_n,\vec{u}_1) & \cdots & K(\vec{u}_n,\vec{u}_n) & f_0(\vec{u}_n) & \cdots & f_L(\vec{u}_n) \\ f_0(\vec{u}_1) & \cdots & f_0(\vec{u}_n) & 0 & \cdots & 0 \\ \vdots & & & \vdots & & \\ f_L(\vec{u}_1) & \cdots & f_L(\vec{u}_n) & 0 & \cdots & 0 \end{bmatrix} \begin{bmatrix} \lambda_1 \\ \vdots \\ \lambda_n \\ \mu_0 \\ \vdots \\ \mu_L \end{bmatrix}$$

$$= \begin{bmatrix} K(\vec{u}_1,\vec{u}_0) \\ \vdots \\ K(\vec{u}_n,\vec{u}_0) \\ f_0(\vec{u}_0) \\ \vdots \\ f_L(\vec{u}_0) \end{bmatrix}. \quad \text{(D.102)}$$

The error variance is estimated as

$$\hat{\sigma}_E^2 = K(\vec{u}_0,\vec{u}_0) - \sum_{i=1}^{n}\lambda_i K(\vec{u}_i,\vec{u}_0) - \sum_{l=0}^{L}\mu_l f_l(\vec{u}_0). \quad \text{(D.103)}$$

The advantage of this approach is that the residual covariance does not need to be modeled. Also, by determining the appropriate trend as part of the process, the type of trend does not need to be known. Although the procedure appears to be elegant, the actual application of the procedure has been questioned.

D.5.4 Kriging with External Drift. This method is an extension of universal kriging, where the trend in the data is assumed to be controlled by a secondary variable. If $Y(\vec{u})$ is the secondary variable, we can write

$$E[X(\vec{u})] = m(\vec{u}) = a_0 + a_1 Y(\vec{u}). \quad \text{(D.104)}$$

The value at the unsampled location is estimated as

$$X^*(\vec{u}_0) = \sum_{i=1}^{n}\lambda_i X(\vec{u}_i). \quad \text{(D.105)}$$

Unbiasedness. By applying the unbiased condition,

$$E\left[X^*(\vec{u}_0) - \sum_{i=1}^{n}\lambda_i X(\vec{u}_i)\right] = 0. \quad \text{(D.106)}$$

By substituting Eq. D.104 in Eq. D.106, we obtain

$$a_0 + a_1 Y(\vec{u}_0) - \sum_{i=1}^{n}\lambda_i[a_0 + a_1 Y(\vec{u}_i)] = 0,$$

and

$$a_0\left[\sum_{i=1}^{n}\lambda_i - 1\right] + a_1\left[\sum_{i=1}^{n}\lambda_i Y(\vec{u}_i) - Y(\vec{u}_0)\right] = 0.$$

This equation can be satisfied by assuming the constraints,

$$\sum_{i=1}^{n}\lambda_i - 1 = 0, \quad \text{(D.107)}$$

and

$$\sum_{i=1}^{n}\lambda_i Y(\vec{u}_i) - Y(\vec{u}_0) = 0. \quad \text{(D.108)}$$

Minimum Variance. An equation for error variance can be written as

$$\hat{\sigma}_E^2 = C(\vec{u}_0,\vec{u}_0) - \sum_{i=1}^{n}\sum_{j=1}^{n}\lambda_i\lambda_j C(\vec{u}_i,\vec{u}_j) - \sum_{i=1}^{n}\lambda_i C(\vec{u}_i,\vec{u}_0). \quad \text{(D.109)}$$

Using the two constraints in Eqs. D.107 and D.108, we can write function F as

$$F = \hat{\sigma}_E^2 = 2\mu_0\left[\sum_{i=1}^{n}\lambda_i - 1\right] + 2\mu_1\left[\sum_{i=1}^{n}\lambda_i Y(\vec{u}_i) - Y(\vec{u}_0)\right]. \quad \text{(D.110)}$$

Taking the derivative with respect to λ_i and μ_i, we can write the final equation in matrix form as

$$\begin{bmatrix} C(\vec{u}_1,\vec{u}_1) & \cdots & 1 & Y(\vec{u}_1) \\ \vdots & & & \\ C(\vec{u}_n,\vec{u}_1) & \cdots & 1 & Y(\vec{u}_n) \\ 1 & \cdots & 0 & 0 \\ Y(\vec{u}_1) & \cdots & 0 & 0 \end{bmatrix} \begin{bmatrix} \lambda_1 \\ \vdots \\ \lambda_n \\ \mu_0 \\ \mu_1 \end{bmatrix} = \begin{bmatrix} C(\vec{u}_1,\vec{u}_0) \\ \vdots \\ C(\vec{u}_n,\vec{u}_0) \\ 1 \\ Y(\vec{u}_0) \end{bmatrix}. \quad \text{(D.111)}$$

This equation can be solved to obtain the values of λ_i and μ_i. The error variance can be calculated as

$$\hat{\sigma}_E^2 = C(\vec{u}_0, \vec{u}_0) - \sum_{i=1}^{n} \lambda_i C(\vec{u}_i, \vec{u}_0) - \mu_0 - \mu_1 Y(\vec{u}_0). \quad \text{(D.112)}$$

One of the advantages of this method is that it does not require cross covariance modeling.

Nomenclature

a = coefficient to describe trend, Eq. D.73
$C(\vec{u}_i, \vec{u}_j)$ = covariance between two points located at $\vec{u}_i$ and $\vec{u}_j$
$\vec{C}_{lm}$ = cross covariance matrix between variables l and m, Eq. D.66
$E(X)$ = expected value of variable X
$f(\vec{u})$ = trend function, Eq. D.73
F = function to be minimized under constraints
$\vec{h}$ = lag distance
$K(\vec{h})$ = generalized covariance function at lag distance $\vec{h}$
L = number of covariables or number of terms used to define trend
m = mean
n = number of sample points
$n(\vec{h})$ = number of sample points at a lag distance $\vec{h}$
$\vec{u}$ = lag distance
$Var(X)$ = variance of variable X
$X(\vec{u})$ = variable at $\vec{u}$
$X^*(\vec{u})$ = estimated value of variable at $\vec{u}$
$x(\vec{u})$ = sample at $\vec{u}$
$\gamma(\vec{h})$ = variogram at distance $\vec{h}$
λ = weight assigned to sample
Λ = vector of weights assigned to samples
μ = Lagrange multiplier
$\hat{\sigma}_E^2$ = error variance

Subscripts

c = cross covariance
i,j = locations
l = covariable
o = unsampled location
v = block/point relationship
vv = block/block relationship
X = variable X
Y = variable Y

References

1. Myers, D.E.: "Matrix Formulation of Co-Kriging," *Math Geology* (1982) **14,** No. 3, 249–257.
2. Matheron, G.: "The Intrinsic Random Functions and Their Applications," *Adv. in Applied Probability* (1973) **5,** 439–468.
3. Delfiner, P.: "Linear Estimation of Non-Stationary Spatial Phenomena," *Advanced Geostatistics in the Mining Industry,* Guarascio *et al.* (eds.) Reidel, Dordrecht, Holland (1976) 49–68.

Appendix E
Mathematical Details for Simulated Annealing

E.1 Estimation of Initial Control Parameter

The initial control parameter has an important effect on the efficiency of simulated annealing. If the initial control parameter is too large, then most of the swaps during the first steps will be accepted because the probability of accepting a swap given by the Metropolis condition is large. The objective function will fluctuate significantly during these first steps, and the net reduction of the global objective function will be small. Therefore, a significant amount of computation can be wasted if the initial control parameter is too large. If the initial control parameter is too small, then most swaps that increase the objective function will be rejected. This situation is similar to the greedy algorithm that only accepts swaps that reduce the objective function. This might appear to be an efficient strategy; however, there is the danger that the solution will be trapped in a local minimum and yield undesirable results.

Aarts and Korst[1] proposed a method to calculate the initial control parameter. The steps of this method are outlined next.

We must calculate objective functions (O^k_{initial}) for M_{initial} cycles. The change of the objective function is given by

$$\Delta O^k_{\text{initial}} = O^k_{\text{initial}} - O_0 \quad \text{(E.1)}$$

for k=1, ..., $M_{\text{initial}}N_s$ and N_s is the number of gridblocks ($N_xN_yN_z$).

1. Calculate the number of swaps that result in negative (m_1) and positive (m_2) changes of the objective function.

2. Calculate the mean change of the objective function for swaps that result in positive changes of the objective function, ($\Delta O^{+,k}_{\text{initial}} \geq 0$),

$$\Delta\overline{O}^{+}_{\text{initial}} = \frac{1}{m_2}\sum_{i=1}^{m_2} \Delta\overline{O}^{+,i}_{\text{initial}}. \quad \text{(E.2)}$$

3. Calculate the initial control parameter,

$$T^0 = \frac{\Delta\overline{O}^{+}_{\text{initial}}}{\ln\left[\dfrac{m_2}{A_{\text{initial}}m_2 - (1 - A_{\text{initial}})m_1}\right]}, \quad \text{(E.3)}$$

where, A_{initial} is the initial acceptance ratio.

The additional parameters required to calculate the initial control parameter have been evaluated for several simulations and different conditions. Guidelines were developed for the values of the parameters required for this approach and are written as

$$A_{\text{initial}} = 0.99$$

and

$$M_{\text{initial}} > 0.2 \text{ cycles.}$$

Fig. E.1 shows the initial control parameter calculated with Eq. E.3 for several simulations with different values of M_{initial} and $A_{\text{initial}} = 0.99$.[2] These simulations demonstrate that the estimated initial control parameter converges to a stable value for the guidelines.

E.2 Estimation of Maximum Number of Swaps per Step

The maximum number of swaps per step (M_t) is the criterion used to reduce the control parameter and move to the next step when the maximum number of accepted swaps per step (M_a) is not reached. At the beginning of the simulation, the number of swaps reached in each step is M_a. The number of accepted swaps decreases and becomes smaller than M_a as the simulation proceeds and the control parameters decreases. In these steps, it is important to perform a sufficient number of swaps to ensure that the solution space is fully searched and the system does not get trapped in a local minima solution. M_t can be set to a very large number to avoid this problem. However, this is not a practical solution because it can make the simulation process very slow and inefficient.

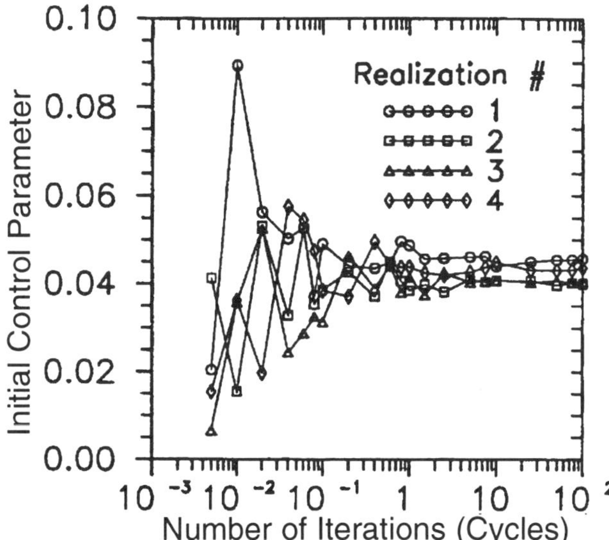

Fig. E.1—Simulation results for estimated initial control parameters for different initial number of iterations ($M_{initial}$).

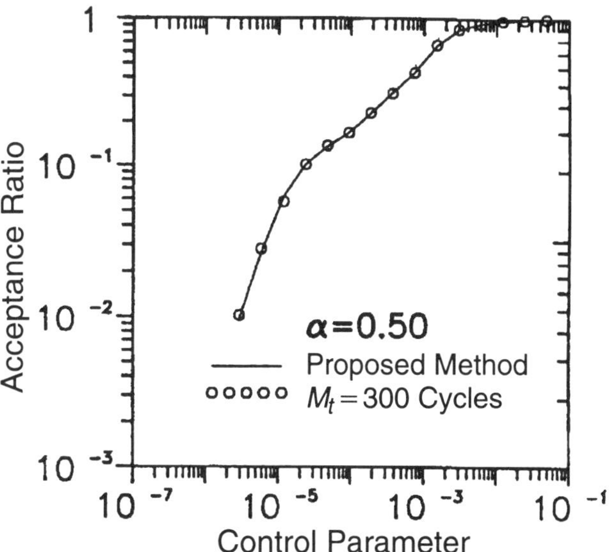

Fig. E.2—Acceptance ratio for simulations generated with Eq. E.9 (proposed method) and $M_t = 300$ cycles.

The method to calculate M_t is based on the relationship between the acceptance ratio (A) and the control parameter (T),[2]

$$\frac{\ln\left(\frac{A^{r+1}}{A^r}\right)}{\ln\left(\frac{T^{r+1}}{T^r}\right)} = \frac{\ln\left(\frac{A^r}{A^{r-1}}\right)}{\ln\left(\frac{T^r}{T^{r-1}}\right)}, \quad \text{(E.4)}$$

where, r^{-1}, r and r^{+1} are the previous, current and next steps, respectively. This relationship can be used to predict A^{r+1} for a subsequent step based on the acceptance ratios from current and previous steps. It is developed empirically by evaluating the relations among simulation parameters for several simulations with large values for M_t.

In Chap. 6, the control parameter is defined as

$$T^{r+1} = \alpha T^r. \quad \text{(E.5)}$$

When combining Eqs. E.4 and E.5, the equation to calculate A becomes

$$A^{r+1} = \frac{(A^r)^2}{A^{r-1}}. \quad \text{(E.6)}$$

Using the definition of A given in Chap. 6, M_t is given by

$$M_t^{r+1} = \frac{m^{r+1}}{A^{r+1}}, \quad \text{(E.7)}$$

where m^{r+1} is the number of accepted swaps in step $r+1$. Because $m^{r+1} \leq M_a$, an upper limit for M_t is

$$M_t^{r+1} = \frac{M_a}{A^{r+1}}. \quad \text{(E.8)}$$

The equation to calculate M_t is obtained by substituting A from Eq. E.6 into Eq. E.8 and is written as

$$M_t^{r+1} = M_a \frac{A^{r-1}}{(A^r)^2}, \quad \text{(E.9)}$$

for $r = 1, 2, \ldots$. For the first step, $r = 1$, it is assumed that $A^0 = A_{initial}$.

Eq. E.9 has been validated with several simulations in which M_t was set to a very large value.[2] **Figs. E.2** and **E.3** compare the acceptance ratio and objective function for simulations with Eq. E.9 and $M_t = 300$ cycles. The simulations with Eq. E.9 were in close agreement with the ones with a large M_t.

There are two advantages of using the approach given by Eq. E.9:

- It does not require us to guess a value of M_t for each simulation case.
- It is more efficient because only the necessary maximum number of swaps is used, and this is calculated at each step.

E.3 Equations To Update a Variogram When Two Values are Interchanged

The key for the efficient application of simulated annealing is that the objective function must be updated rapidly with a small number of operations. The objective function must be calculated for each of the several thousands or millions of swaps during a simulation. Therefore, it is important to update the objective function for each swap in the minimum number of operations. Chap. 6 describes several components that can be included in the objective function to account for different types of constraints in the simulations. The vario-

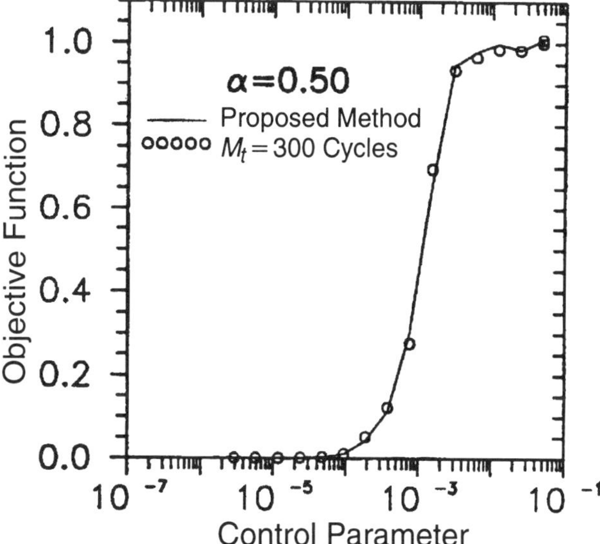

Fig. E.3—Objective function for simulations generated with Eq. E.9 (proposed method) and $M_t = 300$ cycles.

gram component controls the spatial correlation and plays an important role in all the conditional simulation techniques described in this book. This section presents the equations to update the variogram component when two values in the simulation domain are interchangeable. With these equations, it is quicker to update the objective function than to completely recalculate the variogram function for each swap.

The variogram (γ_e) for a lag $\vec{L}_a$, because of swapping two points, is updated with the equation,

$$\gamma_e(\vec{L}_a) = \gamma_e^k(\vec{L}_a) + \lambda(\vec{L}_a), \quad \text{(E.10)}$$

where γ_e^k is the variogram at step k before the swap, and λ is the correction factor. The variogram is updated with Eq. E.10 for each lag included in the objective function. The equations to calculate the correction factor efficiently were derived by tracking the effects of swapping two points on the variogram. The components of the correction factor at the first point located at $\vec{u}_1$, because of a swap with a value from a second point located at $\vec{u}_2$, are

$$\lambda_1^+ = -\left[X(\vec{u}_1) - X(\vec{u}_1 + \vec{L}_a)\right]^2 + \left[X(\vec{u}_2) - X(\vec{u}_1 + \vec{L}_a)\right]^2, \quad \text{(E.11)}$$

and

$$\lambda_1^- = -\left[X(\vec{u}_1) - X(\vec{u}_1 - \vec{L}_a)\right]^2 + \left[X(\vec{u}_2) - X(\vec{u}_1 - \vec{L}_a)\right]^2, \quad \text{(E.12)}$$

where X is the simulation variable. These equations consider the effects on the variogram in the two possible directions around point $\vec{u}_1$. Similar equations can be derived for the point at the second location, ($\vec{u}_2$). Eqs. E.11 and E.12 can be simplified as

$$\lambda_1^+ = d_{1,2}X(\vec{u}_1 + \vec{L}_a) - D_{1,2}, \quad \text{(E.13)}$$

and

$$\lambda_1^- = d_{1,2}X(\vec{u}_1 - \vec{L}_a) - D_{1,2}, \quad \text{(E.14)}$$

where $d_{1,2}$ and $D_{1,2}$ are given by

$$d_{1,2} = 2[X(\vec{u}_1) - X(\vec{u}_2)], \quad \text{(E.15)}$$

and

$$D_{1,2} = [X(\vec{u}_1) + X(\vec{u}_2)][X(\vec{u}_1) - X(\vec{u}_2)]. \quad \text{(E.16)}$$

The components of the correction factor at the location of the second swap point, ($\vec{u}_2$), are given by

$$\lambda_2^+ = -d_{1,2}X(\vec{u}_2 + \vec{L}_a) + D_{1,2}, \quad \text{(E.17)}$$

and

$$\lambda_2^- = -d_{1,2}X(\vec{u}_2 - \vec{L}_a) + D_{1,2}. \quad \text{(E.18)}$$

For each component of the correction factor (Eqs. E.13, E.14, E.17 and E.18), locations $\vec{u}_1 + \vec{L}_a$, $\vec{u}_1 - \vec{L}_a$, $\vec{u}_2 + \vec{L}_a$, and $\vec{u}_2 - \vec{L}_a$ must be checked to ensure they fall within the limits of the simulation region. If one of these points falls outside the simulation region, then the corresponding component must be set to zero.

The correction factor for Eq. E.10, in terms of the components derived, is

$$\lambda(\vec{L}_a) = \frac{1}{2n(\vec{L}_a)}(\lambda_1^+ + \lambda_1^- + \lambda_2^+ + \lambda_2^-), \quad \text{(E.19)}$$

where $n(\vec{L}_a)$ is the number of pairs for the lag $\vec{L}_a$. The number of pairs for each lag is calculated at the beginning of the simulation for the initial variogram, and it is saved to calculate the correction factors during the simulation process.

Similar localized correction factors can be derived for other static constraints.

Nomenclature

A = acceptance ratio
$\vec{L}_a$ = lag distance
M = cycle of swaps
$n(\vec{L}_a)$ = number of pairs at distance $\vec{L}_a$
N = gridblocks
N_S = total number of gridblocks
O = objective function
ΔO = change in objective function
T = control parameter
$X(\vec{u})$ = variable value at $\vec{u}$
α = multiplier for changing control parameter
$\lambda(\vec{L}_a)$ = correction for variogram function for lag distance $\vec{L}_a$
λ^+, λ^- = corrections for variogram function

Subscripts

a = accepted swaps
t = total number of swaps
x = direction
y = direction
z = direction
1,2 = locations of swapped points

Superscripts

k = iteration step
r = step in changing control parameter

References

1. Aarts, E. and Korst, J.: *Simulated Annealing and Boltzmann Machines: A Stochastic Approach to Combinatorial Optimization and Neural Computing,* John Wiley & Sons Ltd., Chichester (1989) 1.
2. Perez, G.: "Stochastic Conditional Simulation for Description of Reservoir Properties," PhD dissertation, U. of Tulsa, Tulsa (1991).

Appendix F
Mathematical Details for Upscaling

Appendix F discusses derivations of various upscaling methods.

F.1 Permeability Upscaling

F.1.1 Incomplete Layer Method.[1] Refer to Fig. 8.5 where a fine-scale permeability description consisting of m blocks in x direction and n blocks in y direction is shown.

To calculate effective permeability in the x direction, we first calculate $(k_{max})_x$. We conduct arithmetic averaging in the y direction. For block i in the x direction, the arithmetic average is given by

$$k_{a,i} = \frac{1}{n}\sum_{j=1}^{n} k_{i,j}. \quad \text{(F.1)}$$

The harmonic average of arithmetically averaged values is given by

$$k_{(\max)x} = \frac{m}{\sum_{i=1}^{m} \frac{1}{k_{a,i}}}. \quad \text{(F.2)}$$

By substituting Eq. F.1 in Eq. F.2, we obtain

$$k_{(\max)x} = \frac{m}{\sum_{i=1}^{m} \frac{n}{\sum_{j=1}^{n} k_{i,j}}}. \quad \text{(F.3)}$$

To calculate $(k_{min})_x$, we first conduct harmonic averaging in the x direction. For block j in the y direction, we calculate the harmonic average as

$$k_{h,j} = \frac{m}{\sum_{i=1}^{m} \frac{1}{k_{i,j}}}. \quad \text{(F.4)}$$

The arithmetic average of these values can be calculated as

$$k_{(\min)x} = \frac{1}{n}\sum_{j=1}^{n} k_{h,j}. \quad \text{(F.5)}$$

Substituting Eq. F.4 in Eq. F.5, we obtain

$$k_{(\min)n} = \frac{1}{n}\sum_{j=1}^{n} \left[\frac{m}{\sum_{i=1}^{m} \frac{1}{k_{i,j}}} \right]. \quad \text{(F.6)}$$

Effective permeability is calculated with the geometric mean of maximum and minimum values,

$$\bar{k}_x = \sqrt{k_{(\max)x} k_{(\min)x}}. \quad \text{(F.7)}$$

For calculating permeability in the y direction, we follow the same procedure. Using an analogy,

$$k_{(\max)y} = \frac{n}{\sum_{j=1}^{n} \frac{m}{\sum_{i=1}^{m} k_{i,j}}}, \quad \text{(F.8)}$$

and

$$k_{(\min)y} = \frac{1}{m}\sum_{i=1}^{m} \left[\frac{n}{\sum_{j=1}^{n} \frac{1}{k_{i,j}}} \right]. \quad \text{(F.9)}$$

With Eqs. F.8 and F.9, we can calculate

$$\bar{k}_y = \sqrt{k_{(\max)y} k_{(\max)x}}. \quad \text{(F.10)}$$

We point out that the geometric average of two extreme values might not always work. Alternately, we can define

$$\bar{k}_x = \frac{(k_{\max})_x + (k_{\min})_x}{2}, \quad \text{(F.11)}$$

and

$$\bar{k}_y = \frac{(k_{\max})_y + (k_{\min})_y}{2}. \quad \text{(F.12)}$$

Numerical experiments indicate that, in the presence of anisotropic reservoirs, an arithmetic average might work better than a geometric average.

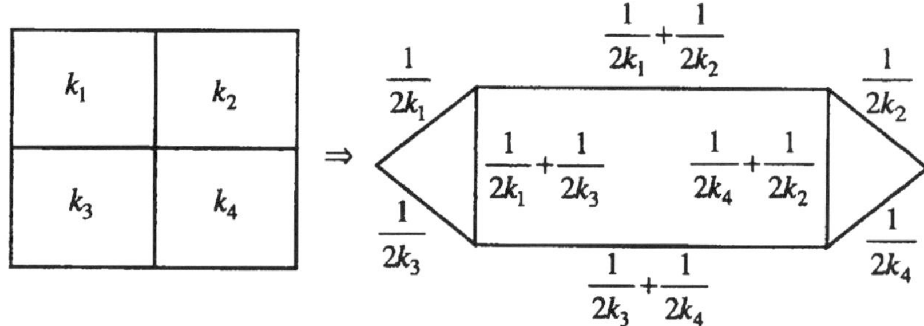

Fig. F.1—Renormalization arrangement.

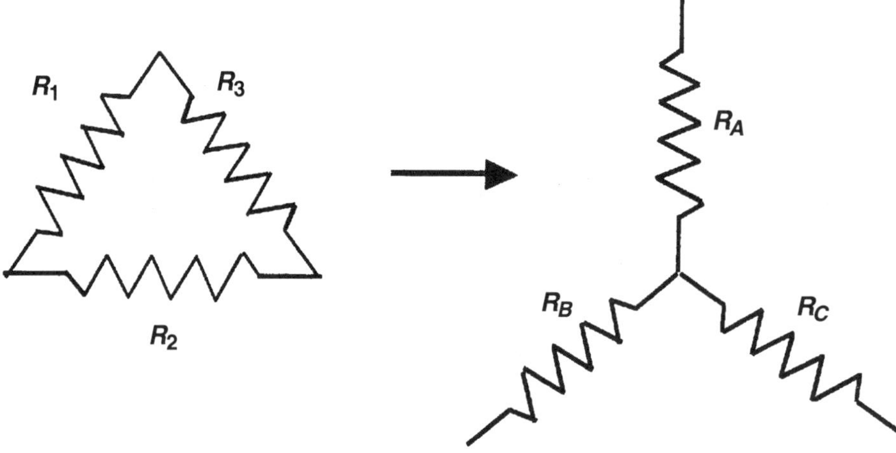

Fig. F.2—Star-Delta Transform.

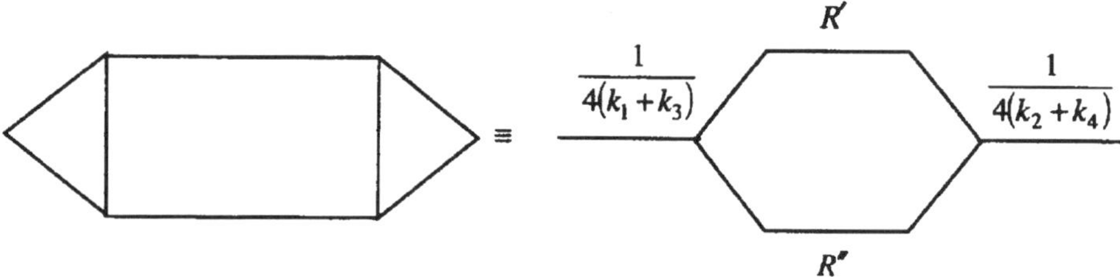

Fig. F.3—Application of Star-Delta Transform.

F.1.2 Renormalization. The method of renormalization is based on the circuit analogs.[2] Examining Fig. 8.6 (**Fig. F.1**), which consists of four gridblocks, we can write the resistances in terms of permeabilities. By defining half a gridblock, which has a resistance of $\frac{1}{2}k_i$ and where k_i is the permeability, we can transform the four gridblocks into an equivalent circuit analog, where we assume flow in the x direction with a no-flow boundary in the y direction. The transformation from delta to star in electrical circuits allows us to show this (see **Fig. F.2**) if

$$R_A = \frac{R_3 R_1}{R_1 + R_2 + R_3}, \qquad \text{(F.13)}$$

$$R_B = \frac{R_1 R_2}{R_1 + R_2 + R_3}, \qquad \text{(F.14)}$$

and

$$R_C = \frac{R_3 R_2}{R_1 + R_2 + R_3}, \qquad \text{(F.15)}$$

where R_i = resistance. Comparing the transform with the figure of permeability resistance shows that (see **Fig. F.3**), where

$$R' = \frac{3}{4}\left(\frac{k_1 + k_2}{k_1 k_2}\right), \text{ and } R'' = \frac{3}{4}\left(\frac{k_3 + k_4}{k_3 k_4}\right).$$

Because R' and R'' are parallel, they can be combined by harmonic average. By combining the result and three resistances,

$$k_x = \frac{4(k_1 + k_3)(k_2 + k_4)C_1}{(k_1 + k_2 + k_3 + k_4)C_1 + 3C_2}, \qquad \text{(F.16)}$$

where

$$C_1 = k_2 k_4 (k_1 + k_3) + k_1 k_3 (k_2 + k_4), \qquad \text{(F.17)}$$

and

$$C_2 = (k_1 + k_2)(k_3 + k_4)(k_1 + k_3)(k_2 + k_4). \qquad \text{(F.18)}$$

Similarly, permeability in the y direction can be calculated.

F.1.3 Permeability Tensor. Consider the four-gridblock system, defined in Fig. 8.6. For convenience, we assume that the gridblock dimensions are identical, and each small-scale gridblock can be represented by principal isotropic components in permeability tensor only. That is, k_{yx} and k_{xy} are zero on a fine scale, and $k_x = k_y$. Both assumptions are not necessary for deriving the equation (see Assum *et al.*[3]).

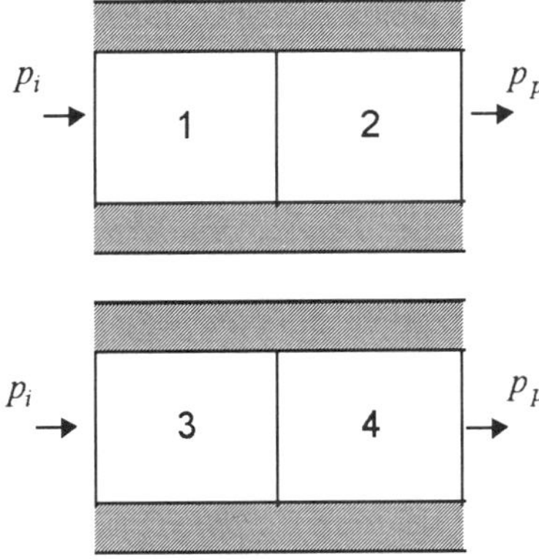

Fig. F.4—Schematic for a no-crossflow case.

We can determine $k_{x\text{app}}$ and $k_{y\text{app}}$ by following a procedure. We outline the six steps in that procedure.

1. Isolate the top layer, blocks one and two in Fig. 8.6 (see **Fig. F.4**). With Darcy's law, determine the apparent permeability in the x direction ($\hat{k}_{x12}$) of blocks one and two combined. Constant potential drop is applied in the principal direction of flow (x direction) with a no-flow condition at the upper and lower boundaries.
2. Isolate the bottom layer, blocks three and four in Fig. 8.6 (see Fig. F.4). With Darcy's law, determine the apparent permeability in the x direction ($\hat{k}_{x34}$) of blocks three and four combined. Constant potential drop is applied in the principal direction of flow (x direction) with a no-flow condition at the upper and lower boundaries.
3. Combine effective apparent permeability in the x direction of the top and bottom layers ($\hat{k}_{x1234}$) as an arithmetic average of $\hat{k}_{x12}$ and $\hat{k}_{x34}$ because these layers are in parallel with respect to the direction of flow.
4. Isolate the first column, blocks one and three in Fig. 8.6 (**Fig. F.5**). With Darcy's law, determine the apparent permeability in the x direction ($\hat{k}_{x13}$) of blocks one and three combined. Constant potential drop is applied in the principal direction of flow (x direction) with a no-flow condition at the upper and lower boundaries.
5. Isolate the second column, blocks two and four in Fig. 8.6 (Fig. F.5). With Darcy's law, determine the apparent permeability in the x direction ($\hat{k}_{x24}$) of blocks two and four combined. Constant potential drop is applied in the principal direction of flow (x direction) with a no-flow condition at the upper and lower boundaries.
6. The effective apparent permeability in the x direction of the first and second columns combined ($\hat{k}_{x1324}$) is a harmonic average of $\hat{k}_{x13}$ and $\hat{k}_{x24}$ because these columns are in series with respect to the direction of flow.

Steps one through six are repeated to determine the effective apparent permeability for cases in the y direction. For these cases, the flow of fluid is along the y axis for each of the separate sections: (top and bottom, and first and second columns). Consequently, the closed boundaries for each case are the left and right vertical boundaries in Fig. 8.6.

Note that no crossflow refers to no communication, and crossflow refers to the maximum communication between blocks/layers in the direction transverse to the principal flow direction.

The expressions for $\hat{k}_{x1234}$ and $\hat{k}_{x1324}$ are

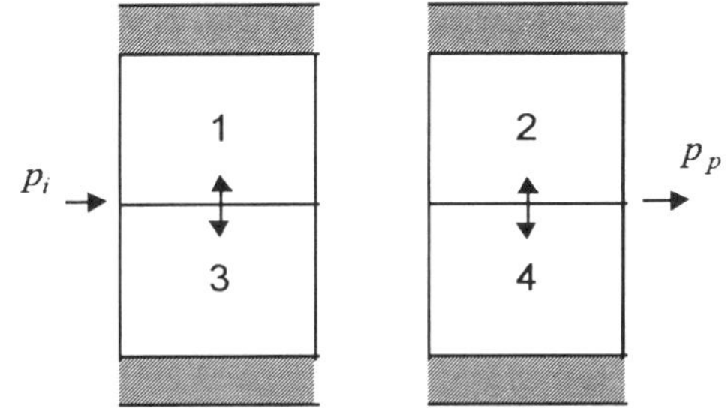

Fig. F.5—Schematic for a crossflow case.

$$\hat{k}_{x1234} = \frac{\hat{k}_{x12} + \hat{k}_{x34}}{2}, \qquad \text{(F.19)}$$

where

$$\hat{k}_{x12} = \frac{2k_1k_2}{k_1 + k_2}, \qquad \text{(F.20)}$$

and

$$\hat{k}_{x34} = \frac{2k_3k_4}{k_3 + k_4}. \qquad \text{(F.21)}$$

Eq. F.19 accounts for the no crossflow condition. For a transverse equilibrium condition,

$$\hat{k}_{x1324} = \frac{2\hat{k}_{x13}\hat{k}_{x24}}{\hat{k}_{x13} + \hat{k}_{x24}}, \qquad \text{(F.22)}$$

where

$$\hat{k}_{x13} = \frac{k_1 + k_3}{2}, \text{ and } \hat{k}_{x24} = \frac{k_2 + k_4}{2}. \qquad \text{(F.23)}$$

With similar arguments, we can also develop equations for ($\hat{k}_{y1324}$) and $\hat{k}_{y1234}$. The apparent permeabilities in the x and y directions are calculated by solving Eqs. F.24 and F.25.

$$\tilde{k}_{x\text{app}} = \frac{\tilde{k}_{x\text{app}}\hat{k}_{x1234} + \tilde{k}_{y\text{app}}\hat{k}_{x1324}}{\tilde{k}_{x\text{app}} + \tilde{k}_{y\text{app}}}, \qquad \text{(F.24)}$$

and

$$\tilde{k}_{y\text{app}} = \frac{\tilde{k}_{x\text{app}}\tilde{k}_{x1324} + \tilde{k}_{y\text{app}}\tilde{k}_{x1234}}{\tilde{k}_{x\text{app}} + \tilde{k}_{y\text{app}}}. \qquad \text{(F.25)}$$

Eqs. F.24 and F.25 can be solved simultaneously for the two unknowns, $\tilde{k}_{x\text{app}}$ and $\tilde{k}_{y\text{app}}$.

If the simulator is not capable of using permeability tensors, Eqs. F.24 and F.25 represent upscaled permeabilities in the x and y directions, respectively.

To develop the permeability tensor equation, we must consider the transverse pressure gradient. Examining **Fig. F.6,** if we assume no crossflow between blocks one and two and blocks three and four, for unit lengths of gridblocks and unit viscosity fluid, we can write

$$q_{x12} = -k_1(p^* - p_i), \qquad \text{(F.26)}$$

and

$$q_{x12} = -k_2(p_p - p^*), \qquad \text{(F.27)}$$

where q_{x12} is the flow rate in the x direction through the top two blocks.

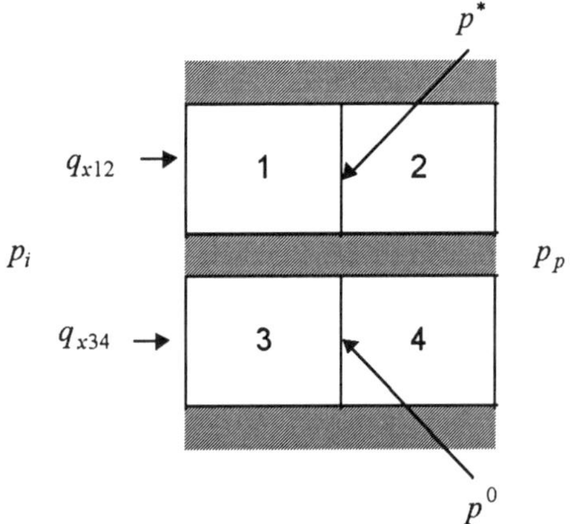

Fig. F.6—Closed upper, middle and lower boundaries.

By combining Eqs. F.26 and F.27,

$$p^* = \frac{q_{x12}}{2}\left(\frac{1}{k_2} - \frac{1}{k_1}\right) + \left(\frac{p_i + p_p}{2}\right). \quad \text{(F.28)}$$

Similarly,

$$q_{x34} = -k_3(p^0 - p_i) = -k_4(p_p - p^0). \quad \text{(F.29)}$$

Rearranging,

$$p^0 = \frac{q_{x34}}{2}\left(\frac{1}{k_4} - \frac{1}{k_3}\right) + \left(\frac{p_i + p_p}{2}\right). \quad \text{(F.30)}$$

By subtracting Eq. F.30 from Eq. F.28, we obtain

$$(\Delta p_t)_{\max} = p^* - p^0 = \frac{q_{x12}}{2}\left(\frac{1}{k_2} - \frac{1}{k_1}\right) - \frac{q_{x34}}{2}\left(\frac{1}{k_4} - \frac{1}{k_3}\right). \quad \text{(F.31)}$$

Eq. F.31 represents the maximum transverse pressure drop across the no-flow, middle boundary. It can be shown that the average transverse pressure drop is half of the maximum.[3]

We can also write

$$q_{x12} = \hat{k}_{x12}(p_i - p_p) = \frac{2k_1k_2}{k_1 + k_2}\cdot(p_i - p_p), \quad \text{(F.32)}$$

and

$$q_{x34} = \hat{k}_{x34}(p_i - p_p) = \frac{2k_3k_4}{k_3 + k_4}\cdot(p_i - p_p). \quad \text{(F.33)}$$

By substituting Eqs. F.33 and F.32 in Eq. F.31, we can calculate the average transverse pressure drop as

$$(\Delta p_t)_{\text{avg}} = \left[\frac{k_1 - k_2}{2(k_1 + k_2)} - \frac{k_3 - k_4}{2(k_3 + k_4)}\right](p_i - p_p). \quad \text{(F.34)}$$

The transverse pressure gradient in the x direction is

$$\left(\frac{\partial p}{\partial y}\right)_x = \frac{(\Delta p_t)_{\text{avg}}}{L_y},$$

and

$$\left(\frac{\partial p}{\partial x}\right)_x = \frac{p_i - p_p}{L_x},$$

for

$$\frac{\left(\frac{\partial p}{\partial y}\right)_x}{\left(\frac{\partial p}{\partial x}\right)_x} = \frac{1}{2}\left(\frac{k_1 - k_2}{k_1 + k_2} - \frac{k_3 - k_4}{k_3 + k_4}\right). \quad \text{(F.35)}$$

A similar expression can be derived for the y direction as well. With the developed equation, we can calculate $\tilde{k}_{xx}$, $\tilde{k}_{yy}$, $\tilde{k}_{xy}$, and $\tilde{k}_{yx}$ with Eqs. 8.31 through 8.34.

We only discussed the method for four gridblocks. This method can be extended to multiple gridblocks, as well as in three dimensions.[4]

F.2 Upscaling of Relative Permeabilities

Upscaling of relative permeabilities is divided into two limiting cases—no crossflow and vertical equilibrium. Both limiting cases are amenable to analytical solutions, if we assume that only vertical heterogeneity exists and individual layers are homogeneous.

F.2.1 No Crossflow. Under this condition, we assume that communication only exists at the wellbore. Each layer is isolated from other layers. Because the layers are isolated from each other, we must first arrange them with the highest displacement velocity at the top and the lowest at the bottom. See **Fig. F.7.**

Each layer has unique properties defined by k_i, ϕ_i, h_i, ΔS_{w_i}, k'_{rw_i}, and k'_{ro_i}, where k is the permeability; ϕ is the porosity; h is the thickness; $\Delta S_w = 1 - S_{rw_i} - S_{o_r}$, and k'_{rw} and k'_{ro} are endpoint relative permeabilities for water and oil, respectively. S_{w_i} is the connate water saturation, and S_{o_r} is the residual oil saturation. The highest-ranking layer (layer one) has the highest value of $k/\phi\Delta S_w$ because it reflects the highest in-situ velocity.

For developing an analytical solution, we assume piston-like displacement and constant injection/production pressure. For convenience, we assume a unit width of the reservoir.

We define the mobility ratio, M_j, for layer j, as

$$M_j = \frac{\dfrac{k'_{rw_j}}{\mu_w}}{\dfrac{k'_{ro_j}}{\mu_o}}. \quad \text{(F.36)}$$

For layer j, where the front has advanced distance x_j, we can write the flow rate as (see **Fig. F.8**)

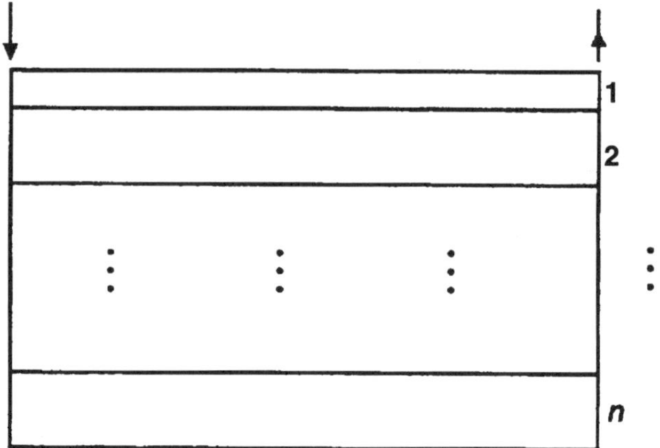

Fig. F.7—Horizontal noncommunicating layers.

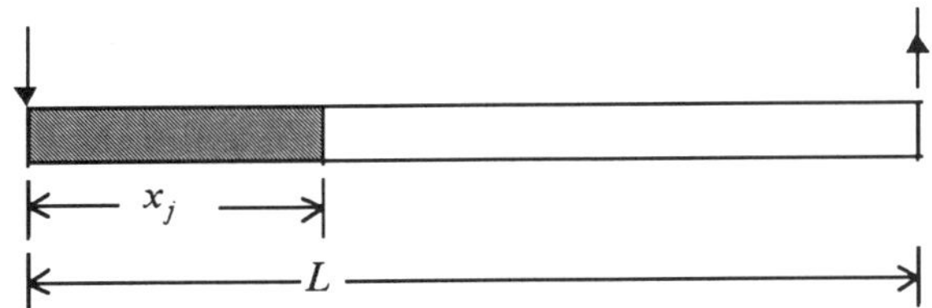

Fig. F.8—Displacement of oil in Layer *j*.

$$q_j = \frac{(kk'_{rw}h)_j}{\mu_w} \cdot \frac{\Delta p}{L\left[x_{Dj} + M_j(1 - x_{Dj})\right]}, \quad \text{(F.37)}$$

where $x_{Dj} = \frac{x_j}{L}$.

The in-situ velocity equation can be written as

$$v_j = \frac{dx_j}{dt} = \frac{q_j}{h_j \phi_j \Delta S_{w_j}}. \quad \text{(F.38)}$$

By substituting Eq. F.37 in Eq. F.38 and slightly modifying it,

$$\frac{dx_{Dj}}{dt} = \left(\frac{kk'_{rw}}{\phi \Delta S_w}\right)_j = \frac{\Delta p}{\mu_w L^2} \cdot \frac{1}{\left[x_{Dj} + M_j(1 - x_{Dj})\right]}. \quad \text{(F.39)}$$

If we assume that layer l has broken through, all the layers above l have also broken through, and all the layers below l have yet to break through. For the layers below l,

$$\left(\frac{\phi \Delta S_w}{kk'_{rw}}\right)_j \frac{\mu_w L^2}{\Delta p} \int_0^{x_{Dj}} \left[x_{Dj} + M_j(1 - x_{Dj})\right] dx_{Dj}$$

$$= \left(\frac{\phi \Delta S_w}{kk'_{rw}}\right)_l \frac{\mu_w L^2}{\Delta p} \int_0^1 \left[x_{Dl} + M_l(1 - x_{Dl})\right] dx_{Dj}. \quad \text{(F.40)}$$

By simplifying Eq. F.40, assuming $M_j = M_l = M$, we define

$$R_{jl} = \left(\frac{\phi \Delta S_w}{kk'_{rw}}\right)_l \Big/ \left(\frac{\phi \Delta S_w}{kk'_{rw}}\right)_j \quad \text{(F.41)}$$

and obtain

$$x_{Dj}^2\left(\frac{1 - M}{2}\right) + Mx_{Dj} - R_{jl}\left(\frac{1 + M}{2}\right) = 0. \quad \text{(F.42)}$$

By solving the quadratic equation, we can write

$$x_{Dj} = \frac{-M + \sqrt{M^2 + R_{jl}(1 - M^2)}}{1 - M}. \quad \text{(F.43)}$$

Note that, by examining Eq. F.42, if $M = 1$,

$$x_{Dj} = R_{jl}. \quad \text{(F.44)}$$

With Eq. F.43, we can calculate the distance traveled by layer j when layer l breaks through.

Because all the layers above l have already broken through, the water production can be calculated as

$$q_w = \sum_{i=1}^{l} \left(\frac{kk'_{rw}h}{\mu_w}\right)_i \frac{\Delta p}{L}. \quad \text{(F.45)}$$

In terms of effective property, we can write

$$q_w = \frac{(\tilde{k}\tilde{k}_{rw})H}{\mu_w} \frac{\Delta p}{L}, \quad \text{(F.46)}$$

where H is the total thickness. If we calculate,

$$\tilde{k}H = \frac{\sum_{i=1}^{n} k_i h_i}{n}. \quad \text{(F.47)}$$

Then, by combining Eqs. F.45 through F.47, we obtain

$$\tilde{k}_{rw_l} = \frac{\sum_{i=1}^{l} (kk'_{rw}h)_i}{\tilde{k}H}. \quad \text{(F.48)}$$

All the layers below l are producing oil.

The total oil production is

$$q_o = \sum_{i=l+1}^{n} \frac{(kk'_{ro}h)_i}{\mu_o} \frac{\Delta p}{L\left[\frac{1}{M}x_{D_i} + \left(1 - x_{D_i}\right)\right]}. \quad \text{(F.49)}$$

With an expression similar to water for effective relative permeability, we can write

$$\tilde{k}_{ro_l} = \frac{\sum_{i=l+1}^{n} \frac{(kk'_{ro}h)_i}{\frac{1}{M}x_{D_i} + \left(1 - x_{D_i}\right)}}{\tilde{k}H}. \quad \text{(F.50)}$$

Note that for the unit mobility ratio, Eq. F.50 reduces to

$$\tilde{k}_{ro_l} = \frac{\sum_{i=l+1}^{n} (kk'_{ro}h)_i}{\tilde{k}H}. \quad \text{(F.51)}$$

The corresponding water saturation, at the time of l bed's breakthrough, can be calculated as

$$\bar{S}_{w_l} = \frac{\sum_{i=1}^{l}\left[h\phi(1 - S_{or})_i\right] + \sum_{i=l+1}^{n}\left\{\left[h\phi(1 - S_{or})\right]_i x_{D_i} + \left(h\phi S_{wi}\right)_i\left(1 - x_{D_i}\right)\right\}}{\sum_{i=1}^{n}(h\phi)_i}. \quad \text{(F.52)}$$

By repeating the calculations, as each bed breaks through, we can calculate the effective relative permeability curves.

F.2.2 Vertical Equilibrium. Vertical equilibrium exists when the sum of all fluid flow driving forces, in the direction perpendicular to the direction of bulk fluid flow, is zero. Typically, three forces are important in establishing vertical equilibrium—viscous, gravity, and capillary. Depending on the dominance of these forces, different analytical expressions can be derived for layered medium.

Viscous Crossflow Domination. Under a case dominated by viscous crossflow, we assume capillary and gravity forces are negligible, and only viscous forces predominate. Under vertical equilibrium conditions, we assume that pressure gradient, in the transverse direction, is negligible. Consider a layered system with n layers. Similar to the no crossflow case, we assume that the layers are arranged from the fastest to the slowest layer, based on the value of $\frac{k}{\phi \Delta S_w}$.

Consider that the lth layer has just broken through. The floodfront location is shown by the crosshatch portion. See **Fig. F.9.** At the time the lth layer has broken through,

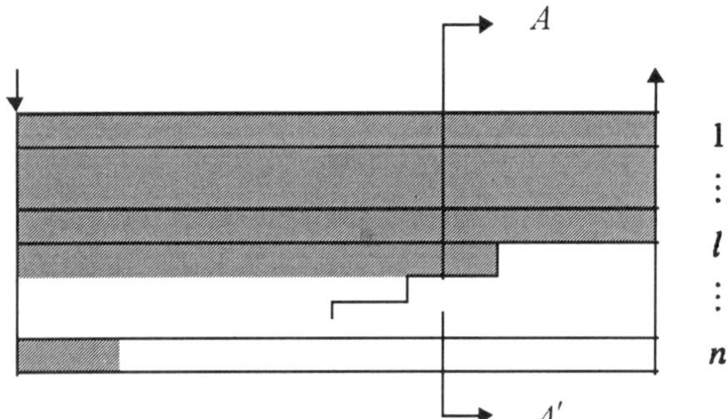

Fig. F.9—Flow under Viscous Crossflow Domination.

$(n-l+1)$ segments exist below layer l, each having different combinations of oil/water fronts (e.g., for section AA' across, the $(l+1)$ layer of water is flowing, and $(n-l-1)$ layers of oil is flowing).

Under the condition of vertical equilibrium, pressure gradient across each of these segments is independent of any layers. That is, same pressure gradient exists across every layer for a given cross section. Consider the last horizontal segment closest to the producer. We can write the total rate as

$$q_t = \left(\frac{dp}{dx}\right)_{n-l+1}\left(\sum_{i=1}^{l}\frac{k_{w_i}h_i}{\mu_w} + \sum_{i=l+1}^{n}\frac{k_{o_i}h_i}{\mu_o}\right), \quad \text{(F.53)}$$

and

$$\left(\frac{dp}{dx}\right)_{n-l+1} = \frac{q_t}{\left(\sum_{i=1}^{l}\frac{k_{w_i}h_i}{\mu_w} + \sum_{i=l+1}^{n}\frac{k_{o_i}h_i}{\mu_o}\right)}. \quad \text{(F.54)}$$

Similarly, pressure gradient across each segment can be calculated. For example, in segment one,

$$\left(\frac{dp}{dx}\right)_1 = \frac{q_t}{\sum_{i=1}^{n-1}\frac{k_{w_i}h_i}{\mu_w} + \frac{k_{o_n}h_n}{\mu_o}}. \quad \text{(F.55)}$$

In addition to pressure gradient, we can also locate the floodfront location in each layer below l. In the absence of crossflow, each layer has a fixed rate. For the layer below l, we can write

$$\frac{u_j}{u_l} = \left(\frac{k}{\phi\Delta S_w}\right)_j \Big/ \left(\frac{k}{\phi\Delta S_w}\right)_l = \frac{x_j}{L}, \quad \text{(F.56)}$$

where x_j is the floodfront location in layer j. Therefore, the length of each segment is

$$\Delta x_j = x_{j+1} - x_j. \quad \text{(F.57)}$$

By knowing the pressure gradient in each segment (Eq. F.54) and the length of each segment, we can calculate the total pressure gradient as

$$\left(\frac{\Delta p}{L}\right)_T = \frac{1}{L}\sum_{j=1}^{n-l+1}\left(\frac{dp}{dx}\right)_j \Delta x_j. \quad \text{(F.58)}$$

The effective relative permeability equation can be written as

$$q_w = \frac{\tilde{k}\tilde{k}_{rw}H}{\mu_w}\left(\frac{\Delta p}{L}\right)_T \quad \text{(F.59)}$$

$$= \sum_{i=1}^{l}\frac{(kk'_{rw})_i}{\mu_w}\left(\frac{dp}{dx}\right)_{n-l+1}. \quad \text{(F.60)}$$

Assuming $\tilde{k}H = \frac{\sum_{i=1}^{n} k_i h_i}{n}$, we can write

$$\tilde{k}_{rw} = \frac{\sum_{i=1}^{l}(kk'_{rw})_i h_i \left(\frac{dp}{dx}\right)_{n-l+1}}{\tilde{k}H\left(\frac{\Delta p}{L}\right)_T}. \quad \text{(F.61)}$$

The oil relative permeability is

$$\tilde{k}_{ro} = \frac{\sum_{i=l+1}^{n}(kk'_{ro})_i h_i \left(\frac{dp}{dx}\right)_{n-l+1}}{\tilde{k}H\left(\frac{\Delta p}{L}\right)_T}. \quad \text{(F.62)}$$

The average water saturation is

$$\bar{S}_{w_l} = \frac{\sum_{i=1}^{l}\left[h\phi(1-S_{or})_i\right] + \sum_{i=l+1}^{n}\left\{\left[h\phi(1-S_{or})\right]_i x_{D_i} + \left(h\phi S_{w_i}\right)_i\left(1-x_{D_i}\right)\right\}}{\sum_{i=1}^{n}(h\phi)_i}. \quad \text{(F.52)}$$

Repeating these steps as each layer breaks through, we can calculate the entire range of relative permeabilities.

Gravity Crossflow Domination. Under gravity-dominated flow, we assume that the water segregates to the bottom of the reservoir immediately. That is, irrespective of the permeability ordering in the reservoir, the bottom-most layer will break through first. See **Fig. F.10.**

Assuming piston-like displacement, we can easily develop the equations for effective relative permeabilities.

$$\tilde{k}_{rw_l} = \frac{\sum_{i=1}^{l}(kk'_{rw})_i h_i}{\tilde{k}H}, \quad \text{(F.63)}$$

where the lth layer has broken through,

$$\text{and } \tilde{k}_{ro_l} = \frac{\sum_{i=l+1}^{n}(kk'_{ro})_i h_i}{\tilde{k}H}. \quad \text{(F.64)}$$

In both Eqs. F.63 and F.64, $\tilde{k}H$ reflects the arithmetic weighted average.

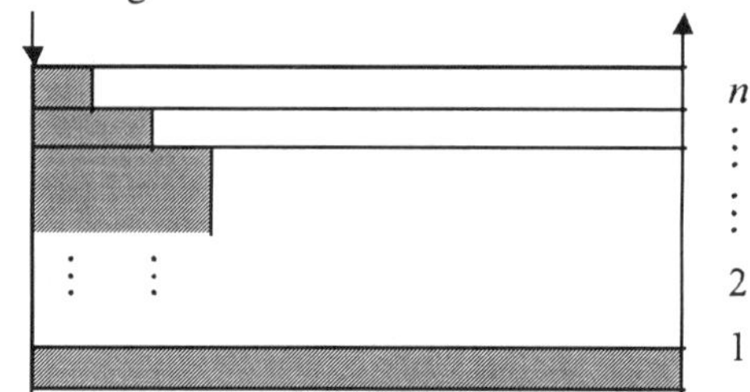

Fig. F.10—Flow under Gravity Flow Domination.

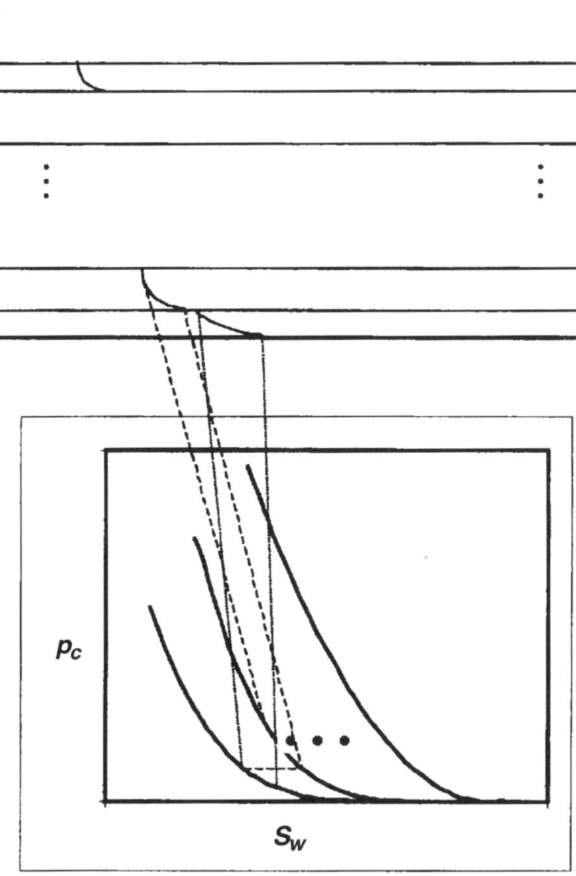

Fig. F.11—Flow under Gravity and Capillary Crossflow Domination.

The saturation can be calculated as

$$\tilde{S}_{w_l} = \frac{\sum_{i=l}^{l}[h\phi(1 - S_{or})]_i + \sum_{i=l+1}^{n}(h\phi S_{wi})_i}{\sum_{i=l}^{n}(h\phi)_i}. \quad \text{(F.65)}$$

With Eqs. F.63 through F.65, the effective relative permeabilities, as a function of saturation, are calculated.

Gravity and Capillary Crossflow Domination. In this case, we assume the reservoir is under vertical equilibrium, and the saturation distribution is governed by capillary pressure.

Assuming that we have capillary pressure curves for each layer, we can calculate the saturation for each layer (see **Fig. F.11**). We have n layers, and each layer has a unique capillary pressure curve.

We arbitrarily pick a water saturation value at the bottom of layer one. Knowing the thickness and relationship between capillary and gravity forces, we write

$$p_c = \Delta\rho g h. \quad \text{(F.66)}$$

We can calculate the saturation distribution in layer one. Assuming pressure continuity, we can calculate the saturation value at the bottom of layer two and, with Eq. F.66, determine the saturation distribution in layer two. The same procedure is repeated for all layers, so we can obtain the saturation distributions for all layers. If the functional relationship between capillary pressure and saturation is known, we can analytically calculate the average water saturation in each layer. Otherwise,

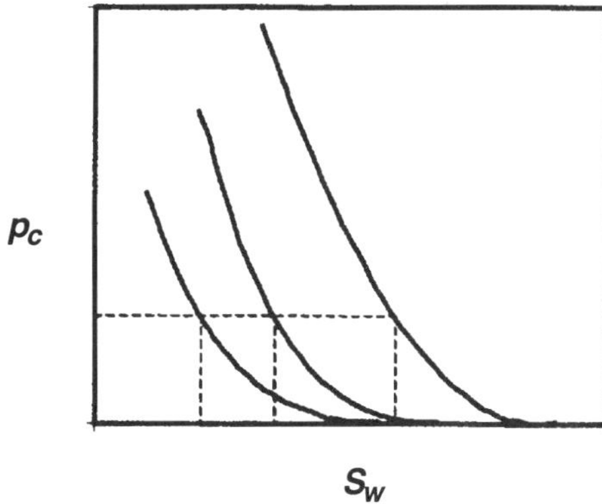

Fig. F.12—Flow under Capillary Force Domination.

$$\overline{S}_{w_l} = \frac{1}{(h\phi)_i}\int_o^{(h\phi)_i} S_w d(h\phi). \quad \text{(F.67)}$$

With Eq. F.67, we can calculate the average water saturation in layer l. Eq. F.67 will simplify, if porosity is assumed constant across the thickness.

Relative permeabilities are defined as a unique function of saturation. Therefore, once the average saturation in each layer is determined, we can calculate the effective relative permeabilities as

$$\tilde{k}_{rw} = \frac{\sum_{i=1}^{n}\left[kk_{rw}(\overline{S}_w)\right]_i h_i}{\sum_{i=1}^{n} k_i h_i}, \quad \text{(F.68)}$$

$$\tilde{k}_{ro} = \frac{\sum_{i=1}^{n}\left[kk_{ro}(\overline{S}_w)\right]_i h_i}{\sum_{i=1}^{n} k_i h_i}, \quad \text{(F.69)}$$

and

$$\tilde{S}_w = \frac{\sum_{i=1}^{n} h_i\phi_i\overline{S}_{w_l}}{\sum_{i=1}^{n} h_i\phi_i}. \quad \text{(F.70)}$$

The procedure can be repeated starting with different water saturations at the bottom of layer one, and the effective relative permeabilities are calculated over the entire range of saturation.

Capillary Force Domination. If we assume that both gravity and viscous crossflow are negligible, we can assume that the saturation profile is uniform in each layer. See **Fig. F.12.**

To calculate the saturation in each layer, pick a value of capillary pressure and calculate the saturations in each layer.

By knowing the relationship between relative permeability and saturation, we can calculate

$$\tilde{k}_{rw} = \frac{\sum_{i=1}^{n}[kk_{rw}(S_w)]_i h_i}{\sum_{i=1}^{n} k_i h_i}, \quad \text{(F.71)}$$

$$\tilde{k}_{ro} = \frac{\sum_{i=1}^{n} [kk_{ro}(S_w)]_i h_i}{\sum_{i=1}^{n} k_i h_i}, \quad \text{(F.72)}$$

and

$$\tilde{S}_w = \frac{\sum_{i=1}^{n} h_i \phi_i S_{wi}}{\sum_{i=1}^{n} h_i \phi_i}. \quad \text{(F.73)}$$

Eqs. F.71 through F.73 can be used for different values of capillary pressures and corresponding saturation distributions.

Nomenclature

C = constant as defined in Eqs. F.17 and F.18
g = gravitational acceleration
h = thickness
H = total thickness
k = permeability
$\bar{k}, \tilde{k}$ = average permeability
k_r = relatively permeability
L = length of the medium
m,n = number of fine-scale gridblocks
M = mobility ratio
p = pressure
p_c = capillary pressure
Δp = pressure drop
q = flow rate
R = resistance
R_{ji} = ratio of velocities, as defined in Eq. F.41
S = saturation
t = time
v = velocity
x = direction, distance
x_D = dimensionless distance
y = direction
ϕ = porosity
μ = viscosity
ρ = density

Subscripts

a = arithmetic average
app = apparent effective permeability
A = resistance as defined in Fig. F.2
B = resistance as defined in Fig. F.2
C = resistance as defined in Fig. F.2
D = dimensionless
h = harmonic average
i = inlet value
j = dummy variable, indicates y direction
l = bed that just broke through
o = oil
or = remaining oil
w = water
wi = irreducible water
x = direction
xy,yx = nondiagonal terms in the permeability tensor matrix
y = direction

References

1. Cardwell, W.T. Jr. and Parsons, R.L.: "Average Permeabilities of Heterogeneous Oil Sands," AIME (March 1945) No. 1852.
2. King, P.R.: "Renormalization Calculations of Immiscible Flow," *Transport in Porous Media* (1993) **12,** 237.
3. Aasum, Y., Kasap, E., and Kelkar, M.: "Analytical Up-Scaling of Small-Scale Permeability Using a Full Tensor," *Petroleum Geoscience* (1995) **1,** 365.
4. Lee, J., Kasap, E., and Kelkar, M.G.: "Analytical Upscaling of Permeability for 3D Gridblocks," *SPEJ* (March 1996) 59.

Author Index

A

B

C

D

E-H

I-J

K

L

M-O

P-R

S-T

V-Z

Subject Index

A

B

C

D

E

F

G

H

I

J

K

L

M

N

O

P

Q

R

S

T

U

V

W

X-Z